THERMODYNAMICS AND STATISTICAL MECHANICS

THERMODYNAMICS AND STATISTICAL MECHANICS

BY

A. H. WILSON

F.R.S.

CAMBRIDGE

AT THE UNIVERSITY PRESS

1966

CAMBRIDGE UNIVERSITY PRESS
Cambridge, New York, Melbourne, Madrid, Cape Town,
Singapore, São Paulo, Delhi, Tokyo, Mexico City

Cambridge University Press
The Edinburgh Building, Cambridge CB2 8RU, UK

Published in the United States of America by Cambridge University Press, New York

www.cambridge.org
Information on this title: www.cambridge.org/9780521093644

First published 1957
Reprinted 1960, 1966
Re-issued 2011

A catalogue record for this publication is available from the British Library

ISBN 978-0-521-09364-4 Paperback

CONTENTS

PREFACE

The subject of thermodynamics seems to present peculiar difficulties to both physicists and chemists, and, while there are many excellent books which help to smooth the path of the chemist, there are relatively few which have been written with the physicist, and particularly the theoretical physicist, in mind. In my opinion any book on thermodynamics which is intended primarily for the chemist is bound to be distinctly unsatisfactory for the theoretical physicist, and the converse is also true. The fundamental ideas of thermodynamics are simple but of very considerable depth, and their proper exposition for any particular class of readers must be closely related to their probable background and habits of thought. My own experience is that many theoretical physicists cannot follow an argument which appeals to chemists, because they are unable to distinguish between what is pure thermodynamics and what is physical chemistry. In addition, many of the mathematical steps are normally given in such detail as to be wearisome to a professional theoretical physicist, and this tends to distract his attention from the physical significance of the important steps in the argument.

The present book is an attempt to give an account of thermodynamics and statistical mechanics intended mainly for theoretical physicists. It covers a considerable range of elementary and advanced topics, but, though it starts at the beginning of thermodynamics, it is not likely to be suitable as a first introduction to the subject since the elementary portions are treated from an advanced standpoint. My primary aim has been to develop the subjects of thermodynamics and of statistical mechanics from starting points which are likely to be acceptable to the majority of the probable readers and which at the same time make it possible to deal with the difficult points which most accounts raise in the minds of critical readers. I have not, however, attempted to push the analysis of fundamental notions, particularly of statistical mechanics, as far as is possible. My secondary aim has been to illustrate the use of the general theory by applications to particular problems, which constitute an introduction, but not more than an introduction, to the equilibrium theory of the gaseous and solid states.

The first three chapters deal with the classical development of thermodynamics on traditional lines, but with more emphasis than is normal upon the special characteristics of a thermodynamic system as contrasted with a dynamical one. The fourth chapter, which is based upon Carathéodory's axiomatic approach, analyses in much greater detail those points which are only cursorily dealt with in the first two chapters. It is probable that many readers will be satisfied to read Chapters 1, 2 and 3 and to omit Chapter 4, or to read Chapter 4 and to take the preceding ones for granted; but those who wish to understand the full significance of Chapter 5 will require to study at least some parts of Chapter 4. Chapter 4 completes thermodynamics proper, and at this stage the logical order of development is broken in order to derive the fundamental formulae of statistical mechanics in Chapter 5. The reason for this inelegance is my wish to illustrate the general theory by specific applications, and this can only be done either by introducing and analysing a considerable amount of experimental data or by discussing thermodynamic functions which have been derived theoretically from atomic models of particular thermodynamic systems. The number of applications using experimental data alone is small if only an elementary knowledge of physical chemistry is assumed. I have therefore felt it necessary to draw the illustrations partly from purely physical phenomena, such as the electrical and magnetic properties of matter, and partly from subjects lying on the border between physics and physical chemistry, such as the specific heats of gases and solids and the behaviour of imperfect gases, which can best be discussed by considering both the experimental data and the atomic theory. It is assumed that the reader is fully familiar with quantum mechanics.

The selection of the applications given in Chapters 6–14 has been made on a somewhat arbitrary basis, and they represent those subjects in which I have been interested at some time or other. One major omission is the theory of liquids, which would require a whole book to itself. I have also omitted any reference to surface phenomena, and to the so-called thermodynamics of irreversible processes, a subject which is best treated as part of the theory of transport phenomena. On the other hand, Chapters 12 and 13, on solutions and electrochemical systems, have been included, after a considerable amount of hesitation, on the grounds that they constitute an introduction to problems of a physical chemical nature and that it is desirable to give some elementary illustrations of how the thermodynamic potentials in multi-component and multi-phase systems can be used and measured. No subject is discussed exhaustively, since this would result in a book of very considerably increased size, and to obtain a full account of any of the subjects the reader will therefore have to refer to more specialized books. In general, I have not given references to the sources of the older experimental data since these are all taken from the *International Critical Tables* and from Landolt & Börnstein's *Physikalisch-Chemische Tabellen*. Also, in view of the extensive bibliographies which are already available, I have not thought it necessary to discuss in any detail the historical aspects of the subject nor in general to give references to any but recent work.

I am much indebted to Dr F. C. Powell for a criticism of Chapters 1–5, to Sir Francis Simon for advice concerning the treatment of the third law of thermodynamics, and to Dr C. H. Bamford for commenting on the more chemical topics.

A. H. W.

LIST OF IMPORTANT SYMBOLS

$\mathbf{a}_i$ $(i = 1, 2, 3)$	Axes of the unit cell of a crystal.
a	Lattice constant.
a, a_i	Generalized coordinate of a thermodynamic system.
A, A_i	Generalized force of a thermodynamic system.
c	Concentration (in gram molecules per litre).
c	Velocity of light.
C_P	Specific heat at constant pressure.
C_V	Specific heat at constant volume.
$\mathscr{E}$, $\boldsymbol{\mathscr{E}}$	Electric field in electrostatic units.
$\mathscr{E}$	Electromotive force.
E	Energy of a particle or element.
$-\epsilon$	Charge on the electron in electrostatic units.
ζ	The thermodynamic potential of the conduction electrons in a metal.
ϑ	Empirical temperature.
ϑ	Reduced temperature.
Θ	The Debye characteristic temperature.
f	Fugacity.
F	Free energy at constant volume.
$\mathrm{F} = F/n$	Free energy at constant volume per gram molecule.
$\mathscr{F}$	Faraday's constant.
G	Free energy at constant pressure.
$\mathrm{G} = G/n$	Free energy at constant pressure per gram molecule.
h	Planck's constant.
$\hbar = h/(2\pi)$	
H	Heat function at constant pressure.
$\mathrm{H} = H/n$	Heat function at constant pressure per gram molecule.
H, $\mathbf{H}$	Magnetic field in magnetic units.
$\mathscr{H}$	Hamiltonian function.
i	Vapour-pressure constant.
I	Moment of inertia.
I, $\mathbf{I}$	Intensity of magnetization.
J	Chemical constant.
$\mathbf{k}$	Wave vector of an electronic wave function in a metal.
k	Boltzmann's constant.
K_c, K_p, K_V, K_x	Equilibrium constants.
L	Latent heat.

$\lambda = \mu/(kT)$	
$\mathscr{L}$	Avogadro's number.
m, M	Mass.
M^*	Molecular weight.
μ_0	Bohr magneton.
μ	Dipole moment.
μ	Joule-Thomson coefficient.
μ	Thermodynamic potential.
n	Number of gram molecules.
N	Number of molecules.
q_i	Generalized coordinate of a dynamical system.
Q	Quantity of heat.
p_i	Generalized momentum of a dynamical system.
p	Partial pressure.
P	Total pressure.
$P^\dagger$	Standard pressure.
$\mathscr{P}$, $\boldsymbol{\mathscr{P}}$	Electric polarization.
π	Reduced pressure.
ϖ	Weight factor.
R	Gas constant.
S	Entropy.
$s = S/n$	Entropy per gram molecule.
T	Absolute temperature.
$T^\dagger$	Standard temperature.
U	Internal energy.
$u = U/n$	Internal energy per gram molecule.
V	Volume.
$v = V/n$	Volume per gram molecule.
$\mathscr{V}$	Potential energy.
W	Mechanical work.
W	Number of complexions.
x_i	Mole fraction.
ϕ	Reduced volume.
χ	Electric or magnetic susceptibility.
ψ, Ψ	Wave function.
Z	Partition function.

Units

In theoretical work it is simplest to use c.g.s. units and to translate the results into practical units at the end of the calculation. Some of the conversion factors are as follows:

1 calorie = 4·184 joules = $4{\cdot}184 \times 10^7$ ergs.

1 joule = 0·239 calorie = 10^7 ergs.

0° C. is equivalent to 273·16° K.

1 atmosphere = $1{\cdot}0133 \times 10^6$ dynes/cm.^{2} = 1·0333 kg./cm.^{2} = 760 mm. Hg.

10^6 dynes/cm.^{2} = 0·9869 atmosphere = 1·0197 kg./cm.^{2} = 750 mm. Hg.

Physical constants

ϵ	Electronic charge	$4{\cdot}802 \times 10^{-10}$ e.s.u.
m	Mass of electron	$9{\cdot}106 \times 10^{-28}$ g.
c	Velocity of light	$2{\cdot}998 \times 10^{10}$ cm./sec.
$\mathscr{L}$	Avogadro's number	$6{\cdot}024 \times 10^{23}$ atoms/gram atom.
k	Boltzmann's constant	$1{\cdot}380 \times 10^{-16}$ erg/degree.
R	Gas constant	8·313 joules/degree/gram molecule (1·987 calories/degree/gram molecule).
h	Planck's constant	$6{\cdot}623 \times 10^{-27}$ erg sec.
$\mathscr{F}$	Faraday's constant	96,494 coulombs/gram equivalent.
μ_0	Bohr magneton	$9{\cdot}223 \times 10^{-21}$ erg/gauss.

Chapter 1

THE CLASSICAL DEVELOPMENT OF THERMODYNAMICS. THE FIRST LAW

THE AIM AND SCOPE OF THERMODYNAMICS

1·1. The methods used in thermodynamics are at the same time more general and less illuminating than those employed in other branches of theoretical physics. A full discussion of the reasons for this would take us too far afield, and so only a few remarks will be made upon the subject. The great generality of the results obtained by thermodynamic methods is a consequence of the theory being based solely upon the three very general principles known as the first, second and third laws of thermodynamics. No attempt is made to derive these principles from equations governing the behaviour of material bodies, and in this respect thermodynamics differs fundamentally from mechanics and electromagnetism. In the latter two subjects the energy equation plays an important part, but it can be derived from the Newtonian equations of motion and the Maxwellian equations of the electromagnetic field. These equations are more important than the equation of energy since the equation of energy may be derived from them, while the converse is not true, and some further principle, such as Hamilton's principle, must be invoked. In thermodynamics, however, we have nothing to work with except the three fundamental principles, and the results obtainable are therefore of the same nature as those that could be obtained in mechanics or electromagnetism if nothing more than the conservation of energy were used. For example, the equation of energy and the principle of virtual work tell us that the force on a system of conductors in an electrostatic field can be found without a knowledge of the surface density at all points of the conductors. We merely need to know the total charge on each conductor and the coefficients of potential as functions of position, but the equation of energy provides no means of determining the values of the coefficients of potential.

A thermodynamic argument normally results in a relation between various physical quantities associated with any type of substance, but thermodynamics provides no means of calculating any of the quantities occurring in the relation. There are a number of ways in which we may use such a thermodynamic relation in investigating the properties of a particular substance. (1) We may measure all the quantities which occur in the relation. The thermodynamic formula then provides a check upon the correctness of the measurements. (2) We may measure all but one of the quantities and use the thermodynamic formula to deduce the remaining quantity. This is particularly valuable when one quantity is much more difficult to measure than the others. For example, it is difficult to measure directly the specific heat at constant volume of a solid or a liquid, but it can be deduced by means of equation (3·212·3) on p. 39 from measurements that can easily be carried out. (3) We may use the general thermodynamic theory to determine which quantities must be measured in order to specify a system completely. This can be done by setting up what is known as a fundamental thermodynamic equation (see § 3·42). The fundamental equation may either be determined experimentally or calculated by means of statistical mechanics. Thermodynamics then enables us to deduce all the properties of the substance from this equation.

We shall not encounter the subject of statistical mechanics until Chapter 5. At the present stage it is sufficient to say that it is the branch of mechanics which attempts to derive the properties of matter in bulk from the equations of classical or quantum mechanics in cases where the number of particles involved is so great that a detailed treatment of the mechanical problem is impossible, and only the average properties of the system can be calculated.

When thermodynamics is supplemented by statistical mechanics it is no longer a general theory independent of the laws governing the detailed behaviour of matter, but while it becomes less general it becomes more informative, since the behaviour of particular systems can be deduced. Unfortunately, the number of systems which can be dealt with by statistical mechanics is very small, and, though rapid progress is being made, one of the main functions of thermodynamics for a long time to come will be to help to interpret the experimental results for systems which are too complicated to be dealt with entirely theoretically.

Another way in which thermodynamics differs from other subjects is that it only deals with equilibrium states and with transitions from one equilibrium state to another. It does not, of course, completely ignore kinetic changes, but since the subject has not been developed sufficiently to deal with phenomena which require an infinite number of macroscopic variables to specify them, it can only give a limited amount of information concerning such phenomena. The same limitation also applies to statistical mechanics, which essentially is concerned with calculating the equilibrium properties of matter in bulk from special models. The discussion of non-equilibrium states, that is, of transport phenomena, is much more difficult than the problems which fall into the sphere of statistical mechanics, and the theory of transport phenomena is less highly developed, its main spheres of applicability being to the motion of not too dense gases and of the electrons in a metal.

Fundamental Notions

1·2. The subject of thermodynamics deals with the extension of ordinary mechanics to those processes in nature in which the concept of temperature cannot be ignored; and to carry out this extension we have to introduce the two fundamental notions of temperature and quantity of heat. It is often assumed that the idea of temperature is a primary one, like the concepts of length and time, which cannot be analysed into anything simpler, and this is the viewpoint adopted in the first two chapters. In Chapter 4, however, we discuss this point more thoroughly, because, unless we can define temperature in terms of more fundamental properties, there is no hope of setting up a logical formulation of thermodynamics in terms of atomic dynamics, which is one of the principal aims of statistical mechanics.

1·21. *Temperature.* If we assume that temperature is a primary idea, the methods of measuring it must be to a certain extent empirical. We make use of the fact that a number of properties of substances depend in a continuous manner upon the temperature, such as the volume of a liquid, the volume of a gas under constant pressure, the pressure of a gas kept at constant volume, or the electrical resistance of a metal at constant pressure. We take advantage of these properties to construct test-bodies known as thermometers, and measure the temperature of a given body by bringing a thermometer into thermal contact with it, and by noticing the 'reading' on the thermometer when equilibrium has been established, that is, when the reading on the thermometer has reached a steady value. By comparing the results obtained by using different thermometers we can reject as unsuitable those substances, such as water, whose behaviour is anomalous, and, by taking suitable precautions, which are too well known to need mention here, we can set up a number of perfectly reproducible temperature scales. The commonest one, the centigrade scale, is obtained by taking as 0° C. the melting-point of pure ice and as 100° C. the boiling-point of pure water, both at standard atmospheric pressure. The subdivision of the temperature scale into degrees so that temperatures between 0 and 100° may be obtained by interpolation, and those outside that range by extrapolation, is not, however, unambiguous. For example, we may divide up the standard 100° interval by saying that every temperature interval must be proportional to the change in volume which it produces in a given mass of mercury. We then obtain the centigrade mercury scale. On the other hand, we may obtain another subdivision by taking the temperature interval to be proportional to the change in volume of a given mass of hydrogen gas at a given pressure. We then obtain the centigrade hydrogen scale at the given pressure, and the readings on such a hydrogen thermometer will not agree exactly with those on a mercury thermometer, except of course at 0 and 100°. There is, however, one scale, the centigrade thermodynamic or absolute scale, which is independent of the properties of any particular substance. This scale is defined in § 2·31, and it is the scale which is meant when we refer to the centigrade scale without qualification. But, for the moment, the existence of such an absolute scale is quite irrelevant. All that we require of a substance to be used in constructing a thermometer is that the property to be measured, such as length, volume, electrical resistance, etc., should be a strictly increasing function of the temperature. It is then possible to assign to every temperature, in a given scale, a definite numerical value. We do not need to inquire whether the size of the degree is the same between say 15 and 16° and between 420 and 421°, for the question has no meaning unless some criterion is laid down for determining whether two different temperature intervals are of the same magnitude. The only comparison that has any meaning is the comparison of two different scales of temperature, and until we have established the existence of the absolute scale of temperature, there is no reason for preferring one scale to another. We therefore choose any convenient scale by specifying the substance to be used in constructing the thermometer and the property which is to be measured. The temperature so defined will be called an empirical temperature and will be denoted by ϑ.

The variables required to specify the state of a body are the ordinary mechanical variables supplemented by the temperature. For example, for a single gas or liquid the pressure P, the volume V and the temperature ϑ suffice, whereas for mixtures we should require, in addition, variables specifying the composition, and for solids we should need to know the strain and stress tensors. These variables are not all independent but are connected by a number of relations known as the equations of state. For simple gases and liquids there is only one equation of state, of the form $f(P, V, \vartheta) = 0$, and we can use this relation to reduce the number of independent variables to two, which may be chosen to suit the particular problem under discussion. Any quantity which is a single-valued function of the independent variables chosen is called a function of the state. For most purposes it is sufficient to confine our attention to cases in which there is only one equation of state. More complicated cases can be considered as and when necessary.

1·22. *Heat.* The concept of quantity of heat is more difficult than the concept of temperature, since we do not normally observe quantities of heat directly but infer them from observing changes in temperature. The original experiments which led to an unambiguous definition of a quantity of heat were carried out by Joule, and his experiments deserve a more searching analysis than they usually receive. Joule determined the amount of mechanical work required to raise the temperature of a given mass of water through a given range, the water being stirred by paddles and the kinetic energy generated being dissipated by the viscosity of the water. He also carried out experiments in which the temperature of mercury was raised either by stirring with an iron paddle or by friction between two iron rings. The result of these and other experiments was to show that the amount of work required to raise the temperature of a body through a given range is independent of the exact manner in which the work is applied.

It is usually held that these results establish the equivalence of heat and work, but the adoption of this interpretation would make unnecessarily complicated the formulation of the first law of thermodynamics, namely, that heat is a form of energy and that energy is conserved. For example, if we adopted the above interpretation, we should be forced to say that when the temperature of a gas is raised by compression we have added heat to it, and this is by no means the usual method of describing what happens in this case. The simpler and better way of interpreting Joule's experiments is as follows. Joule showed that the amount of mechanical work required to change the state of a thermally insulated body is independent of how the work is applied and only depends upon the initial and final states of the body. It is therefore consistent to say that a body in a given state possesses a definite amount of energy, and the change in this energy from one state to another can be measured by the amount of mechanical work required to make the body pass from the initial to the final state when the body is thermally insulated. We call the energy of the body the internal energy and denote it by the symbol U. The function U contains an arbitrary constant, since only changes in U are directly measurable.

In all this there is no mention of heat, and to proceed any further we must have recourse to further experiments. It is found that there are processes in which the state of a body can change without work being done, and we then say that heat is absorbed or emitted by the body. For example, if we place two liquids in thermal contact which are at different temperatures ϑ_1 and ϑ_2 ($\vartheta_1 > \vartheta_2$), the final equilibrium state has temperature ϑ lying between ϑ_1 and ϑ_2 if no external work is done. It is found that the mechanical work required to raise the temperature of the first liquid from ϑ to ϑ_1 is equal to the mechanical work required to raise the temperature of the second liquid from ϑ_2 to ϑ when both liquids are thermally insulated, and we describe this phenomenon by saying that the first liquid has given out a certain amount of heat Q while the second liquid has absorbed the same amount of heat. We can also consider more general processes in which mechanical work $-W$ is communicated to the body but in which $-W$ is not equal to the increase in the internal energy. It is found that in any such process it is consistent to say that an amount of heat Q has been imparted to the body of such amount that

$$Q = \Delta U + W, \tag{1·22·1}$$

where ΔU is the increase in the internal energy.

Equation (1·22·1) is the definition of the quantity of heat, and it first requires the determination of the internal energy U for all states of the body under consideration. The next section is devoted to making these ideas more precise, while a still deeper analysis will be found in Chapter 4.

The First Law of Thermodynamics

1·3. We first require a number of definitions. If a body is in equilibrium in a 'vessel' and if its state can only be altered from outside either by action at a distance or by moving the 'wall' of the vessel, the vessel is said to be an adiabatic vessel.‡ By action at a distance we mean through the agency of such forces as gravitational and electromagnetic forces which do not need a material medium through which to be transmitted. For the moment we shall not consider such forces, so that in their absence the state of a body in an adiabatic vessel can only be changed by moving the wall. (Note that a vessel which transmits radiation is not adiabatic.)

We can state the definition of the word adiabatic in more mathematical terms as follows. If we have any general dynamical system, whose state is a function of certain parameters a_i, we can take the a_i's as generalized coordinates, corresponding to which there are generalized forces A_i which are such that in an infinitesimal displacement the work done by the system is $dW = \Sigma A_i da_i$. A_i is therefore the generalized force exerted by the system on its surroundings. If the state of the system can only be changed by varying at least one of the a_i's, the system is an adiabatic one.

An adiabatic process is one which takes place in an adiabatic vessel. A diathermal wall is one which is not adiabatic, that is, if a body is enclosed in a vessel, the whole or part of which is diathermal, changes can take place in the state of the body without the wall being moved.

‡ Unless the contrary is stated, we suppose that all vessels are impervious to matter.

The definition of an adiabatic process has been worded so as to cover not only such processes as the compression and expansion of a gas but also the stirring of a liquid and the rubbing together of two solid surfaces. When a liquid is being stirred by a paddle we must include the paddle as part of the 'wall'; for a solid the vessel and the wall usually consist of the surface of the solid itself so that the rubbing together of two solid surfaces can be considered as the tangential motion of the wall. Stirring and rubbing are the main processes concerned in Joule's experiments, and they are the only important ways in which the internal energy of liquids and solids can be changed by adiabatic processes. In one form of Joule's experiment a weight is attached to a paddle by means of pulleys. The weight falls slowly, and the temperature of the liquid (and of the calorimeter containing it, and of the paddle) increases. The result of the experiment is a decrease in the potential energy of the weight and an increase in the internal energy of the liquid, etc. According to the definition given above, this is an adiabatic process. For gases, expansions and compressions are as important as stirring, and are somewhat more familiar, though the only difference is that, if the expansions and compressions are slow enough, viscosity plays no part, whereas viscosity can never be neglected when the adiabatic process consists in stirring.

We can now state the results of Joule's experiments in the following way. If a body (or system of bodies) is brought by an adiabatic process from one state to another, the mechanical work necessary is independent of the intermediate states of the body and depends only upon the initial and final states. If, for example, we consider a fluid, the state of which can be specified by the parameters P, V, and if we keep the initial state fixed (defined by P_0, V_0) and vary the final state, we can define a function $U(P, V)$, called the internal energy of the body, by the relation

$$U(P, V) = U(P_0, V_0) - W, \qquad (1\cdot3\cdot1)$$

where $-W$ is the mechanical work which must be done by outside agencies to change the state from (P_0, V_0) to (P, V). Alternatively we may say that

$$\oint dW = 0 \qquad (1\cdot3\cdot2)$$

for all adiabatic cyclic processes, from which the existence of U as a function of the state follows at once.

The function U is a measurable quantity and we suppose that it is known for all values of the parameters of state P, V (see §4·32). We now consider processes in which part of the wall of the vessel containing the body is diathermal. We define the heat Q absorbed by the body by the equation

$$Q = \Delta U + W, \qquad (1\cdot3\cdot3)$$

where ΔU is the increase in the internal energy (and is known since U is known for all states) and where W is the mechanical work done by the body, ΔU depends only upon the initial and final states, whereas Q and W depend upon the particular way in which the change is carried out. Thus the heat absorbed and the mechanical work done are not functions of the state.

1·31. Having defined quantity of heat in a way which only involves the mechanical variables P, V and not the temperature, we can now give the following criterion for determining which of two bodies is the hotter, without being involved in a circular argument. We define the temperature ϑ_A of a body A to be greater than the temperature ϑ_B of a body B if, when the bodies are put in contact through a diathermal wall and neither body performs any mechanical work, heat flows from A to B, i.e. U_A decreases and U_B increases. This definition is necessary to ensure that the temperature scale used is in accordance with our intuitive ideas about temperature, and in effect it places a limitation upon the properties and substances which can usefully be employed in constructing thermometers. Thus it prevents us from taking ϑ to be the volume of a certain mass of water if the temperature ranges considered include the region in which water has its maximum density (at 4° C.). It excludes our using an inverse power of the pressure of a gas kept at constant volume, though it allows us to use any strictly increasing function of the pressure, and it excludes, of course, any multi-valued functions of the pressure. In practical cases there is never any doubt as to whether a given temperature scale is a legitimate one or not.

In statistical mechanics, however, an inverse temperature scale $1/\vartheta$ (or even $-1/\vartheta$) occurs naturally. There is no objection to using such scales so long as we do not mind having one body hotter than another when the temperature of the first is less than that of the second (or if we are prepared to have all temperatures negative instead of positive), but we must then change all the definitions accordingly. If we keep to the normal convention, the temperature scales allowable are such that ϑ is positive and that $(\partial U/\partial\vartheta)_V > 0$ for any stable substance.‡

‡ Since any two of P, V and ϑ can be treated as the independent variables it is always necessary in any partial derivative to specify the variable (or variables) that are being kept constant. The usual way of indicating this in thermodynamics is to place brackets round the derivative and to add the variable which is being kept constant as a suffix. Thus $(\partial U/\partial\vartheta)_V$ means that V is to be considered constant in the differentiation.

In pure mathematics it is normal to use a different functional symbol for the dependent variable when the independent variables are changed. Thus if we change from the independent variables P, V to ϑ, V we would normally write the energy as

$$U(P, V) = U\{P(V, \vartheta), V\} = \mathscr{U}(\vartheta, V).$$

In physics, on the other hand, it is inconvenient to use different functional symbols

Since heat has been defined so as to make the conservation of energy valid, we can use the erg as the unit of heat as well as of energy. It has, however, been customary to use the calorie, which is defined as the amount of energy required to raise 1 gram of water from $14\frac{1}{2}$ to $15\frac{1}{2}°$ C. The calorie is found by measurement to be equal to $41\cdot84 \times 10^6$ ergs $= 4\cdot184$ joules, but the modern tendency is to avoid the use of the calorie as a separate unit. Further, it is impossible for any thermodynamic equation to contain a symbol for the 'mechanical equivalent of heat'. Such a symbol can only appear when two different units of measurement are used for the physical quantities which occur in the equation, and the equation is not then invariant when the units are changed.

Quasi-static Processes

1·4. Consider a gas contained in a cylinder closed by a movable piston, and suppose that the gas is expanding and at the same time receiving heat from outside. When the gas is in equilibrium its state is completely determined by its volume and the pressure which it exerts upon the piston, but this is no longer true in general when the gas is in a state of motion. We then require an infinite number of variables to specify the behaviour of a moving gas, for we must know the pressure tensor and the velocity at every point in the gas. The problem of describing such an expansion is therefore a hydrodynamical one of great complexity, and no progress is possible if we try to follow the motion in all its details. We usually have to content ourselves with determining the difference in the properties of the system before and after a dynamical change, the initial and final states of the system being equilibrium states which can therefore be described by a finite number of parameters.

There is, however, one type of process which can easily be described. Considering once more the expanding gas, suppose that the motion of the piston is resisted by a force P^* per unit area, while the force exerted on the piston per unit area by the gas is P. If $P = P^*$ there can be no motion, but if P^* is infinitesimally less than P the gas will expand; but the kinetic energy generated will be infinitesimal and the gas will still be effectively in an equilibrium state. If we adjust P^* so that it is always infinitesimally less than P, the gas will undergo a finite expansion while always remaining effectively in equilibrium. Such an ideal expansion is called a quasi-static expansion and cannot be realized in practice, though we may approximate to it in favourable circumstances. Two important properties of a quasi-

for the same physical quantity, and therefore by $U(P, V)$ and $U(\vartheta, V)$ we mean the internal energy expressed as a function of P, V and of ϑ, V respectively, whatever the particular functional dependence on the variables may be.

static expansion are, first, that during the expansion the state of the body can be described by the parameters, finite in number, which are sufficient to define an equilibrium state, and secondly, that the motion can be reversed at any stage by an infinitesimal increase in the external forces. If heat is absorbed in the direct process the same amount of heat must be abstracted in the reversed process. For these reasons a quasi-static process is often referred to as a reversible process.

The formal definition of a quasi-static process is as follows:

A quasi-static change in which a body passes from the equilibrium state 1 *to the equilibrium state* 2 *is such that there is a linear continuum of equilibrium states to which* 1 *and* 2 *belong, and such that the body will successively take the states connecting* 2 *to* 1 *by a simple reversal of sign of the displacements, of the work done and of the heat absorbed.*

We now consider the work done in a quasi-static process. If a gas expands quasi-statically from volume V_1 to volume V_2 ($V_2 > V_1$), the work done is $\int_{V_1}^{V_2} P(V)\,dV$, where $P(V)$ is the pressure when the volume is V. This is obvious when the gas expands in a cylinder of constant cross-section. When the movable part of the wall is not plane the result can be proved as follows. Consider an element dS of the bounding surface and let it be displaced an infinitesimal (vector) distance $\mathbf{\delta s}$. Then if $\mathbf{n}$ is the unit vector in the direction of the outward normal to the surface, the work done by the gas in the infinitesimal displacement is $\int P\,\mathbf{\delta s}\,.\,\mathbf{n}\,dS$, the integral being taken over the bounding surface. But, since the expansion is quasi-static, P has the same value at all points at a given time. Also $\mathbf{\delta s}\,.\,\mathbf{n}\,dS$ is the volume of the slant cylinder formed by the displacement of the element of surface dS, and hence $\int \mathbf{\delta s}\,.\,\mathbf{n}\,dS$ is the change in volume δV in the displacement. The work done is therefore $P(V)\,\delta V$, and the result follows. For a general dynamical system the work done is $\Sigma A_i\,da_i$.

The work done in a quasi-static expansion is the maximum work that can be obtained if V_1 and V_2 are given, since the expansion cannot take place unless the external pressure is less than or equal to $P(V)$, while if the external pressure is less than $P(V)$ kinetic energy will be generated in the gas, and this energy will ultimately be dissipated by the viscosity of the gas. Any amount of mechanical work between 0 and $\int_{V_1}^{V_2} P(V)\,dV$ can be obtained in the expansion from volume V_1 to volume V_2 by allowing the gas to expand against a suitable external pressure. When the work done is not the maximum attainable, the final internal energy is greater than that for the quasi-static process since the total energy must be conserved. Also, since U and P increase together at constant volume for any gas, the final pressure is greater than that for the quasi-static process. Similarly, if we consider a compression from volume V_2 to volume V_1, $\int_{V_1}^{V_2} P(V)\,dV$ is the minimum work required to effect the compression. Any greater amount of work can be applied to effect the compression, and then the final pressure and internal energy are greater than those for the quasi-static process. (It is assumed that, in comparing the work done and the change in internal energy in different processes, the heat absorbed is the same.)

In general, there is no simple expression for the work done in an expansion of a gas in terms of its pressure, which in any case is not a well-defined quantity if the expansion is irreversible. The work done may be obtained in terms of the external pressure, but this is independent of the pressure distribution in the gas. There are, however, two cases in which the work done can be calculated for irreversible processes. The first is when the process takes place at constant volume; the work done is then zero. The second case is when the expansion takes place at constant pressure; the work done is then $P\Delta V$, where ΔV is the increase in volume.

The above results may be summarized as follows. In an infinitesimal quasi-static change involving a single gas

$$dQ = dU + P\,dV, \tag{1·4·1}$$

where dQ is the infinitesimal amount of heat absorbed. In any change (finite or infinitesimal) at constant volume

$$\Delta Q = \Delta U, \tag{1·4·2}$$

while in any change at constant pressure

$$\Delta Q = \Delta U + P\Delta V. \tag{1·4·3}$$

For a general dynamical system (1·4·1) must be replaced by

$$dQ = dU + \Sigma A_i\,da_i. \tag{1·4·4}$$

Applications of the First Law

1·5. In dealing with equation (1·4·1) it is essential to remember that although dQ, the heat absorbed by the body, is an infinitesimal, it is not a perfect differential, that is, it is not the differential of a quantity Q which is a function of the parameters defining an equilibrium state of the body. On the other hand, U is a function of these parameters, i.e. U is a function of the state, and so dU is a perfect differential.

Let us take ϑ and V as the independent variables defining the state of a body. Then, in an infinitesimal change at constant volume

$$dQ = \frac{\partial U(\vartheta, V)}{\partial \vartheta} d\vartheta \quad (dV = 0). \tag{1·5·1}$$

We define C_V, the specific heat at constant volume, as the amount of heat required to raise the temperature of unit mass of the substance by unity, the volume being kept constant. The value of the specific heat therefore depends upon the unit of mass, which may be chosen to be either the gram or the gram molecule (or mole). The gram has the advantage that the unit of mass is the same for all substances, but the gram molecule is in many ways more convenient, especially when dealing with gases. In general, we shall use the gram molecule, and denote the mass of a body in gram molecules by the symbol n. (The mass m in grams is nM^*, where M^* is the molecular weight.) Hence, if U is the internal energy and V the volume of n gram molecules of a substance, and if we put

$$v = U/n, \quad \mathrm{v} = V/n, \tag{1·5·2}$$

we have

$$C_V = \frac{\partial v(\vartheta, V)}{\partial \vartheta} = \left(\frac{\partial v}{\partial \vartheta}\right)_V. \tag{1·5·3}$$

Owing to the restrictions placed upon the permissible temperature scales in § 1·31, C_V is necessarily positive.

Now consider an infinitesimal reversible change. We have

$$dQ = \left(\frac{\partial U}{\partial \vartheta}\right)_V d\vartheta + \left\{\left(\frac{\partial U}{\partial V}\right)_\vartheta + P\right\} dV. \tag{1·5·4}$$

We define L_V, the latent heat of expansion, as the amount of heat required to increase the volume of unit mass of the substance by unity in a reversible manner, the temperature being kept constant. Then

$$L_V = P + (\partial U/\partial V)_\vartheta, \tag{1·5·5}$$

and in a reversible change (1·4·1) can be written

$$dQ = nC_V d\vartheta + L_V dV. \tag{1·5·6}$$

Similar equations hold at constant pressure. In any infinitesimal change at constant pressure we have, taking ϑ and P as the independent variables,

$$dQ = \left(\frac{\partial U(\vartheta, P)}{\partial \vartheta} + P\frac{\partial V(\vartheta, P)}{\partial \vartheta}\right) d\vartheta \quad (dP = 0). \tag{1·5·7}$$

Hence, if we define C_P, the specific heat at constant pressure, as the amount of heat required to raise the temperature of unit mass of the substance by unity, the pressure remaining constant, we have

$$C_P = \left(\frac{\partial v}{\partial \vartheta}\right)_P + P\left(\frac{\partial \mathrm{v}}{\partial \vartheta}\right)_P = \frac{\partial}{\partial \vartheta}(v + P\mathrm{v})_P. \tag{1·5·8}$$

Also in any reversible change we have

$$dQ=\left\{\left(\frac{\partial U}{\partial \vartheta}\right)_P+P\left(\frac{\partial V}{\partial \vartheta}\right)_P\right\}d\vartheta+\left\{\left(\frac{\partial U}{\partial P}\right)_\vartheta+P\left(\frac{\partial V}{\partial P}\right)_\vartheta\right\}dP=nC_P d\vartheta+nL_P dP, \tag{1·5·9}$$

and L_P, the heat required to raise the pressure of unit mass of the substance by unity in a reversible manner, the temperature remaining constant, is given by

$$L_P=\left(\frac{\partial U}{\partial P}\right)_\vartheta+P\left(\frac{\partial v}{\partial P}\right)_\vartheta. \tag{1·5·10}$$

We can obtain a relation between C_V and C_P which is sometimes useful. If we treat v as a function of ϑ and P we have

$$\left(\frac{\partial U}{\partial \vartheta}\right)_P=\frac{\partial}{\partial \vartheta}U\{\vartheta, v(\vartheta,P)\}=\left(\frac{\partial U}{\partial \vartheta}\right)_V+\left(\frac{\partial U}{\partial v}\right)_\vartheta\left(\frac{\partial v}{\partial \vartheta}\right)_P. \tag{1·5·11}$$

Substituting this in (1·5·8) we obtain

$$C_P=C_V+\left\{\left(\frac{\partial U}{\partial v}\right)_\vartheta+P\right\}\left(\frac{\partial v}{\partial \vartheta}\right)_P=C_V+L_V\left(\frac{\partial v}{\partial \vartheta}\right)_P. \tag{1·5·12}$$

Theorems on Partial Differentiation

1·6. The variables P, V, ϑ which specify the state of a simple substance are not all independent, and it is a matter of convenience which pair we regard as the independent variables. We are therefore often faced with the problem of changing from one pair of independent variables to another. This is discussed in any book on differential calculus, but it is convenient to collect here the formulae most frequently used. The most important formulae are (1·6·4) and (1·6·6); those derived in subsection (iii) are only necessary for complicated systems, and are given here mainly for ease of reference.

(i) Let x, y, z be three variables satisfying the functional relation $f(x,y,z)=0$. Then we may regard any one of x, y, z as being a function of the other two and we can define partial differential coefficients accordingly. There are six such derivatives, but only two are independent. The four relations between them can be found as follows.

We have

$$\frac{\partial f}{\partial x}dx+\frac{\partial f}{\partial y}dy+\frac{\partial f}{\partial z}dz=0. \tag{1·6·1}$$

Hence, taking z constant, we find

$$\left(\frac{\partial x}{\partial y}\right)_z=-\frac{\partial f}{\partial y}\bigg/\frac{\partial f}{\partial x}, \tag{1·6·2}$$

where the brackets round $\partial x/\partial y$ and the suffix z denote that z is to be kept constant in the differentiation, that is, that y and z are the independent variables. Also

$$\left(\frac{\partial y}{\partial x}\right)_z = -\frac{\partial f}{\partial x}\Big/\frac{\partial f}{\partial y}, \tag{1·6·3}$$

and so we have the relation

$$\left(\frac{\partial x}{\partial y}\right)_z = 1\Big/\left(\frac{\partial y}{\partial x}\right)_z \tag{1·6·4}$$

and two similar relations.

From (1·6·3) we find the fourth relation

$$\left(\frac{\partial x}{\partial y}\right)_z\left(\frac{\partial y}{\partial z}\right)_x = \frac{\partial f}{\partial z}\Big/\frac{\partial f}{\partial x} = -\left(\frac{\partial x}{\partial z}\right)_y. \tag{1·6·5}$$

By (1·6·4) this can be written in the symmetrical form

$$\left(\frac{\partial x}{\partial y}\right)_z\left(\frac{\partial y}{\partial z}\right)_x\left(\frac{\partial z}{\partial x}\right)_y = -1. \tag{1·6·6}$$

(ii) Suppose that we have a number of quantities $y_1, y_2, \ldots, y_n$ which are all functions of two variables u and v. Then we often have to calculate derivatives when one of the quantities $y_1, \ldots, y_n$ is kept constant, or we may have to consider derivatives of u or v with respect to one of the other quantities. The formulae required are as follows.

We have

$$dy_i = \left(\frac{\partial y_i}{\partial u}\right)_v du + \left(\frac{\partial y_i}{\partial v}\right)_u dv, \quad dy_j = \left(\frac{\partial y_j}{\partial u}\right)_v du + \left(\frac{\partial y_j}{\partial v}\right)_u dv. \tag{1·6·7}$$

To treat u as a function of y_i and v, we put $dv=0$ and $dy_i=0$ successively in the first of these equations; we obtain

$$\left(\frac{\partial u}{\partial y_i}\right)_v = 1\Big/\left(\frac{\partial y_i}{\partial u}\right)_v \tag{1·6·8}$$

and

$$\left(\frac{\partial u}{\partial v}\right)_{y_i} = 1\Big/\left(\frac{\partial v}{\partial u}\right)_{y_i} = -\left(\frac{\partial y_i}{\partial v}\right)_u\Big/\left(\frac{\partial y_i}{\partial u}\right)_v \tag{1·6·9}$$

These are merely particular cases of (1·6·4) and (1·6·5).

To calculate a derivative such as $(\partial y_i/\partial u)_{y_j}$ we put $dy_j=0$ in the second equation of (1·6·8) and eliminate dv. Using the standard notation for Jacobians we obtain

$$\left(\frac{\partial y_i}{\partial u}\right)_{y_j} = 1\Big/\left(\frac{\partial u}{\partial y_i}\right)_{y_j} = \frac{\partial(y_i, y_j)}{\partial(u, v)}\Big/\left(\frac{\partial y_j}{\partial v}\right)_u. \tag{1·6·10}$$

To calculate $(\partial y_i/\partial y_j)_{y_k}$ we eliminate du and dv from the equations (1·6·7) and the equation

$$dy_k = \left(\frac{\partial y_k}{\partial u}\right)_v du + \left(\frac{\partial y_k}{\partial v}\right)_u dv. \tag{1·6·11}$$

The result is
$$\left(\frac{\partial y_i}{\partial y_j}\right)_{y_k} = \frac{\partial(y_i, y_k)}{\partial(u,v)} \bigg/ \frac{\partial(y_j, y_k)}{\partial(u,v)}. \tag{1·6·12}$$

(iii) The formulae of the preceding subsection can be generalized to the case in which there are n independent variables $x_1, x_2, \ldots, x_n$. Let the dependent variables be $y_1, y_2, \ldots, y_m$. Then

$$\left.\begin{aligned} dy_1 &= \frac{\partial y_1}{\partial x_1} dx_1 + \ldots + \frac{\partial y_1}{\partial x_n} dx_n, \\ &\vdots \\ dy_m &= \frac{\partial y_m}{\partial x_1} dx_1 + \ldots + \frac{\partial y_m}{\partial x_n} dx_n. \end{aligned}\right\} \tag{1·6·13}$$

We can define a differential coefficient such as $\partial y_1/\partial x_1$ subject to $r-1$ of the y's, say $y_2, \ldots, y_r$ and $n-r$ of the x's, say $x_{r+1}, \ldots, x_n$ remaining constant, the total number of variables having constant values being always $n-1$. To calculate such a derivative we consider the r equations of the set (1·6·13) which involve $y_1, y_2, \ldots, y_r$ and put

$$dy_2 = dy_3 = \ldots = dy_r = 0, \quad dx_{r+1} = \ldots = dx_n = 0.$$

We then obtain a set of r linear inhomogeneous equations for the r differentials $dx_1, \ldots, dx_r$ which can be solved by Cramer's rule. The result is

$$\frac{dx_1}{\dfrac{\partial(y_2, y_3, \ldots, y_r)}{\partial(x_2, x_3, \ldots, x_r)}} = \ldots = \frac{(-1)^{r-1} dx_r}{\dfrac{\partial(y_2, y_3, \ldots, y_r)}{\partial(x_1, x_2, \ldots, x_{r-1})}} = \frac{dy_1}{\dfrac{\partial(y_1, y_2, \ldots, y_r)}{\partial(x_1, x_2, \ldots, x_r)}}. \tag{1·6·14}$$

Hence
$$\left(\frac{\partial y_1}{\partial x_1}\right)_{y_2, \ldots, y_r, x_{r+1}, \ldots, x_n} = 1 \bigg/ \left(\frac{\partial x_1}{\partial y_1}\right)_{y_2, \ldots, y_r, x_{r+1}, \ldots, x_n}$$
$$= \frac{\partial(y_1, y_2, \ldots, y_r)}{\partial(x_1, x_2, \ldots, x_r)} \bigg/ \frac{\partial(y_2, y_3, \ldots, y_r)}{\partial(x_2, x_3, \ldots, x_r)}. \tag{1·6·15}$$

The value of the derivative $\partial x_i/\partial x_j$, subject to $y_2, \ldots, y_r$ and $x_{r+1}, \ldots, x_n$ remaining constant, can be found directly from (1·6·14) by disregarding the last expression involving dy_1, which is now irrelevant. We find, for $i, j \neq r+1, r+2, \ldots, n$,

$$\left(\frac{\partial x_i}{\partial x_j}\right)_{y_2, \ldots, y_r, x_{r+1}, \ldots, x_n} = 1 \bigg/ \left(\frac{\partial x_j}{\partial x_i}\right)_{y_2, \ldots, y_r, x_{r+1}, \ldots, x_n}$$
$$= (-1)^{i+j} \frac{\dfrac{\partial(y_2, \ldots, y_r)}{\partial(x_1, \ldots, x_{j-1}, x_{j+1}, \ldots, x_r)}}{\dfrac{\partial(y_2, \ldots, y_r)}{\partial(x_1, \ldots, \partial x_{i-1}, \partial x_{i+1}, \ldots, x_r)}}. \tag{1·6·16}$$

In a similar manner we can find the derivative $\partial y_1/\partial y_2$ subject to $y_3, \ldots, y_r$ and $x_r, \ldots, x_n$ remaining constant. In this case we have, from (1·6·14),

$$\frac{dy_1}{\dfrac{\partial(y_1, y_3, \ldots, y_r)}{\partial(x_1, x_2, \ldots, x_{r-1})}} = \frac{dx_1}{\dfrac{\partial(y_3, \ldots, y_r)}{\partial(x_2, \ldots, x_{r-1})}}$$

and

$$\frac{dy_2}{\dfrac{\partial(y_2, \ldots, y_r)}{\partial(x_1, \ldots, x_{r-1})}} = \frac{dx_1}{\dfrac{\partial(y_3, \ldots, y_r)}{\partial(x_2, \ldots, x_{r-1})}}$$

Hence

$$\left(\frac{\partial y_1}{\partial y_2}\right)_{y_3, \ldots, y_r, x_r, \ldots, x_n} = 1 \Big/ \left(\frac{\partial y_2}{\partial y_1}\right)_{y_3, \ldots, y_r, x_r, \ldots, x_n}$$

$$= \frac{\partial(y_1, y_3, \ldots, y_r)}{\partial(x_1, x_2, \ldots, x_{r-1})} \Big/ \frac{\partial(y_2, y_3, \ldots, y_r)}{\partial(x_1, x_2, \ldots, x_{r-1})} \qquad (1·6·17)$$

Other differential coefficients can be found in a similar way.

Chapter 2

THE CLASSICAL DEVELOPMENT OF THERMODYNAMICS. THE SECOND LAW

Carnot's Theorem and the Second Law of Thermodynamics

2·1. By itself, the first law of thermodynamics is not very illuminating since it merely shows that the internal energy is a function of the state and provides a definition of heat. Neither the heat absorbed by a system nor the work done by it is a function of the state, and on this account it is impossible to make much progress by using the first law alone, since the only physical quantities whose changes can be calculated in irreversible processes by thermodynamic methods are those which are functions of the state. The second law of thermodynamics, which we now consider, proves the existence of a second function of the state, the entropy; and with two functions of the state at our disposal we can deal with a much wider class of phenomena than can be discussed by means of the first law above.

2·11. In a quasi-static process the infinitesimal variation of a function of the state is necessarily a perfect differential, and so far we have only found one, namely, dU. But although the infinitesimal amount of heat dQ absorbed in such a process is not a perfect differential it is a multiple of a perfect differential. This result is known as Carnot's theorem, of which a formal statement is as follows:

There exist two functions of the state S and T, where T is a positive function of the empirical temperature ϑ only, such that, in any infinitesimal quasi-static change of a body or system of bodies, $dQ = T\,dS$.

The function T is called the absolute temperature, and the function S is called the entropy. There is a further theorem that, in any spontaneous process, the entropy always increases.

These two theorems are consequences of the very general principle known as the second law of thermodynamics, which states that certain processes are impossible. The second law differs from the usual laws of physics in that it is of a purely negative nature and this makes it difficult to grasp its full import. In its simplest form the principle merely states that irreversible processes exist (generalizations C and C′, pp. 78 and 81), but it is usual to adopt a less abstract approach and to specify certain processes which cannot be reversed. The two most common formulations are due to Clausius and Kelvin, who took as the examples of irreversible process, first, the conduction of heat and, secondly, the transformation of mechanical work into the internal energy of a body. Based upon these two processes, we have the following alternative statements of the second law.

Clausius's principle. No cyclic process exists which has as its sole effect the transference of heat from a colder to a hotter body.

Kelvin's principle. No cyclic process exists which produces no other effect than the extraction of heat from a body and its conversion into an equivalent amount of mechanical work.

The above statements do not exclude the possibility of transferring heat from a colder to a hotter body or of converting heat into work, but they deny that these effects can be produced by cyclic processes. Alternative statements which indicate when these possibilities can occur are as follows: (1) It is impossible to transfer heat from a colder to a hotter body without the expenditure of mechanical work or unless some other compensating change takes place in the systems concerned. (2) It is impossible to abstract heat from a body at a given temperature and convert it into work unless a positive amount of heat is given out at a lower temperature or another compensating change takes place in the systems concerned. In these alternative statements of Clausius's and Kelvin's principles, the nature of the compensating changes is necessarily left vague since they can be of almost any kind, depending upon the nature of the system. (Typical compensating changes are changes in volume and changes in composition.) Such compensating changes cannot occur in cyclic processes, since all the systems return to their initial states at the end of a cycle, and it is therefore advantageous in some ways to formulate the principles in terms of cyclic processes.

The second law is often stated in the form that no perpetual motion machine of the second kind exists, that is, a machine which will convert the whole of the internal energy of a body into mechanical work. (A perpetual motion machine of the first kind is one which obtains work from nothing, that is, a machine which contravenes the principle of the conservation of energy.) This is, in effect, Kelvin's principle; and Clausius's principle is equivalent to it, since if Clausius's principle were false it would be possible to construct a perpetual motion machine of the second kind in the following way. We construct a heat engine which takes in a certain amount of heat from a source, converts part of it into work, and gives out the remainder of the heat at a lower temperature. We now use a cyclic process to transfer this heat to the original source, and by hypothesis this can be done without expending any work. By repeating the process the whole of the internal energy of the source could be transformed into work, and since this is impossible Clausius's principle cannot be false.

We now proceed to prove Carnot's theorem by means of reversible cycles. In chapter 4 an entirely different proof is given.

Carnot's Cycle

2·2. Consider a system whose state is defined by the two variables P and V, and suppose that we have two large sources or sinks of heat at temperatures ϑ_1 and ϑ_2 ($\vartheta_1 > \vartheta_2$). Let the system initially have temperature ϑ_1, and be in a state A with parameters P_A and V_A. Let the system be put in thermal contact with the source whose temperature is ϑ_1, and let it expand quasi-statically and isothermally (that is, at constant temperature) until it reaches the state B, with parameters P_B and V_B. In this isothermal process an amount of heat Q_1 is abstracted from the source, and the system performs a certain amount of external work. Now insulate the system thermally and let it undergo a quasi-static adiabatic expansion until it reaches the state C, the temperature then being ϑ_2 and the parameters of state being P_C, V_C. Next put the system in thermal contact with the source whose temperature is ϑ_2, and compress the system quasi-statically at constant temperature ϑ_2, until it reaches the state D, with parameters P_D, V_D, such that the system can return to the initial state A by a quasi-static adiabatic compression. In this isothermal process work is done on the system and it gives up an amount of heat Q_2 to the source. Finally, isolate the system thermally and make it return to the initial state A by an adiabatic compression.

The cycle, which is known as Carnot's cycle, is shown diagrammatically in fig. 2·1, which gives what is known as the P, V diagram of the cycle. The curves connecting the points A, B, C and D show the succession of states occupied by the system, in the two isothermal and two adiabatic processes.

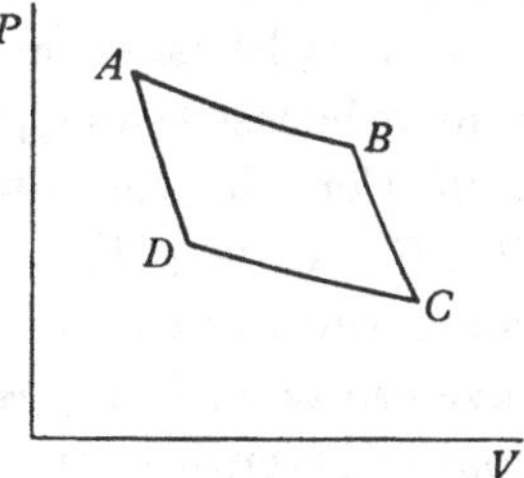

Fig. 2·1. The Carnot cycle in the P, V diagram.

Since all the processes take place quasi-statically, and since heat is only taken in and given out when the temperature of the system is the same as that of a source, the cycle is clearly reversible. Also the work done in any of the processes is $\int P\,dV$ taken between the appropriate limits, and the net work done by the system is therefore the area of the curvilinear quadrilateral $ABCD$ in the P, V diagram. Let this work be W. Then by the first law we must have

$$W = Q_1 - Q_2. \tag{2·2·1}$$

The efficiency of the system is defined to be the ratio of the work done by the system to the heat taken in, that is, the efficiency is

$$W/Q_1 = 1 - Q_2/Q_1. \tag{2·2·2}$$

We now prove a number of theorems concerning Carnot cycles.

THEOREM 1. *The efficiency of a Carnot cycle depends only upon ϑ_1 and ϑ_2.*

In the first place the efficiency can only depend upon the nature of the cycle and not upon the amount of the working substance. For, by combining a number of identical cycles, we can increase W, Q_1 and Q_2 in the same proportions, and it is clear that it is only the ratios that matter. Further, W, Q_1 and Q_2 are all positive in the direct cycle or all negative in the reversed cycle. We take them to be positive. We give two proofs of the theorem based upon Clausius's and Kelvin's principles respectively.

(*a*) Consider two independent systems which can undergo Carnot cycles between the same temperatures ϑ_1 and ϑ_2. We can suppose that their sizes are adjusted so that $W = W'$, where undashed letters refer to the first system and dashed letters to the second system. If $Q_1 \neq Q_1'$, suppose that $Q_1 < Q_1'$.

We now make a composite system in which the first system carries out its Carnot cycle while the second system carries out its Carnot cycle reversed. The composite system takes in a negative amount of heat $Q_1 - Q_1'$ at the temperature ϑ_1 and gives out a negative amount of heat

$$Q_2 - Q_2' = Q_1 - Q_1'$$

at temperature ϑ_2, while no work is performed. Thus heat has been conveyed by a cyclic process from a lower to a higher temperature, and there are no compensating processes. This contradicts Clausius's principle, and hence Q_1 cannot be less than Q_1'. By a similar argument we can prove that Q_1' cannot be less than Q_1, and it therefore follows that $Q_1 = Q_1'$.

(*b*) Consider the same two systems as before and adjust them so that $Q_1 = Q_1'$. If $W \neq W'$, suppose that $W > W'$, and hence that $Q_2 < Q_2'$. The composite system consisting of the first system and the second system reversed takes in a positive amount of heat $Q_2' - Q_2$ at temperature ϑ_2 and converts it entirely into work. This contradicts Kelvin's principle and hence Q_2 cannot be less than Q_2'. By a similar argument Q_2' cannot be less than Q_2, and hence $Q_2 = Q_2'$.

We have now proved that the efficiency of a Carnot cycle is independent of the particular substance undergoing the cycle. It therefore follows that

$$Q_1/Q_2 = f(\vartheta_1, \vartheta_2), \tag{2·2·3}$$

where $f(\vartheta_1, \vartheta_2)$ is a universal function of ϑ_1 and ϑ_2. This is Theorem 1.

THEOREM 2. *In any Carnot cycle $Q_1/Q_2 = \phi(\vartheta_1)/\phi(\vartheta_2)$, where $\phi(\vartheta)$ is a universal function of the temperature ϑ.*

The theorem states that the universal function $f(\vartheta_1, \vartheta_2)$, whose existence is shown by theorem 1, is of an especially simple kind. To prove it we consider two cycles. The first takes in an amount of heat Q_1 at temperature ϑ_1 and gives out an amount of heat Q_2 at temperature ϑ_2. The second takes in an amount of heat Q_2 at temperature ϑ_2 and gives out an amount of heat Q_3 at temperature ϑ_3. By theorem 1 we have

$$Q_1/Q_2 = f(\vartheta_1, \vartheta_2), \quad Q_2/Q_3 = f(\vartheta_2, \vartheta_3). \tag{2·2·4}$$

But, if we consider the two cycles combined to form a third cycle, the third cycle takes in an amount of heat Q_1 at temperature ϑ_1 and gives out an amount of heat Q_3 at temperature ϑ_3. Hence

$$Q_1/Q_3 = f(\vartheta_2, \vartheta_3). \tag{2·2·5}$$

Since the two equations in (2·2·4) must imply equation (2·2·5) we have

$$f(\vartheta_1, \vartheta_2) = f(\vartheta_1, \vartheta_3)/f(\vartheta_2, \vartheta_3) \tag{2·2·6}$$

for all values of ϑ_1, ϑ_2 and ϑ_3. Since the left-hand side of (2·2·6) does not contain ϑ_3, ϑ_3 cannot occur on the right-hand side, and the solution of the functional equation is

$$f(\vartheta_1, \vartheta_2) = \phi(\vartheta_1)/\phi(\vartheta_2). \tag{2·2·7}$$

Since Q_1 and Q_2 are either both positive or both negative, we can take $\phi(\vartheta)$ to be a positive function.

THEOREM 3. *There exists a positive function T of the empirical temperature ϑ such that the efficiency of a Carnot cycle is $1 - T_2/T_1$.*

To prove this we have only to take T equal to $\alpha\phi(\vartheta)$, where α is a positive constant.

THEOREM 4. *In any Carnot cycle, $\Sigma Q/T = 0$, where Q is the (positive or negative) amount of heat taken in at temperature T.*

This follows at once from theorem 2 by using the definition of T given by theorem 3.

THE PROOF OF CARNOT'S THEOREM

2·3. We can now prove the following theorem which is equivalent to Carnot's theorem:

In any reversible cycle $\oint \frac{dQ}{T} = 0$, where dQ is the infinitesimal (positive or negative) amount of heat taken in by the system undergoing the cycle when its temperature is T.

If we are dealing with a system specified by two independent variables, we can represent the cycle on a diagram equivalent to the P, V diagram. We can then divide up the cycle into a number of infinitesimal Carnot cycles by drawing, in the above diagram, a network of isothermal and adiabatic lines. The theorem can then be proved by applying theorem 4 of § 2·2 to each infinitesimal Carnot cycle. In general, however, more than two independent variables will be required to specify the state of the system, and it is not easy to carry out the above procedure. We therefore prove the theorem by introducing a number of auxiliary Carnot cycles.

Consider any general cycle and let dQ be the heat absorbed by the system during an infinitesimal change during which the temperature can be supposed to be constant and equal to T. Without loss of generality we can suppose that this heat is taken from an auxiliary body which performs a Carnot cycle between the temperatures T and T_0, where T_0 is the fixed temperature of a large reservoir M_0 of heat. The heat lost by M_0 in this infinitesimal Carnot cycle is dQ_0, where by theorem 4 of § 2·2, $dQ_0 = T_0 dQ/T$. If now the cycle is split up into infinitesimal changes in each of which the heat is supplied by an auxiliary Carnot cycle connected with M_0, the total amount of heat withdrawn from M_0 in a cycle is

$$Q_0 = \oint dQ_0 = T_0 \oint \frac{dQ}{T}. \tag{2·3·1}$$

If Q_0 is positive, the net effect of the cycle and all the auxiliary Carnot cycles is that a quantity of heat has been completely converted into work, while no heat has been given out, and this contradicts Kelvin's principle. It is possible for Q_0 to be negative without contradicting Kelvin's principle since this would mean that work has been converted into heat. This is, however, an irreversible process and cannot occur if all the cycles are reversible. For, if $Q_0 < 0$ for a reversible cycle, then by reversing the cycle and the auxiliary Carnot cycles, we could convert a positive amount of heat into work. Hence for a reversible cycle we must have $\oint \frac{dQ}{T} = 0$.

Carnot's theorem follows as an immediate corollary of the above result, since the statement that $\oint \frac{dQ}{T} = 0$ for every closed path in the parameter space is the necessary and sufficient condition that dQ/T should be a perfect differential dS of a function S of the state. Thus in an infinitesimal reversible process we have $dQ = T\,dS$, and, for any finite reversible change connecting the states A and B,

$$S_B - S_A = \int_A^B \frac{dQ}{T}. \tag{2·3·2}$$

It should be noted that $S_B - S_A$ can be defined by the integral taken along any reversible path connecting the states A and B, since all such paths necessarily give the same value for the integral. The entropy difference $S_B - S_A$ has, however, no relation to the integral (2·3·2) taken along an irreversible path, and, in fact, $S_B - S_A$ cannot be defined at all unless a reversible process can be found connecting the states A and B.

2·31. *The thermodynamic temperature scale.* The absolute temperature T, the existence of which is shown by Carnot's theorem, is independent of the properties of any particular substance and is unique except for a multiplying constant which determines the size of the degree. The absolute temperature scale is therefore the most fundamental one, and is the one always used in thermodynamics. It is often called the thermodynamic temperature scale. To determine the thermodynamic centigrade scale we need to know the absolute temperatures corresponding to 100 and 0° C. The ratio of these absolute temperatures is the ratio of the heat taken in to the heat given out by a Carnot cycle working between 100 and 0° C. If the ratio $T_{100°\,\text{C.}}/T_{0°\,\text{C.}}$ can be found directly or indirectly, the thermodynamic centigrade scale is completely established. It is found, for example, from the properties of gases (§ 2·72) that $T_{100°\,\text{C.}}/T_{0°\,\text{C.}} = 373{\cdot}16/273{\cdot}16$, and hence that 0° C. corresponds to 273·16° on the absolute scale. Temperatures on the absolute scale are usually distinguished by the letter K, after Kelvin. Thus 0° C. is the same as 273·16° K., and 0° K. is the same as $-273{\cdot}16$° C.

Irreversible Cycles

2·4. We can apply the arguments of § 2·2 to prove that an irreversible cycle working between the temperatures ϑ_1 and ϑ_2 is less efficient than a Carnot cycle working between the same temperatures. Let W, Q_1, Q_2 refer to the irreversible cycle and let W', Q_1', Q_2' refer to the reversible Carnot cycle. By considering the first cycle and the second cycle reversed, we can show, exactly as in the proof of theorem 1 of § 2·2, that if $Q_1 = Q_1'$ we cannot have $W > W'$. Thus $W \leqslant W'$, and in general $W < W'$, since, for example, any changes which are not quasi-static must result in the work done being less than that ideally possible. The irreversible cycle is therefore less efficient than the reversible one.

If we now consider any general cycle, reversible or not, and repeat the argument of § 2·3, we obtain the result that the amount of heat Q_0 withdrawn from the reservoir M_0 must be negative or zero. If the cycle is reversible Q_0 is zero, while if the cycle is irreversible Q_0 is negative. This result is known as Clausius's inequality. It can be stated as follows:

In any cycle $\oint \frac{dQ}{T} \leqslant 0$, *where* dQ *is the heat absorbed in an infinitesimal process when the temperature is* T. *The equality sign holds if and only if the cycle is reversible.*

2·41. *The increasing property of entropy.* Clausius's inequality tells us nothing directly about the change of entropy in an irreversible process, since there is no connexion between the heat absorbed and the change in entropy unless the process is reversible. We can, however, use the inequality to prove that the entropy always increases in a spontaneous process.

Consider a closed system and suppose that it is capable of undergoing a spontaneous change which connects the states A and B. This change is, of course, irreversible. To define the change in entropy we need a reversible process connecting the states, and this will require the use of outside agencies to add heat to or subtract heat from the system and to do work on it. If we now consider the spontaneous change from A to B followed by the reversible process restoring the system back to A, we have an irreversible cycle, and in this cycle $\oint \frac{dQ}{T} < 0$. This can be written

$$\int_A^B \frac{dQ}{T} + \int_B^A \frac{dQ}{T} < 0,$$

where the first integral refers to the spontaneous change and the second to the reversible process. The second integral is $S_A - S_B$, and we therefore have

$$S_B - S_A > \int_A^B \frac{dQ}{T}, \qquad (2{\cdot}41{\cdot}1)$$

where dQ is the heat absorbed at temperature T during the spontaneous change. But the spontaneous change takes place in a closed system, and therefore no heat is absorbed during the process (and no work is done). Hence $S_B - S_A > 0$, and the entropy always increases in a spontaneous process.

The inequality (2·41·1) provides an alternative way of stating Clausius's inequality as follows. In any infinitesimal process $dQ \leqslant T\,dS$, where dQ is the heat absorbed by the system, T is the temperature and dS is the increase in entropy.

If we consider a system consisting of several bodies and consider the separate heat exchanges between the bodies, we arrive at the same result. The total change in entropy of the system still satisfies the inequality (2·41·1), but the right-hand side of the inequality now represents irreversible heat exchanges between the various bodies. The integrand therefore consists of terms such as $-\delta Q/T + \delta Q/T'$, where $T > T'$, and hence it must be positive or zero. Thus $S_B - S_A > 0$ as before.

Remarks upon the Proof of Carnot's Theorem

2·5. The proof of Carnot's theorem has been based on the principle that certain processes are irreversible. This assumption is more than is necessary to prove the existence of the entropy and the absolute temperature, and the theorems concerning entropy can be split up into two. The first of these, Carnot's theorem, requires only the consideration of reversible processes, while the second, the increasing property of entropy, involves irreversible processes (Ehrenfest-Afanassjewa, 1925; Ruark, 1925).

We can prove Carnot's theorem by postulating the opposite of Clausius's and Kelvin's principles, but applied to reversible processes. Let us postulate, for example, that no reversible cyclic process exists which produces no other effect than the conversion of mechanical work into an equivalent amount of heat. Suppose that we have two reversible Carnot cycles, and suppose them adjusted so that $Q_1 = Q_1'$. If $W \neq W'$, suppose that $W > W'$, and hence that $Q_2 < Q_2'$. The composite system consisting of the second system and the first system reversed does an amount of work $W' - W$ and gives out an amount of heat $Q_2' - Q_2$. That is, a positive amount of work $W - W'$ is done on the system and converted entirely into heat. This contradicts the postulate and hence $W \leqslant W'$. By a similar argument we can show that $W' \leqslant W$, so that we must have $W = W'$. This proves that the efficiency of all reversible cycles is the same, and we can now prove all the theorems of § 2·2 which are required for the proof of Carnot's theorem.

It is clear from the above that the only postulates that are required to prove Carnot's theorem are that certain effects cannot be produced by reversible processes. This is also a logical necessity, since Carnot's theorem only deals with reversible processes and has nothing to say about what happens in irreversible ones. We can therefore restate Clausius's and Kelvin's principles in the following restricted forms:

Clausius's principle for reversible changes: No reversible cyclic process exists which has as its sole effect either (*a*) *the transference of heat from a colder to a hotter body or* (*b*) *the transference of heat from a hotter to a colder body.*

Kelvin's principle for reversible changes: No reversible cyclic process exists which produces no other effect than either (*a*) *the extraction of heat from a body and its conversion into mechanical work or* (*b*) *the disappearance of mechanical work and its conversion into the internal energy of a single body.*

To bring out more clearly the content of these postulates we restate them in a form which gives conditions under which heat transfer and the conversion of work into heat can occur reversibly:

(1) No reversible cyclic process exists which can transfer heat from a colder to a hotter body unless a certain amount of work is converted into heat.

(2) No reversible cyclic process exists which can transfer heat from a hotter to a colder body unless a certain amount of heat is converted into work.

(3) No reversible cyclic process exists which can convert heat into work unless a certain amount of heat is transferred from a hotter to a colder body.

(4) No reversible cyclic process exists which can convert work into heat unless a certain amount of heat is transferred from a colder to a hotter body.

The statements (3) and (4) are the converses of (2) and (1), while (2) and (4) follow from (1) and (3) since the cycles considered are reversible.

Although any of the postulates (1)–(4) is sufficient to prove Carnot's theorem, they do not tell us anything about what happens in irreversible processes. To prove the increasing property of entropy we need Clausius's and Kelvin's principles in their original form and not merely for reversible changes. If the word 'reversible' is omitted, the postulates (1) and (3) are still true while (2) and (4) are false.

Available Energy and Entropy

2·6. Another way of expressing the results of the second law is in terms of what is known as available energy. It is impossible to utilize all the internal energy of a body for the production of mechanical work, since work can only be obtained by abstracting heat from the body and giving it to a reversible engine, whose efficiency is necessarily less than unity. Suppose that the temperature of a body is T and that the reversible engine works between T and T_0, which is the temperature of a large reservoir to which the engine can give up heat and is the lowest temperature available. Then, if the body gives an amount of heat dQ to the engine, only part of it, namely, $dQ(1-T_0/T)$, can be converted into mechanical work. If the engine is irreversible, still less work can be obtained. The available energy is defined to be $dQ(1-T_0/T)$, while the unavailable energy is $T_0 dQ/T$.

The available energy has a more immediate physical significance than the entropy, but it is not a well-defined quantity, owing to its dependence on the temperature T_0, which is completely arbitrary. Corresponding to the theorem that the entropy always increases in a spontaneous process, there is the theorem that the available energy always decreases in an irreversible cycle. (Alternatively, we can assume as a general principle that there exists no process which can increase the available energy in the universe, and we can then deduce Clausius's inequality from it.) If a system at temperature T receives a small quantity of heat dQ and performs a small amount of work dW, the increase in the available energy of the system is

$$d\psi = dQ(1-T_0/T) - dW. \qquad (2\cdot6\cdot1)$$

By the first law, $dQ = dU + dW$, and so

$$d\psi = dU - T_0 dQ/T. \qquad (2\cdot6\cdot2)$$

If the system performs a cycle, the net gain in available energy by the system is

$$\oint d\psi = \oint dU - T_0 \oint \frac{dQ}{T} = -T_0 \oint \frac{dQ}{T}. \qquad (2\cdot6\cdot3)$$

But after the cycle the system is in the same state as it was at the beginning, and its available energy must be unaltered. If it has gained any available energy from outside, this available energy must have been dissipated owing to the occurrence of irreversible processes in the system. The net dissipation of available energy in the cycle is therefore $-T_0\oint\frac{dQ}{T}$, and this is positive by Clausius's inequality. Only if the cycle is reversible is there no dissipation of available energy.

2·61. *The physical significance of entropy.* It is not easy to explain the significance of entropy in ordinary physical terms. Perhaps the most immediate meaning to be ascribed to entropy is by means of its connexion with available and unavailable energy. We have seen that the whole of the internal energy of a body cannot be utilized to perform mechanical work, and that when mechanical work is transformed into internal energy part of it becomes unavailable. When an infinitesimal amount of heat dQ is imparted to a body reversibly its entropy is increased by dQ/T and its unavailable energy is increased by T_0dQ/T, where T is the temperature of the body and T_0 is the lowest temperature available. The increasing property of entropy is therefore equivalent to saying that energy is always being degraded into forms which are more and more difficult to utilize for the production of work.

The significance of entropy becomes much clearer when we consider heat to be the energy of the molecular motion of the ultimate constituents of matter. On this well-established hypothesis, the lack of complete availability of heat energy for the production of work is due to the molecular motion being a random motion. It is impossible to reduce the motion of each molecule simultaneously to zero by the action of large-scale forces acting upon the body as a whole, and so it is impossible to abstract all the heat energy from a body. From this point of view the increasing property of entropy means that the molecular motion of an isolated system always tends to become more random, and the entropy can be thought of as a measure of the 'randomness' of the internal motion of a system. This qualitative connexion between entropy and randomness can be made quantitative by means of the theory of statistical mechanics (see § 5·33).

The Thermodynamics of a Perfect Gas

2·7. It was found by Boyle that, for any of the so-called permanent gases, PV is constant at constant temperature. We can therefore take PV as an empirical temperature. Further, if we consider 1 gram molecule of a gas, the product PV is the same for all the permanent gases at the same temperature. If there are n gram molecules of gas present, we write

$$PV = nR\vartheta, \tag{2·7·1}$$

where R is a constant whose value is determined by the convention that $\vartheta_{100^\circ\text{C.}} - \vartheta_{0^\circ\text{C.}} = 100$, that is, that

$$R = \tfrac{1}{100}\{(PV)_{100^\circ\text{C.}} - (PV)_{0^\circ\text{C.}}\}.$$

R is called the universal gas constant and is found by measurement to have the value $83{\cdot}2 \times 10^6$ ergs/° C., which is approximately 1·98 calories/° C. The equation (2·7·1) is not rigorously true for any real gas for all P, V and ϑ, but it applies to any gas at sufficiently low pressures and sufficiently high temperatures. The permanent gases are those for which (2·7·1) is a good approximation at ordinary temperatures and at pressures of the order of a few atmospheres. A perfect gas is an ideal gas for which (2·7·1) is valid for all values of P, V and ϑ. If we measure the mass in grams instead of in gram molecules, (2·7·1) has to be replaced by

$$PV = m(R/M^*)\,\vartheta, \tag{2·7·2}$$

where m is the mass in grams and M^* is the molecular weight of the gas.

To determine the properties of a perfect gas we require to know the internal energy as well as the equation of state. It is found that for the monatomic permanent gases such as helium and argon, the internal energy is given to a high degree of accuracy by $U = A\vartheta + \text{constant}$, where A is a constant. Hence $nC_V = (\partial U/\partial\vartheta)_V = A$, so that A/n is the specific heat at constant volume, and

$$U = nC_V\vartheta + U_0, \tag{2·7·3}$$

where C_V and U_0 are constants. We take this to represent the internal energy of a perfect gas.

The linear dependence of U upon ϑ is not true for all the permanent gases, and for many purposes the assumption that U is a linear function of ϑ is unduly restrictive. The first experiments which determined the internal energy of a gas were carried out by Joule. He compressed air to about 20 atmospheres in a large vessel which was connected by a stopcock to an evacuated vessel. On opening the cock, the air rushed into the vacuum, and after a time an equilibrium state was reached. It was found that no perceptible change in temperature was produced by the expansion. In this irreversible process no external work is done by the gas, and the process is an adiabatic one. The internal energy therefore remains unaltered, and the result of the experiment was that, to a first approximation, $(\partial\vartheta/\partial V)_U = 0$. By (1·6·5) we have

$$\left(\frac{\partial\vartheta}{\partial V}\right)_U = -\left(\frac{\partial U}{\partial V}\right)_\vartheta \Big/ \left(\frac{\partial U}{\partial\vartheta}\right)_V,$$

and hence $$(\partial U/\partial V)_\vartheta = 0. \tag{2·7·4}$$

The internal energy is therefore a function of ϑ only, and so is C_V.

It is found experimentally that C_V is constant over very wide ranges for the monatomic permanent gases, while for the diatomic permanent gases, such as oxygen, nitrogen and hydrogen, it is constant for moderate temperature ranges but varies with ϑ if the temperature range considered is large. In general, we shall assume that perfect gases are such that (2·7·3) is true, but it is sometimes convenient to define perfect gases by the relations (2·7·1) and (2·7·4). We shall then refer to them as perfect gases in the wide sense. It will be shown in § 3·21 that the relation $(\partial U/\partial V)_\vartheta = 0$ is not independent of the equation of state $PV = nR\vartheta$, but is a direct consequence of it.

2·71. *The specific heats.* Equation (1·5·12) together with (2·7·1) and (2·7·4) gives for the difference of the specific heat at constant pressure and constant volume

$$C_P - C_V = \frac{P}{n}\left(\frac{\partial V}{\partial \vartheta}\right)_P = R. \tag{2·71·1}$$

This is true for perfect gases in the wide sense. The ratio C_P/C_V is denoted by γ. It is greater than unity and is constant for perfect gases in the strict sense. If γ is constant, or can be taken to be constant over the temperature range involved, we can easily obtain the expansion law for a quasi-static adiabatic process. We have

$$dQ = nC_V d\vartheta + P dV = n(C_V + R)\, d\vartheta - V dP, \tag{2·71·2}$$

and since $dQ = 0$ we obtain, on eliminating $d\vartheta$,

$$\frac{dP}{P} + \gamma \frac{dV}{V} = 0, \tag{2·71·3}$$

which gives on integration

$$PV^\gamma = \text{constant}. \tag{2·71·4}$$

This family of curves in the P, V plane is known as the family of adiabatic curves. From the definition of entropy they are curves of constant entropy.

2·72. *The entropy and the absolute temperature.* We can obtain the entropy of a perfect gas and the relation between the gas temperature ϑ and the absolute temperature T in various ways. The shortest method is to find an integrating factor for dQ. In a quasi-static process we have

$$dQ = nC_V d\vartheta + P dV = nC_V d\vartheta + nR\vartheta\, dV/V.$$

An obvious integrating factor which is a function of ϑ only is $1/\vartheta$, for

$$\frac{dQ}{\vartheta} = nC_V \frac{d\vartheta}{\vartheta} + nR\frac{dV}{V} = nd(C_V \log \vartheta + R \log V).$$

It follows from the general theory that ϑ must be identical with T except for a multiplying constant which merely determines the size of the degree, and, further, that $dQ/\vartheta = dS$. Hence, putting $\vartheta = T$, we have

$$S = nC_V \log T + nR \log V + S_0', \tag{2·72·1}$$

where S_0' is an arbitrary constant. We can express S in the alternative forms

$$S = nC_P \log T - nR \log P + S_0' + nR \log R \tag{2·72·2}$$

and

$$S = nC_P \log V + nC_V \log P + S_0' - nC_V \log R \tag{2·72·3}$$

$$= nC_V \log (PV^\gamma) + S_0' - nC_V \log R, \tag{2·72·4}$$

from the latter of which it is clear that S is constant along an adiabatic curve.

The quantity $S_0 = S_0' + nR \log R$ is the value of the entropy when $T = 1$ and $P = 1$. Now the reference state $T = 1$, $P = 1$ is not an absolute state since it depends upon the units in which T and P are measured, and it can be made to be any state we please by choosing the units correctly. For most purposes this arbitrariness does not matter, and the entropy constant can be put equal to zero since we are normally concerned with entropy differences, but we run into difficulties if we ignore the entropy constants when the entropy difference is measured between a gas and its solid. Also the quantity S_0' by itself has not the dimensions of an energy. To avoid such an inelegancy, we shall either consider entropy differences explicitly or write

$$S = nC_P \log T - nR \log P + S_0 - nC_P \log T^\dagger + nR \log P^\dagger, \tag{2·72·5}$$

where $T^\dagger$ and $P^\dagger$ denote a standard temperature and a standard pressure. It would of course be preferable to have an absolute state from which to measure S, but apart from $T = 0$, $P = 0$, which is impossible on account of the logarithms in (2·72·5), we have at present no reference state which has any advantages over any other. We shall, however, see later (§§ 5·4, 6·311 and 6·41) that, owing to the fact that an ideal gas cannot remain perfect in the sense used here down to $T = 0$, $P = 0$, an absolute choice of $T^\dagger$ and $P^\dagger$ is in fact possible, and that $T^\dagger$ and $P^\dagger$ can be defined so as to make S_0 the entropy of the gas at $T = 0$.

If C_P is not constant but is a function of T, we write

$$C_P(T) = C_P^0 + C_P^1(T), \quad C_P^0 = C_P(0), \tag{2·72·6}$$

and

$$dQ = T\,dS = nC_P\,dT - V\,dP = n\{C_P^0 + C_P^1(T)\}\,dT - nRT\,dP/P, \tag{2·72·7}$$

by (2·71·2). Hence

$$S = nC_P^0 \log T + n\int_0^T \frac{C_P^1(T')}{T'}\,dT' - nR \log P + S_0 - nC_P^0 \log T^\dagger + nR \log P^\dagger. \tag{2·72·8}$$

Similarly we may write

$$S = nC_V^0 \log T + n\int_0^T \frac{C_V^1(T')}{T'}\,dT' + nR\log V + S_0' - nC_V^0 \log T^\dagger - nR\log V^\dagger, \tag{2·72·9}$$

where $C_P^0 - C_V^0 = R$, $C_P^1(T) = C_V^1(T)$. The quantity $V^\dagger$ is proportional to the number of gram molecules present, and it is sometimes more convenient to replace $V^\dagger$ by $nv^\dagger$, where $v^\dagger$ is the volume of 1 gram molecule of the gas in the standard state. The parameters $P^\dagger$, $v^\dagger$, $T^\dagger$ of the standard state are then independent of the amount of the gas. It is also often convenient to write $S_0 = ns_0$, $S_0' = ns_0'$ in order to obtain entropy constants which refer to 1 gram molecule of the gas.

2·721. We have now shown that the perfect gas temperature, defined as PV, is the same as the absolute temperature. We can therefore determine the relation between the absolute scale and the centigrade scale by measuring the ratio $(PV)_{100^\circ \mathrm{C.}}/(PV)_{0^\circ \mathrm{C.}}$ for a gas which is at a sufficiently low pressure to behave as a perfect gas. In practice this ratio is measured for a number of low pressures and the results are extrapolated to zero pressure. In this way it is found that 0° C. is the same as 273·16° K. In a similar way we can use the perfect gas temperature to subdivide the centigrade scale into degrees. From now onwards we identify the perfect gas scale and the absolute scale and use the symbol T for both temperatures.

We have also found expressions for the entropy of a perfect gas which determine the entropy for all states of the gas. We can therefore calculate the change in entropy in any process, reversible or irreversible.

REFERENCES

Ehrenfest-Afannassjewa, T. (1925). The axioms of the second law of thermodynamics. *Z. Phys.* **33**, 933; **34**, 638.

Ruark, A. E. (1925). Proof of the corollary of Carnot's theorem. *Phil. Mag.* (6), **49**, 584.

Chapter 3

THERMODYNAMIC FUNCTIONS

THE INTERNAL ENERGY, THE HEAT FUNCTION AND THE FREE ENERGY

3·1. We have so far developed the subject by taking the internal energy as the fundamental quantity. There are, however, three other quantities which are of equal importance, which, together with the internal energy, are known as the principal thermodynamic functions. In the present chapter we derive the main properties of the thermodynamic functions that are required in the applications discussed later.

It is convenient to begin by considering substances which have only two independent parameters of state, a generalized coordinate a and a generalized force A. Such substances are called simple substances, and for definiteness we shall usually consider a fluid or a solid subject to an isotropic pressure. The generalized coordinate is then the volume V and the generalized force is the pressure P. We shall afterwards extend the theory to include substances which are characterized by more than one generalized coordinate and also substances whose composition varies.

(i) *The internal energy.* The internal energy U is measured by the amount of work required to bring the substance into any given state from a fixed initial state by an adiabatic process. In any general process

$$Q = \Delta U + W, \tag{3·1·1}$$

where Q is the heat absorbed by the substance and W is the mechanical work done by the system. When no external work is done, the heat absorbed by the system is equal to the increase in the internal energy. For a fluid of invariable composition the external work is zero if the volume is constant, and the specific heat at constant volume is given by

$$nC_V = (\partial U/\partial T)_V,$$

where n is the number of gram molecules of the substance. For an infinitesimal quasi-static process equation (3·1·1) becomes

$$T\,dS = dU + P\,dV, \tag{3·1·2}$$

and so

$$T = (\partial U/\partial S)_V, \quad P = -(\partial U/\partial V)_S. \tag{3·1·3}$$

(ii) *The heat function at constant pressure.* The heat function H at constant pressure is defined by

$$H = U + PV, \tag{3·1·4}$$

and it plays the same part for processes at constant pressure as the internal energy U does for processes at constant volume. For any process at constant pressure

$$(\Delta H)_P = \Delta U + P\Delta V = Q - W + W_{\text{exp.}}, \tag{3·1·5}$$

where $W_{\text{exp.}} = P\Delta V$ is the work done by the substance in expanding against the constant pressure. For a fluid of invariable composition the only work that can be done is work of expansion, so that $W = W_{\text{exp.}}$, and the heat absorbed at constant pressure is equal to the increase in the heat function. Hence the specific heat at constant pressure of a simple substance is given by

$$nC_P = (\partial H/\partial T)_P. \tag{3·1·6}$$

For an infinitesimal quasi-static process we have, by combining equations (3·1·2) and (3·1·4),

$$dH = T\,dS + V\,dP, \tag{3·1·7}$$

and so

$$T = (\partial H/\partial S)_P, \quad V = (\partial H/\partial P)_S. \tag{3·1·8}$$

An important property of H is that, in the steady adiabatic flow of gas under the influence of a pressure gradient, $\textsc{h} + \frac{1}{2}q^2$ is constant along a stream-line, where $\textsc{h} = H/n$ is the heat function per gram molecule and where q is the velocity of the gas. To prove this, consider a tube of flow of infinitesimal cross-section, and consider the energy flowing across a plane at right angles to the direction of flow. Let A be the cross-section of the tube of flow and let ρ be the molar density of the fluid. Then in time δt a volume $Aq\,\delta t$ of the gas passes over the given cross-section. The mass of this element of gas is $A\rho q\,\delta t$, its kinetic energy is $\frac{1}{2}A\rho q^3\,\delta t$ and its internal energy is $A\rho q \textsc{u}\,\delta t$, where $\textsc{u} = U/n$ is the internal energy per gram molecule. Also, the work done by the pressure is $APq\,\delta t$. Thus the energy which crosses the section in time δt in the direction of flow is $Aq\,\delta t(\frac{1}{2}\rho q^2 + \rho \textsc{u} + P)$, and this must be the same for every cross-section since the flow is steady. Also the mass crossing the section must be the same for all cross-sections, so that $A\rho q$ is constant. Hence $\frac{1}{2}q^2 + \textsc{u} + P/\rho = \frac{1}{2}q^2 + \textsc{h}$ is constant along a stream-line.

A particular case of great importance is when the pressure drop occurs abruptly, as when the gas passes through a constriction in a pipe. The process is then known as a throttling process.

(iii) *The free energy at constant volume.* The free energy F at constant volume is defined by

$$F = U - ST. \tag{3·1·9}$$

For any process at constant temperature

$$(\Delta F)_T = \Delta U - T\Delta S = Q - T\Delta S - W. \tag{3·1·10}$$

If, in addition, the process is reversible, then $Q = T\Delta S$, and $\Delta F = -W$, so that F is the force function for isothermal changes. Now most quasi-static processes in mechanics are carried out with no precautions being taken to insulate the system from its surroundings. The force function that is used in mechanics and in electromagnetic phenomena is therefore in general F and not U. The force function is U when changes take place so rapidly that heat exchanges between the system and its surroundings are negligible, provided that no irreversible processes take place. An example is provided by sound waves in air. The velocity of propagation of sound waves is determined by the adiabatic compressibility, and the entropy can be considered as constant for very small amplitudes of vibration; for large amplitudes the phenomenon is much more complicated and there is an irreversible increase in entropy.

For a simple substance ΔF is zero for a reversible change in which T and V are constant, but, for substances in which chemical reactions, for instance, can occur, $-(\Delta F)_{T,V}$ is the work obtainable from an isothermal reversible reaction at constant volume. This gives rise to the name free energy at constant volume. If the reaction is not reversible less work is obtainable, since, by Clausius's inequality, $Q \leqslant T\Delta S$, and so

$$W = -(\Delta F)_{T,V} + Q - T\Delta S \leqslant -(\Delta F)_{T,V}. \tag{3·1·11}$$

Thus $-(\Delta F)_{T,V}$ is the maximum work that can be obtained from an isothermal process at constant volume.

For a fluid of invariable composition and for an infinitesimal quasi-static change we have

$$dF = dU - S\,dT - T\,dS = -S\,dT - P\,dV, \tag{3·1·12}$$

and so

$$S = -(\partial F/\partial T)_V, \quad P = -(\partial F/\partial V)_T. \tag{3·1·13}$$

The first of the relations (3·1·13) enables us to obtain a very important relation between U and F which does not involve the entropy. It can be put into either of the forms

$$U = F - T\left(\frac{\partial F}{\partial T}\right)_V, \quad U = -T^2\left(\frac{\partial}{\partial T}\frac{F}{T}\right)_V, \tag{3·1·14}$$

and it is known as the Gibbs-Helmholtz equation. We may therefore write (3·1·12) in the alternative form

$$d\frac{F}{T} = -\frac{U}{T^2}dT - \frac{P}{T}dV. \tag{3·1·15}$$

(iv) *The free energy at constant pressure.* The free energy G at constant pressure is defined by

$$G = U - ST + PV = H - ST. \tag{3·1·16}$$

For any process at constant temperature and pressure

$$(\Delta G)_{T,P} = \Delta U - T\Delta S + P\Delta V = Q - T\Delta S - W + W_{\text{exp.}}, \tag{3·1·17}$$

where $W_{\text{exp.}} = P\Delta V$ is the work done by the substance in expanding against the constant pressure. Now $Q \leqslant T\Delta S$, the equality sign holding when the process is reversible. Hence

$$W - W_{\text{exp.}} = -(\Delta G)_{T,P} + Q - T\Delta S \leqslant -(\Delta G)_{T,P}, \tag{3·1·18}$$

and $-(\Delta G)_{T,P}$ is the maximum work obtainable from the substance in a given change at constant temperature and pressure, over and above the external work necessarily used in expanding the substance against the external pressure. This is the origin of the name 'free energy at constant pressure'. For a simple substance, $(\Delta G)_{T,P} = 0$ for a reversible change.

For a fluid of invariable composition and for an infinitesimal quasi-static change we have

$$dG = dU - S\,dT - T\,dS + P\,dV + V\,dP = -S\,dT + V\,dP, \tag{3·1·19}$$

and so
$$S = -(\partial G/\partial T)_P, \quad V = (\partial G/\partial P)_T. \tag{3·1·20}$$

We can obtain a Gibbs-Helmholtz equation connecting H and G by eliminating the entropy. It can be put into either of the forms

$$H = G - T\left(\frac{\partial G}{\partial T}\right)_P, \quad H = -T^2\left(\frac{\partial}{\partial T}\frac{G}{T}\right)_P. \tag{3·1·21}$$

We may therefore write
$$d\frac{G}{T} = -\frac{H}{T^2}dT + \frac{V}{T}dP. \tag{3·1·22}$$

(v) Of the four thermodynamic functions the two which occur naturally in experimental work are H and G, since it is much easier to keep the pressure constant than the volume. On the other hand, the functions U and F are the ones which arise naturally in a theoretical treatment based upon statistical mechanics. Also, the functions F and G are somewhat easier to deal with than the functions U and H, since the temperature is a less recondite independent variable than the entropy.

(vi) When we deal with systems with more than one generalized coordinate the number of thermodynamic functions is greater than four. If there are s generalized coordinates a_i and s generalized forces A_i, which are such that the force exerted on‡ the system by external bodies is $-A_i$ in the direction tending to increase a_i, we may write

$$dU = T\,dS - \sum_{i=1}^{s} A_i\,da_i \tag{3·1·23}$$

for an infinitesimal reversible change. Then instead of the two functions U and H we have the 2^s functions

$$U + \Sigma' A_i a_i, \tag{3·1·24}$$

‡ The sign of A_i is fixed in this way so that the pressure exerted by a gas on the containing vessel is a generalized force.

where the summation is taken over any set of the coordinates and forces. Similarly, instead of the two functions F and G we have the 2^s functions

$$U - ST + \Sigma' A_i a_i. \tag{3·1·25}$$

The properties of these functions can easily be found in any particular case from their definitions and from equation (3·1·23).

Maxwell's Equations and their Applications

3·2. The differentials of the four thermodynamic functions of a simple substance are given by

$$\left.\begin{aligned} dU &= T\,dS - P\,dV, \qquad & dH &= T\,dS + V\,dP, \\ dF &= -S\,dT - P\,dV, \qquad & dG &= -S\,dT + V\,dP. \end{aligned}\right\} \tag{3·2·1}$$

Now dU is a perfect differential, and so we must have

$$(\partial T/\partial V)_S = \partial^2 U/\partial S\,\partial V = -(\partial P/\partial S)_V.$$

There are three similar equations arising from the conditions that dH, dF, dG should be perfect differentials. These equations are known as Maxwell's equations and are explicitly as follows:

$$\left.\begin{aligned} (\partial P/\partial S)_V &= -(\partial T/\partial V)_S, \qquad & (\partial V/\partial S)_P &= (\partial T/\partial P)_S, \\ (\partial S/\partial V)_T &= (\partial P/\partial T)_V, \qquad & (\partial S/\partial P)_T &= -(\partial V/\partial T)_P. \end{aligned}\right\} \tag{3·2·2}$$

We may also add the equations

$$\left(\frac{\partial U}{\partial V}\right)_T = T^2\left(\frac{\partial}{\partial T}\frac{P}{T}\right)_V, \quad \left(\frac{\partial H}{\partial P}\right)_T = -T^2\left(\frac{\partial}{\partial T}\frac{V}{T}\right)_P, \tag{3·2·3}$$

which are derived from (3·1·15) and (3·1·22). The four Maxwell equations, together with the formulae (1·6·4) and (1·6·5), are very useful in eliminating the entropy from thermodynamic formulae and obtaining results which contain only quantities which are easily measurable. The way in which this is done is best illustrated by a few simple examples.

For systems with more than one generalized coordinate there is a set of Maxwell's equations similar to the above, but the number of equations is greater.

3·21. *The specific heat equations.* We can use the Maxwell equations to obtain relations for the specific heats which are more useful in practice than those obtained so far.

The specific heat at constant volume is given by‡

$$C_V = (\partial u/\partial T)_V = T(\partial s/\partial T)_V. \tag{3·21·1}$$

Hence
$$\left(\frac{\partial C_V}{\partial v}\right)_T = T\frac{\partial^2 s}{\partial T\,\partial v} = T\left(\frac{\partial^2 P}{\partial T^2}\right)_V, \tag{3·21·2}$$

by the third Maxwell equation. This formula shows us that we can always determine $(\partial C_V/\partial v)_T$ when the equation of state is known.

The latent heat of expansion at constant temperature L_V is defined by

$$dQ = T\,dS = nC_V\,dT + L_V\,dV. \tag{3·21·3}$$

Therefore
$$L_V = T(\partial S/\partial V)_T = T(\partial P/\partial T)_V, \tag{3·21·4}$$

and (3·21·3) can be written as

$$dS = n\frac{C_V}{T}dT + \left(\frac{\partial P}{\partial T}\right)_V dV. \tag{3·21·5}$$

This enables us to find the entropy if the specific heat at constant volume and the equation of state are known.

Similarly, by (3·2·3), we have

$$dU = \left(\frac{\partial U}{\partial T}\right)_V dT + \left(\frac{\partial U}{\partial V}\right)_T dV = nC_V\,dT + T^2\left(\frac{\partial}{\partial T}\frac{P}{T}\right)_V dV, \tag{3·21·6}$$

from which we can find U if the specific heat at constant volume and the equation of state are known.

In § 2·7 we defined a perfect gas in the wide sense by the equations $Pv = RT$ and $(\partial U/\partial V)_T = 0$. We now see that these equations are not independent, for the second follows from the first on account of equation (3·2·3). The specific heat of a perfect gas is therefore necessarily a function of the temperature only.

To obtain S and U we have to integrate equations (3·21·5) and (3·21·6). Since S and U are both functions of the state, the integrations can be carried out along any path in the V, T plane, and it is simplest to integrate along lines parallel to the axes. In this way we obtain the following results:

$$U(V,T) - U(V_0,T_0) = n\int_{T_0}^{T} C_V(V_0,T')\,dT' + \int_{V_0}^{V} T^2\frac{\partial}{\partial T}\frac{P(V',T)}{T}\,dV', \tag{3·21·7}$$

$$S(V,T) - S(V_0,T_0) = n\int_{T_0}^{T} C_V(V_0,T')\frac{dT'}{T'} + \int_{V_0}^{V}\frac{\partial P(V',T)}{\partial T}\,dV', \tag{3·21·8}$$

$$F(V,T) - F(V_0,T_0) = n\int_{T_0}^{T} C_V(V_0,T')\left(1-\frac{T}{T'}\right)dT' - \int_{V_0}^{V} P(V',T)\,dV' - (T-T_0)\,S(V_0,T_0). \tag{3·21·9}$$

‡ The internal energy U, the volume V and the entropy S are examples of extensive variables, which are defined as those parameters of a body whose values are proportional to the mass of the substance present. It is convenient to use capital

It is perhaps somewhat illogical to express the thermodynamic functions in terms of the specific heat, since calorimetric measurements result in a knowledge of $U(V_0, T) - U(V_0, T_0)$ rather than that of C_V itself. If, therefore, we take (3·1·15) as our starting point we readily find that

$$F(V,T) = -T\int_{T_0}^{T} \{U(V_0,T') - U(V_0,T_0)\}\frac{dT'}{T'^2} - \int_{V_0}^{V} P(V',T)\,dV' + U(V_0,T_0) - TS(V_0,T_0). \quad (3\cdot21\cdot10)$$

If we express $U(V_0, T') - U(V_0, T_0)$ in terms of C_V, this can be written as

$$F(V,T) = -T\int_{T_0}^{T}\frac{dT'}{T'^2}\int_{T_0}^{T'} nC_V(V_0,T'')\,dT'' - \int_{V_0}^{V} P(V',T)\,dV' + U(V_0,T_0) - TS(V_0,T_0), \quad (3\cdot21\cdot11)$$

which can be used as an alternative to (3·21·9). Alternative expressions for U and S can be obtained from (3·21·10) and (3·21·11) by the appropriate differentiations.

3·211. There are corresponding formulae involving C_P. They are as follows:

$$C_P = (\partial \textsc{h}/\partial T)_P = T(\partial \textsc{s}/\partial T)_P, \quad (3\cdot211\cdot1)$$

$$(\partial C_P/\partial P)_T = -T(\partial^2 \textsc{v}/\partial T^2)_P, \quad (3\cdot211\cdot2)$$

$$L_P = (\partial \textsc{h}/\partial P)_T - V = T(\partial \textsc{s}/\partial P)_T = -T(\partial \textsc{v}/\partial T)_P. \quad (3\cdot211\cdot3)$$

The proof of these is left to the reader.

The formula (3·211·2) is often used to determine the specific heat at high pressures where direct calorimetric measurements are difficult to carry out. Some calculated results for nitrogen are shown in fig. 8·15, p. 236.

If the specific heat at constant pressure and the equation of state are known, S and H can be found from the formulae

$$dS = n\frac{C_P}{T}dT - \left(\frac{\partial V}{\partial T}\right)_P dP, \quad (3\cdot211\cdot4)$$

$$dH = nC_P dT - T^2\left(\frac{\partial}{\partial T}\frac{V}{T}\right)_P dP. \quad (3\cdot211\cdot5)$$

It is difficult in practice to carry out thermal measurements at high pressures, whereas the determination of the equation of state is a much simpler matter. If, therefore, it is desired to obtain any of the thermodynamic functions in the high-pressure region, it is desirable to integrate first with respect to T at a low pressure and then to integrate with respect to P.

letters for extensive variables when we are considering an arbitrary amount of the substance, and to use small capitals to denote the values of an extensive variable per unit mass (measured in grams or in gram molecules) of the substance. Parameters such as the temperature and the pressure which are not extensive variables are called intensive variables.

The formulae corresponding to (3·21·7)–(3·21·9) are as follows:

$$H(P,T)-H(P_0,T_0)=n\int_{T_0}^{T} C_P(P_0,T')\,dT'-\int_{P_0}^{P} T^2\frac{\partial}{\partial T}\frac{V(P',T)}{T}\,dP', \quad (3·211·6)$$

$$S(P,T)-S(P_0,T_0)=n\int_{T_0}^{T} C_P(P_0,T')\frac{dT'}{T'}-\int_{P_0}^{P}\frac{\partial V(P',T)}{\partial T}\,dP', \quad (3·211·7)$$

$$G(P,T)-G(P_0,T_0)=n\int_{T_0}^{T} C_P(P_0,T')\left(1-\frac{T}{T'}\right)dT'+\int_{P_0}^{P} V(P',T)\,dP' -(T-T_0)\,S(P_0,T_0). \quad (3·211·8)$$

The alternative expressions for G, namely,

$$G(P,T)=-T\int_{T_0}^{T}\{H(P_0,T')-H(P_0,T_0)\}\frac{dT'}{T'^2}+\int_{P_0}^{P} V(P',T)\,dP' +H(P_0,T_0)-TS(P_0,T_0) \quad (3·211·9)$$

and

$$G(P,T)=-T\int_{T_0}^{T}\frac{dT'}{T'^2}\int_{T_0}^{T'} nC_P(P_0,T'')\,dT''+\int_{P_0}^{P} V(P',T)\,dP' +H(P_0,T_0)-TS(P_0,T_0), \quad (3·211·10)$$

are the analogues of (3·21·10) and (3·21·11). The equivalence of (3·211·8) and (3·211·10) can be verified directly by integrating by parts.

3·212. A relation of a different type between C_P and C_V can be found by eliminating L_V between equations (1·5·12) and (3·21·4). This gives

$$C_P-C_V=\left(\frac{\partial v}{\partial T}\right)_P L_V=T\left(\frac{\partial v}{\partial T}\right)_P\left(\frac{\partial P}{\partial T}\right)_V. \quad (3·212·1)$$

We may also deduce this result directly by writing

$$C_P-C_V=T\left(\frac{\partial s}{\partial T}\right)_P-T\left(\frac{\partial s}{\partial T}\right)_V=T\left(\frac{\partial s}{\partial v}\right)_P\left(\frac{\partial v}{\partial T}\right)_P, \quad (3·212·2)$$

where $\partial s(P,T)/\partial T$ has been evaluated by treating P as a function of v and T. Since $(\partial s/\partial v)_T=(\partial P/\partial T)_V$ by (3·2·2), we regain (3·212·1).

Equation (3·212·1) can be put into a more useful form by using the fact that

$$\left(\frac{\partial P}{\partial T}\right)_V=-\left(\frac{\partial V}{\partial T}\right)_P\Big/\left(\frac{\partial V}{\partial P}\right)_T,$$

which is a particular case of (1·6·5). If we introduce the isothermal compressibility κ_T defined by

$$\kappa_T=-\frac{1}{V}\left(\frac{\partial V}{\partial P}\right)_T,$$

(3·212·1) becomes

$$C_P-C_V=\frac{T}{v\kappa_T}\left(\frac{\partial v}{\partial T}\right)_P^2, \quad (3·212·3)$$

which shows that $C_P \geqslant C_V$, since κ_T is necessarily positive for any stable substance. If the substance has a point of maximum density for a given pressure, then $C_P = C_V$ at the temperature at which the density is a maximum.

The formula (3·212·1) for $C_P - C_V$ is much more useful than (1·5·12), since the former only involves the isothermal compressibility and the coefficient of thermal expansion at constant pressure, which can be directly and easily measured. One of its main uses is to find C_V from the measured values of C_P. For a perfect gas $C_P - C_V = R$, but for a real gas $C_P - C_V$ may be much greater than R if the pressure is high and the temperature is low. Some calculated values of $C_P - C_V$ for nitrogen are shown in fig. 8·16, p. 236.

3·213. Quantities such as the adiabatic compressibility κ_S can be obtained in terms of directly measurable quantities by using the formulae of § 1·6 (ii). For example, we have

$$\kappa_S = -\frac{1}{V}\left(\frac{\partial V}{\partial P}\right)_S.$$

By (1·6·10) this becomes

$$\kappa_S = -\frac{1}{v}\left\{\left(\frac{\partial v}{\partial P}\right)_T\left(\frac{\partial s}{\partial T}\right)_P - \left(\frac{\partial v}{\partial T}\right)_P\left(\frac{\partial s}{\partial P}\right)_T\right\}\bigg/\left(\frac{\partial s}{\partial T}\right)_P.$$

Using the fourth Maxwell equation we can transform this into

$$\kappa_S = \kappa_T - \frac{T}{vC_P}\left(\frac{\partial v}{\partial T}\right)_P^2, \qquad (3·213·1)$$

and this shows that $\kappa_T \geqslant \kappa_S$.

A large number of equations of the type considered in §§ 3·21, 3·211, 3·212 and 3·213 can be derived, and many of them are of value in enabling us to calculate quantities which cannot be directly measured. A number of applications of such equations are given at appropriate places in the text. A systematic method of obtaining all the possible relations has been given by Shaw (1935).

The Specific Heat of a Perfect Gas

3·3. In order to apply the preceding results to real gases it is necessary to know the equation of state of the gas and the specific heat at a sufficiently low pressure for the gas to be considered as perfect. The equation of state is discussed in Chapter 8, while a detailed account of the theory of the specific heat of a perfect gas is given in § 6·1.

A knowledge of the simple empirical facts concerning the specific heat is sufficient for many purposes, and a brief account of these is given in the following sections.

3·31. The internal energy of a perfect gas is the sum of the energies of the individual molecules. In a monatomic gas the atoms possess translational energy only, whereas in a polyatomic gas we must also take into account the possible contributions of the rotations of the molecules and of the internal motions of the individual atoms in the molecules relative to one another, there being two rotational degrees of freedom for linear molecules and three for non-linear molecules. According to the statistical theory, to be given in § 6·1, the internal energy of n gram molecules of a monatomic gas is $\frac{3}{2}nRT$, while the specific heat associated with the rotational degrees of freedom is strongly temperature-dependent and varies from zero at $T=0$ to $\frac{1}{2}R$ for each degree of freedom at sufficiently high temperatures, which in practice are considerably less than room temperature. The rotational contribution to the internal energy at ordinary temperatures is therefore nRT for linear and $\frac{3}{2}nRT$ for non-linear molecules. The contribution of the vibrational degrees of freedom to the specific heat is also strongly temperature-dependent and is small for diatomic molecules at ordinary temperatures but may be considerable for polyatomic molecules containing a number of heavy atoms. These matters are discussed further in § 6·14, but for the present it is sufficient to note that, if we write

$$C_V(T)=C_V^0+C_V^1(T), \quad C_V^0=C_V(0), \tag{3·31·1}$$

then

$$C_V^0=\tfrac{3}{2}R, \tag{3·31·2}$$

and that normally $C_V^1(T)$ will be about R for linear molecules and $\frac{3}{2}R$ for non-linear molecules at ordinary temperatures. Higher values of $C_V^1(T)$ will be attained if there is any appreciable contribution from the vibrations.

3·311. The measurement of C_V for a gas is, in principle, simple, since we have merely to enclose the gas in a calorimeter and measure the energy required to change its temperature by a given amount. The practical difficulties are, however, considerable, and it is easier to measure C_P. The specific heat C_P at constant pressure is usually measured by a continuous flow method in which, for example, a steady stream of gas is heated electrically in a tube, and the rate of flow, the rate of supply of electrical energy and the rise in temperature are determined.

An alternative method is based upon the fact that we can find both C_V and C_P for a perfect gas if we can measure $\gamma=C_P/C_V$, since we have $C_P-C_V=R$ for any type of perfect gas. Now if c is the velocity of sound in a gas and if ρ is the density, we have $c^2=\partial P/\partial\rho$, where the differentiation must be carried out at constant entropy since the expansions and rarefactions take place so rapidly in a sound wave that they must be considered to be adiabatic. Therefore

$$c^2=(\partial P/\partial\rho)_S. \tag{3·311·1}$$

But, putting $dQ=0$ and $d\vartheta=0$ successively in (1·5·6) and (1·5·9), we have

$$\left(\frac{\partial P}{\partial V}\right)_S=\frac{C_P}{C_V}\frac{L_V}{nL_P}, \quad \left(\frac{\partial P}{\partial V}\right)_T=\frac{L_V}{nL_P}=\frac{1}{\gamma}\left(\frac{\partial P}{\partial V}\right)_S, \tag{3·311·2}$$

and so
$$c^2=\gamma(\partial P/\partial\rho)_T, \tag{3·311·3}$$

which becomes
$$c^2=\gamma P/\rho \tag{3·311·4}$$

for a perfect gas. This gives a method of determining γ. Some experimental results for C_P/R and γ are given in table 3·1. Helium and argon are monatomic, ammonia, methane and ethylene have non-linear molecules, while the remaining gases have linear molecules. If there is no vibrational contribution to the specific heat, we should have $\gamma=1·4$ for linear molecules and $\gamma=1·33$ for non-linear molecules, but it will be seen that there are considerable variations from these values. For the present, the only facts about C_V and C_P for a perfect gas that we need to use are that they are functions of ϑ only, but we shall resume the discussion of the specific heat data in § 6·1, where we establish the requisite theoretical formulae.

Table 3·1. *Values of C_P/R and γ for various gases at* 15° C.

	He	A	H_2	N_2	O_2	CO	NO
C_P/R	2·51	2·54	3·45	3·51	3·51	3·52	3·64
γ	1·67	1·67	1·41	1·4	1·4	1·4	1·38

	C_2H_2	CO_2	N_2O	NH_3	CH_4	C_2H_4
C_P/R	5·20	4·44	4·61	4·28	4·28	5·11
γ	1·24	1·29	1·28	1·30	1·30	1·24

Homogeneous Substances of Variable Composition

3·4. It is an experimental fact that, if surface effects are neglected, the specific heat of a homogeneous substance is independent of the amount of the substance present. This fact can conveniently be expressed in the form that the heat function $H(T,P,n)$ of n gram molecules of a substance at the temperature T and the pressure P can be written as

$$H(T,P,n)=n\text{H}(T,P), \tag{3·4·1}$$

where $\text{H}(T,P)$, the heat function of 1 gram molecule, is a function of T and P only. Similarly, by (3·1·21), we may write

$$G(T,P,n)=n\text{G}(T,P). \tag{3·4·2}$$

If (3·4·1) and (3·4·2) do not hold, either the substance is not homogeneous or surface effects must be taken into account. If the surface of a body makes a significant contribution to the thermodynamic functions we have to set up a more elaborate theory in which the volume and the surface of the body are considered as separate entities. The work done by the body in a small displacement can still be written in the form $\Sigma A_i da_i$, with the surface area as one of the generalized coordinates and the negative of the surface tension as the corresponding generalized force. The general theory is therefore applicable to cases where surface effects are important, but we shall not pursue this topic here.‡

From equation (3·4·2) we have

$$dG(T,P,n)=n\,d\sigma(T,P)+\sigma(T,P)\,dn \tag{3·4·3}$$

as the equation giving the variation in the free energy when we do not keep the mass of the system constant. But equation (3·1·19) applied to 1 gram molecule of the substance is

$$d\sigma(T,P)=-s(T,P)\,dT+v(T,P)\,dP. \tag{3·4·4}$$

Also, corresponding to (3·4·1) and (3·4·2) we have

$$S(T,P,n)=ns(T,P), \quad V(T,P,n)=nv(T,P), \tag{3·4·5}$$

and so (3·4·3) can be written as

$$dG=-S\,dT+V\,dP+\mu\,dn, \tag{3·4·6}$$

where

$$\mu=G(T,P,n)/n. \tag{3·4·7}$$

The quantity μ is known as the thermodynamic potential or the chemical potential. It may be defined by the relation

$$\mu=(\partial G/\partial n)_{T,P}. \tag{3·4·8}$$

By a similar argument we have

$$U(T,P,n)=nu(T,P), \tag{3·4·9}$$

and, since $G=U+ST-PV$, equation (3·4·6) gives

$$dU=T\,dS-P\,dV+\mu\,dn \tag{3·4·10}$$

as the generalization of (3·1·2). We therefore have

$$\mu=(\partial U/\partial n)_{S,V}. \tag{3·4·11}$$

Also, if we eliminate T and P from the relations (3·4·5) and (3·4·9) and consider U to be a function of S, V and n, $U(S,V,n)$ must be a homogeneous function of degree unity, since, if S, V and n are changed to kS, kV and kn, where k is a constant, U is changed to kU.

3·41. In order to specify the state of a homogeneous substance containing r constituents whose proportions can be varied we require to know,

‡ The reader who wishes to go more deeply into many of the topics only briefly mentioned here is referred to Gibbs's classical dissertation on the equilibrium of heterogeneous substances (1875–8).

in addition to the usual variables, the proportions of the various constituents. We denote the constituents by $\mathscr{C}_i$ $(i=1, \ldots, r)$, and we write n_i for the number of gram molecules of $\mathscr{C}_i$ present. Then, for a fluid or for a solid under isotropic pressure we can generalize equations (3·4·6) and (3·4·10) by writing

$$dG = -S\,dT + V\,dP + \sum_{i=1}^{r} \mu_i\,dn_i \tag{3·41·1}$$

and

$$dU = T\,dS - P\,dV + \sum_{i=1}^{r} \mu_i\,dn_i, \tag{3·41·2}$$

where

$$\mu_i = \left(\frac{\partial G}{\partial n_i}\right)_{T,P} = \left(\frac{\partial U}{\partial n_i}\right)_{S,V}. \tag{3·41·3}$$

The variation in equation (3·41·1) or equation (3·41·2) is supposed to be the most general possible variation provided that it refers to a quasi-static process and that the substance remains homogeneous. The total mass of the system need not remain constant.

For a given mass of a substance of invariable composition the internal energy is determinate when the entropy and the volume are given, and hence the internal energy of any homogeneous substance is determinate when the entropy, the volume and the masses of the constituents are given. Now the condition for a substance to be homogeneous is that, if the extensive variables S, V and n_i are changed to kS, kV and kn_i, where k is a constant, the internal energy is changed from U to kU. This is the condition that U should be a homogeneous function of degree unity in S, V and n_i. Hence, by Euler's theorem on homogeneous functions,

$$U = S\frac{\partial U}{\partial S} + V\frac{\partial U}{\partial V} + \sum_{i=1}^{r} n_i \frac{\partial U}{\partial n_i} = ST - PV + \sum_{i=1}^{r} \mu_i n_i. \tag{3·41·4}$$

(Alternatively, we may put $dU = kU$, $dS = kS$, $dV = kV$, $dn_i = kn_i$ in (3·41·2), where k is any infinitesimal constant, and we obtain (3·41·4) at once.)

The r quantities μ_i defined by (3·41·1) or (3·41·2) are intensive variables since they are independent of the amount of matter present, and like P and T they are homogeneous functions of S, V and n_i of degree zero. They are called the thermodynamic potentials (sometimes the chemical potentials) of the constituents, and they may, if desired, be defined in terms of H or F (see equations (3·42·3) and (3·42·6)).

3·42. *Fundamental thermodynamic equations.* A homogeneous fluid is completely specified by $2r+5$ variables, namely, the $r+3$ extensive variables U, S, V, n_i, and the $r+2$ intensive variables P, T, μ_i. Of these variables only $r+2$ are independent, since a knowledge of the entropy, the volume and the masses of the constituents is sufficient to determine the state of the substance. There must therefore be $r+3$ relations between the $2r+5$ variables, and any equation from which all these relations can be deduced is called a fundamental thermodynamic equation. Once a fundamental equation has been determined all the properties of the substance can be found from it. The $r+2$ variables S, V, n_i are not the only ones that can be used to specify the state of the substance, nor are they always the most convenient ones. If we take one variable from each of the pairs (S,T), (V,P), (n_1,μ_1), (n_2,μ_2), ..., (n_r,μ_r) as the independent variables, we can set up a fundamental equation by choosing the dependent variable in the fundamental equation properly.

(i) $U=U(S,V,n_i)$ *is a fundamental equation.* We have the general equation (3·41·2), from which it follows that

$$T=(\partial U/\partial S)_{V,n_i}, \quad P=-(\partial U/\partial V)_{S,n_i}, \quad \mu_i=(\partial U/\partial n_i)_{S,V,n_j}. \quad (3\cdot42\cdot1)$$

These $r+2$ equations, together with the equation $U=U(S,V,n_i)$, give $r+3$ equations from which U, T, P and μ_i can be determined when S, V and n_i are given.

It should be noted that $U=U(T,V,n_i)$ would not be a fundamental equation because the only relations that are given by the first derivatives are

$$(\partial U/\partial T)_{V,n_i}=T(\partial S/\partial T)_{V,n_i}=nC_V$$

and $$(\partial U/\partial V)_{T,n_i}=-P+T(\partial S/\partial V)_{T,n_i}=-P+T(\partial P/\partial T)_{V,n_i}.$$

These equations are not sufficient to determine P and S, but only give

$$\left(\frac{\partial S}{\partial T}\right)_{V,n_i} \quad \text{and} \quad \left(\frac{\partial}{\partial T}\frac{P}{T}\right)_{V,n_i}.$$

(ii) $H=H(S,P,n_i)$ *is a fundamental equation.* From the definition of H as $U+PV$ and from equation (3·41·2) we have

$$dH=T\,dS+V\,dP+\sum_{i=1}^{r}\mu_i\,dn_i, \quad (3\cdot42\cdot2)$$

and so $$T=(\partial H/\partial S)_{P,n_i}, \quad V=(\partial H/\partial P)_{S,n_i}, \quad \mu_i=(\partial H/\partial n_i)_{S,P,n_j}. \quad (3\cdot42\cdot3)$$

The $r+2$ equations (3·42·3) together with the equation $H=H(S,P,n_i)$ give $r+3$ equations from which H (and hence U), T, V and μ_i can be determined when S, P and n_i are given. We also have, from (3·41·4),

$$H=ST+\sum_{i=1}^{r}\mu_i n_i. \quad (3\cdot42\cdot4)$$

(iii) $F=F(T,V,n_i)$ *is a fundamental equation.* Since F is defined as $U-ST$ we have

$$dF=-S\,dT-P\,dV+\sum_{i=1}^{r}\mu_i\,dn_i, \quad (3\cdot42\cdot5)$$

and so $S = -(\partial F/\partial T)_{V,n_i}, \quad P = -(\partial F/\partial V)_{T,n_i}, \quad \mu_i = (\partial F/\partial n_i)_{T,V,n_j}.$ (3·42·6)

The $r+2$ equations (3·42·6) together with the equation $F = F(T, V, n_i)$ give $r+3$ equations from which F (and hence U), S, P and μ_i can be determined when T, V and n_i are given. An explicit expression for F is

$$F = -PV + \sum_{i=1}^{r} \mu_i n_i. \tag{3·42·7}$$

Instead of differentiating F itself we may differentiate F/T. Then, by (3·1·15), equations (3·42·5) and (3·42·6) are replaced by

$$d\frac{F}{T} = -\frac{U}{T^2}dT - \frac{P}{T}dV + \sum_{i=1}^{r} \frac{\mu_i}{T} dn_i \tag{3·42·8}$$

and $$U = -T^2\left(\frac{\partial}{\partial T}\frac{F}{T}\right)_{V,n_i}, \quad P = -\left(\frac{\partial F}{\partial V}\right)_{T,n_i}, \quad \mu_i = \left(\frac{\partial F}{\partial n_i}\right)_{T,V,n_j}. \tag{3·42·9}$$

(iv) $G = G(T, P, n_i)$ *is a fundamental equation*. The definition of G is $G = U - ST + PV$, and so

$$dG = -S\,dT + V\,dP + \sum_{i=1}^{r} \mu_i dn_i. \tag{3·42·10}$$

Hence $S = -(\partial G/\partial T)_{P,n_i}, \quad V = (\partial G/\partial P)_{T,n_i}, \quad \mu_i = (\partial G/\partial n_i)_{T,P,n_j}.$ (3·42·11)

The $r+2$ equations (3·42·11) together with the equation $G = G(T, P, n_i)$ give $r+3$ equations from which G (and hence U), S, V and μ_i can be determined when T, P and n_i are given. An explicit expression for G is

$$G = \sum_{i=1}^{r} \mu_i n_i. \tag{3·42·12}$$

The equations (3·42·10) and (3·42·11) may be replaced by

$$d\frac{G}{T} = -\frac{H}{T^2}dT + \frac{V}{T}dP + \sum_{i=1}^{r} \frac{\mu_i}{T} dn_i, \tag{3·42·13}$$

and $$H = -T^2\left(\frac{\partial}{\partial T}\frac{G}{T}\right)_{P,n_i}, \quad V = \left(\frac{\partial G}{\partial P}\right)_{T,n_i}, \quad \mu_i = \left(\frac{\partial G}{\partial n_i}\right)_{T,P,n_j}. \tag{3·42·14}$$

(v) $P = P(T, \mu_i)$ *is a fundamental equation*. The $r+3$ equations (3·41·4) and (3·42·1) are general equations independent of the properties of any particular substance. If we are given the fundamental equation

$$U = U(S, V, n_i)$$

for a certain substance we can, in principle, eliminate from these $r+4$ equations the $r+3$ variables U, S, V and n_i, and obtain an equation between P, T and μ_i. This equation is shown below to be a fundamental equation, but it is of a somewhat different nature from the four fundamental equations already found. A knowledge of the intensive variables P, T and μ_i is insufficient to determine completely the extensive variables, but is sufficient to determine their ratios, say U/V, S/V, n_i/V. We must therefore be given the value of one of the extensive variables, for example the volume or the total mass, in order to specify the system completely.

To show that $P=P(T,\mu_i)$ is a fundamental equation we differentiate the identity (3·41·4), obtaining

$$dU=SdT+TdS-PdV-VdP+\sum_{i=1}^{r}(\mu_i dn_i+n_i d\mu_i).$$

We now eliminate dU by means of (3·41·2) and divide by V. The result is

$$dP=\frac{S}{V}dT+\sum_{i=1}^{r}\frac{n_i}{V}d\mu_i. \tag{3·42·15}$$

Hence $$S/V=(\partial P/\partial T)_{\mu_i},\quad n_i/V=(\partial P/\partial \mu_i)_{T,\mu_j}. \tag{3·42·16}$$

The $r+1$ equations (3·42·16), together with the equation $P=P(T,\mu_i)$, determine the ratios S/V and n_i/V, and also the pressure, when T and μ_i are given. The ratio U/V can then be found from (3·41·4). If we are given the volume V in addition, all the extensive variables are then fixed. If we are given the total mass instead of the volume, we determine V from the relation

$$\Sigma n_i=V\Sigma(\partial P/\partial \mu_i)_{T,\mu_j}. \tag{3·42·17}$$

(vi) There are numerous other fundamental equations of less importance in which some of the n's and some of the μ's are taken as the independent variables. For example,

$$U-\sum_{i=1}^{r'}\mu_i n_i=f(S,V,\mu_1,\ldots,\mu_{r'},n_{r'+1},\ldots,n_r)$$

is a fundamental equation. For we have

$$d\left(U-\sum_{i=1}^{r'}\mu_i n_i\right)=TdS-PdV-\sum_{i=1}^{r'}n_i d\mu_i+\sum_{i=r'+1}^{r}\mu_i dn_i,$$

and so $$T=\frac{\partial}{\partial S}\left(U-\sum_{i=1}^{r'}\mu_i n_i\right),\quad P=-\frac{\partial}{\partial V}\left(U-\sum_{i=1}^{r'}\mu_i n_i\right),$$

$$n_i=-\frac{\partial}{\partial \mu_i}\left(U-\sum_{j=1}^{r'}\mu_j n_j\right)\quad (i=1,2,\ldots,r'),$$

$$\mu_i=\frac{\partial}{\partial n_i}\left(U-\sum_{j=1}^{r'}\mu_j n_j\right)\quad (i=r'+1,r'+2,\ldots,r).$$

These equations are just sufficient to determine all the dependent variables. There are 2^r such equations associated with U, of which the most important is $U=U(S,V,n_i)$ which we have already discussed. The only other one that is ever likely to be useful is the one involving $U-\sum_{i=1}^{r}\mu_i n_i$, and this can be written as $ST - PV = f(S, V, \mu_i)$. There are also 2^r equations associated in the same way with each of the functions H, F and G. Of the 2^r equations associated with G, one of them, namely, that referring to $G - \sum_{i=1}^{r} \mu_i n_i$, vanishes identically. Its place is taken by the $P(T, \mu_i)$ equation.

The equation associated with $F - \sum_{i=1}^{r} \mu_i n_i = -PV$ is of a relatively simple form and is often useful, since it occurs naturally in theoretical work. It is

$$PV = f(T, V, \mu_i), \tag{3·42·18}$$

and its properties are as follows. We have

$$-d(PV) = d\left(F - \sum_{i=1}^{r} \mu_i n_i\right) = -SdT - PdV - \sum_{i=1}^{r} n_i d\mu_i. \tag{3·42·19}$$

Hence

$$S = \{\partial(PV)/\partial T\}_{V,\mu_i}, \quad P = \{\partial(PV)/\partial V\}_{T,\mu_i}, \quad n_i = \{\partial(PV)/\partial \mu_i\}_{T,V,\mu_j}. \tag{3·42·20}$$

This fundamental equation is essentially the same as the $P(T, \mu_i)$ equation.

3·421. *The fundamental equations of a general thermodynamic system.* For a general system with s deformation coordinates and r constituents we have

$$dU = TdS - \sum_{i=1}^{s} A_i da_i + \sum_{j=1}^{r} \mu_j dn_j \tag{3·421·1}$$

as the generalization of (3·1·23) and (3·41·2). There are then numerous fundamental equations. For example, we can take $U + \sum_{i=1}^{s'} A_i a_i - \sum_{j=1}^{r'} \mu_j n_j$ and express it as a function of $S, A_1, \ldots, A_{s'}, a_{s'+1}, \ldots, a_s, \mu_1, \ldots, \mu_{r'}, n_{r'+1}, \ldots, n_r$ and obtain a fundamental equation. There are 2^{r+s} equations of this type. Similarly, there are 2^{r+s} fundamental equations with U replaced by $F = U - ST$, and so there are 2^{r+s+1} equations in all, the properties of any of which can readily be found. If we consider in particular

$$F - \sum_{j=1}^{r} \mu_j n_j \equiv -\sum_{i=1}^{s} A_i a_i, \tag{3·421·2}$$

we have

$$d\left(F - \sum_{j=1}^{r} \mu_j n_j\right) \equiv -d\sum_{i=1}^{s} A_i a_i = -SdT - \sum_{i=1}^{s} A_i da_i - \sum_{j=1}^{r} n_j d\mu_j, \tag{3·421·3}$$

and so

$$S = \left(\frac{\partial}{\partial T}\sum_{i=1}^{s} A_i a_i\right)_{a_i,\mu_j}, \quad A_i = \left(\frac{\partial}{\partial a_i}\sum_{i=1}^{s} A_i a_i\right)_{T,\mu_j}, \quad n_j = \left(\frac{\partial}{\partial \mu_j}\sum_{i=1}^{s} A_i a_i\right)_{T,a_i}. \tag{3·421·4}$$

3·422. Each of the fundamental equations has its own merits, and the particular one we choose in any special case depends upon the object in view. The $U(S, V, n_i)$ equation is most appropriate in discussing problems concerning heterogeneous substances such as the coexistence of a liquid and its vapour, since it contains only the extensive variables which are particularly well adapted to problems in which the amounts of matter in the liquid or the vapour states are variable. On the other hand, the $G(T, P, n_i)$ equation only contains variables whose significance is obvious intuitively, while the $P(T, \mu_i)$ equation, involving only the intensive variables, is in many ways the simplest of all, since, as shown in § 3·62, the parameters P, T and μ_i must be the same for all substances in equilibrium with one another. The uses of the various fundamental equations will become apparent when special cases are considered in subsequent chapters.

In general, the additive constants which occur in the definitions of the internal energy and the entropy are unimportant, but when we are dealing with the thermodynamic potentials it is essential that these constants should be fixed once and for all, and they must not be considered as parameters whose values can be varied in different parts of a calculation to simplify the formulae.

The Fundamental Equations for a Simple Substance

3·5. While the determination of the fundamental equations for a substance of variable composition presents considerable difficulties, a knowledge of the specific heat and of the equation of state is sufficient to determine the fundamental equations of any simple substance. Equation (3·21·9), or the equivalent equations (3·21·10) and (3·21·11), gives $F(T, V)$ if P is known explicitly as a function of T and V, while equation (3·211·8) (or (3·211·9) or (3·211·10)) gives $G(T, P)$ if V is known explicitly as a function of T and P. Further, since $G = n\mu$ for a simple substance, equation (3·211·8) is also the $P(T, \mu)$ equation. To obtain the $U(S, V)$ and the $H(S, P)$ equations we have to eliminate the temperature between equations (3·21·7) and (3·21·8) and between (3·211·6) and (3·211·7) respectively, but in general it is impossible to carry out the eliminations explicitly.

3·51. *A perfect gas.* The simplest fundamental equations are those for a perfect gas, but even then the equations are distinctly complicated. The fundamental equations can be found from the expressions for the internal energy and the entropy which have already been given in § 2·72. The energy and entropy constants are proportional to the number of gram molecules present, and to show this explicitly we write

$$U_0 = nv_0, \quad S_0 = ns_0, \quad S_0' = ns_0', \tag{3·51·1}$$

where v_0, s_0 and s_0' are independent of T, P, V and n. The internal energy and the entropy are then given by

$$U = n(C_V T + v_0), \tag{3·51·2}$$

and $$S=n(C_V\log T+R\log V+s_0'-C_V\log T^\dagger-R\log V^\dagger),\tag{3·51·3}$$
where v_0 is the internal energy of 1 gram molecule when $T=0$, and s_0' is the entropy of 1 gram molecule when its temperature is $T^\dagger$ and its volume is $v^\dagger=V^\dagger/n$. Alternatively, we may write
$$S=n(C_P\log T-R\log P+s_0-C_P\log T^\dagger+R\log P^\dagger),\tag{3·51·4}$$
where s_0 is the entropy of 1 gram molecule when its temperature is $T^\dagger$ and its pressure is $P^\dagger$. We also require the equation of state
$$PV=nRT\tag{3·51·5}$$
and the relation $C_P=C_V+R$.

(i) The $U(S,V,n)$ equation is obtained by eliminating T between (3·51·2) and (3·51·3). It is
$$U=nC_VT^\dagger\left(\frac{nv^\dagger}{V}\right)^{R/C_V}\exp\frac{S-ns_0'}{nC_V}+nv_0.\tag{3·51·6}$$
The relations (3·42·1) then give
$$T=T^\dagger\left(\frac{nv^\dagger}{V}\right)^{R/C_V}\exp\frac{S-ns_0'}{nC_V},\quad P=\frac{nRT^\dagger}{V}\left(\frac{nv^\dagger}{V}\right)^{R/C_V}\exp\frac{S-ns_0'}{nC_V}\tag{3·51·7}$$
and $$\mu=\left(C_P-\frac{S}{n}\right)T^\dagger\left(\frac{nv^\dagger}{V}\right)^{R/C_V}\exp\frac{S-ns_0'}{nC_V}+v_0.\tag{3·51·8}$$

(ii) The $H(S,P,n)$ equation is obtained by eliminating T from
$$H=U+PV=n(C_PT+v_0)$$
and equation (3·51·4). It is
$$H=nC_PT^\dagger\left(\frac{P}{P^\dagger}\right)^{R/C_P}\exp\frac{S-ns_0}{nC_P}+nv_0.\tag{3·51·9}$$
Then $$T=T^\dagger\left(\frac{P}{P^\dagger}\right)^{R/C_P}\exp\frac{S-ns_0}{nC_P},\quad V=\frac{nRT^\dagger}{P^\dagger}\left(\frac{P^\dagger}{P}\right)^{C_V/C_P}\exp\frac{S-ns_0}{nC_P}\tag{3·51·10}$$
and $$\mu=\left(C_P-\frac{S}{n}\right)T^\dagger\left(\frac{P}{P^\dagger}\right)^{R/C_P}\exp\frac{S-ns_0}{nC_P}+v_0.\tag{3·51·11}$$

(iii) The $F(T,V,n)$ equation can be obtained at once by combining (3·51·2) and (3·51·3). It is
$$F=nC_V\{T-T\log(T/T^\dagger)\}-nRT\log(V/nv^\dagger)+n(v_0-Ts_0').\tag{3·51·12}$$
By differentiation we obtain
$$S=n\{C_V\log(T/T^\dagger)+R\log(V/nv^\dagger)+s_0'\},\quad P=nRT/V\tag{3·51·13}$$
and $$\mu=C_PT-C_VT\log(T/T^\dagger)-RT\log(V/nv^\dagger)+v_0-Ts_0'.\tag{3·51·14}$$

(iv) The $G(T,P,n)$ equation is obtained by substituting (3·51·4) into the relation $G=H-ST=n(C_PT+v_0)-ST$. It is
$$G=nC_P\{T-T\log(T/T^\dagger)\}+nRT\log(P/P^\dagger)+n(v_0-Ts_0).\tag{3·51·15}$$

By differentiation we obtain

$$S = n\{C_P \log (T/T^\dagger) - R \log (P/P^\dagger) + s_0\}, \quad V = nRT/P \qquad (3\cdot51\cdot16)$$

and

$$\mu = C_P\{T - T \log (T/T^\dagger)\} + RT \log (P/P^\dagger) + v_0 - Ts_0. \qquad (3\cdot51\cdot17)$$

(v) The $P(T, \mu)$ equation is given at once by (3·51·17). If we solve explicitly for P we obtain

$$P = P^\dagger \exp [(s_0 - C_P)/R] (T/T^\dagger)^{C_P/R} \exp [(\mu - v_0)/RT], \qquad (3\cdot51\cdot18)$$

and hence

$$\frac{S}{V} = \frac{P^\dagger}{RT^\dagger} \exp [(s_0 - C_P)/R] \left(\frac{T}{T^\dagger}\right)^{C_V/R} \left(C_P - \frac{\mu - v_0}{T}\right) \exp [(\mu - v_0)/RT], \qquad (3\cdot51\cdot19)$$

$$\frac{n}{V} = \frac{P^\dagger}{RT^\dagger} \exp [(s_0 - C_P)/R] \left(\frac{T}{T^\dagger}\right)^{C_V/R} \exp [(\mu - v_0)/RT]. \qquad (3\cdot51\cdot20)$$

3·52. When we try to formulate the fundamental equations of a perfect gas in the wide sense, we find that it is impossible to carry out all the eliminations required explicitly. The $F(T, V, n)$ and the $G(T, P, n)$ equations can, however, easily be found from equations (2·72·8), (2·72·9), the relations $F = U - ST$, $G = H - ST$ and

$$U(V, T) = n\int_0^T C_V(T')\,dT' + nv_0, \quad H(P, T) = n\int_0^T C_P(T')\,dT' + nv_0. \quad (3\cdot52\cdot1)$$

The expressions for F and G are

$$F(T, V, n) = n\left[\int_0^T C_V(T')\,dT' - C_V^0 T \log \frac{T}{T^\dagger} - T\int_0^T \frac{C_V^1(T')}{T'}\,dT' - RT \log \frac{V}{nV^\dagger} + v_0 - Ts_0'\right] \quad (3\cdot52\cdot2)$$

$$= n\left[C_V^0\left(T - T\log\frac{T}{T^\dagger}\right) - T\int_0^T \frac{dT'}{T'^2}\int_0^{T'} C_V^1(T'')\,dT'' - RT \log \frac{V}{nV^\dagger} + v_0 - Ts_0'\right], \quad (3\cdot52\cdot3)$$

and

$$G(T, P, n) = n\mu(T, P) = n\left[\int_0^T C_P(T')\,dT' - C_P^0 T \log \frac{T}{T^\dagger} - T\int_0^T \frac{C_P^1(T')}{T'}\,dT' + RT \log \frac{P}{P^\dagger} + v_0 - Ts_0\right], \quad (3\cdot52\cdot4)$$

$$= n\left[C_P^0\left(T - T\log\frac{T}{T^\dagger}\right) - T\int_0^T \frac{dT'}{T'^2}\int_0^{T'} C_P^1(T'')\,dT'' + RT \log \frac{P}{P^\dagger} + v_0 - Ts_0\right], \quad (3\cdot52\cdot5)$$

where C_V^0, C_P^0 are the constant parts of the specific heats and $C_V^1(T) = C_P^1(T)$ are the parts which vary with the temperature. The alternative forms (3·52·3) and (3·52·5) are obtained from (3·52·2) and (3·52·4) by integrating by parts.

Heterogeneous Equilibrium

3·6. *The general conditions of equilibrium.* It was shown in § 2·41 that, in any spontaneous process, the entropy always increases. If, therefore, the entropy of an isolated system is a maximum, no spontaneous change is possible and the system is in a state of stable equilibrium. An isolated system is one which is enclosed by a rigid adiabatic wall, and not only is the internal energy of such a system constant, but the volume and the total mass of each of its independent constituents are also constant. We can therefore state the conditions for equilibrium as follows:

For the equilibrium of any isolated system it is necessary and sufficient that in all possible variations of the state of the system which do not alter its internal energy, volume or the masses of any of its independent constituents, the variation of its entropy shall either vanish or be negative.

If we use d to denote an infinitesimal of the first order, the higher order variations being neglected, the condition of equilibrium can be written as

$$(dS)_{U,V,n_i} = 0. \qquad (3·6·1)$$

Now let Δ denote a general variation in which infinitesimals of higher order than the first are not neglected. Then the stability conditions are as follows. For stable equilibrium we must have

$$(\Delta S)_{U,V,n_i} < 0. \qquad (3·6·2)$$

For neutral equilibrium there must be some variations for which

$$(\Delta S)_{U,V,n_i} = 0,$$

while in general $(\Delta S)_{U,V,n_i} \leqslant 0$. For unstable equilibrium there must be some variations for which $(\Delta S)_{U,V,n_i} > 0$, while in general $(dS)_{U,V,n_i} = 0$.

3·61. The conditions (3·6·1) and (3·6·2) are the necessary and sufficient conditions for the equilibrium and stability of an isolated system, but it is often convenient to consider the stability of systems which are not isolated but which form part of a larger thermodynamic system. There are then four stability conditions each associated with one of the functions U, H, F and G.

(i) To discuss these stability conditions we first put (3·6·1) into a slightly different form, which can be stated as follows:

For the equilibrium of any isolated system it is necessary and sufficient that in all possible variations of the state of the system which do not alter its entropy, volume or the masses of any of its independent constituents, the variation of its internal energy shall either vanish or be positive.

To prove this, suppose that there is a variation for which $dS=0$ and $dU<0$. Now we can increase both S and U simultaneously while keeping V and n_i constant by adding heat reversibly to the system. There is therefore a variation of the system such that $dS>0$ and $dU=0$, which means that the system is not in equilibrium since the variation violates condition (3·6·1). We therefore see that the condition (3·6·1) implies the condition

$$(dU)_{S,V,n_i}=0, \tag{3·61·1}$$

and the converse of this can be proved by the same method. Similarly, the stability condition (3·6·2) is equivalent to

$$(\Delta U)_{S,V,n_i}>0. \tag{3·61·2}$$

(ii) If we have a system contained in an adiabatic envelope which is not rigid, the pressure must be the same as that in the medium surrounding the envelope. If this medium is so large that its pressure is unaffected by movements of the envelope, any changes in the thermodynamic system under consideration take place at constant pressure, and the change in energy will be given by $dH=dU+PdV$ and not by dU. The conditions (3·61·1) and (3·61·2) have therefore to be replaced by the conditions

$$(dH)_{S,P,n_i}=0 \tag{3·61·3}$$

for equilibrium, and

$$(\Delta H)_{S,P,n_i}>0 \tag{3·61·4}$$

for stability.

(iii) If a thermodynamic system is contained in a rigid diathermal envelope and surrounded by a large reservoir of heat, any changes will take place at constant volume and constant temperature. Now when T and V are constant, the maximum work that the system can perform in a given process is $-\Delta F$ (equation (3·1·11)), so that if ΔF is positive no process can take place. The equilibrium condition is therefore

$$(dF)_{T,V,n_i}=0, \tag{3·61·5}$$

and the stability condition is

$$(\Delta F)_{T,V,n_i}>0. \tag{3·61·6}$$

(iv) If a thermodynamic system is contained in an envelope which is neither adiabatic nor rigid and is surrounded by a large medium whose temperature and pressure are constant, any changes that take place are at constant temperature and pressure. Now $-\Delta G$ is the maximum work that can be obtained from the system in a given process over and above the necessary work of expansion or compression (equation (3·1·18)), and if ΔG is positive no process can take place. The equilibrium condition is therefore

$$(dG)_{T,P,n_i}=0, \tag{3·61·7}$$

and the stability condition is

$$(\Delta G)_{T,P,n_i} > 0. \tag{3·61·8}$$

The equilibrium conditions (3·61·1), (3·61·3), (3·61·5) and (3·61·7) must all be equivalent, since if a system is in equilibrium when it is considered as part of another system it must also be in equilibrium when considered as an isolated system. The corresponding stability conditions are not, however, necessarily equivalent for finite variations, since they refer to different physical conditions, and finite changes may well be possible in a system forming part of a larger system which are impossible in an isolated system.

3·62. *The equilibrium of heterogeneous masses.* The different homogeneous states of aggregation of a substance are commonly called phases (e.g. the solid phase, the liquid phase and the vapour phase), and in the most general type of equilibrium it is necessary to determine the conditions for the coexistence of more than one phase. We regard all bodies which merely differ in quantity and shape as different examples of the same phase, and a phase is therefore defined by its composition and thermodynamic state. We consider the general case in which there are r independent constituents $\mathscr{C}_i$ and p phases. We exclude for the moment cases in which chemical reactions take place between the constituents (see § 11·4). If we use the index α $(\alpha = 1, 2, \ldots, p)$ to distinguish the various phases, and consider an isolated system enclosed in a rigid, impermeable, adiabatic envelope, we can write

$$U = \Sigma U^{(\alpha)}, \quad S = \Sigma S^{(\alpha)}, \quad V = \Sigma V^{(\alpha)}, \quad n_i = \Sigma n_i^{(\alpha)} \quad (i = 1, 2, \ldots, r), \tag{3·62·1}$$

where $U^{(\alpha)}$, $S^{(\alpha)}$ and $V^{(\alpha)}$ are the internal energy, the entropy and the volume of the phase α, and $n_i^{(\alpha)}$ is the number of gram molecules of the ith constituent in the phase α. The general condition of equilibrium (3·6·1) can most easily be written down by introducing undetermined multipliers ϑ, ξ, λ_i and replacing (3·6·1) by the condition

$$d \sum_{\alpha} (S^{(\alpha)} - \vartheta U^{(\alpha)} - \xi V^{(\alpha)} + \sum_{i} \lambda_i n_i^{(\alpha)}) = 0, \tag{3·62·2}$$

where the quantities to be varied are $S^{(\alpha)}$, $U^{(\alpha)}$, $V^{(\alpha)}$, $n_i^{(\alpha)}$ subject to the constancy of the internal energy U, the volume V, and the number of gram molecules n_i of each of the constituents in the whole system.

Each phase is homogeneous, and therefore equation (3·41·2) is applicable to each separate phase. Therefore, substituting for $dS^{(\alpha)}$ from (3·41·2), we can write (3·62·2) as

$$\sum_{\alpha} \left[\left(\frac{1}{T^{(\alpha)}} - \vartheta \right) dU^{(\alpha)} + \left(\frac{P^{(\alpha)}}{T^{(\alpha)}} - \xi \right) dV^{(\alpha)} + \sum_{i} \left(\lambda_i - \frac{\mu_i^{(\alpha)}}{T^{(\alpha)}} \right) dn_i^{(\alpha)} \right] = 0, \tag{3·62·3}$$

and $dU^{(\alpha)}$, $dV^{(\alpha)}$, $dn_i^{(\alpha)}$ can be treated as independent differentials. Hence

$$T^{(\alpha)} = 1/\vartheta, \quad P^{(\alpha)} = \xi T^{(\alpha)}, \quad \mu_i^{(\alpha)} = \lambda_i T^{(\alpha)} \quad (\text{all } \alpha), \tag{3·62·4}$$

i.e.
$$T^{(\alpha)} = T, \quad P^{(\alpha)} = P, \quad \mu_i^{(\alpha)} = \mu_i \quad (\text{all } \alpha). \tag{3·62·5}$$

Thus the temperature, the pressure and the thermodynamic potential of each constituent must be the same for all phases. The first two conditions give nothing new, since the temperature is defined so as to be the same for all bodies in thermal equilibrium and the pressure must be uniform for hydrostatic equilibrium. The requirement that the thermodynamic potential of each constituent must be constant throughout the system is, however, new.

We still have to verify that the conditions of equilibrium (3·62·4), together with a knowledge of U, V and n_i for the whole system, are sufficient to determine the state of the system. Now since $U^{(\alpha)}$ $(S^{(\alpha)}, V^{(\alpha)}, n_i^{(\alpha)})$ is a fundamental equation for a homogeneous mass, the state of the phase α is determined if the values of $S^{(\alpha)}$, $V^{(\alpha)}$ and $n_i^{(\alpha)}$ are given. There are therefore $p(r+2)$ independent variables to be determined. But there are $(p-1)(r+2)$ equations for $T^{(\alpha)} = \partial U^{(\alpha)}/\partial S^{(\alpha)}$, $P^{(\alpha)} = -\partial U^{(\alpha)}/\partial V^{(\alpha)}$, $\mu_i^{(\alpha)} = \partial U^{(\alpha)}/\partial n_i^{(\alpha)}$, and there are $r+2$ further equations since we are given the values of U, V and n_i. Hence we have $p(r+2)$ equations for $p(r+2)$ quantities and the problem is determinate unless there are special relations between the equations. Singular cases, though mathematically possible, do not arise in practice.

3·621. *The effect of additional conditions of constant.* If there are additional constraints upon the system, the number of conditions of equilibrium will be correspondingly reduced. Consider for example, the effect of dividing the mass into two by a rigid diathermal membrane which is permeable to all the constituents except $\mathscr{C}_1$. If we distinguish the two portions of the system by single and double accents, the volumes V' and V'' of the two parts and the amounts n_1' and n_1'' of the constituent $\mathscr{C}_1$ in them are constant, and equation (3·62·3) must be replaced by

$$\sum_\alpha \left[\left(\frac{1}{T^{(\alpha)}} - \vartheta \right) dU^{(\alpha)} + \left(\frac{P'^{(\alpha)}}{T^{(\alpha)}} - \xi' \right) dV'^{(\alpha)} + \left(\frac{P''^{(\alpha)}}{T^{(\alpha)}} - \xi'' \right) dV''^{(\alpha)} \right.$$
$$\left. + \left(\lambda_1' - \frac{\mu_1'^{(\alpha)}}{T^{(\alpha)}} \right) dn_1'^{(\alpha)} + \left(\lambda_1'' - \frac{\mu_1''^{(\alpha)}}{T^{(\alpha)}} \right) dn_1''^{(\alpha)} + \sum_{i \geqslant 2} \left(\lambda_i - \frac{\mu_i^{(\alpha)}}{T^{(\alpha)}} \right) dn_i^{(\alpha)} \right] = 0. \tag{3·621·1}$$

We therefore obtain all the conditions (3·62·5) except that

$$P' \neq P'', \quad \mu_1' \neq \mu_1''.$$

We now lose two of the equations (3·62·5), but since we know V', V'' and n_1', n_1'', there are two less independent variables, so that the problem is still soluble.

3·63. *The phase rule.* In considering the equilibrium of a number of coexistent phases we are not interested in general in the amount of each phase present. A set of coexistent phases is therefore determined by the temperature, the pressure and the concentrations of the various constituents, i.e. by $p(r-1)+2$ quantities. But the $r(p-1)$ conditions $\mu_i^{(\alpha)}=\mu_i$ (all α) must be satisfied, and there are therefore $r-p+2$ independent variables or degrees of freedom. This is known as the phase rule.

A simple substance has $3-p$ degrees of freedom. Therefore, if there is only one phase present, both the temperature and the pressure can be chosen arbitrarily. If there are two phases present either the temperature or the pressure can be chosen arbitrarily, but not both, while three phases can only exist together at one definite temperature and pressure known as the triple point. Sulphur is an example of a substance which can exist in two different crystalline modifications. It is, however, impossible for both these modifications to exist simultaneously in contact with liquid and gaseous sulphur, since this is at variance with the phase rule. More complicated examples of the phase rule will be encountered in Chapter 12.

THE STABILITY OF HOMOGENEOUS SUBSTANCES

3·7. In hydrostatics the usual criterion for the stability of a fluid is that $\partial P/\partial V$ should be negative, and this applies whether we consider isothermal or adiabatic conditions. Therefore, since $P=-\partial U(S,V)/\partial V$, a necessary condition for stability is that, for a fixed value of S, $\partial^2 U/\partial V^2$ should be positive, i.e. that U should be convex downwards. We may state this in the alternative form that the curve $U(V)$ must lie above all its tangents. Now the equation of the tangent at $V=V_1$ is

$$(U-U_1)/(V-V_1)=(\partial U/\partial V)_{V=V_1}=-P_1,$$

and hence $U(V)$ lies above all its tangents if $U-U_1+P_1(V-V_1)>0$.

To derive a similar condition in the general case we consider the expression

$$\Phi(S,V,n_i;\,S_1,V_1,n_{i,1})=U(S,V,n_i)-T_1S+P_1V-\Sigma\mu_{i,1}n_i, \quad (3\cdot7\cdot1)$$

relating to a given homogeneous mass of a fluid or an isotropic solid, where

$$T_1=\partial U/\partial S, \quad P_1=-\partial U/\partial V, \quad \mu_{i,1}=\partial U/\partial n_i,$$

evaluated at $S=S_1$, $V=V_1$, $n_i=n_{i,1}$. By (3·41·4) we may write Φ as

$$\Phi(S,V,n_i;\,S_1,V_1,n_{i.1})=U-U_1-T_1(S-S_1)+P_1(V-V_1)-\Sigma\mu_{i,1}(n_i-n_{i,1}), \quad (3\cdot7\cdot2)$$

and we see that the geometrical significance of Φ is that it is the distance, measured parallel to the U axis, of the point S, V, n_i on the surface $U(S,V,n_i)$ from the tangent plane at the point $S_1, V_1, n_{i,1}$. We shall now show that, if

$\Phi(S, V, n_i; S_1, V_1, n_{i,1}) > 0$ for all values of S, V, n_i other than S_1, V_1, $n_{i,1}$, the fluid is stable in the state S_1, V_1, $n_{i,1}$.

To discuss the stability of the fluid we have to compare the values of U for all the states with $S = S_1$, $V = V_1$, $n_i = n_{i,1}$. These states, other than the state whose stability is in question, are necessarily heterogeneous, and for these states Φ is the sum of a number of Φ's referring to homogeneous masses α satisfying the conditions $\Sigma S^{(\alpha)} = S_1$, $\Sigma V^{(\alpha)} = V_1$, $\Sigma n_i^{(\alpha)} = n_{i,1}$. Now if $\Phi > 0$ for every homogeneous mass, we must have $\Sigma \Phi^{(\alpha)} > 0$. But, by (3·7·1), $\Sigma \Phi^{(\alpha)} = \Sigma U^{(\alpha)} - T_1 S_1 + P_1 V_1 - \Sigma \mu_{i,1} n_{i,1}$, and this is equal to $\Sigma U^{(\alpha)} - U_1$. Hence, if $\Phi(S, V, n_i; S_1, V_1, n_{i,1}) > 0$ for all S, V, n_i, the internal energy $U(S_1, V_1, n_{i,1})$ for the given homogeneous mass is less than that for any heterogeneous mass with the same values of S_1, V_1, $n_{i,1}$. Therefore, by (3·61·1), the substance is stable in the given state.

3·71. From the geometrical significance of Φ it is clear that a necessary condition for the stability of the state S_1, V_1, $n_{i,1}$ is that the U surface should be convex downwards at S_1, V_1, $n_{i,1}$. A sufficient condition for stability is that the U surface should be convex downwards everywhere, and in this case every state is stable. If, however, the U surface has portions which are not convex downwards, unstable states are bound to occur. These are of two types, namely, those which are stable with respect to small variations and those which are essentially unstable. The unstable states of the first type (the metastable or locally stable states) occur when the U surface is convex downwards but has a tangent plane which intersects the surface again at a finite distance from the point of contact. The regions in which U is not convex downwards are composed of essentially unstable states.

The determination of the states which are stable for small but unstable for large variations depends upon the large-scale geometry of the U surface and is not amenable to general treatment. The essentially unstable states are, however, determined by the differential geometry of the U surface, which we now investigate. In the remainder of this section we shall use the word stable to mean 'locally stable' and unstable to mean 'locally (or essentially) unstable'.

If we have two neighbouring states S_1, V_1, $n_{i,1}$ and S_2, V_2, $n_{i,2}$ which are both stable, the tangent plane at the first point on the U surface passes below the second point and the tangent plane at the second point passes below the first point. We therefore have

$$\Phi(S_2, V_2, n_{i,2}; S_1, V_1, n_{i,1}) > 0, \quad \Phi(S_1, V_1, n_{i,1}; S_2, V_2, n_{i,2}) > 0. \quad (3·71·1)$$

Hence by addition and by using (3·7·2) we have

$$(T_2 - T_1)(S_2 - S_1) - (P_2 - P_1)(V_2 - V_1) + \Sigma(\mu_{i,2} - \mu_{i,1})(n_{i,2} - n_{i,1}) > 0. \quad (3·71·2)$$

We may write this as

$$\Delta T\,\Delta S - \Delta P\,\Delta V + \Sigma \Delta\mu_i \Delta n_i > 0, \tag{3·71·3}$$

and we can use this to deduce all the conditions for stability, which are

$$\partial T/\partial S > 0, \quad \partial P/\partial V < 0, \quad \partial\mu_i/\partial n_i > 0. \tag{3·71·4}$$

There is, however, a considerable choice as to the variables which are to be kept constant in these differentiations, since $\partial T/\partial S$, for example, must be positive if either P or V and if either μ_i or n_i (for all i) are kept constant. There are therefore 2^{r+1} possible values of $\partial T/\partial S$, all of which must be positive. There is a similar choice of independent variables in the conditions $\partial P/\partial V < 0$ and $\partial\mu_i/\partial n_i > 0$.

All the above stability conditions are not independent, and there are in fact only $r+1$ separate conditions. For most purposes the most convenient set of conditions is

$$(\partial S/\partial T)_{P,n_i} > 0, \quad (\partial V/\partial P)_{T,n_i} < 0, \quad (\partial\mu_i/\partial n_i)_{T,P,n_j} > 0, \tag{3·71·5}$$

since the independent variables are then T, P and n_i. If, however, we wish to determine the limits of stability it is desirable to choose the independent variables in another way. Before discussing the general case further it is convenient to consider in some detail the stability of a fluid of invariable composition.

3·72. In dealing with a fluid of invariable composition it is simplest to keep the total mass constant. The stability conditions (3·71·4) can then be written as

$$(\partial T/\partial S)_{P,n} > 0, \quad (\partial T/\partial S)_{V,n} > 0, \quad (\partial P/\partial V)_{T,n} < 0, \quad (\partial P/\partial V)_{S,n} < 0, \tag{3·72·1}$$

and since, by (3·212·3) and (3·213·1), $C_P \geqslant C_V$ and $\kappa_T \geqslant \kappa_S$ for any stable substance, we see that the first condition implies the second and that the third condition implies the fourth. The limits of stability are therefore defined by

$$(\partial T/\partial S)_{P,n} = 0, \quad (\partial P/\partial V)_{T,n} = 0. \tag{3·72·2}$$

The same results are contained in equations (3·72·6) below.

The general behaviour of U as a function of S and of V is shown in fig. 3·1. If we consider U as a function of S for fixed V and n, the tangent at every point lies below the whole of the curve except for points on the portion $ABCD$. Further, the states defined by the portions AB and CD are metastable while those defined by the portion BC are essentially unstable. Similar considerations apply to U as a function of V for fixed S and n. The portion $KLMN$ of the curve represents unstable states, the portions KL and MN referring to metastable and the portion LM to essentially unstable states.

The stable states between A and D and between K and N are represented by points on the double tangents AD and KN. A state such as X (or Y) is a heterogeneous state whose temperature is equal to the gradient of AD and whose pressure is equal to minus the gradient of KN. It is a mixture of the phases represented by A and D (or by K and N) and if n_A and n_D (or n_K and n_N) are the amounts present of the phases A and D (or of K and N), then $n_A/n_D = XD/AX$ and $n_K/n_N = YN/KN$.

In determining the exact stability conditions it is not sufficient to consider only the sections of the U surface by the planes S = constant and V = constant, and we have to determine the relation of the U surface to its

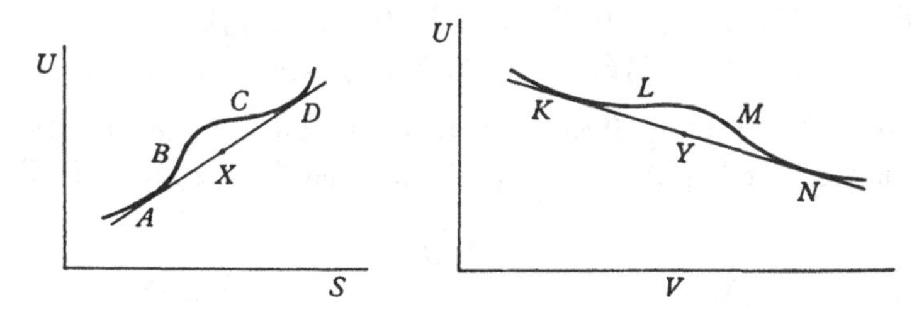

Fig. 3·1. The internal energy U as a function of S and V.

tangent planes. To discuss the conditions for local stability or instability it is sufficient to expand Φ in powers of $S - S_1$ and of $V - V_1$. Then, omitting the symbol n since it is constant, we have, to the second order,

$$2\Phi(S, V; S_1, V_1) = (S - S_1)^2 \frac{\partial^2 U_1}{\partial S_1^2} + 2(S - S_1)(V - V_1)\frac{\partial^2 U_1}{\partial S_1 \partial V_1} + (V - V_1)^2 \frac{\partial^2 U_1}{\partial V_1^2}, \tag{3·72·3}$$

and if the right-hand side is a positive-definite form the state S_1, V_1 will be at least locally stable. The conditions for this are that (dropping the suffix 1)

$$\frac{\partial^2 U}{\partial S^2}\frac{\partial^2 U}{\partial V^2} - \left(\frac{\partial^2 U}{\partial S \partial V}\right)^2 > 0, \quad \frac{\partial^2 U}{\partial S^2} > 0, \quad \frac{\partial^2 U}{\partial V^2} > 0. \tag{3·72·4}$$

Only two of these conditions are independent, since the first condition and one of the other two imply the third.

Now

$$\frac{\partial^2 U}{\partial S^2}\frac{\partial^2 U}{\partial V^2} - \left(\frac{\partial^2 U}{\partial S \partial V}\right)^2 = -\frac{\partial(T, P)}{\partial(S, V)} = -\left(\frac{\partial T}{\partial S}\right)_P \left(\frac{\partial P}{\partial V}\right)_S = -\left(\frac{\partial P}{\partial V}\right)_T \left(\frac{\partial T}{\partial S}\right)_V \tag{3·72·5}$$

by (1·6·10), and so we may write the conditions (3·72·4) in the form

$$-\left(\frac{\partial T}{\partial S}\right)_P \left(\frac{\partial P}{\partial V}\right)_S = -\left(\frac{\partial T}{\partial S}\right)_V \left(\frac{\partial P}{\partial V}\right)_T > 0, \quad \left(\frac{\partial T}{\partial S}\right)_V > 0, \quad -\left(\frac{\partial P}{\partial V}\right)_S > 0, \tag{3·72·6}$$

i.e.
$$\frac{1}{C_P\kappa_S}=\frac{1}{C_V\kappa_T}>0,\quad C_V>0,\quad \kappa_S>0. \tag{3·72·7}$$

At the limit of the metastable phases the first of the conditions (3·72·4) breaks down, and then $C_P=\infty$, $\kappa_T=\infty$ while C_V and κ_S are still finite. The essentially unstable phases are characterized by having $C_P<0$, $\kappa_T<0$, while C_V and κ_S may be positive or negative according to circumstances.

3·73. The stability conditions can be obtained by considering any of the thermodynamic functions. For example, (3·7·1) with $S=S_2$, $V=V_2$, $n_i=n_{i,2}$ can be written as

$$\begin{aligned}\Phi(T_2,P_2,n_{i,2};\,T_1,P_1,n_{i,1})&=G(T_2,P_2,n_{i,2})-G(T_1,P_1,n_{i,1})\\&\quad-(T_1-T_2)S_2+(P_1-P_2)V_2+\Sigma\mu_{i,1}(n_{i,1}-n_{i,2})>0.\end{aligned} \tag{3·73·1}$$

To see the geometrical significance of this inequality we must split it up into two parts. First, put $n_{i,1}=n_{i,2}$ (all i). Then $\Phi(T_2,P_2,n_{i,1};\,T_1,P_1,n_{i,1})$

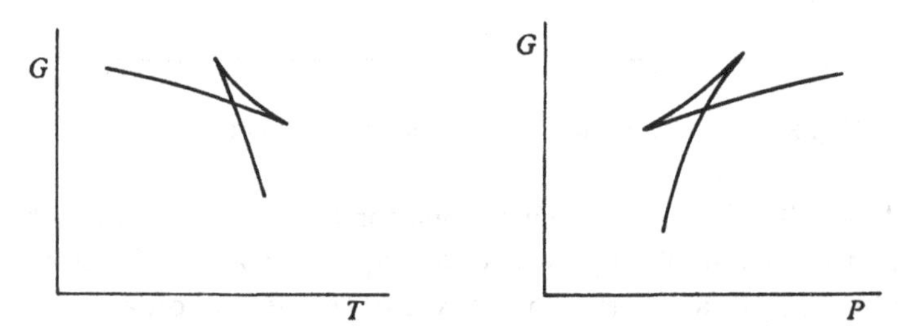

Fig. 3·2. The free energy G as a function of T and P.

is the distance, measured parallel to the G axis, of the point T_1, P_1, $n_{i,1}$ on the $G(T,P,n_i)$ surface below the tangent plane to the surface at the point T_2, P_2, $n_{i,1}$. Thus one of the conditions for the state T_1, P_1, $n_{i,1}$ to be stable is that the tangent plane at every point T_2, P_2, $n_{i,1}$ of the $G(T,P,n_{i,1})$ surface should pass above the point T_1, P_1, $n_{i,1}$. The general form of G as a function of T and P for fixed n_i is shown in fig. 3·2 for a substance for which the internal energy has the characteristics shown in fig. 3·1. It will be seen that in the unstable regions G is a multi-valued function of T or of P. It is obvious that, if there are two values of G for given T and P, namely, G_1 and G_2, where $G_1>G_2$, the tangent plane at G_2 passes below the point G_1, and therefore that the state of a substance is unstable if, for given T, P and n_i, there is another state with a lower value of G. This in accordance with the stability condition (3·61·8).

Next put $T_1=T_2$, $P_1=P_2$. Then $\Phi(T_1,P_1,n_{i,2};\,T_1,P_1,n_{i,1})$ is the distance, measured parallel to the G axis, of the point T_1, P_1, $n_{i,2}$ on the $G(T,P,n_i)$ surface above the tangent plane to the surface at the point T_1, P_1, $n_{i,1}$.

Thus the state T_1, P_1, $n_{i,1}$ is stable if the tangent plane to the $G(T_1, P_1, n_i)$ surface (where T_1, P_1 are fixed) at that point lies wholly below the surface.

If we consider the behaviour of G as a function of the mass n_1 of one of the constituents, T, P and the masses of the other constituents remaining constant, instability will occur when $G(n_1)$ is of the form shown in fig. 3·3. The states lying on ab and cd, though unstable for large variations, are metastable, while those lying on bc are essentially unstable. The stable states between a and d are heterogeneous and are represented by points lying on the double tangent ad. They are mixtures of the phases corresponding to a and d, and the composition of the state represented by the point x is given by $n_{1,a}/n_{1,d} = xd/ax$.

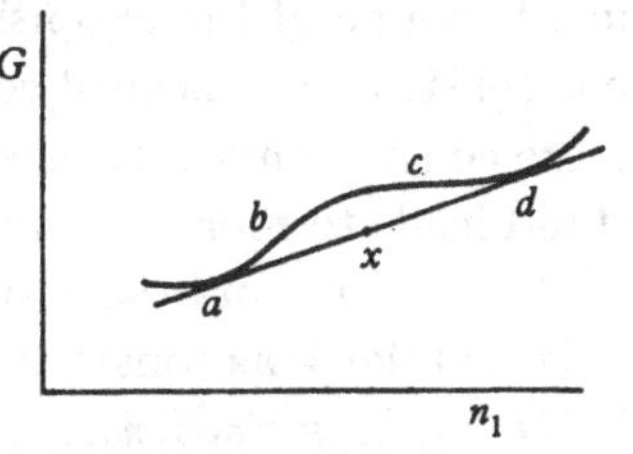

Fig. 3·3. The free energy G as a function of the composition.

3·74. We now return to the general case of r constituents. If we put $\delta S = S - S_1$, $\delta V = V - V_1$, $\delta n_i = n_i - n_{i,1}$, and expand Φ in powers of δS, δV and δn_i, neglecting terms of the third and higher orders, we obtain (3·63·2) in the form

$$2\Phi = \frac{\partial^2 U_1}{\partial S_1^2}(\delta S)^2 + \frac{\partial^2 U_1}{\partial V_1^2}(\delta V)^2 + \sum_{i,j} \frac{\partial^2 U_1}{\partial n_{i,1}\,\partial n_{j,1}}\,\delta n_i\,\delta n_j$$

$$+ 2\frac{\partial^2 U_1}{\partial S_1\,\partial V_1}\,\delta S_1\,\delta V_1 + 2\sum_i \frac{\partial^2 U_1}{\partial S_1\,\partial n_{i,1}}\,\delta S\,\delta n_i + 2\sum_i \frac{\partial^2 U_1}{\partial V_1\,\partial n_{i,1}}\,\delta V\,\delta n_i, \quad (3·74·1)$$

and we have to determine under what conditions Φ is a positive definite form. The discriminant of (3·74·1) is (omitting the suffix 1)

$$\begin{vmatrix} \dfrac{\partial T}{\partial S} & \dfrac{\partial T}{\partial V} & \cdots & \dfrac{\partial T}{\partial n_r} \\ -\dfrac{\partial P}{\partial S} & -\dfrac{\partial P}{\partial V} & \cdots & -\dfrac{\partial P}{\partial n_r} \\ \vdots & & & \vdots \\ \dfrac{\partial \mu_r}{\partial S} & \dfrac{\partial \mu_r}{\partial V} & \cdots & \dfrac{\partial \mu_r}{\partial n_r} \end{vmatrix} = \frac{\partial(T, -P, \mu_1, \ldots, \mu_r)}{\partial(S, V, n_1, \ldots, n_r)}. \quad (3·74·2)$$

Now the $P(T, \mu_i)$ fundamental equation for the substance is

$$S\,dT - V\,dP + \Sigma n_i\,d\mu_i = 0.$$

Hence, multiplying the successive rows of the determinant (3·74·2) by $S, V, n_1, \ldots, n_r$ and adding, we see that the discriminant is identically zero whatever the properties of the system may be. This is a consequence of the fact that for any homogeneous substance U is a homogeneous function of degree unity in S, V, n_i, and that a mere increase in the mass of the system with fixed ratios of S, V and n_i is without effect upon the properties of the system. To obtain a significant result we must therefore consider variations in which one of the extensive variables has a fixed value. If there is only one constituent it is most convenient to take n constant, but when there are more constituents than one the choice of V as the variable to be kept constand leads to more symmetrical formulae. We may, however, fix any one of S, V and n_i, and the resulting sets of formulae are all equivalent.

If we take V as constant, the discriminant of the quadratic form (3·74·1) is $\partial(T,\mu_1,\ldots,\mu_r)/\partial(S,n_1,\ldots,n_r)$, and, if the quadratic form is to be positive-definite, this determinant and all the determinants formed by striking out successive rows and their associated columns must also be positive. We may strike out the rows and columns in any order, but (see, for example, Ferrar, 1941, p. 138) there are only $r+1$ independent conditions which can be taken to be

$$\frac{\partial(T,\mu_1,\ldots,\mu_r)}{\partial(S,n_1,\ldots,n_r)}>0,\quad \frac{\partial(T,\mu_1,\ldots,\mu_{r-1})}{\partial(S,n_1,\ldots,n_{r-1})}>0,\quad \ldots,\quad \frac{\partial T}{\partial S}>0. \tag{3·74·3}$$

If we divide each of the above expressions by the succeeding one and use the relation (1·6·15), we see that the stability conditions can be written as

$$\left(\frac{\partial\mu_r}{\partial n_r}\right)_{V,T,\mu_1,\ldots,\mu_{r-1}}>0,\quad \left(\frac{\partial\mu_{r-1}}{\partial n_{r-1}}\right)_{V,T,\mu_1,\ldots,\mu_{r-1},n_r}>0,\quad \ldots,$$
$$\left(\frac{\partial\mu_1}{\partial n_1}\right)_{V,T,n_2,\ldots,n_r}>0,\quad \left(\frac{\partial T}{\partial S}\right)_{V,n_1,\ldots,n_r}>0. \tag{3·74·4}$$

The limit of stability occurs when the discriminant of the quadratic form vanishes, i.e. when the first, and in general only the first, of the conditions (3·74·3) ceases to hold. In this event the first of the inequalities (3·74·4) also ceases to hold, and, by symmetry, we must also have

$$\left(\frac{\partial T}{\partial S}\right)_{V,\mu_1,\ldots,\mu_r}=0,\quad \left(\frac{\partial\mu_i}{\partial n_i}\right)_{V,T,\mu_1,\ldots,\mu_{i-1},\mu_{i+1},\ldots,\mu_r}=0\quad (\text{all } i). \tag{3·74·5}$$

As mentioned above, it is possible to obtain other equivalent sets of stability conditions by keeping constant any one of the extensive variables $V,S,n_1,\ldots,n_r$. For example, if we make n_r constant in (3·74·2), the stability conditions can be written as

$$\frac{\partial(-P,T,\mu_1,\ldots,\mu_{r-1})}{\partial(V,S,n_1,\ldots,n_{r-1})}>0,\quad \frac{\partial(-P,T,\mu_1,\ldots,\mu_{r-2})}{\partial(V,S,n_1,\ldots,n_{r-2})}>0,\quad \ldots,\quad -\frac{\partial P}{\partial V}>0, \tag{3·74·6}$$

or equivalently,

$$\left(\frac{\partial \mu_{r-1}}{\partial n_{r-1}}\right)_{P,T,\mu_1,\ldots,\mu_{r-1},n_r} > 0, \quad \left(\frac{\partial \mu_{r-2}}{\partial n_{r-2}}\right)_{P,T,\mu_1,\ldots,\mu_{r-3},n_{r-1},n_r} > 0, \quad \ldots,$$

$$\left(\frac{\partial T}{\partial S}\right)_{P,n_1,\ldots,n_r} > 0, \quad -\left(\frac{\partial P}{\partial V}\right)_{S,n_1,\ldots,n_r} > 0. \qquad (3\cdot74\cdot7)$$

Numerous singular cases can arise when the rank of the quadratic form (3·74·1) is less than $r+1$, but these are best dealt with as they arise.

REFERENCES

Ferrar, W. L. (1941). *Algebra.* Oxford.

Gibbs, J. W. (1875–8). On the equilibrium of heterogeneous substances. *Trans. Conn. Acad. Arts Sci.* **3**, 108, 343; reprinted in *Collected Works*, vol. **1**. New York, 1928.

Shaw, A. N. (1935). The derivation of thermodynamic relations for a simple system. *Phil. Trans.* A, **334**, 299.

Chapter 4

THE AXIOMATIC FOUNDATION OF THERMODYNAMICS

Introduction

4·1. In the preceding chapters the fundamental results of thermodynamics are derived by the traditional methods used by Carnot, Clausius and Kelvin. The treatment goes a little deeper than most accounts of the subject in that the logical difficulties concerning temperature and quantity of heat are not completely ignored, and the account given in Chapters 1 and 2 is logically consistent. In the present chapter a more elaborate discussion is given of several points merely touched upon in Chapter 1, and in addition an entirely different method is given of proving Carnot's theorem. This method, due to Carathéodory (1909),‡ does not use the properties of Carnot cycles, which, although necessary for the theory of heat engines and by tradition deeply rooted in the classical method of exposition of thermodynamics, are somewhat inelegant tools for discussing and critically examining the fundamental principles of the subject. By the use of Carathéodory's method the subject gains greatly in mathematical elegance but becomes much more difficult and abstract, and it has therefore been thought advisable to illustrate the main points by concrete examples, provided mainly by the properties of gases.

To begin with, we consider only simple systems. We use the term 'simple system' in a technical sense to denote substances such as gases and liquids, and solids subject to uniform hydrostatic pressures, whose properties are determined by two variables P and V (the parameters of state). In more general systems there are s generalized coordinates a_i and s generalized forces A_i which are such that the work done by the system in an infinitesimal displacement is $dW = \Sigma A_i da_i$. (The generalized coordinates a_i are often called 'deformation coordinates'.) For a simple system there is only one deformation coordinate and one generalized force, and it is immaterial so far as the general theory is concerned whether we consider a fluid, in which case $a = V$ and $A = P$, or whether we consider other physical systems with similar properties such as a spring under tension or compression, or a magnetic body in a magnetic field (with the magnetic moment and field parallel to one another). We merely use the pressure and volume of a gas

‡ Simplified versions have been given by Born (1921) and Carathéodory (1925).

or of a liquid, or of a solid, as convenient and familiar examples of the types of parameters that occur. The extension of the theory to systems for which $s > 1$ is straightforward and is given in § 4·9.

The Empirical Temperature

4·2. It was pointed out in § 1·11 that the temperature is always defined and measured in terms of one or other of the mechanical variables specifying the state of the thermometer (or of a combination of both variables). We now intend to give a formal proof that temperature is not a primitive concept, but can be defined and given a numerical value in terms of the mechanical variables. It is, in fact, possible to develop the whole subject of thermodynamics without introducing the temperature explicitly as a separate variable until the absolute temperature is defined, and this was done by Carathéodory in his original paper. We shall adopt a less abstract method of exposition which is more in harmony with our physical picture of the world.

The reader of the following pages must be on his guard against using technical phrases such as 'adiabatic' and 'diathermal', etc., in the loose manner in which they are so often employed. For example, 'adiabatic' is usually defined more or less as 'insulated so as to prevent the passage of heat', and if defined thus it obviously implies that we already know what heat is, and how to measure it. The definition of 'adiabatic' that is given on p. 6, namely, that a substance is enclosed in an adiabatic vessel if the parameters of state of the substance cannot be altered unless the deformation coordinates a_i are changed (or unless there is action at a distance), comes to exactly the same thing in the end, but it is expressed in terms of the mechanical variables alone and can be used later to define what heat is without our becoming involved in a circular argument. Similarly, the definition of a diathermal vessel as one which is not adiabatic does not involve the notion of heat as yet, and merely states the experimental fact that substances can be put into certain vessels and have their parameters of state altered without moving the wall of the vessel.

The definitions give a simple test whether the vessel in which a substance is enclosed is adiabatic or diathermal. We have merely to bring any other thermodynamic system into contact with the vessel and then change the parameters of the second system. If all the parameters of the substance inside the vessel being investigated are unaffected by the changes in the second system, whatever the changes are (provided that the wall is not moved) and whatever the system may be, then the vessel is adiabatic. If, however, the parameters can be affected by changes in the external system (which do not entail moving the wall of the vessel), then the vessel is diathermal. In practice, of course, every wall is diathermal to some extent, and an adiabatic wall is a theoretical abstraction which can never be completely realized.

The mode of exposition adopted is to state certain experimental facts and to draw logical conclusions from them. In order not to be involved in circular arguments it is sometimes necessary to state the facts in an unfamiliar fashion. To make clearer the content of the experimental facts, their meaning has been explained in terms of familiar concepts, including temperature, even before temperature has been logically defined. The reader must therefore make a sharp distinction between the definitions and logical deductions on the one hand and the illustrations on the other. In dealing with the first we must use very precise language and avoid using undefined quantities, whereas in the illustrations we may draw upon our everyday experience and use familiar but not very well-defined concepts.

STATEMENT 1. *An equilibrium state of a given mass of a simple system is defined if the values of both of its parameters of state are given. A knowledge of the value of the deformation coordinate alone is not sufficient.*

This is merely a statement of the way in which a thermodynamic system differs from those normally considered in mechanics. In statics the state of a system in equilibrium is completely defined by a knowledge of the coordinates specifying the positions of the various bodies. (We also need to know the 'physical constants' of the bodies, such as their mass, compressibility, etc.) A knowledge of the forces is not required; they can be deduced from the equations of equilibrium. For example, to specify the state of a given piece of wire under tension we normally need only know the length of the wire. The tension can then be deduced by Hooke's law (assuming that we know the unstretched length and cross-section and the Young's modulus of the wire, which are constants for a given piece of wire). If, however, we consider wide ranges of temperature, the wire must be considered as a thermodynamic system and not as a mechanical one, since we can alter the length of the wire without altering the tension simply by raising the temperature. We must then specify both the tension and the extension. To sum up: a thermodynamic system possesses more degrees of freedom than the corresponding mechanical system, and in fact one more degree of freedom. It is customary to employ the temperature as this extra variable, but as we have not yet defined temperature we use the two independently variable mechanical parameters P and V to define the state of a thermodynamic system, whereas the single variable V is sufficient to describe the state of the corresponding mechanical system.

STATEMENT 2. *If two simple systems which are otherwise enclosed in adiabatic vessels are put into contact through a fixed diathermal wall, then in equilibrium they cannot have arbitrary values of their four parameters of state.*

To illustrate this statement, consider two gases contained in a cylinder closed by two movable pistons, the gases being kept separate by a fixed rigid wall D, as shown in fig. 4·1. The cylinder and the pistons are supposed to be adiabatic. Let the parameters of state of the two gases be P_1, V_1 and P_2, V_2. If the wall D is adiabatic we can alter the parameters P_1, V_1 of the first gas by moving the piston 1 without in any way affecting the state of the second gas. If, however, the wall D is diathermal, any motion of the piston 1 will affect the states of both gases. In familiar terms, compressing the first gas causes its temperature to rise, and, since the wall is diathermal, heat will flow from the first gas to the second until both are once more at the same temperature. Thus the temperatures of both gases are raised by compressing either, and consequently a change in the parameters P_1, V_1 of the first gas entails a corresponding change in the parameters P_2, V_2 of the second gas. When two substances are in equilibrium with one another through a diathermal wall, they are said to be in thermal equilibrium with one another.

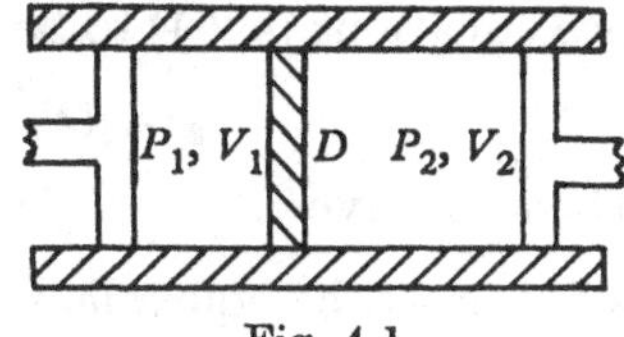

Fig. 4·1

Leaving aside the description of the mechanism by which equilibrium is established, we can state the result of the above idealized experiment in precise terms as follows. When the wall D is diathermal there is a definite functional relation $F(P_1, V_1; P_2, V_2) = 0$ between the four parameters of the two gases, and this functional relation can be found by direct experiment.

The function F (or, to be precise, a table giving the values of the parameters P_1, V_1, P_2, V_2 which satisfy $F = 0$) can be obtained experimentally by carrying out the above idealized experiment a number of times, starting each time from a different initial state. It is found in every case that the function F is of a particularly simple form, namely, the difference of a function of P_1, V_1 and a function of P_2, V_2. This can be proved by using the experimental fact contained in the following statement.

STATEMENT 3. *If two simple systems are each in thermal equilibrium with a third simple system then they are in thermal equilibrium with one another.*

If the systems 1 and 2, with parameters P_1, V_1 and P_2, V_2, are each in thermal equilibrium with a third system with parameters P_3, V_3, then there must be two relations of the form

$$F_2(P_1, V_1; P_3, V_3) = 0, \quad F_1(P_2, V_2; P_3, V_3) = 0 \tag{4·2·1}$$

between the parameters. Statement 3 then means that, if the two functional equations (4·2·1) are simultaneously true, then there is a third functional equation

$$F_3(P_1, V_1; P_2, V_2) = 0 \tag{4·2·2}$$

between P_1, V_1, P_2, V_2. (Note that the functions F_1, F_2, F_3 are not arbitrary functions, but can be found by direct measurement.) Now suppose the two equations (4·2·1) to be solved for P_3, so that

$$P_3 = f_1(P_1, V_1; V_3) \quad \text{and} \quad P_3 = f_2(P_2, V_2; V_3).$$

Then we have

$$f_1(P_1, V_1; V_3) = f_2(P_2, V_2; V_3), \tag{4·2·3}$$

which must be equivalent to (4·2·2). Since V_3 does not occur in (4·2·2) it cannot occur in (4·2·3), and hence (4·2·2) and (4·2·3) must both be equivalent to

$$f_1(P_1, V_1) = f_2(P_2, V_2). \tag{4·2·4}$$

Now write

$$\vartheta_1 = f_1(P_1, V_1), \quad \vartheta_2 = f_2(P_2, V_2).$$

Then the condition (4·2·4) for thermal equilibrium to exist between the two systems is that $\vartheta_1 = \vartheta_2$, i.e. that the two quantities ϑ_1 and ϑ_2, each characteristic of one system, should have the same value ϑ, say. The quantities ϑ_1 and ϑ_2 are called empirical temperatures of the systems since they have the properties usually ascribed to temperature. ϑ_1 is not unique, since instead of $f_1(P_1, V_1)$ we could choose any arbitrary single-valued function $\phi\{f_1(P_1, V_1)\}$ of $f_1(P_1, V_1)$, but we should then have to take the corresponding function $\phi\{f_2(P_2, V_2)\}$ instead of ϑ_2. Thus although the condition (4·2·4) gives one definite value for say P_1 when V_1, P_2 and V_2 are given, the functions f_1 and f_2 are not unique, and we therefore speak of ϑ_1 as an empirical temperature of system 1 and not as the empirical temperature.

The result of the preceding argument can be stated as follows:

Generalization A. *Associated with any simple system there is a function ϑ of the variables defining the state such that ϑ must have the same value for all substances in thermal equilibrium with one another. ϑ is known as an empirical temperature and it not unique since any arbitrary single-valued function $\phi(\vartheta)$ will serve equally as an empirical temperature. Once a particular function ϕ is chosen for any one substance the empirical temperature is determined for all substances.*

It should be noted that the processes required for setting up an empirical temperature scale are just those normally used to determine the equation of state of a substance. The only difference, and it is more a difference in outlook than in method of measurement, is that we have been able to define and measure both an empirical temperature and the equation of state simultaneously, whereas usually temperature is taken for granted and only the equation of state is obtained.

The states for which ϑ is constant are states of constant temperature and can be represented by a family of curves $f(P, V) = \vartheta = \text{constant}$ in the P, V diagram. They are known as isothermal curves, or simply as isothermals. Through every point in that part of the P, V plane which corresponds to possible states of the system under consideration there passes one and only one isothermal curve. For, if two isothermals were to intersect, the parameters of the point of intersection would define two different states of the system, which could not therefore be a simple one but would require more than two variables to specify its state. Further, $(\partial P/\partial V)_\vartheta$ must be negative in order for the system to be mechanically stable, and so the isothermals have the general shape shown in fig. 4·2.

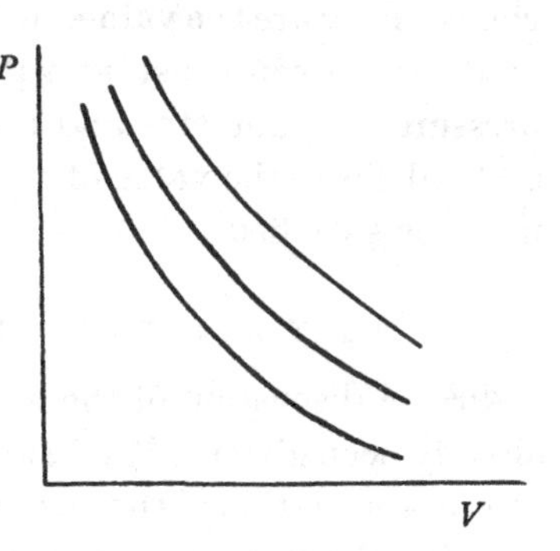

Fig. 4·2. The isothermals of a stable system in the P, V diagram.

The whole family of isothermal curves is clearly independent of the choice of the particular function used to define the empirical temperature, and the arbitrary function at our disposal can only be used to vary the way in which the parameter ϑ is associated with the curves. It is essential that there should be a one-one correspondence between the isothermal curves and the parameter ϑ, and in general this restricts the choice of the arbitrary function to be such that the equation $f(P, V) = \vartheta$ has the property that $(\partial\vartheta/\partial V)_P \neq 0$, $(\partial\vartheta/\partial P)_V \neq 0$. These partial derivatives must both be positive or both be negative since $(\partial V/\partial P)_\vartheta < 0$, and the usual convention is to take them positive. Unless the contrary is stated we shall therefore assume that any empirical temperature scale is so chosen that

$$(\partial\vartheta/\partial V)_P > 0, \quad (\partial\vartheta/\partial P)_V > 0. \tag{4·2·5}$$

There are, however, certain exceptional substances such as water and liquid helium whose isothermal curves form two distinct families which overlap in part of the P, V diagram. Such substances have a maximum density and there is a line in the P, V diagram along which $(\partial V/\partial\vartheta)_P$ and $(\partial P/\partial\vartheta)_V$ are either zero or are discontinuous. In the case of water $(\partial V/\partial\vartheta)_P$ and $(\partial P/\partial\vartheta)_V$ are negative between 0 and 4° C., become zero at 4° C. and are positive above 4° C. (at atmospheric pressure). In order to avoid having two isothermals passing through the same point it is necessary to consider the P, V diagram to consist of two separate sheets which are joined along the line where $(\partial V/\partial\vartheta)_P = (\partial P/\partial\vartheta)_V = 0$. In the case of liquid helium, which is considered in detail in § 9·4, the differential coefficients are discontinuous along a line in the P, V plane; they are positive for normal liquid helium, known as liquid helium I, and negative for the modification, known as liquid helium II, which exists at low temperatures (below 2·19° K.). These abnormal substances cannot be treated as simple systems.

It must again be emphasized that ϑ is not a mere qualitative symbol distinguishing one isothermal from another, but is a continuously variable parameter which ascribes a number to every isothermal. The simplest choices for ϑ are the values of V for a given constant value of P or the values of P for a given constant value of V. These choices lead to the constant-pressure thermometer and the constant-value thermometer; to make the scales definite the value of the constant pressure or of the constant volume must be specified.

Heat and the First Law of Thermodynamics

4·3. A discussion of the way in which heat is defined and measured has already been given in §§ 1·12 and 1·2, and the present section consists mainly of a recapitulation of the argument given there. We take as our fundamental principle the conservation of energy, which can be stated in the following form.

Generalization B. *In any process whatever, the sum of the energies of all the bodies taking part remain constant.*

A discussion of the status of the principle of the conservation of energy belongs more to the philosophy of science than to thermodynamics. Nevertheless, the subject can scarcely be passed over without a few general remarks being made. In mechanics kinetic energy is the fundamental concept, and potential energy can be considered as a theoretical abstraction which is only introduced because the kinetic energy of a body is not in general constant. Essentially we define potential energy so that the sum of the kinetic energy and the potential energy of a system is constant for certain systems known as conservative systems, and we call the sum the total energy of the system. The justification for this procedure is distinctly complex, since, if we have, for example, a particle attached to a spring, the primary measurements that can be carried out only determine the position $\mathbf{r}$ of the particle as a function of the time t, and the extension or compression of the spring. From these we have to infer the existence of a scalar quantity, the mass m, associated with the particle, and a vector quantity, the force $\mathbf{F}$, associated with the spring, which are such that $m d^2\mathbf{r}/dt^2 = \mathbf{F}$. We have to generalize this for arbitrary force fields, and we then find, as a generalization from experience, that for a large class of fields (the conservative fields) $\oint dW \equiv -\oint \mathbf{F}.d\mathbf{r} = 0$ for all cyclic paths. From this we can deduce that $\mathscr{H} = \mathscr{T} + W$, where $\mathscr{T}$ is the kinetic energy, is constant in time.

When we consider systems which interact with electromagnetic fields it is not in general true that the energy of a material system is constant. We can, however, extend the definition of the energy function so that an energy is ascribed to the electromagnetic field itself, and it is then found that the total energy is conserved for a closed system consisting of both material bodies and an electromagnetic field. We can therefore widen the fundamental notions of the conservation of energy in such a way as to lead to a logically satisfactory description of the physical phenomena and to a very considerable simplification in our mode of thought.

Perhaps the most extreme example is provided by the β-active elements, that is, those substances whose atomic nuclei emit positive or negative electrons. When a β-active nucleus decays it can emit an electron with any energy from zero up to a maximum energy E_m characteristic of the nucleus. We should therefore expect the energy difference between the original and the final nuclei to be determined by the energy of the electron emitted. It is, however, found that this energy difference is constant and equal to the maximum energy E_m with which an electron can be emitted, and this means that there is an apparent breakdown in the conservation of energy. To get over the difficulty the hypothesis is made that a neutral particle, called a neutrino, is emitted together with the electron when a nucleus decays and that the neutrino carries off the excess energy, the sum of the energies of the electron and the neutrino always being equal to E_m. The neutrino is supposed to have zero (or very small) mass and to be undetectable except by the part it plays in β-disintegrations. Its function is therefore to conserve the energy (and also the momentum) of the nuclear system, and it may well be asked whether the introduction of such a hypothetical, unobservable particle is preferable to giving up the conservation of energy. Nevertheless, it has been found possible to set up a satisfactory theory of β-decay by means of the neutrino hypothesis, whereas attempts to do so by abandoning the conservation of energy have failed, and the neutrino hypothesis therefore plays a useful part in nuclear physics, though its value may be transitory and it may be replaced at a later date by some other more satisfying postulate.

4·31. According to the views expressed above, the role that heat plays in the conservation of energy is as follows. There are processes in which the energy, computed from the known forms of mechanical and electrical energy, is not conserved, but which are accompanied by thermal changes in the systems concerned, that is, by changes in the parameters defining their states, which cannot be ascribed solely to mechanical or electrical effects. We therefore postulate a new form of energy known as heat, and we show that we can define the quantity of heat so as to make the conservation of energy to be true for these more general processes. It is clear from this that heat, like temperature, can be defined in purely mechanical terms, and we do not need to inquire into the mechanism by which mechanical energy is transformed into thermal energy. Of course, we gain a great insight into the physical phenomena when we identify thermal energy with the energy associated with the random motions of the molecules composing a body, but this is irrelevant to the definition and measurement of a quantity of heat. The attitude taken here is that the two fundamental concepts are the internal energy of a body and the mechanical work it does in changing from one state to another, both of which concepts can be expressed in purely mechanical terms. Heat is a secondary concept, introduced to balance the energy equation.

With these preliminaries we can now restate the results of § 1·2 in the following form:

STATEMENT 4. *The internal energy of a body (or of a system of bodies) in a given state is measured by the mechanical work required to bring the body into the given state from some fixed initial state by means of an adiabatic process. The work required is independent of the intermediate states through which the body may pass and depends only upon the initial and final states provided that the processes considered are adiabatic. The internal energy is therefore a function of the state and is uniquely defined except for an additive constant which is determined by the internal energy ascribed to the initial state.*

It must again be emphasized that the adiabatic processes envisaged must not be limited to be quasi-static. (A discussion of how far we can get by considering only quasi-static processes is given in § 4·8.) Further, even in the case of gases we must not limit ourselves merely to compressions and expansions, but we must also include stirring, and for solids we must consider the rubbing of two solids together as an adiabatic process.

The internal energy $U(P, V)$ can be found as a function of P, V by direct measurement from the definition

$$U(P, V) - U(P_0, V_0) = -W, \tag{4·31·1}$$

where P_0, V_0 are the parameters defining the initial state and $-W$ is the work done by outside agencies in changing the state from P_0, V_0 to P, V by an adiabatic process. There is, however, one proviso. It must be possible to connect the states P_0, V_0 and P, V by an adiabatic process, and it is not obvious that this is always true. It is, in fact, impossible to attain all states by adiabatic processes starting from a given initial state, but we shall see later (§ 4·61) that if the state P_1, V_1 is unattainable by an adiabatic process starting from P_0, V_0 it is possible to attain the state P_0, V_0 by an adiabatic process starting from P_1, V_1. We may therefore suppose that the internal energy is known for all values of P and V for which the substance exists.

If we have a number of systems coupled together and forming a larger closed system, then in any process we have $\Sigma\Delta U + \Sigma W = 0$, but it does not necessarily follow that $\Delta U + W = 0$ for each of the individual systems. Therefore, there are processes for which (4·31·1) is untrue. (Note that it is W, not ΔU, which has a different value in a general change from the value it has in an adiabatic change.) When the equation (4·31·1) does not hold we say that heat has been absorbed or emitted by the system and we define the amount of heat absorbed by the body by the equation

$$Q = \Delta U + W, \tag{4·31·2}$$

where ΔU is the increase in the internal energy of the system and W is the mechanical work done by the system. Equation (4·31·2) expresses the first law of thermodynamics in mathematical form.

4·32. The possibility of measuring the internal energy U of a system by adiabatic processes depends upon the fact that the work obtainable from an adiabatic change of a thermodynamic system (unlike the work obtainable from a mechanical system) does not depend merely upon the change in the deformation coordinates. For example, the maximum work that can be obtained by expanding a gas adiabatically from the initial state P_1, V_1 to a state with volume $V_2 > V_1$ is $\int_{V_1}^{V_2} P\,dV$, where P is the pressure corresponding to the volume V when the expansion is quasi-static. For given P_1 and V_1 this maximum work is uniquely defined by the final volume V_2, but it is possible to obtain a less amount of work by allowing the gas to acquire kinetic energy during the expansion, the kinetic energy being subsequently reconverted into internal energy by means of the internal friction of the fluid. For given V_2, the minimum final pressure $P_{2,\,\text{min.}}$ is obtained when the expansion is quasi-static and the maximum final pressure $P_{2,\,\text{max.}}$ is obtained when no external work is done. Any final pressure lying between these two limits can be obtained by a suitable adiabatic process, and the final pressure is uniquely determined by the final volume and the external work done (that is, by the change in internal energy). Similarly, the minimum work required to effect an adiabatic compression to volume $V_3 < V_1$ is the integral $\int_{V_3}^{V_1} P\,dV$ evaluated for the quasi-static compression, but any greater amount of work may be used. We therefore see that the states which are attainable

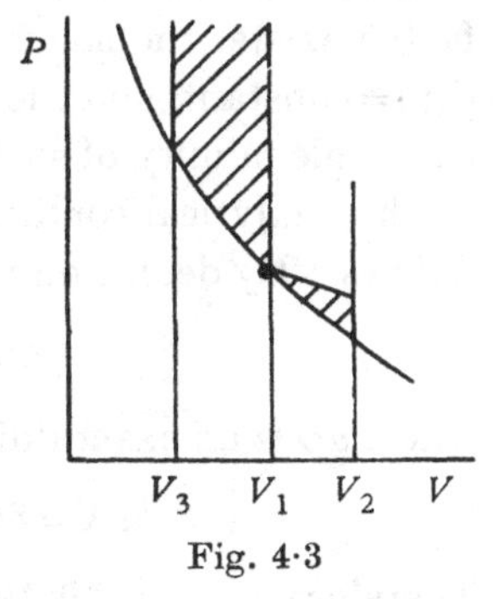

Fig. 4·3

from the state P_1, V_1 by a single adiabatic compression or expansion (and whose volumes lie between V_3 and V_2) are as shown in fig. 4·3. Other states are attainable by more than one adiabatic compression or expansion or by stirring the gas at constant volume, and we can attain in this way any state which lies on or above the quasi-static adiabatic curve through P_1, V_1. We cannot obtain, however, any state lying below this curve by means of an adiabatic process; this is a particular case of the second law of thermodynamics.

The Nature of Linear Differential Forms

4·4. In an infinitesimal quasi-static process the heat absorbed can be expressed as a linear combination of the differentials of the independent variables. We now investigate those properties of linear differential forms which have applications to thermodynamics, and for simplicity we consider only three independent variables, though the theorems extend at once to n variables.

Consider the linear differential form

$$df = X(\mathbf{r})\,dx + Y(\mathbf{r})\,dy + Z(\mathbf{r})\,dz, \tag{4·4·1}$$

where the use of the notation df does not necessarily imply that the form is the differential of a function of $\mathbf{r} = (x, y, z)$. The condition $df = 0$ defines a double infinity of curves, there being an infinite number of curves passing through a given point $\mathbf{r}_0$, since the differential equation merely imposes the condition that the tangent to a curve through $\mathbf{r}_0$ should be perpendicular to the direction whose direction ratios are $X(\mathbf{r}_0)$, $Y(\mathbf{r}_0)$, $Z(\mathbf{r}_0)$. In general, the expression df is not an exact differential, nor does the equation $df = 0$ admit of an integrating factor. That is, no function $M(\mathbf{r})$ exists such that $M\,df$ is the differential of a function $\phi(\mathbf{r})$. If, however, an integrating factor exists, the solution of the differential equation $df = 0$ is given by $\phi(\mathbf{r}) = \text{constant}$, and the double infinity of curves defined by $df = 0$ lie upon the single infinity of surfaces given by $\phi(\mathbf{r}) = \text{constant}$.

The analytical condition for the existence of an integrating factor is as follows. By definition we have

$$d\phi = MX\,dx + MY\,dy + MZ\,dz,$$

where $d\phi$ is an exact differential. Hence

$$MX = \partial\phi/\partial x, \quad MY = \partial\phi/\partial y, \quad MZ = \partial\phi/\partial z.$$

Therefore $$\partial(MY)/\partial x = \partial^2\phi/\partial x\,\partial y = \partial(MX)/\partial y,$$

which gives $$M\left(\frac{\partial Y}{\partial x} - \frac{\partial X}{\partial y}\right) = X\frac{\partial M}{\partial y} - Y\frac{\partial M}{\partial x}.$$

Similarly $$M\left(\frac{\partial Z}{\partial y}-\frac{\partial Y}{\partial z}\right)=Y\frac{\partial M}{\partial z}-Z\frac{\partial M}{\partial y}$$

and $$M\left(\frac{\partial X}{\partial z}-\frac{\partial Z}{\partial x}\right)=Z\frac{\partial M}{\partial x}-X\frac{\partial M}{\partial z}.$$

On eliminating M we obtain

$$X\left(\frac{\partial Z}{\partial y}-\frac{\partial Y}{\partial z}\right)+Y\left(\frac{\partial X}{\partial z}-\frac{\partial Z}{\partial x}\right)+Z\left(\frac{\partial Y}{\partial x}-\frac{\partial X}{\partial y}\right)=0, \qquad (4\cdot4\cdot2)$$

which is the necessary condition for an integrating factor to exist. It can also be shown that the condition is sufficient for the existence of an integrating factor.

The condition (4·4·2) can easily be generalized to apply to a linear differential form $df=\Sigma X_i dx_i$ in n variables. We have merely to replace X, x, Y, y, Z, z by X_i, x_i, X_j, x_j, X_k, x_k, where i, j and k are all different, and we obtain $\frac{1}{6}n(n-1)(n-2)$ conditions of which only $\frac{1}{2}(n-1)(n-2)$ are independent. The case of two dimensions is peculiar, since a differential equation of the first order with one independent variable is always integrable. This is in accordance with the geometrical interpretation that is given above, since a curve and a surface in n-dimensions are the same thing when $n=2$. It should be noted that, if an integrating factor exists, it is not unique. For if $M\,df$ is an exact differential $d\phi$, then so is $M\psi'(\phi)\,df$, where $\psi(\phi)$ is any differentiable function of ϕ, since $M\psi'(\phi)\,df=d\psi(\phi)$.

4·41. We shall not pursue the analytical theory of linear differential forms any further since our interest lies in their geometrical properties. It is clear that, if the differential equation $df=0$ possesses an integrating factor, it is impossible to join two points by a curve satisfying the differential equation unless one of the integral surfaces passes through both points. The converse of this statement is also true, and it is this theorem which primarily interests us. We can state it as follows:

THEOREM. *If a linear differential form $df=X\,dx+Y\,dy+Z\,dz$, where X, Y, Z are continuous functions of x, y, z, is such that in every neighbourhood of an arbitrary point P there exist points which cannot be attained from P by any curve satisfying the differential equation $df=0$, then the linear differential form possesses an integrating factor.*

To prove this, take a point Q in the neighbourhood of P and unattainable from P by a curve satisfying $df=0$. Let l be a straight line through P whose direction does not satisfy $df=0$. Then we first prove that there are points on l which are as close as we please to P and which are unattainable from P along a curve for which $df=0$.

Draw the plane through Q and l, and consider the curve through Q which lies in this plane and satisfies $df=0$. There is one and only one such curve, since the curve satisfying $df=0$ and lying in the surface whose parametric representation is $x=x(u, v)$, $y=y(u, v)$, $z=z(u, v)$ must satisfy the equation

$$X\,dx+Y\,dy+Z\,dz=0,$$

where $\quad dx=\frac{\partial x}{\partial u}du+\frac{\partial x}{\partial v}dv, \quad dy=\frac{\partial y}{\partial u}du+\frac{\partial y}{\partial v}dv, \quad dz=\frac{\partial z}{\partial u}du+\frac{\partial z}{\partial v}dv.$

Hence the differential equation determining the curve is

$$\left(X\frac{\partial x}{\partial u}+Y\frac{\partial y}{\partial u}+Z\frac{\partial z}{\partial u}\right)du+\left(X\frac{\partial x}{\partial v}+Y\frac{\partial y}{\partial v}+Z\frac{\partial z}{\partial v}\right)dv=0,$$

and since this is an equation in two variables only there is one and only one curve in the parametric space (u, v) which satisfies the equation and passes through a given point (u_0, v_0).

Now the curve passing through Q, lying in the plane determined by l and Q, and satisfying $df=0$, must cut the line in a point R (fig. 4·4). This point R is unattainable from P along a curve for which $df=0$, since otherwise Q would be attainable from P. Further, since the direction of l does not satisfy $df=0$, the tangent at Q to the curve cannot be parallel to l, and the point R can be made to lie as close as we please to P by choosing Q close enough to P.

We now use this lemma to prove the main theorem. Take any line l' parallel to l and join the lines by a cylindrical surface C, and consider the curve satisfying $df=0$ which starts from P and lies on C. Let it cut l' in the point M. Further, join l and l' by another cylindrical surface C' and continue the curve PM (satisfying $df=0$) in the surface C' until it cuts l again in some point N (fig. 4·5). Then N and P must coincide. For if not, deform the surface C' continuously into the surface C. In this process N must pass continuously into P, and, unless N and P always coincide, every point in a certain neighbourhood of P and lying on the line l is attainable from P. This contradicts the lemma, and hence N and P must coincide. (We must, however, consider the possibility that all the points on l attainable from P lie on one side of P, while all the points on the other side of P, and sufficiently near to P, are unattainable from P. Now since the coefficients in the linear differential form are continuous functions, the properties of the system of integral curves must depend in a continuous manner on the coordinates of P. If, therefore, all points such as N are attainable from P while a point N' on the opposite side of P is unattainable, displace P so as to take up the position P' coincident with N'. Then by continuity the set of points attainable from P' lies on the same side of P' as P, and includes P if P' is near enough to P. Thus N' is attainable from P, and the possibility considered cannot arise.)

Further, by deforming the surface C' we make the curve PM trace out a surface which contains all the curves which pass through P and which satisfy $df=0$. If the equation of the surface so described is $\psi(x,y,z)=$ constant, the linear differential form $df=0$ has the integral $\psi(x,y,z)=$ constant and possesses an integrating factor.

The above theorem can be extended at once to n-dimensions. We have merely to take for the surfaces C and C' two-dimensional cylindrical surfaces whose parametric representations are of the form $x_i=x_i(u,v)$ $(i=1,2,\ldots,n)$. The argument is then exactly the same as before, and it is

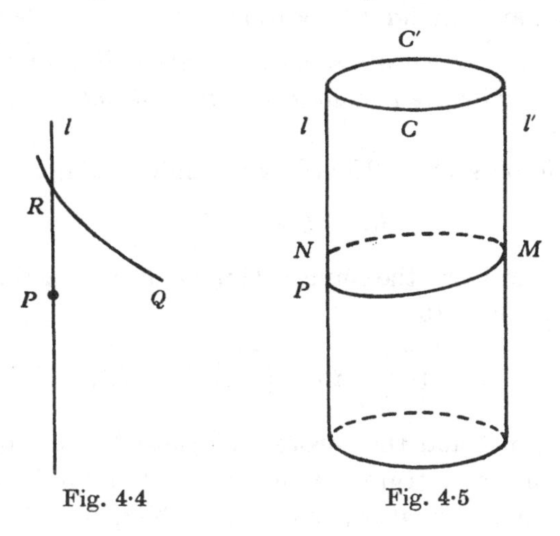

Fig. 4·4 Fig. 4·5

clear that by deforming the surface C' we make the curve PM trace out an $(n-1)$-dimensional surface, whose equation is of the form $\psi(x_1,x_2,\ldots,x_n)=$ constant. This family of surfaces provides the integral of the linear differential form.

The Second Law of Thermodynamics and Carnot's Theorem

4·5. The proof of Carnot's theorem given in Chapter 2 is somewhat indirect, and it seems strange at first sight that anything so definite as Carnot's theorem could be deduced from such a negative principle as the second law of thermodynamics. However, the theorem proved in the preceding section shows that the existence or non-existence of an integrating factor of a linear differential form depends solely upon whether any two neighbouring points can be joined by one of the integral curves of the linear differential form. The criterion for a linear differential form to be integrable is therefore of the same negative form as the second law of thermodynamics, and we can make the content of Carnot's theorem more readily understandable by making the proof depend upon the properties of linear differential forms.

We state the second law of thermodynamics in the form of Carathéodory's principle, but we divide it into two parts. It was pointed out in § 2·5 that the existence of the entropy function depends only upon the properties of quasi-static processes, whereas the increasing property of entropy depends upon the properties of irreversible processes, and that these two results are really independent of one another. We therefore first state Carathéodory's principle in a restricted form which enables us to deduce Carnot's theorem, and extend it later to include irreversible processes.

GENERALIZATION C. *There are states of a system, differing infinitesimally from a given state, which are unattainable from that state by any quasi-static adiabatic process.*

Consider one simple system. Then for an infinitesimal quasi-static process the first law gives

$$dQ = dU + P\,dV.$$

If we use the volume V and the empirical temperature ϑ as the independent variables, this can be written as

$$dQ = \left(\frac{\partial U}{\partial \vartheta}\right)_V d\vartheta + \left\{\left(\frac{\partial U}{\partial V}\right)_\vartheta + P\right\} dV.$$

Now generalization C and the theorem of § 4·41 tell us that this linear differential form has an integrating factor. But this result is trivial, since any form in two variables is always integrable. (Note that Carnot's theorem applied to one simple system is not trivial, since it states that the integrating factor is a function of ϑ only. However, we cannot prove Carnot's theorem by considering one system only.) We therefore obtain a non-trivial result only by considering at least two systems in thermal equilibrium with one another.

On account of the properties of the empirical temperature we can specify the state of a compound system consisting of two simple systems in thermal equilibrium by the two volumes V_1 and V_2 and the empirical temperature ϑ. Then, for an infinitesimal quasi-static process, we have

$$dQ = \left\{\left(\frac{\partial U_1}{\partial \vartheta}\right)_{V_1} + \left(\frac{\partial U_2}{\partial \vartheta}\right)_{V_2}\right\} d\vartheta + \left\{\left(\frac{\partial U_1}{\partial V_1}\right)_\vartheta + P_1\right\} dV_1 + \left\{\left(\frac{\partial U_2}{\partial V_2}\right)_\vartheta + P_2\right\} dV_2, \quad (4·5·1)$$

and the quasi-static adiabatic curves are given by $dQ = 0$. This differential form must possess an integrating factor, and so we can write (4·5·1) as

$$dQ = M\,d\sigma, \qquad (4·5·2)$$

where M and σ are functions of ϑ, V_1 and V_2. Also, by applying generalization C to each of the two systems separately we have

$$dQ_1 = \left(\frac{\partial U_1}{\partial \vartheta}\right)_{V_1} d\vartheta + \left\{\left(\frac{\partial U_1}{\partial V_1}\right)_{\vartheta} + P_1\right\} dV_1 = M_1 \, d\sigma_1 \quad (4\cdot5\cdot3)$$

and

$$dQ_2 = \left(\frac{\partial U_2}{\partial \vartheta}\right)_{V_2} d\vartheta + \left\{\left(\frac{\partial U_2}{\partial V_2}\right)_{\vartheta} + P_2\right\} dV_2 = M_2 \, d\sigma_2, \quad (4\cdot5\cdot4)$$

where M_1 and σ_1 are functions of ϑ and V_1 only, while M_2 and σ_2 are functions of ϑ and V_2 only. Hence

$$M(\vartheta, V_1, V_2)\, d\sigma(\vartheta, V_1, V_2) = M_1(\vartheta, V_1)\, d\sigma_1(\vartheta, V_1) + M_2(\vartheta, V_2)\, d\sigma_2(\vartheta, V_2). \quad (4\cdot5\cdot5)$$

4·51. *The absolute temperature.* The integrating factors M, M_1 and M_2 are not unique, and we have now to prove the most important part of Carnot's theorem, that the integrating factors can be chosen to be functions of ϑ only. The simplest way of doing this is to change the independent variables from ϑ, V_1, V_2 to ϑ, σ_1, σ_2. Equation (4·5·5) then becomes

$$M(\vartheta, \sigma_1, \sigma_2)\, d\sigma(\vartheta, \sigma_1, \sigma_2) = M_1(\vartheta, \sigma_1)\, d\sigma_1 + M_2(\vartheta, \sigma_2)\, d\sigma_2. \quad (4\cdot51\cdot1)$$

Therefore

$$(\partial\sigma/\partial\vartheta)_{\sigma_1,\sigma_2} = 0, \quad (\partial\sigma/\partial\sigma_1)_{\vartheta,\sigma_2} = M_1/M, \quad (\partial\sigma/\partial\sigma_2)_{\vartheta,\sigma_1} = M_2/M \quad (4\cdot51\cdot2)$$

and

$$\frac{\partial}{\partial\vartheta}\frac{M_1}{M} = \left(\frac{\partial^2\sigma}{\partial\vartheta\,\partial\sigma_1}\right)_{\sigma_2} = 0, \quad \frac{\partial}{\partial\vartheta}\frac{M_2}{M} = \left(\frac{\partial^2\sigma}{\partial\vartheta\,\partial\sigma_2}\right)_{\sigma_1} = 0.$$

Hence

$$\frac{\partial}{\partial\vartheta}\log M_1(\vartheta, \sigma_1) = \frac{\partial}{\partial\vartheta}\log M_2(\vartheta, \sigma_2) = \frac{\partial}{\partial\vartheta}\log M(\vartheta, \sigma_1, \sigma_2). \quad (4\cdot51\cdot3)$$

Now the first of the expressions in (4·51·3) is independent of σ_2 while the second is independent of σ_1. They can therefore only be equal if they are functions of ϑ only, and hence all the three expressions in (4·51·3) are equal to $g(\vartheta)$, where $g(\vartheta)$ is a function of the empirical temperature only and is independent of the nature of the particular systems under consideration. On integrating with respect to ϑ we obtain

$$M_1 = \Sigma_1(\sigma_1)\exp\left[\int g(\vartheta)\, d\vartheta\right], \quad M_2 = \Sigma_2(\sigma_2)\exp\left[\int g(\vartheta)\, d\vartheta\right],$$

$$M = \Sigma(\sigma_1, \sigma_2)\exp\left[\int g(\vartheta)\, d\vartheta\right], \quad (4\cdot51\cdot4)$$

where the Σ's are functions of the variables shown.

To obtain that integrating factor which is a function of ϑ only we have now merely to replace M (and also M_1 and M_2) by the function

$$T = C\exp\left[\int g(\vartheta)\, d\vartheta\right], \quad (4\cdot51\cdot5)$$

where C is a constant. The function T is the absolute temperature and is unique except for the constant C which determines the size of the degree. C must be positive if we wish to make the absolute temperature positive.

4·52. *The entropy.* We have now proved that the linear differential form (4·5·1) can be written in the form

$$dQ = T\Sigma(\sigma_1, \sigma_2)\, d\sigma/C, \tag{4·52·1}$$

and from the method of proof it is clear that $\Sigma(\sigma_1, \sigma_2)\, d\sigma$ must be a perfect differential. An explicit proof of this fact is as follows.

Consider the system 1 above. Then

$$dQ = T\Sigma_1(\sigma_1)\, d\sigma_1/C, \tag{4·52·2}$$

and we can define a function S_1 of the state by

$$S_1 = \frac{1}{C}\int \Sigma_1(\sigma_1)\, d\sigma_1. \tag{4·52·3}$$

S_1 is the entropy, and (4·52·2) can be written as

$$dQ_1 = T\, dS_1. \tag{4·52·4}$$

Similar results hold for system 2, while for the combined system we have equation (4·51·1), which, by the help of (4·51·4), can be written in the form

$$\Sigma(\sigma_1, \sigma_2)\, d\sigma = \Sigma_1(\sigma_1)\, d\sigma_1 + \Sigma_2(\sigma_2)\, d\sigma_2,$$

which gives

$$\Sigma(\sigma_1, \sigma_2)\, \partial\sigma/\partial\sigma_1 = \Sigma_1(\sigma_1), \quad \Sigma(\sigma_1, \sigma_2)\, \partial\sigma/\partial\sigma_2 = \Sigma_2(\sigma_2).$$

Hence, differentiating the first of these equations with respect to σ_2 and the second with respect to σ_1, we have

$$\frac{\partial \Sigma}{\partial \sigma_2}\frac{\partial \sigma}{\partial \sigma_1} + \Sigma\frac{\partial^2\sigma}{\partial\sigma_2\partial\sigma_1} = 0, \quad \frac{\partial \Sigma}{\partial \sigma_1}\frac{\partial \sigma}{\partial \sigma_2} + \Sigma\frac{\partial^2\sigma}{\partial\sigma_1\partial\sigma_2} = 0.$$

On subtraction these give

$$\frac{\partial(\Sigma, \sigma)}{\partial(\sigma_1, \sigma_2)} = 0. \tag{4·52·5}$$

This is the condition for $\Sigma(\sigma_1, \sigma_2)$ to be a function of σ only, and we can therefore define the entropy S of the combined system by the formula

$$S = \frac{1}{C}\int \Sigma(\sigma_1, \sigma_2)\, d\sigma. \tag{4·52·6}$$

We then have $dS = dS_1 + dS_2$, and we can choose the additive constants so that $S = S_1 + S_2$. Thus the entropy of a combined system is the sum of the entropies of the separate systems of which it is composed.

THE INCREASING PROPERTY OF ENTROPY

4·6. We have been able to define the absolute temperature and the entropy by the consideration of quasi-static processes alone. To find out what happens in more general processes we need further experimental evidence which is provided by the following form of Carathéodory's principle:

GENERALIZATION C′. *There are states of a system, differing infinitesimally from a given state, which are unattainable from that state by any adiabatic process whatever.*

We have seen in §4·32 that, when a system in the state P_0, V_0 undergoes an adiabatic change to a state in which its volume is V_1, the final pressure is not unique, but depends upon the external work done. The parameters of the final state therefore form a continuum, and consequently the values of the entropy of the final state completely fill an interval (the entropy being a function of the parameters of state). The initial value S_0 of the entropy must lie in this interval, since if the adiabatic expansion is quasi-static the entropy remains unchanged. We now prove that the initial value S_0 cannot be an interior point of the interval.

Consider, for example, a system composed of two simple systems in thermal contact. The state of the system is described by the parameters ϑ, V_1, V_2, but it can equally well be described by S, V_1, V_2. Let the initial values of the parameters be S_0, V_1^0, V_2^0. Then, if S_0 is an interior point of the interval composed of the values of S after an adiabatic expansion in which the volumes become V_1 and V_2, every state in a sufficiently small neighbourhood of S_0, V_1^0, V_2^0 is attainable from the initial state by an adiabatic process. For if S_0 is an interior point we can first change S by a positive or negative amount, and we can then change the volumes in a quasi-static adiabatic process. In this last process S remains constant, while the volumes can be varied arbitrarily. (A quasi-static adiabatic process is characterized by the equation $S(\vartheta, V_1, V_2) = \text{constant}$, and since ϑ is arbitrary, V_1 and V_2 can be varied arbitrarily.) This argument extends at once to more general systems. We therefore arrive at a contradiction to generalization C , and so the initial value S_0 of the entropy must be either the greatest or the least value that the entropy can have after an adiabatic process. This is equivalent to saying that the entropy either always increases or always decreases in an adiabatic process.

We cannot determine which of the last two statements is true without a further appeal to experiment. It is sufficient to measure the change in entropy in one adiabatic process which is not quasi-static, and it is found that the entropy always increases. Only if the adiabatic process is quasi-static does the entropy remain constant. We may state this result in the following form:

GENERALIZATION C″. *In any adiabatic process the entropy changes monotonically. If the absolute temperature is defined to be positive, the entropy always increases or remains constant in such a process.*

4·61. One way of stating the results of the preceding section is to say that, if in any process the entropy increases, there is no adiabatic process which will transform the final state into the initial state. In other words, processes which involve a change of entropy are irreversible. We now prove the complementary theorem that, if in any process the entropy increases, the final state can be attained from the initial state by an adiabatic process.

To prove this, consider, for example, a system composed of two simple systems and take as the parameters of state the quantities S, V_1, V_2. By means of a quasi-static adiabatic process, change the values of V_1 and V_2 into their final values. Then, keeping the volumes constant, change the entropy by an adiabatic process, for example, by stirring. It is clear that the entropy can be increased by any desired amount by such an adiabatic process, and this proves the theorem. The extension to more general systems is obvious.

This is the result that was quoted on p. 72 and which is required in order to show that the internal energy of a system can be determined for all values of the parameters by means of adiabatic processes.

THE STABILITY OF A THERMODYNAMIC SYSTEM

4·7. We have seen that in any irreversible change the entropy always increases. We can go further than this and state the following result as a generalization from experience:

GENERALIZATION D. *If any small changes in any isolated system are possible which increase the entropy, the system will actually undergo one or other of these changes.*

This can be stated in the following alternative form:

GENERALIZATION D′. *The necessary and sufficient condition for the stable equilibrium of an isolated thermodynamic system is that the entropy is a maximum.*

The restriction to small changes in generalization D is necessary because, if the entropy could be increased by a large change in the state of the system, the change will not take place if it entails the system passing through intermediate states with smaller entropies than the entropy of the initial state. Similarly, in the statement of generalization D′ the condition that the entropy is a maximum is to be understood in the sense in which the term is used in the differential calculus. There may be other states with greater values of the entropy, but if these are not adjacent states the system will still be stable.

The Determination of the Internal Energy, the Absolute Temperature and the Entropy by Reversible Processes

4·8. The proof of Carnot's theorem given in § 4·5 depends only upon the properties of reversible adiabatic processes. However, we have implicitly used the properties of irreversible processes since we have assumed that the internal energy is known for all states, and the method of measuring U given in § 4·3 depends upon the use of irreversible adiabatic processes. We now investigate how far it is possible to develop the theory without using irreversible processes at all.

Consider the system illustrated in fig. 4·1, p. 67. By confining our measurements to quasi-static processes in which the wall separating the two gases is either adiabatic or diathermal, we can determine the (quasi-static) adiabatic curves and the isothermal curves of system 1. Let the system of adiabatic curves be given by $\sigma=\sigma(P, V)=$ constant, while the isothermals are given by $\vartheta=\vartheta(P, V)=$ constant. The function $\sigma(P, V)$ is not unique, since any arbitrary function $\phi\{\sigma(P, V)\}$ of σ would do equally well. One and only one adiabatic curve goes through each point in the P, V plane, and in order for there to be a one-one correspondence between the values of σ and the adiabatic curves it is necessary that the derivatives of σ with respect to P and V should not vanish. We can therefore impose the restrictions

$$(\partial\sigma/\partial P)_V > 0, \quad (\partial\sigma/\partial V)_P > 0$$

on the function σ. By analogy with ϑ we can call σ the empirical entropy.

We take σ and ϑ as the independent variables, and the problem is whether we can determine functions U, S and T such that in any infinitesimal quasi-static process

$$dU = T\,dS - P\,dV,$$

where S is a function of σ only and T is a function of ϑ only. Rewrite the equation in the form

$$dU = \left\{T\frac{dS}{d\sigma} - P\left(\frac{\partial V}{\partial\sigma}\right)_\vartheta\right\} d\sigma - P\left(\frac{\partial V}{\partial\vartheta}\right)_\sigma d\vartheta. \tag{4·8·1}$$

Now dU must be a perfect differential, and we must therefore have

$$\frac{\partial}{\partial\vartheta}\left(T\frac{dS}{\partial\sigma} - P\frac{\partial V}{\partial\sigma}\right) = \frac{\partial^2 U}{\partial\vartheta\,\partial\sigma} = -\frac{\partial}{\partial\sigma}\left(P\frac{\partial V}{\partial\vartheta}\right),$$

which reduces to $$\frac{dT}{d\vartheta}\frac{dS}{d\sigma}=\frac{\partial(P,V)}{\partial(\vartheta,\sigma)}. \qquad (4\cdot8\cdot2)$$

Hence, if we calculate the expression on the right-hand side from the measured quantities it must split up into a product $\Theta(\vartheta)\,\Sigma(\sigma)$ of a function of ϑ and of a function of σ. We then have

$$T=C\int\Theta(\vartheta)\,d\vartheta+C',\quad S=\frac{1}{C}\int\Sigma(\sigma)\,d\sigma+C'', \qquad (4\cdot8\cdot3)$$

where C, C' and C'' are constants. Further,

$$dU=\Sigma(\sigma)\,d\sigma\int\Theta(\vartheta)\,d\vartheta+\frac{C'}{C}\Sigma(\sigma)\,d\sigma-P\,dV,$$

and, if we use σ and V as the independent variables and integrate, we obtain an expression of the form

$$U=f(S,V)+C'S. \qquad (4\cdot8\cdot4)$$

The constant C' is not arbitrary since the absolute temperature is unique except for the size of the degree. However, we cannot determine C' by means of reversible processes alone, and we need to measure the changes in U and S for one irreversible process. For, since S changes in an irreversible process, we can use equation (4·8·4) to determine C'. The simplest process to consider is an adiabatic process in which no external work is done, since then U remains constant.

4·81. We may illustrate the preceding argument by considering a perfect gas for which the empirical temperature and entropy are given by

$$\vartheta=PV,\quad \sigma=PV^{\gamma}, \qquad (4\cdot81\cdot1)$$

where γ is a constant. We then have, by (4·8·2),

$$\frac{dT}{d\vartheta}\frac{dS}{d\sigma}=\frac{1}{(\gamma-1)\,PV^{\gamma}}=\frac{1}{(\gamma-1)\,\sigma}. \qquad (4\cdot81\cdot2)$$

Hence $$T=C\vartheta+C',\quad S=\frac{1}{(\gamma-1)\,C}\log\sigma+C''. \qquad (4\cdot81\cdot3)$$

Also, substituting (4·81·3), $\vartheta=\sigma/V^{\gamma-1}$ and $P=\sigma/V^{\gamma}$ into $dU=T\,dS-P\,dV$, we have

$$dU=\frac{d\sigma}{(\gamma-1)\,V^{\gamma-1}}-\frac{\sigma\,dV}{V^{\gamma}}+\frac{C'}{C}\frac{d\sigma}{(\gamma-1)\,\sigma}, \qquad (4\cdot81\cdot4)$$

which gives

$$U=U_0+\frac{\sigma}{(\gamma-1)\,V^{\gamma-1}}+\frac{C'}{C(\gamma-1)}\log\sigma=U_0+\frac{\vartheta}{\gamma-1}+\frac{C'}{C(\gamma-1)}\log\sigma. \qquad (4\cdot81\cdot5)$$

To determine the constant C' we use the result of Joule's experiment, namely, that when σ changes and U remains constant ϑ is unchanged. This shows that $C'=0$. Finally, to obtain (4·81·5) in its usual form we write $\gamma-1=R/C_V$ and $\vartheta=nRT$, where n is the number of gram molecules of the gas. We then have

$$U=U_0+nC_V T.$$

4·82. It should be noted that the method of calculating T and S from the isothermal and adiabatic curves will only work if the Jacobian $\partial(P,V)/\partial(\vartheta,\sigma)$ does not vanish or become infinite at any point. This is the condition that the parameters ϑ and σ can be chosen so that there is a one-one correspondence between points in the P, V and ϑ, σ planes. Restrictions which are sufficient to ensure that the empirical temperature and entropy scales satisfy this condition have already been noted on several occasions, but it does not follow that these are always satisfied. The condition for the Jacobian to vanish (or to become infinite) defines a curve in the P, V plane which divides the thermodynamic states into two types, and when such a boundary curve exists there is no reason for the two sets of states to possess the same characteristics. For example, one set might be such that the entropy always increased in an irreversible process, while for the other set the entropy might always decrease. Until recently, the possibility of a thermodynamic system possessing two sets of states with essentially different characteristics was considered to be a theoretical abstraction, but, as described in § 10·423, the atomic nuclei of certain solids can be obtained in orientational states which can best be described as having negative absolute temperatures. In general, however, we shall not consider such exceptional systems.

4·83. It is clear from the physical properties of the quantities concerned that the results of the preceding section cannot depend upon the particular choice of the arbitrary functions which occur in the definitions of the empirical temperature and the empirical entropy. A formal proof is as follows.

If instead of σ and ϑ we use σ^* and ϑ^*, where σ^* is a function of σ and ϑ^* a function of ϑ such that the derivatives are strictly positive, then by repeating the calculation of the preceding section we arrive at the result that

$$\frac{dT}{d\vartheta^*}\frac{dS}{d\sigma^*}=\frac{\partial(P,V)}{\partial(\vartheta^*,\sigma^*)}=\Theta^*(\vartheta^*)\,\Sigma^*(\sigma^*). \qquad (4·83·1)$$

But
$$\frac{\partial(P,V)}{\partial(\vartheta^*,\sigma^*)}=\frac{\partial(P,V)}{\partial(\vartheta,\sigma)}\frac{\partial(\vartheta,\sigma)}{\partial(\vartheta^*,\sigma^*)}=\frac{\partial(P,V)}{\partial(\vartheta,\sigma)}\frac{d\vartheta}{d\vartheta^*}\frac{d\sigma}{d\sigma^*},$$

and so
$$\frac{dT}{d\vartheta^*}=C\Theta^*(\vartheta^*)=C\Theta(\vartheta)\frac{d\vartheta}{d\vartheta^*},\quad \frac{dS}{d\sigma}=\frac{1}{C}\Sigma^*(\sigma^*)=\frac{1}{C}\Sigma(\sigma)\frac{d\sigma}{d\sigma^*}. \qquad (4·83·2)$$

Integrating these equations we have

$$\left.\begin{aligned} T&=C\int\Theta^*(\vartheta^*)\,d\vartheta^*+C'=C\int\Theta(\vartheta)\,d\vartheta+C',\\ S&=\frac{1}{C}\int\Sigma^*(\sigma^*)\,d\sigma^*+C''=\frac{1}{C}\int\Sigma(\sigma)\,d\sigma+C'',\end{aligned}\right\} \tag{4·83·3}$$

which proves that the functions S and T, and consequently U, are independent of the particular empirical temperature and entropy used.

Extensions to Complex Systems

4·9. We now consider briefly the extensions of the theory that are required to make it applicable to general thermodynamical systems. To begin with, consider the composite system formed by two simple systems in thermal equilibrium and depicted schematically in fig. 4·1 on p. 67. This system is described by four parameters P_1, V_1, P_2, V_2, but these are not all independent. For the condition for thermal equilibrium is expressed by an equation of the form $F(P_1, V_1; P_2, V_2)=0$, or, more conveniently in terms of the empirical temperature, by two equations of the form $\vartheta=f_1(P_1, V_1)$, $\vartheta=f_2(P_2, V_2)$. Thus there are only three independent parameters, which we may take to be P_1, V_1, V_2, or, more symmetrically ϑ, V_1, V_2. Similarly, if we have s simple systems in equilibrium the conditions for thermal equilibrium give us equations of state of the form $\vartheta=f_i(P_i, V_i)$ $(i=1, 2, \ldots, s)$, and of the $2s$ variables P_i, V_i only $s+1$ are independent. The most symmetrical set of variables to choose is ϑ, V_i $(i=1, 2, \ldots, s)$, and it is clear that we can always choose the independent variables to consist of the s geometrical coordinates which are sufficient to specify the state of the corresponding mechanical system, together with one of the generalized forces corresponding to the geometrical coordinates or, alternatively, the empirical temperature.

We can extend the above argument to solids in which the stresses are not isotropic. The state of a stressed solid is determined by the strain and stress tensors $x_{\mu\nu}$ and $p_{\mu\nu}$ $(\mu, \nu=1, 2, 3)$. Since these tensors are symmetrical there are twelve quantities to consider, but the equations of state reduce the number of independent variables to seven. By putting the solid into thermal contact with a simple system and varying the strains we can establish a number of equations of state, which, so long as the elastic limit is not overstepped, are of the form

$$p_{\mu\nu}=\sum_{\rho,\sigma}\beta_{\mu\nu,\rho\sigma}x_{\rho\sigma}, \tag{4·9·1}$$

where the $\beta_{\mu\nu,\rho\sigma}$ are the elastic constants and are functions of the temperature only. We may therefore choose as our independent variables the six $x_{\mu\nu}$ together with ϑ.

We may now state the following general result. The state of a general thermodynamic system is specified by one more variable than is required to specify the corresponding mechanical system. This extra variable may be taken to be the empirical temperature ϑ. The other variables a_i required to specify the state of the system may always be chosen to be deformation coordinates (volumes, masses, strains, etc.) such that the work done by the system in an infinitesimal quasi-static adiabatic process is of the form

$$dW = \sum_{i=1}^{s} A_i da_i.$$

The existence of the internal energy function $U(\vartheta, a_i)$ can be demonstrated by the same argument as was used in § 4·3, and the heat absorbed by the system in an infinitesimal quasi-static process is given by

$$dQ = dU(\vartheta, a_i) + \sum_{i=1}^{s} A_i da_i. \tag{4·9·2}$$

4·91. We now indicate how the proofs given in §§ 4·5–4·8 have to be modified to apply to general systems, leaving the reader to fill in the details.

(*a*) To prove Carnot's theorem, consider as in § 4·5 two systems in thermal equilibrium, and let them be specified by the variables $\vartheta, a_i^{(1)}$ and $\vartheta, a_i^{(2)}$. Then the three linear differential forms (4·9·2) associated with the two individual systems and with the combined systems possess integrating factors M_1, M_2 and M, and the generalized form of equation (4·5·5) is

$$M(\vartheta, a_i^{(1)}, a_i^{(2)})\, d\sigma(\vartheta, a_i^{(1)}, a_i^{(2)}) = M_1(\vartheta, a_i^{(1)})\, d\sigma_1(\vartheta, a_i^{(1)}) + M_2(\vartheta, a_i^{(2)})\, d\sigma_2(\vartheta, a_i^{(2)}). \tag{4·91·1}$$

Now change the independent variables from $\vartheta, a_i^{(1)}, a_i^{(2)}$ to $\vartheta, \sigma_1, \sigma_2, y_i^{(1)}, y_i^{(2)}$ $(i = 1, 2, \ldots, s-1)$, where $y_i^{(1)} = a_i^{(1)}$, $y_i^{(2)} = a_i^{(2)}$ for $i = 1, 2, \ldots, s-1$. Then with these variables we have

$$\frac{\partial \sigma}{\partial \vartheta} = 0, \quad \frac{\partial \sigma}{\partial y_i^{(1)}} = 0, \quad \frac{\partial \sigma}{\partial y_i^{(2)}} = 0, \quad \frac{\partial \sigma}{\partial \sigma_1} = \frac{M_1}{M}, \quad \frac{\partial \sigma}{\partial \sigma_2} = \frac{M_2}{M},$$

and hence

$$\left.\begin{aligned} \frac{\partial}{\partial \vartheta}\frac{M_1}{M} = 0, \quad \frac{\partial}{\partial y_i^{(1)}}\frac{M_1}{M} = 0, \quad \frac{\partial}{\partial y_i^{(2)}}\frac{M_1}{M} = 0, \\ \frac{\partial}{\partial \vartheta}\frac{M_2}{M} = 0, \quad \frac{\partial}{\partial y_i^{(1)}}\frac{M_2}{M} = 0, \quad \frac{\partial}{\partial y_i^{(2)}}\frac{M_2}{M} = 0. \end{aligned}\right\} \tag{4·91·2}$$

It follows from these equations that M_1 can only depend upon ϑ and σ_1, M_2 upon ϑ and σ_2, and M upon ϑ, σ_1 and σ_2, so that we obtain equation (4·51·4) once more. The remainder of the proof then proceeds as before.

(*b*) The changes required in the argument of § 4·8 are as follows. Let a thermodynamic system be described by the empirical temperature ϑ and s deformation coordinates a_i $(i=1,2,\ldots,s)$, and let A_i be the generalized force corresponding to a_i. Then by reversible processes we can determine the s isothermal relations (equations of state)

$$A_i = A_i(\vartheta, a_1, \ldots, a_s), \tag{4·91·3}$$

and the family of adiabatics

$$\sigma = \sigma(\vartheta, a_1, \ldots, a_s) = \text{constant}. \tag{4·91·4}$$

Rewriting equation (4·9·2) in the form

$$dU(\vartheta, a_i) = T\,dS - \sum_{i=1}^{s} A_i\,da_i, \tag{4·91·5}$$

we have to determine the functions U, S and T from (4·91·3) and (4·91·4), where S depends only on σ, and T upon ϑ. Change the independent variables from ϑ, a_i $(i=1,2,\ldots,s)$ to ϑ, σ, y_i $(i=1,2,\ldots,s-1)$, where $y_i = a_i$ for $i=1,2,\ldots,s-1$. Then (4·91·5) becomes

$$dU(\vartheta,\sigma,y_i) = \left\{T\frac{dS}{d\sigma} - A_s\left(\frac{\partial a_s}{d\sigma}\right)_{\vartheta,y_i}\right\}d\sigma - A_s\left(\frac{\partial a_s}{\partial\vartheta}\right)_{\sigma,y_i}d\vartheta - \sum_{i=1}^{s-1} A_i\,dy_i. \tag{4·91·6}$$

There are a number of relations to be satisfied in order to make dU a perfect differential. The only one that we are concerned with at the moment is

$$\frac{\partial}{\partial\vartheta}\left\{T\frac{dS}{d\sigma} - A_s\left(\frac{\partial a_s}{\partial\sigma}\right)_{y_i}\right\} = \left(\frac{\partial^2 U}{\partial\vartheta\,\partial\sigma}\right)_{y_i} = -\frac{\partial}{\partial\sigma}\left\{A_s\left(\frac{\partial a_s}{\partial\vartheta}\right)_{y_i}\right\}.$$

This reduces to

$$\frac{dT}{d\vartheta}\frac{dS}{d\sigma} = \left(\frac{\partial(A_s, a_s)}{\partial(\vartheta,\sigma)}\right)_{y_i}, \tag{4·91·7}$$

which is the generalization of equation (4·8·2). The calculation now proceeds as before.

There are s equations like (4·91·7), but they must all be the same on account of the conditions required to make dU a perfect differential. The easiest way to show this is to obtain (4·91·7) in an alternative form. Take $\vartheta, a_1, a_2, \ldots, a_s$ as the independent variables and rewrite (4·91·5) as

$$dU = T\frac{dS}{d\sigma}\frac{\partial\sigma}{\partial\vartheta}d\vartheta + \sum_{i=1}^{s}\left(T\frac{dS}{d\sigma}\frac{\partial\sigma}{\partial a_i} - A_i\right)da_i.$$

Since dU is a perfect differential we must have

$$\frac{\partial}{\partial a_i}\left(T\frac{dS}{d\sigma}\frac{\partial\sigma}{\partial\vartheta}\right) = \frac{\partial}{\partial\vartheta}\left(T\frac{dS}{d\sigma}\frac{\partial\sigma}{\partial a_i} - A_i\right) \quad (i=1,2,\ldots,s),$$

i.e. $$\frac{dT}{d\vartheta}\frac{dS}{d\sigma}=\left(\frac{\partial A_i}{\partial \vartheta}\right)_{a_1,\ldots,a_s}\bigg/\left(\frac{\partial \sigma}{\partial a_i}\right)_{\vartheta, a_1,\ldots,a_{i-1},a_{i+1},\ldots,a_s} \quad (i=1,2,\ldots,s), \quad (4·91·8)$$

which is a more symmetrical form than (4·91·7). The conditions that the s quantities on the right-hand sides of the s equations (4·91·8) should be equal are s of the conditions for dU to be a perfect differential. The other conditions are

$$\partial A_i/\partial a_j = \partial A_j/\partial a_i \quad (i,j=1,2,\ldots,s),$$

the independent variables being $\vartheta, a_1, a_2, \ldots, a_s$.

REFERENCES

Born, M. (1921). Critical reflections on the traditional exposition of thermodynamics. *Phys. Z.* **22**, 218, 249, 282.

Carathéodory, C. (1909). Investigations on the foundations of thermodynamics. *Math. Ann.* **67**, 355.

Carathéodory, C. (1925). The determination of the internal energy and the absolute temperature by means of reversible processes. *S.B. preuss. Akad. Wiss.* 39.

Chapter 5

STATISTICAL MECHANICS

Fundamental Concepts

5·1. In any dynamical system the equations of motion can be formulated exactly, and the motion of each individual particle can, in theory, be calculated. In practice, however, the integration of the equations of motion is an impossible task if the number of particles is large, and we have to be content with determining the average behaviour of the system. Even if we could integrate the equations of motion exactly, our information concerning, for example, the initial conditions to be assumed would be far from complete, and in view of our lack of knowledge of the microscopic state of the system we would have to average over all states which are compatible with our partial information. In the averaging process much of the detail of the motion is lost, and the properties of the system have to be described by a few parameters only, instead of by the coordinates and momenta of all the individual particles which make up the system. It is through this averaging process that the concept of temperature is introduced, a concept which is quite foreign to the detailed treatment of a dynamical system.

In dealing with the motion of the atoms, etc., which are the constituents of matter in bulk we shall use the language of quantum rather than of classical mechanics even when quantum effects are negligible. Therefore, to characterize a dynamical system we require to know the stationary states of the system, i.e. the eigenvalues of the energy and the corresponding wave functions. The fundamental assumption of statistical mechanics is that the value of any dynamical variable of a macroscopic system is obtained by taking the average of the variable over all the possible stationary states of the system. (This assumption is often stated in the form that every distinct stationary state of a system has the same *a priori* probability.) It should, however, be mentioned that there are other ways of formulating the fundamental principles of statistical mechanics, but a discussion of them would take us too far afield. Which approach one regards as the most fundamental is largely a matter of taste. The basic assumption from which the present treatment starts has the merit of physical simplicity and therefore of plausibility, and since it is expressed in terms of the stationary states, which have an invariant character, it is obvious that it is consistent with the laws of quantum dynamics and can be considered to be a logical extension of them. The reader who wishes to pursue a deeper analysis of the fundamental assumptions is referred to the books by von Neumann (1932) and Tolman (1938).

5·11. Consider a system consisting of a large number N of 'practically independent' identical particles or 'elements'. The meaning of the phrase 'practically independent' is that there is only a very weak coupling between the elements, so that the stationary states and the energy levels of any element can be calculated without reference to the other elements, although there are slight interactions which ensure that, over a sufficiently long interval of time, the system will assume a state of statistical equilibrium. The elements are not necessarily particles or atoms, though it is often convenient to consider them to be so, but they may be complex dynamical systems which can in their turn be considered, when necessary, as assemblies of simpler systems. Now let us suppose that the stationary states and the energy levels of any one of the N elements are known. Then the state of the system is determined if we know that there are n_1 elements in the stationary state with energy E_1, n_2 elements in the state with energy E_2, etc. Each such state of the system is known as a 'complexion', and for given values of $n_1, n_2, \ldots$, there are many complexions. If we denote the number of these complexions by $W(n_1, n_2, \ldots)$, then, by the fundamental assumption of statistical mechanics, the average value of any physical quantity $F(n_1, n_2, \ldots)$ is given by

$$\bar{F} = \frac{\Sigma F(n_1, n_2, \ldots)\, W(n_1, n_2, \ldots)}{\Sigma W(n_1, n_2, \ldots)}, \tag{5·11·1}$$

where the summation is taken over all the possible distributions of the n's which are consistent with the constraints on the system. To calculate $\bar{F}$ exactly is equivalent to solving completely the dynamical equations of motion (normally, by solving the Schrödinger equation for the type of system under consideration), but, in the limit of infinitely large assemblies, $W(n_1, n_2, \ldots)$ has a sharp maximum for a certain distribution of the n's, and $\bar{F}$ can be replaced by the value of $F(n_1, n_2, \ldots)$ for this distribution.

5·12. An elementary but instructive example of the procedure to be adopted is the following. The series $(1+x)^N$, where $x>0$, $N>0$, has the expansion

$$(1+x)^N = \sum_{n=0}^{N} \frac{N!}{n!\,(N-n)!}\, x^n = \sum_{n=0}^{N} W(n). \tag{5·12·1}$$

By using Stirling's theorem in the form

$$\log m! \sim m \log m - m, \tag{5·12·2}$$

the logarithm of the general term in the expansion (5·12·1) can be written as

$$\log W(n) = N \log N + n \log x - n \log n - (N-n) \log (N-n), \tag{5·12·3}$$

provided that both N and n are sufficiently large. If we treat n as a continuous variable, (5·12·3) is a maximum when $\log x - \log n + \log(N-n) = 0$. The maximum term in (5·12·1) is therefore the one for which

$$n = Nx/(x+1), \tag{5·12·4}$$

and, on inserting this value of n into (5·12·3), we find that the value of the maximum term, to the approximation implied in the preceding calculations, is $(1+x)^N$. In this case, therefore, the maximum term, evaluated by the use of Stirling's theorem in the form given above, gives the exact value of the series.

While the procedure just outlined is correct in that it can readily be generalized to prove that $\bar{F}$ in (5·11·1) is the value of $F(n_1, n_2, \ldots)$ for the set of n's which make $W(n_1, n_2, \ldots)$ a maximum, it is nevertheless untrue that the series (5·12·1) can be replaced by its maximum term. There are, in fact, about $\sqrt{N}$ terms comparable in order of magnitude with the maximum term, and a more accurate calculation is required to find the sum of the series by considering the maximum term. If we use Stirling's theorem in the more exact form

$$\log m! \sim (m+\tfrac{1}{2})\log m - m + \tfrac{1}{2}\log(2\pi) + O(m^{-1}), \tag{5·12·5}$$

the maximum term is the one for which $n = n_{\text{max.}}$, where

$$n_{\text{max.}} = \frac{Nx}{1+x} - \frac{1-x}{2(1+x)} + O(N^{-1}), \tag{5·12·6}$$

and its value, to the same order of approximation, is

$$W_{\text{max.}} = (2\pi Nx)^{-\frac{1}{2}}(1+x)^{N+1}. \tag{5·12·7}$$

Also, if we put $n = n_{\text{max.}} + t$, then, to the second order in t/N, we have

$$W(n) = \exp\{-\tfrac{1}{2}t^2(1+x)^2/(Nx)\}\, W_{\text{max.}}. \tag{5·12·8}$$

Therefore, as stated above, the number of terms for which $W(n)$ is comparable with $W_{\text{max.}}$ is of the order of $\sqrt{N}$, while the others are smaller by a factor of the order of $\exp(-\alpha N)$, where α is a constant of order unity. (It should be noted that we only require the extended form of Stirling's theorem to find the value of $W_{\text{max.}}$. The ratio $W(n)/W_{\text{max.}}$ is correctly given by using the simpler form of Stirling's theorem.) We may now calculate $\Sigma W(n)$ by writing

$$\begin{aligned}\sum_0^N W(n) &= W_{\text{max.}}\int_{-\infty}^{\infty} \exp\{-\tfrac{1}{2}t^2(1+x)^2/(Nx)\}\,dt \\ &= (2\pi Nx)^{\frac{1}{2}}(1+x)^{-1}\, W_{\text{max.}} = (1+x)^N, \end{aligned}\tag{5·12·9}$$

which reproduces correctly the value of the series.‡

‡ Note that $\int_{-\infty}^{\infty} e^{-at^2}\,dt = \left(\frac{\pi}{a}\right)^{\frac{1}{2}}, \quad \int_{-\infty}^{\infty} t^2 e^{-at^2}\,dt = \frac{1}{2}\left(\frac{\pi}{a^3}\right)^{\frac{1}{2}}.$

It should be noted that

$$\log W_{\text{max.}} = N\log(1+x) - \tfrac{1}{2}\log N + \tfrac{1}{2}\log\{(1+x)^2/(2\pi x)\}$$
$$= N\log(1+x) + O(\log N),$$

and therefore that, if we are not interested in including terms which are of the order of $N^{-1}\log N$ compared with the leading term, it is immaterial whether we use the simpler or the more elaborate form of Stirling's theorem and whether we include only the maximum term or the $\sqrt{N}$ terms of comparable magnitude. Since N is usually of the order of 10^{20} at least, the terms in $\log W_{\text{max.}}$ involving $\log N$ are negligible compared with the leading term.

We may also use (5·12·8) to calculate the average value of a function $F(n)$ to a higher degree of accuracy than that given by putting $\overline{F} = F(n_{\text{max.}})$. For example, we have the exact expressions

$$\sum_{n=0}^{N} nW(n) = Nx\sum_{n=1}^{N}\frac{(N-1)!}{(n-1)!\,(N-n)!}x^{n-1} = x\frac{\partial}{\partial x}\sum_{0}^{N}W(n) = Nx(1+x)^{N-1}, \tag{5·12·10}$$

$$\sum_{n=0}^{N} n(n-1)\,W(n) = N(N-1)\,x^2\sum_{n=2}^{N}\frac{(N-2)!}{(n-2)!\,(N-n)!}x^{n-2} = x^2\frac{\partial^2}{\partial x^2}\sum_{0}^{N}W(n)$$
$$= N(N-1)\,x^2(1+x)^{N-2}, \tag{5·12·11}$$

and hence $\quad \overline{n} = Nx/(1+x), \quad \overline{n(n-1)} = N(N-1)\,x^2/(1+x)^2, \qquad$ (5·12·12)

which give $\quad \overline{n^2} = \overline{n(n-1)} + \overline{n} = \overline{n}^2 - \overline{n}^2/N + \overline{n} = \overline{n}^2\{1+O(N^{-1})\}. \qquad$ (5·12·13)

We have already established the relation (5·12·12) for $\overline{n}$ by the approximate method. To obtain a similar approximate expression for $\overline{n^2}$ we write

$$\sum_{n=0}^{N} n^2W(n) = \sum_{n=0}^{N}(\overline{n}^2 + 2\overline{n}t + t^2)\,W(n) \sim W_{\text{max.}}\int_{-\infty}^{\infty} t^2\exp\{-\tfrac{1}{2}t^2(1+x)^2/(Nx)\}\,dt$$
$$= \overline{n}^2(1+x)^N + \tfrac{1}{2}\sqrt{\pi}\,(2Nx)^{\frac{3}{2}}\,(1+x)^{-3}\,W_{\text{max.}}$$
$$= \overline{n}^2(1+x)^N + Nx(1+x)^{N-2},$$

which gives

$$\overline{n^2} = \overline{n}^2 + \frac{Nx}{(1+x)^2} = \overline{n}^2 + \frac{Nx}{1+x} - \frac{Nx^2}{(1+x)^2} = \overline{n}^2 + \overline{n} - \overline{n}^2/N. \tag{5·12·14}$$

This is the same as (5·12·13) and it shows that

$$\frac{(\overline{n^2} - \overline{n}^2)^{\frac{1}{2}}}{\overline{n}} = \frac{1}{\overline{n}^{\frac{1}{2}}}\left(1 - \frac{\overline{n}}{N}\right)^{\frac{1}{2}} = O(N^{-\frac{1}{2}}). \tag{5·12·15}$$

The results of the preceding calculations can be summed up as follows. The use of Stirling's theorem in the form (5·12·2) to calculate the maximum value of $W(n)$ gives correctly the average value of any function $F(n)$ to the zero order. The same procedure leads accidentally to the correct value for the sum of the series $\Sigma W(n)$, though there are in fact about $\sqrt{N}$ terms of the same order of magnitude as the maximum term. To obtain the value of the series by a correct procedure we require to use Stirling's theorem in the form (5·12·5). On the other hand, the simpler form of Stirling's theorem is sufficiently accurate to give the ratio $W(n)/W_{\text{max.}}$ and thus enable us to calculate not only the average value of any function but also the mean square deviation of the function from its mean value. If, in addition, we do not wish to retain terms in $\log \Sigma W(n)$ which are of order $N^{-1} \log N$ compared with the leading term, it is immaterial which form of Stirling's theorem we use and whether we include only the maximum term or all the $\sqrt{N}$ terms of comparable magnitude. In the sections which follow we shall normally adopt the procedure based upon the approximation

$$\log m! \sim m \log m - m$$

and the replacement of the series by its maximum term, since the calculations are then much simplified and invariably lead to the correct result because the important physical quantities depend upon the ratio of two series. The reader can easily verify in any particular case that the full procedure supports the results obtained by the shortened version.

The Effect of the Symmetry Properties of the Wave Functions on the Statistics

5·2. We base the exposition on quantum mechanics partly in order to have discrete energy levels and partly to introduce at the outset one of the fundamental quantal principles, namely, the fact that the total wave function of a number of identical particles must be either symmetrical or antisymmetrical with respect to interchanges of each pair of particles. If we analyse the wave functions into those relating to the ultimate constituents of matter, which we take to be electrons, protons and neutrons, the wave functions must be antisymmetrical with respect to the coordinates (including the spin coordinate) of each pair of electrons, of each pair of protons and of each pair of neutrons. If we consider more complex elements and wish to treat them as permanent units, the wave functions will be symmetrical or antisymmetrical according as the elements contain an even or an odd number of fundamental particles. For example, if we have a system consisting of hydrogen atoms, the wave function of the system must be symmetrical with respect to the interchange of two hydrogen atoms, since such an interchange permutes two protons and two electrons. The deuterium atom, on the other hand, consists of a proton, a neutron and an electron, and the wave function of a system consisting of deuterium atoms must belong to the antisymmetrical class. The wave functions of a system consisting of deuterium molecules, however, belong to the symmetrical class.

If we consider the Schrödinger equation for a number of N loosely coupled elements, we can obtain a 'primitive' wave function as a simple product of N separate wave functions:

$$\Psi = \psi_1(x_1)\,\psi_2(x_2)\,\ldots,$$

where $x_1, x_2, \ldots$ are the generic coordinates of the elements $1, 2, \ldots$. Such a wave function is called an unsymmetrized wave function. The number of wave functions with the correct symmetry which can be obtained from any particular unsymmetrized wave functions depends, *inter alia*, upon whether the wave functions are localized or not, i.e. upon whether the wave functions $\psi_1, \psi_2, \ldots$ can be considered to be associated with distinct and distinguishable regions I, II, ... in space, or whether they must be considered as being spread out over the whole of the space occupied by all the elements. For the present we shall consider only unlocalized elements, which include, for example, gases, since the molecules of a gas clearly cannot be confined to any particular portion of the containing vessel, while we shall deal with localized elements such as the atoms on the surface of a solid in § 5·51.

If there are just two elements present, the wave functions $\psi_1(x_1)\,\psi_2(x_2)$ and $\psi_1(x_2)\,\psi_2(x_1)$ are clearly solutions of the Schrödinger equation

$$[\nabla_1^2 + \nabla_2^2 + (8\pi^2 m/h^2)\{E - V(x_1) - V(x_2)\}]\,\Psi(x_1, x_2) = 0$$

belonging to the same energy level, since the potential energy $V(x_1) + V(x_2)$ is necessarily symmetrical in x_1 and x_2. They cannot, however, be considered to give rise to two separate complexions, and for this purpose they must be considered to be equivalent and to be replaced either by the single symmetrical combination $\psi_1(x_1)\,\psi_2(x_2) + \psi_2(x_1)\,\psi_1(x_2)$ or by the single antisymmetrical combination $\psi_1(x_1)\,\psi_2(x_2) - \psi_2(x_1)\,\psi_1(x_2)$, according as the symmetry requirements are $\Psi(x_2, x_1) = \Psi(x_1, x_2)$ or $\Psi(x_2, x_1) = -\Psi(x_1, x_2)$. In the general case we can derive one and only one symmetrical wave function from a primitive unsymmetrized wave function, namely, by permuting in all possible ways the coordinates of the elements and adding the resulting wave functions. Similarly, we can construct exactly one antisymmetrical wave function, which is the determinant

$$\begin{vmatrix} \psi_1(x_1) & \psi_1(x_2) & \ldots & \psi_1(x_N) \\ \psi_2(x_1) & \psi_2(x_2) & \ldots & \psi_2(x_N) \\ \vdots & & & \\ \psi_N(x_1) & \psi_N(x_2) & \ldots & \psi_N(x_N) \end{vmatrix}$$

If, however, any two of the ψ's are the same, the antisymmetrical wave function is identically zero. Therefore, each unsymmetrized wave function is associated with one symmetrical wave function and with either one antisymmetrical wave function (if all the ψ's are different) or with none (if any two ψ's are the same).

The state of a system of weakly interacting elements is characterized by the numbers of elements $n_1, n_2, \ldots$ with energies $E_1, E_2, \ldots$, where the energies E_r and wave functions $\psi_r(x)$ of a single element are given by the solutions of $[\nabla^2 + (8\pi^2 m/h^2)\{E_r - V(x)\}]\psi_r(x) = 0$. Now, according to the arguments given above, the symmetry requirements of the wave functions of the system place no restrictions on the n_r's for systems with symmetrical wave functions but restrict the n_r's to be either 1 or 0 for systems with antisymmetrical wave functions, and it is not easy to find the number of complexions directly when the n_r's are subject to these conditions. We therefore consider the energy levels to be grouped together into a number of sets, each comprising many levels having approximately the same energy. If M_r is the number of energy levels in the group whose energy is E_r, and if N_r is the number of elements whose energy levels fall into the group M_r, then, provided that both M_r and N_r are large we can determine the average value of N_r.

The calculations must be carried out separately for symmetrical and for antisymmetrical wave functions. They can also be carried out for unsymmetrized wave functions, and, although the use of unsymmetrized wave functions cannot strictly lead to the correct number of complexions for an assembly of unlocalized elements, the symmetry properties of the wave functions of the assembly are irrelevant to the distribution of elements over the energy levels when the average number of elements in every group of energy levels is very small compared with the number of levels. This is, in fact, the case for nearly all macroscopic assemblies, and we are then led to 'classical statistics'. We may always obtain the 'classical' formulae by considering them to be particular cases of the formulae derived by taking the symmetry requirements fully into account, and we shall do so from time to time, but it is desirable to derive the 'classical' formulae independently by disregarding the full effect of symmetry requirements from the outset. We must, however, take account of the symmetry requirements in so far as they affect the number of independent wave functions. That is, we must count $\psi_1(x_1)\psi_2(x_2)$ and $\psi_1(x_2)\psi_2(x_1)$ as being the same wave function, even although we do not need to consider their correct linear combinations. With this proviso we have to write down the number of complexions of a system when the chance of double occupation of any level is negligible.

5·21. *The number of complexions for wave functions which are not explicitly symmetrized.* If we have N_r identical elements and M_r different wave functions to choose from, a wave function of the assembly which is a product of the wave functions of single elements can be constructed in $M_r^{N_r}$ ways, but these are not all different, since permutations of the N_r elements amongst themselves do not result in different wave functions of the groups of systems. We must therefore divide by $N_r!$, and the number of complexions for the whole system, for given $N_1, N_2, \ldots$, is

$$W(N_1, N_2, \ldots) = \frac{M_1^{N_1} M_2^{N_2} \ldots}{N_1! \, N_2! \ldots} \quad (M_r \gg N_r \gg 1). \tag{5·21·1}$$

We now determine the average state of the system when the number of complexions is given by (5·21·1), and show how the laws of thermodynamics can be derived. In § 5·6 we consider the same problem when $W(N_1, N_2, \ldots)$ has its correct quantal form.

THE LAWS OF THERMODYNAMICS

5·3. *The empirical temperature.* In the axiomatic formulation of the laws of thermodynamics given in Chapter 4, it was found impossible to define the temperature without considering at least two bodies in thermal equilibrium with one another. We therefore begin our derivation of the laws of thermodynamics from statistical mechanics by determining the average properties of two loosely coupled systems of elements, which consist respectively of $N^{(1)}$ and $N^{(2)}$ practically independent elements, the combined system being isolated from its surroundings so that its total energy U is a fixed quantity. Let $E_1^{(1)}, E_2^{(1)}, \ldots$ and $E_1^{(2)}, E_2^{(2)}, \ldots$ be the energy levels of the two sets of elements respectively. We suppose that the energy levels are grouped together into large sets, the rth set in each system containing $M_r^{(1)}$ and $M_r^{(2)}$ levels respectively, and the number of elements with energy levels lying in the rth set being $N_r^{(1)}$ and $N_r^{(2)}$. Then the number of complexions of the combined system is the product of the number of complexions for the two separate systems, i.e., by (5·21·1),

$$W(N_r^{(1)}, N_s^{(2)}) = \prod_r \prod_s \frac{M_r^{(1)\,N_r^{(1)}}}{N_r^{(1)}!} \frac{M_s^{(2)\,N_s^{(2)}}}{N_s^{(2)}!} \quad (M_r^{(1)} \gg N_r^{(1)},\ M_s^{(2)} \gg N_s^{(2)}). \tag{5·3·1}$$

We shall now prove that the set of $N_r^{(1)}$'s and $N_s^{(2)}$'s which maximize W are such that $N_r^{(1)} \propto \exp(-\vartheta E_r^{(1)})$ and $N_s^{(2)} \propto \exp(-\vartheta E_s^{(2)})$, where ϑ is a parameter which is the same for both systems.

We can determine the maximum value of W by the method outlined in § 5·12, but in doing so we must take into account the conditions

$$\Sigma N_r^{(1)} = N^{(1)}, \quad \Sigma N_s^{(2)} = N^{(2)} \tag{5·3·2}$$

and the relation $$\Sigma(N_r^{(1)} E_r^{(1)} + N_s^{(2)} E_s^{(2)}) = U, \tag{5·3·3}$$

which is the condition for the energy of the combined system to be constant and equal to U. (Note that we make no assumption about the constancy or otherwise of the energies U_1, U_2 of the separate systems, since any such assumption would be inconsistent with the systems being loosely coupled together.) In order to determine the maximum value of $W(N_r^{(1)}, N_s^{(2)})$ subject to the conditions (5·3·2) and (5·3·3), it is convenient to introduce undetermined multipliers λ_1, λ_2 and ϑ and to maximize

$$\log W + \lambda_1 N^{(1)} + \lambda_2 N^{(2)} - \vartheta U,$$

where $N^{(1)}$, $N^{(2)}$ and U are given by (5·3·2) and (5·3·3). If we treat $N_r^{(1)}$ and $N_s^{(2)}$ as continuous variables we then have

$$\begin{aligned}\delta(\log W + \lambda_1 N_1^{(1)} + \lambda_2 N^{(2)} - \vartheta U)\\ = \Sigma\delta N_r^{(1)}(\log M_r^{(1)} - \log N_r^{(1)} + \lambda_1 - \vartheta E_r^{(1)})\\ + \Sigma\delta N_s^{(2)}(\log M_s^{(2)} - \log N_s^{(2)} + \lambda_2 - \vartheta E^{(2)}),\end{aligned} \tag{5·3·4}$$

and $\delta N_r^{(1)}$ and $\delta N_s^{(2)}$ can be considered to be independent differentials. Hence the maximum value of (5·3·1) occurs when

$$N_r^{(1)} = M_r^{(1)} \exp(\lambda - \vartheta E_r^{(1)}), \quad N_s^{(2)} = M_s^{(2)} \exp(\lambda_2 - \vartheta E_s^{(2)}). \tag{5·3·5}$$

At this point it is convenient to revert to the number of elements in each energy level rather than in each group of levels. If the energy levels are non-degenerate, the occupation numbers $\overline{n_r^{(1)}}$ and $\overline{n_s^{(2)}}$ of the levels $E_r^{(1)}$ and $E_s^{(2)}$ are then

$$\overline{n_r^{(1)}} = N_r^{(1)}/M_r^{(1)} = \exp(\lambda_1 - \vartheta E_r^{(1)}), \quad \overline{n_s^{(2)}} = N_s^{(2)}/M_s^{(2)} = \exp(\lambda_2 - \vartheta E_s^{(2)}), \tag{5·3·6}$$

and the M's, which are introduced in order to apply Stirling's theorem, no longer appear explicitly provided that we now sum over all levels instead of over groups of levels.

The parameters λ_1 and λ_2 are determined by the relations

$$\Sigma\overline{n_r^{(1)}} = e^{\lambda_1}\Sigma\exp(-\vartheta E_r^{(1)}) = N^{(1)}, \quad \Sigma\overline{n_s^{(2)}} = e^{\lambda_2}\Sigma\exp(-\vartheta E_s^{(2)}) = N^{(2)}, \tag{5·3·7}$$

while the parameter ϑ is determined from the energy U by the condition

$$U = \Sigma[n_r^{(1)} E_r^{(1)} + n_s^{(2)} E_s^{(2)}] = \Sigma[e^{\lambda_1} E_r^{(1)} \exp(-\vartheta E_r^{(1)}) + e^{\lambda_2} E_s^{(2)} \exp(-\vartheta E_s^{(2)})]. \tag{5·3·8}$$

The energies of the separate systems are

$$U_1 = e^{\lambda_1}\Sigma E_r^{(1)} \exp(-\vartheta E_r^{(1)}), \quad U_2 = e^{\lambda_2}\Sigma E_s^{(2)} \exp(-\vartheta E_s^{(2)}). \tag{5·3·9}$$

We have therefore been led to infer the existence of a parameter ϑ which is the same for both systems when they are in statistical equilibrium with one another (and by a simple extension of the argument the parameter ϑ would be the same for any number of systems in equilibrium with one another). In accordance with generalization A, p. 68, we may identify ϑ with the empirical temperature, and it is determined in terms of the generalized coordinates of the combined system by the relation (5·3·8).

When the formulae of the present section are applicable, the elements are said to obey classical or Maxwell-Boltzmann statistics. The distribution function $e^{-\vartheta E}$ is called the Maxwell or Boltzmann distribution function.

5·31. We may generalize the above results to include cases where the energy levels of the elements are degenerate. If the multiplicity (or weight) of the energy level $E_r^{(1)}$ is $\varpi_r^{(1)}$ and that of the energy level $E_s^{(2)}$ is $\varpi_s^{(2)}$, then by allowing the appropriate number of energy levels to coalesce, we see that the average numbers of elements with energies $E_r^{(1)}$ and $E_s^{(2)}$ are

$$\overline{n_r^{(1)}} = \varpi_r^{(1)} \exp(\lambda_1 - \vartheta E_r^{(1)}), \quad \overline{n_s^{(2)}} = \varpi_s^{(2)} \exp(\lambda_2 - \vartheta E_s^{(2)}) \qquad (5·31·1)$$

respectively. If we define the functions Z_1, Z_2, known as the partition functions of the two sets of elements, by the relations

$$Z_1 = \Sigma \varpi_r^{(1)} \exp(-\vartheta E_r^{(1)}), \quad Z_2 = \Sigma \varpi_s^{(2)} \exp(-\vartheta E_s^{(2)}), \qquad (5·31·2)$$

the parameters λ_1, λ_2 are determined by the generalizations of equations (5·3·7), namely,

$$N^{(1)} = e^{\lambda_1} Z_1, \quad N^{(2)} = e^{\lambda_2} Z_2, \qquad (5·31·3)$$

and so (5·31·1) may be rewritten in the form

$$\overline{n_r^{(1)}} = (N_1/Z_1) \exp(-\vartheta E_r^{(1)}), \quad \overline{n_s^{(2)}} = (N_2/Z_2) \exp(-\vartheta E_s^{(2)}). \qquad (5·31·4)$$

Finally, the parameter ϑ is determined by the condition

$$U = -e^{\lambda_1} \frac{\partial Z_1}{\partial \vartheta} - e^{\lambda_2} \frac{\partial Z_2}{\partial \vartheta} = -N^{(1)} \frac{\partial \log Z_1}{\partial \vartheta} - N^{(2)} \frac{\partial \log Z_2}{\partial \vartheta}. \qquad (5·31·5)$$

5·32. *The definition of the quantity of heat.* We consider for the moment only the first of the two systems discussed in the preceding sections and drop the index 1. The energy of this system is given by

$$U = -e^{\lambda} \partial Z / \partial \vartheta = -N \partial \log Z / \partial \vartheta,$$

but U is not constant since it may be varied by the exchange of energy between the two systems. We suppose that the energy levels of any element are functions of a number of parameters a_i. We can take the a_i's as generalized coordinates of the system, and when an element is in the rth stationary state, the generalized force corresponding to a_i, exerted on the element by external bodies, is $-A_i^{(r)}$ in the direction tending to increase a_i, where

$$A_i^{(r)} = -\partial E_r / \partial a_i, \qquad (5·32·1)$$

provided that the E_r's are independent of N for fixed a_i. Hence the average generalized force on the whole system is

$$\bar{A}_i = \Sigma \bar{n}_r A_i^{(r)} = -N \frac{\Sigma \varpi_r (\partial E_r / \partial a_i) e^{-\vartheta E_r}}{\Sigma \varpi_r e^{-\vartheta E_r}} = \frac{N}{\vartheta} \frac{\partial \log Z}{\partial a_i}. \qquad (5·32·2)$$

Now in any adiabatic infinitesimal change the increase in the internal energy plus the work done by the system on external bodies is zero. But there are changes for which $dU + \Sigma \bar{A}_i da_i$ is not zero, and for these changes we define the (infinitesimal) heat absorbed dQ by the equation

$$dQ = dU + \Sigma \bar{A}_i da_i = -N d \frac{\partial \log Z}{\partial \vartheta} + \frac{N}{\vartheta} \Sigma \frac{\partial \log Z}{\partial a_i} da_i \qquad (5·32·3)$$

(provided that N is constant), which may be written in the alternative form

$$dQ = -d\left(e^{\lambda} \frac{\partial Z}{\partial \vartheta}\right) + \frac{e^{\lambda}}{\vartheta} \Sigma \frac{\partial Z}{\partial a_i} da_i. \qquad (5·32·4)$$

5·33. *The second law of thermodynamics.* In order to establish the existence of the absolute temperature and of the entropy function, we have merely to follow the procedure of §§ 2·72 and 4·5, and show that, provided the number of elements is constant, the linear differential form (5·32·4) for dQ possesses an integrating factor, which is obviously ϑ. For we have

$$\begin{aligned} \vartheta(dU + \Sigma \bar{A}_i da_i) &= d(\vartheta U) - U d\vartheta + \vartheta \Sigma \bar{A}_i da_i \\ &= d(\vartheta U) + e^{\lambda} (\partial Z / \partial \vartheta) d\vartheta + e^{\lambda} \Sigma (\partial Z / \partial a_i) da_i \\ &= d(\vartheta U) + e^{\lambda} dZ = d(\vartheta U + e^{\lambda} Z) - Z e^{\lambda} d\lambda. \end{aligned} \qquad (5·33·1)$$

The last term can be written as $-N d\lambda = -d(N\lambda) + \lambda dN$, and so

$$\vartheta(dU + \Sigma \bar{A}_i da_i) = d(\vartheta U - N\lambda + N) + \lambda dN. \qquad (5·33·2)$$

Hence, if N is constant, ϑdQ is a perfect differential and we must have $1/\vartheta = kT$, where T is the absolute temperature and k is a constant with the dimensions of an energy. The entropy function S is given by

$$S = U/T - kN\lambda + kN = U/T + k(N \log Z - N \log N + N), \qquad (5·33·3)$$

and the free energy F by

$$F = U - ST = -kT(N \log Z - N \log N + N). \qquad (5·33·4)$$

There is an important connexion between the entropy S and the maximum number of complexions. For

$$\begin{aligned} k \log W_{\text{max.}} &= k\Sigma(\bar{N}_r \log M_r - \bar{N}_r \log \bar{N}_r + \bar{N}_r) \\ &= -k\Sigma \bar{N}_r (\lambda - \vartheta E_r) + kN = k(\vartheta U - N\lambda + N), \end{aligned}$$

and so

$$S = k \log W_{\text{max.}}. \qquad (5·33·5)$$

Alternatively we may write

$$F = -kT \log (W_{\text{max.}} e^{-U/kT}). \tag{5·33·6}$$

It should be noted that we may write (5·21·1) in the form

$$\log W = -\Sigma(n_r \log n_r - n_r), \tag{5·33·7}$$

where the summation is taken over all the energy states. This is often taken to be equivalent to writing

$$W = \prod_r 1/n_r!, \tag{5·33·8}$$

but the use of unsymmetrized wave functions is only permissible if the n_r's are small compared with unity, and (5·33·8) is then illusory. It is, of course, possible to write W in the form (5·33·8) provided that it is interpreted as being a convenient notation for (5·33·7), but it is better to avoid such a formal expression for W since it is liable to lead to confusion. In particular, the use of the factorial notation would suggest that the n_r's are integers and therefore that S is negative, which is impossible.

5·34. *The parameter* λ. It will be noted that (5·33·3) determines the dependence of S upon N, the number of elements present, as well as upon T and the geometrical coordinates. However, in obtaining (5·33·3) from (5·33·2) we have tacitly assumed that the arbitrary additive function of N is zero. This is in fact true, but to prove it we must consider a system which includes some mechanism which allows the number of elements to vary. The simplest way to do this is to consider the distribution of a fixed number of elements between two enclosures in which the energy levels are different, the enclosures being separated from each other by a rigid 'wall' which is permeable to the elements.‡ We are therefore led to a reconsideration of the problem of § 5·3, with the proviso that here the two types of element are the same, and so there is only one relation, namely,

$$\Sigma(N_r^{(1)} + N_r^{(2)}) = N^{(1)} + N^{(2)} = N, \tag{5·34·1}$$

for the conservation of the number of elements, instead of the two independent conditions (5·3·2) for $N^{(1)}$ and $N^{(2)}$ separately. The number of complexions is still given by (5·3·1) and the condition for the constancy of

‡ When the elements are confined to one of the two enclosures we have two separate sets of wave functions, say $\psi_r^{(1)}(x)$ and $\psi_s^{(2)}(x)$, and two separate sets of energy levels $E_r^{(1)}$ and $E_s^{(2)}$. When the two systems are loosely coupled together we have only one set of wave functions and one set of energy levels. However, the wave functions will be of two types, those which have large amplitudes in the first enclosure and very small amplitudes in the second, and those whose amplitudes are large in the second enclosure and very small in the first. The series of energy levels of the elements in the combined system may therefore be considered to be the double series $E_r^{(1)} + E_s^{(2)}$ $(r, s = 1, 2, \ldots)$ formed from the energy levels of the elements in the two separate systems.

the energy is still given by (5·3·3), and to find the maximum value of $\log W$ subject to (5·3·3) and (5·34·1) we have to maximize

$$\log W + \lambda\Sigma(N_r^{(1)} + N_s^{(2)}) - \vartheta\Sigma(N_r^{(1)}E_r^{(1)} + N_s^{(2)}E_s^{(2)}),$$

there being only one parameter λ in addition to ϑ. We therefore obtain all the equations (5·3·5), (5·3·6) and (5·3·8) with λ_1, λ_2 replaced by λ, while the two equations (5·3·7) are replaced by the single equation

$$N = N^{(1)} + N^{(2)} = e^{\lambda}(Z_1 + Z_2). \tag{5·34·2}$$

Hence, for the combined system, we have, corresponding to (5·32·4),

$$dU + \Sigma\bar{A}_i da_i = -e^{\lambda} d\left(\frac{\partial Z_1}{\partial\vartheta} + \frac{\partial Z_2}{\partial\vartheta}\right) + \frac{e^{\lambda}}{\vartheta}\Sigma\frac{\partial}{\partial a_i}(Z_1 + Z_2)\,da_i,$$

and

$$\begin{aligned}\vartheta(dU + \Sigma\bar{A}_i da_i) &= d(\vartheta U) - U\,d\vartheta + \vartheta\Sigma\bar{A}_i da_i \\ &= d\{\vartheta(U_1 + U_2) - (N^{(1)} + N^{(2)})\lambda + N^{(1)} + N^{(2)}\} + \lambda(dN^{(1)} + dN^{(2)}),\end{aligned} \tag{5·34·3}$$

that is, $$k\vartheta(dU + \Sigma\bar{A}_i da_i) = dS_1 + dS_2 + k\lambda(dN^{(1)} + dN^{(2)}), \tag{5·34·4}$$

where $$S_1 = k\vartheta U_1 - kN^{(1)}\lambda + kN^{(1)}, \quad S_2 = k\vartheta U_2 - kN^{(2)}\lambda + kN^{(2)}. \tag{5·34·5}$$

If we consider the whole system, the last term in (5·34·4) is zero, since N is constant, and we see that ϑ is an integrating factor of the linear differential form. (Note that there is no logical inconsistency in allowing the energy U of the composite system to vary while treating the total number N of elements as constant, since we can consider the system to be loosely coupled to a third system with which it can exchange energy.) We can, however, write (5·34·4) in the form

$$k\vartheta(dU + \Sigma\bar{A}_i da_i) = dS_1 + dS_2 + k\lambda\,d\phi(N) + k\lambda\{dN^{(1)} + dN^{(2)} - d\phi(N)\},$$

where $\phi(N)$ is an arbitrary function of N. But, if we impose the requirement that S_1 shall be independent of N_2 and S_2 of N_1, then $\phi(N)$ can only be a linear function, and the introduction of an additive linear function of N is trivial since it can be eliminated by a redefinition of λ. We therefore see that the entropies of the separate systems are correctly given by (5·34·5). In addition, we can consider variations of U_1, for example, in which the number of elements does not remain constant, and we then have, dropping the index 1, the equation

$$dU = T\,dS - \Sigma\bar{A}_i da_i + kT\lambda\,dN, \tag{5·34·6}$$

which, on comparison with the thermodynamic relation (3·4·10), namely,

$$dU = T\,dS - \Sigma\bar{A}_i da_i + \mu\,dN,$$

where μ is the thermodynamic potential, shows that

$$\mu = kT\lambda. \tag{5·34·7}$$

5·341. *The fundamental equation associated with* $\Sigma\bar{A}_i a_i$. Having identified $kT\lambda$ with μ, we can now set up the fundamental equation for

$$\Sigma\bar{A}_i a_i \equiv N\mu - F.$$

We have, from (5·33·4) and (5·34·7),

$$\Sigma\bar{A}_i a_i = N\mu + NkT\log(Z/N) + NkT, \tag{5·341·1}$$

and this must be expressed as a function of T, μ and the a_i's. We have therefore to eliminate N from this expression by using the relation $N = Z\,e^{\mu/kT}$. The result is that

$$\Sigma\bar{A}_i a_i = kT\,e^{\mu/kT}\,Z. \tag{5·341·2}$$

5·35. *The partition function.* The results of the preceding sections show that all the thermodynamic formulae can be expressed in terms of the number of elements N and the partition function Z, where

$$Z = \sum_r \exp(-E_r/kT). \tag{5·35·1}$$

The energy levels of most elements form a multiple infinity, and it is often possible to separate these into simpler independent sets. For example, the molecules of a gas may have internal as well as translational motions, and these will be independent of one another. If, therefore, the energy levels form a double infinity and if they can be written as

$$E_{ij} = E_i^{(1)} + E_j^{(2)}, \tag{5·35·2}$$

then
$$Z = \sum_{i,j} \exp(-E_{ij}/kT) = \sum_i \exp(-E_i^{(1)}/kT) \sum_j \exp(-E_j^{(2)}/kT)$$
$$= Z_1 Z_2, \tag{5·35·3}$$

and the partition function can be written as a product of two separate partition functions. The generalization to a separable multiple infinity of energy levels is obvious.

5·351. The free energy F of N weakly coupled elements can, by (5·33·4), be written in the form

$$F = -kT\log(Z^N/N!), \tag{5·351·1}$$

where Z is the partition function for one element. Let us suppose that these elements are in translational motion in an enclosure and that each element has a number of internal degrees of freedom. Then we may write

$$Z = Z_{\text{trans.}} Z_{\text{int.}}, \tag{5·351·2}$$

and the free energy of the system is

$$F = -kT\log(Z_{\text{trans.}}^N/N!) - NkT\log Z_{\text{int.}}. \tag{5·351·3}$$

But the free energy for the internal motion of one single element cannot depend upon the number of elements present, and it is

$$F_{\text{element}} = -kT \log Z_{\text{int.}}. \tag{5·351·4}$$

Now there is no restriction upon how complex the element may be, and it may even consist of a large number of other elements, which may be almost independent elements or which may be strongly coupled together. Therefore, if we consider the 'element' to be a 'system', we see that (5·351·4) can be written as

$$F_{\text{system}} = -kT \log Z_{\text{system}}, \tag{5·351·5}$$

where

$$Z_{\text{system}} = \sum_i \exp(-U_i^{\text{system}}/kT). \tag{5·351·6}$$

Here the U_i's are the possible values of the energy of the system, and when the free energy is expressed in this form there is no need to restrict the system to consist of loosely coupled elements. It is, however, necessary to ensure that each energy level of the system is counted just as often as it gives rise to a distinguishable stationary state of the system. It is therefore often convenient to rewrite (5·351·6) in the equivalent form

$$Z_{\text{system}} = \sum_i W(U_i) \exp(-U_i^{\text{system}}/kT), \tag{5·351·7}$$

where $W(U_i)$ is the number of complexions of the system when its energy is U_i. If we insert (5·351·7) into (5·351·5) we obtain the analogue of equation (5·33·6). Equation (5·351·5) is one of the most important equations in statistical mechanics.

If the system consists of N identical unlocalized loosely coupled elements, (5·351·6) is the same as (5·351·1). For the elements are independent, and therefore the partition function for the system is the product of the partition functions for each element. But the partition function of each element must be taken as referring to the whole of the space occupied by the system since the elements are unlocalized, and there is nothing to distinguish the Z for one element from the Z for any other element. Hence in order to avoid the multiple counting of complexions we must divide by the number of permutations of the N elements amongst themselves. We therefore have

$$Z_{\text{system}} = (Z_{\text{element}})^N/N!, \tag{5·351·8}$$

and (5·351·1) and (5·351·5) are equivalent.

5·36. We have so far assumed that we can replace sums over the various microscopic states by the maximum term in such sums with a negligible error. We shall investigate this point in detail in § 5·7 by calculating not only the average values but also the mean-square deviations of the physical quantities associated with a system, but an elementary discussion can be given at this point.

If $W(N_1, N_2, \ldots)$ is given by (5·21·1) and if we write

$$N_r = \overline{N}_r + \delta N_r, \tag{5·36·1}$$

then $\log W(N_1, N_2, \ldots) = \Sigma(N_r \log M_r - N_r \log N_r + N_r)$

$$= \log W_{\max.} - \Sigma(\delta N_r)^2/\overline{N}_r + O\{(\delta N_r)^3/\overline{N}_r^2\}. \tag{5·36·2}$$

This shows that $W(N_1, N_2, \ldots)$ has a true maximum at $N_r = \overline{N}_r$, but that $W(N_1, N_2, \ldots)$ is of the same order as $W_{\max.}$ if δN_r is of the order of $\overline{N}_r^{\frac{1}{2}}$ for all r. Now the number of such terms is of the order of $N^{\frac{1}{2}}$, whereas $W_{\max.}$ is of the order of e^N (cf. equations (5·33·3) and (5·33·5)). But in the limit of large N there is no significant difference between e^N and $N^{\frac{1}{2}} e^N$, so that we are justified in taking only the maximum term into account.

5·37. *The increasing property of entropy.* It was pointed out in § 2·5 that the second law of thermodynamics contains two independent principles, namely, the existence of the entropy function and the fact that the entropy increases in irreversible processes. We have now to deduce the second of these principles from our postulates.

Suppose that we have two separate systems with entropies S_1 and S_2. Then if they are coupled together, the number of complexions W_{12} of the composite assembly must be at least equal to the product $W_1 W_2$ of the number of complexions of the two separate assemblies. Hence

$$S_{12} = k \log (W_{12})_{\max.} \geqslant k \log (W_1 W_2)_{\max.}$$
$$\geqslant k \log (W_1)_{\max.} + k \log (W_2)_{\max.} = S_1 + S_2,$$

which proves the principle.

5·38. *Other thermodynamic functions.* Having obtained the free energy F, we may derive expressions for any of the other thermodynamic functions by using the transformations given in Chapter 3. For example, if, apart from the mass, there is only one extensive variable, which we take to be the volume V, the free energy G at constant pressure is given by

$$G(T, P, N) = F(T, V, N) + PV, \quad P = -\partial F/\partial V. \tag{5·38·1}$$

If $Z(T, V, N)$ is the partition function for a system containing N elements, whether quasi-independent or not, we have, by (5·351·5),

$$F = -kT \log Z(T, V, N) \tag{5·38·2}$$

and

$$G(T, P, N) = -kT \log \{Z(T, V, N)\, e^{-PV/kT}\}, \quad P = kT\, \partial \log Z(T, V, N)/\partial V, \tag{5·38·3}$$

where V is expressed in terms of T, P and N by the second of the equations (5·38·3). The elimination of V from the above equations is often troublesome, but, on the other hand, the explicit form of $G(T, P, N)$ is sometimes simpler than that of $F(T, V, N)$. Under certain conditions it is convenient to replace (5·38·3) by

$$G(T,P,N) = -kT\log\left[\int_0^\infty Z(T,V,N)\,e^{-PV/kT}\frac{P\,dV}{kT}\right]. \qquad (5·38·4)$$

When this formula is valid, $e^{-G(T,P,N)/kT}$ and $Z(T, V, N)$ are Laplace transforms of one another, and, if one is known, the other can be obtained by standard methods.

The conditions necessary for the equivalence of (5·38·3) and (5·38·4) have not been established, though sufficient conditions can readily be given. If we write the integral in (5·38·4) as

$$\int_0^\infty e^{\Phi(V)-PV/kT}\frac{P\,dV}{kT}, \qquad (5·38·5)$$

where $\Phi(V) = \log Z(T, V, N)$, and if $\Phi(V) - PV/kT$ has a sufficiently sharp single maximum, we can evaluate the integral approximately as follows. If the maximum occurs at $V = \bar{V}$ we write

$$\Phi(V) - PV/kT = \Phi(\bar{V}) - P\bar{V}/kT + \tfrac{1}{2}(V-\bar{V})^2\,\Phi''(\bar{V}) - \ldots, \qquad (5·38·6)$$

where
$$P/kT = \Phi'(\bar{V}) = \partial\log Z(T,\bar{V},N)/\partial\bar{V}, \qquad (5·38·7)$$

and where $\Phi''(\bar{V}) < 0$. We insert (5·38·6) into (5·38·5) and, if the maximum is sufficiently sharp, we can write

$$\int_0^\infty \exp\left[\Phi(V) - PV/kT\right]\frac{P\,dV}{kT} = \exp\left[\Phi(\bar{V}) - P\bar{V}/kT\right]$$
$$\times\int_{-\infty}^\infty \exp\left[-\tfrac{1}{2}(V-\bar{V})^2\,|\,\Phi''(\bar{V})\,|\right]\frac{P\,d(V-\bar{V})}{kT} \qquad (5·38·8)$$

with a negligible error. With these approximations (5·38·4) becomes

$$G(T,P,N) = -kT\log\{Z(T,\bar{V},N)\,e^{-P\bar{V}/kT}\} - kT\log\left[\left(\frac{2\pi}{|\,\Phi''(\bar{V})\,|}\right)^{\frac{1}{2}}\frac{P}{kT}\right]. \qquad (5·38·9)$$

Now since $\bar{V}$ is given by (5·38·7), the above expression for $G(T, P, N)$ is effectively the same as (5·38·3) if the second term in (5·38·9) is of smaller order than the first for large values of N.

In general we should expect $Z(T, V, N)$ to be of order V^N, and the first term in (5·38·9) is then of order $N\log V$, while the second term is of order $\log(V/N^{\frac{1}{2}})$, which is negligible compared with the first term. In this case, therefore, (5·38·3) and (5·38·4) are equivalent.

It is shown in § 5·4 that for a perfect gas we have

$$Z(T,V,N) = (BV)^N/N! \qquad (5·38·10)$$

exactly, and the integral (5·38·4) can be evaluated at once to give

$$G(T,P,N) = -NkT\log(BkT/P). \tag{5·38·11}$$

This is the same expression as is obtained from (5·38·3) by substituting $V = NkT/P$ and using Stirling's theorem in the form $\log N! \sim N\log N - N$.

When there are more extensive variables than one, we can use the Laplace transformation to pass from any number of extensive variables to the corresponding intensive variables. It is, however, necessary to verify in each particular case that the partition function is of such a form that the transformation is justified. In practice this method of obtaining the thermodynamic functions involving the intensive variables is seldom used.

5·381. *The thermodynamic function* $\Sigma A_i a_i$. If we consider a general system characterized by s deformation coordinates a_i and if we write

$$F = -kT\log Z(T,a_i,N), \quad \mu = \partial F/\partial N = -kT\,\partial\log Z(T,a_i,N)/\partial N, \tag{5·381·1}$$

we have
$$\Sigma A_i a_i \equiv N\mu - F = kT\log[Z(T,a_i,N)\,e^{N\mu/kT}], \tag{5·381·2}$$

where N is to be eliminated from this equation by means of the second of the equations (5·381·1). In general, this expression for $\Sigma A_i a_i$ as a function of T, a_i and μ has no particular virtue over the corresponding expression for $F(T,a_i,N)$ on account of the difficulty of eliminating N in all but simple cases. (The situation is different for perfect gases obeying quantal statistics. See §§ 6·6 and 6·4.) In certain circumstances, however, (5·381·6) can be replaced by

$$\Sigma A_i a_i = kT\log[\sum_N Z(T,a_i,N)\,e^{N\mu/kT}]. \tag{5·381·3}$$

This is true when $Z(T,a_i,N)$ depends upon N in such a manner that the summation in (5·381·3) with respect to N can be replaced by its maximum term (or more strictly by its $\sqrt{N}$ terms which are comparable with the maximum term). The maximum term, when it exists, occurs at $N = \bar{N}$, where $\bar{N}$ is the root of the equation

$$\mu = -kT\,\partial\log Z(T,a_i,\bar{N})/\partial\bar{N}, \tag{5·381·4}$$

which is the same as the second of the equations (5·381·1). The quantity

$$\Xi(T,a_i,\mu) = \sum_N Z(T,a_i,N)\,e^{N\mu/kT} \tag{5·381·5}$$

is often called the grand partition function, and (5·381·3) is written as

$$\Sigma A_i a_i = kT\log\Xi(T,a_i,\mu). \tag{5·381·6}$$

If $Z(T,a_i,N)$ is of the form

$$Z(T,a_i,N) = Z(T,a_i)^N_{\text{element}}/N!, \tag{5·381·7}$$

as occurs, for example, for a perfect gas, the relations (5·381·2) and (5·381·3) are identical if we use Stirling's theorem in the form $\log N! \sim N \log N - N$. From (5·381·1) and (5·381·2) we have

$$\Sigma A_i a_i = N\mu + NkT \log (Z_{\text{element}}/N) + NkT \tag{5·381·8}$$

and

$$\mu = -kT \log (Z_{\text{element}}/N), \tag{5·381·9}$$

so that

$$\Sigma A_i a_i = NkT = kT\, e^{\mu/kT} Z_{\text{element}}. \tag{5·381·10}$$

If, on the other hand, we start from (5·381·3), we obtain (5·381·10) at once, and we see that the grand partition function is given by

$$\log \Xi(T, a_i, \mu) = e^{\mu/kT} Z_{\text{element}}. \tag{5·381·11}$$

5·382. It is often assumed that we may define $\Sigma A_i a_i$ indiscriminately either by (5·381·2) or by (5·381·3), but this is not so. We can always define $\Sigma A_i a_i$ by (5·381·2), but in several important cases the series

$$\Sigma Z(T, a_i, N)\, e^{N\mu/kT}$$

has a singularity, and the use of the grand partition function then requires careful investigation.

The grand partition function (5·381·5) is a particular case of the power series $f(z) = \Sigma a_n z^n$, where z is a complex variable and where the a_n's are all positive. Such a power series either represents an integral function or has a singularity on the positive real axis at a finite distance from the origin (see, for example, Titchmarsh, 1939, p. 214). If $f(z)$ is an integral function, then, for sufficiently large values of $|z|$, we have $|a_0| < |a_1 z| < |a_2 z^2| < \ldots$. But, since $|a_n z^n| \to 0$ as $n \to \infty$ for all z, the series $\Sigma |a_n z^n|$ must have a maximum term $m(|z|)$ for any fixed value of $|z|$. Now it is a well-known property of integral functions of finite order that $m(|z|)$ and $M(z)$, the maximum modulus of $f(z)$, are effectively the same for large values of $|z|$, i.e. that

$$\log m(|z|) \sim \log M(z) \tag{5·382·1}$$

(see, for example, Valiron, 1923, p. 32). We may therefore use $m(|z|)$ or $M(z)$ indiscriminately.

If, on the other hand, $f(z)$ has a finite radius of convergence, the singularity on the positive real axis may be either a pole or a branch-point, or even conceivably an isolated essential singularity. In this case it does not necessarily follow that $M(z)$ can be replaced by a maximum term or that there is a maximum term at all (apart from $|a_0|$). In general, all that we can deduce is Cauchy's inequality $|a_n z^n| < M(z)$ for all n, and this by itself is insufficient to tell us anything about the relation of the maximum term to the maximum modulus.

The conclusion to be drawn from the preceding discussion is that the grand partition function and the ordinary partition function will lead to the same results when the former is an integral function of μ/kT. If the grand partition function has a singularity for a finite value of μ/kT a special investigation is required. An obvious example occurs when

$$Z(T, a_i, N) = \{Z(T, a_i)_{\text{element}}\}^N,$$

which gives

$$\Xi(T, a_i, \mu) = \Sigma\, e^{N\mu/kT}\, Z(T, a_i, N) = \frac{1}{1 - e^{\mu/kT}\, Z(T, a_i)_{\text{element}}}. \quad (5\cdot382\cdot2)$$

Since the series only converges if $|\, e^{\mu/kT}\, Z_{\text{element}}\,| < 1$, there is no maximum term. In addition, the relation

$$\overline{N} = \frac{\partial}{\partial\mu} kT \log \Xi(T, a_i, \mu)$$

gives

$$\overline{N} = \frac{e^{\mu/kT}\, Z_{\text{element}}}{1 - e^{\mu/kT}\, Z_{\text{element}}}, \quad (5\cdot382\cdot3)$$

which differs from the correct relation $\overline{N} = e^{\mu/kT}\, Z_{\text{element}}$. The grand partition function cannot therefore be used for a problem of this type (see also §§ 5·51 and 6·2).

5·39. *Other methods of deriving the fundamental equations of statistical mechanics.* In the treatment adopted in this book, the basic assumption is that the value of any macroscopic property of a system of elements is given by averaging that property over every quantum state of the system, each non-degenerate quantum state being given the same *a priori* probability. In order to avoid introducing any further assumptions, it is necessary to consider two loosely coupled systems whose combined energy is given. These assumptions are sufficient to enable us to determine the average energy of the individual elements. The calculation involves the use of Stirling's theorem, but this can be avoided by a method due to Darwin and Fowler, based upon the multinomial theorem and Cauchy's theorem. This latter method is founded essentially upon the same physical assumptions as is the preceding exposition, but the mathematical details differ considerably. A complete account of the Darwin-Fowler method is given in Fowler's book (1936).

A substantially different standpoint was adopted by Gibbs (1902). He considered from the outset a single assembly of a large number of systems which are such that the number of systems with energies lying between E and $E + dE$ is proportional to $e^{-E/\Theta}$, where Θ is a parameter characteristic of the assembly. Such an assembly is called a canonical assembly, and Gibbs assumed without proof that the properties of macroscopic bodies can be derived from the properties of the appropriate canonical assemblies. Once this assumption is made, the subsequent treatment is mathematically elegant, and it can be shown that there are close analogies between a typical system of an assembly and a thermodynamic body, by which it is possible to identify the parameter Θ with the empirical temperature and to give physical meanings to the various quantities that occur in the theory. To many physicists, however, Gibbs's exposition is aesthetically unsatisfactory in that it lays too much stress upon mathematical form and too little upon physical principles. Those who take this view prefer to take the starting point of the theory some way back in the line of argument, and, either explicitly or implicitly, to derive the canonical assembly as a useful description of a macroscopic thermodynamic body rather than to postulate it. This is the view taken in the present book, but once equation (5·351·5) for the free energy of a general system has been derived, the exposition could be framed entirely in terms of canonical assemblies.

A similar but not identical line of argument was pursued by Gibbs as subsidiary to his main investigation of canonical assemblies. He discussed the properties of a microcanonical assembly, which consists of a large number of systems each having the same energy, and he showed that, in certain circumstances, portions of a microcanonical assembly can be considered to be canonical assemblies. This method of introducing the assumptions explicitly through the consideration of a microcanonical assembly is in my view less easy to follow than the method adopted here of dealing with two loosely coupled systems whose combined energy is fixed and each of which contains a large number of elements. But the difference between the methods is mainly one of emphasis and arrangement of the argument, and the choice between them is largely a matter of taste.

In addition to the microcanonical and canonical assemblies, Gibbs considered what he called a grand canonical assembly which consists of systems whose composition is not identical but which may be made up of 1, 2, ... elements. If the energy of a particular system is E and if it contains N elements, the weight to be ascribed to the system is $\exp[-(E-N\mu)/\Theta]$, and the macroscopic value of any physical quantity is to be found by multiplying the quantity by the above weight factor and integrating over all values of E and summing over all values of N. The normal justification given for using the grand canonical assembly is that, just as the canonical assembly leads to a description of a general thermodynamic body whose temperature is given and whose energy is subject to fluctuations, so the grand canonical assembly might be expected to correspond to a thermodynamic body whose thermodynamic potential is given but in which the number of elements is not fixed but is subject to fluctuations. If, however, we proceed in the manner prescribed, we eventually derive equation (5·381·3) for the thermodynamic function $\Sigma A_i a_i$, and, as we have seen in

§ 5·381, the series $\Sigma Z(T, a_i, N) \exp(N\mu/kT)$ is not necessarily one which is dominated by a single term, or by a small number of terms centred round a particular value $\bar{N}$ of N. The grand canonical assembly will therefore give us the average value of a physical quantity over a set of systems each containing N elements, where N can vary from one to infinity, the average being taken in a particular way; but in certain cases this will not correspond to the value of the physical quantity for a thermodynamic body containing a definite large number $\bar{N}$ of elements. This limitation makes the use of the grand canonical assembly, or equivalently of the grand partition function, much less attractive than it would otherwise be, and I shall not have occasion to use it in any of the subsequent applications of the theory. If the reader prefers to use the grand partition function, the summation with respect to N can always be very easily carried out in those cases in which the method is at all useful, and provided that the resulting grand partition function is an integral function of μ/kT, no difficulties will arise. In other cases, a special investigation will be required.

The advantages of using the thermodynamic function $\Sigma A_i a_i$ and of using the grand partition function should be clearly distinguished. The primary definition of $\Sigma A_i a_i$ is that it is given by $N\mu - F(T, a_i, N)$ (where $\mu = \partial F/\partial N$), expressed as a function of T, a_i and μ. This thermodynamic function is often very convenient, especially when the partition function $Z(T, a_i, N)$ is such that it cannot be written down as a simple explicit function of N. This occurs, for example, in the quantum statistics of a perfect gas (see §§ 6·3 and 6·4). The grand partition function, which is defined by (5·381·5), may or may not be a simple way of deriving the expression for $\Sigma A_i a_i$, and it may or may not be valid.

It is possible to define other canonical assemblies. If we consider the general case in which there are s deformation coordinates a_i and r species of components, we may introduce weight factors

$$\exp[(-E + \Sigma A_i a_i + \Sigma N_j \mu_j)/kT], \tag{5·39·1}$$

where the summations are taken over any arbitrary numbers of the deformation coordinates and of the components. By integrating with respect to those a_i's and summing over those N_j's which occur explicitly in (5·39·1), and integrating with respect to E, we can derive directly an expression for the thermodynamic function appropriate to any desired set of variables. It is easy to give sufficient conditions for this procedure to be valid, but I am not aware of any discussion of the necessary conditions. In general, the direct calculation of any particular thermodynamic function using the weight factor (5·39·1) offers little advantage over the indirect method of first deriving $F(T, a_i, N)$ or $\Sigma A_i a_i$ and subsequently changing the variables as detailed in Chapter 3. Special cases can always be dealt with on their own merits.

Even when these various canonical assemblies give the same values of the macroscopic thermodynamic quantities, the results only agree to order N^{-1}. The mean-square (and higher) deviations of the physical quantities from their average values will therefore not in general be the same for the different kinds of assemblies. The general investigation of fluctuation phenomena is extremely complex, and no full discussion of all the possible cases has ever been given. A treatment of some of the simpler cases is given in § 5·7, and it will be seen from the formulae given explicitly there how the fluctuations in the properties of a system can be influenced by its environment. In dealing with fluctuations it is therefore necessary to specify rigorously the exact manner in which a system is coupled to its surroundings. Much confusion has been caused by the comparison of the fluctuations in certain assemblies calculated in different ways. When these calculations implicitly necessitate different environments, it is not surprising that they lead to different results.

The Thermodynamics of a Perfect Gas

5·4. We consider the motion of a number of structureless particles moving under no forces in an enclosure defined by

$$0 \leqslant x \leqslant L_1, \quad 0 \leqslant y \leqslant L_2, \quad 0 \leqslant z \leqslant L_3,$$

where L_1, L_2, L_3 are large. The motion is entirely translatory, and the Schrödinger equation for one of the particles is $\nabla^2\psi + 8\pi^2 mE/h^2 = 0$, where m is the mass and E the energy of a particle and h is Planck's constant. The simplest boundary condition to impose upon the wave functions is that they should vanish on the boundaries of the enclosure. It is, however, slightly more convenient to impose the periodicity conditions

$$\psi(0,y,z) = \psi(L_1,y,z), \quad \psi(x,0,z) = \psi(x,L_2,z), \quad \psi(x,y,0) = \psi(x,y,L_3)$$

on the wave functions since they can then be taken to be progressive waves instead of standing waves. Both boundary conditions necessarily lead to the same result when all the dimensions of the enclosure tend to infinity.

The wave functions are of the form

$$\psi = \exp\left[2\pi i(p_1 x + p_2 y + p_3 z)/h\right], \tag{5·4·1}$$

where

$$p_1 = hn_1/L_1, \quad p_2 = hn_2/L_2, \quad p_3 = hn_3/L_3, \tag{5·4·2}$$

n_1, n_2 and n_3 being positive or negative integers when the 'periodic' boundary conditions are used. Also, the energy levels are given by

$$E = \tfrac{1}{2}(p_1^2 + p_2^2 + p_3^2)/m. \tag{5·4·3}$$

We shall suppose that the number of particles present is so small that we can apply classical statistics, and we have therefore to calculate the partition function

$$Z(\vartheta)=\varpi \sum_{n_1,n_2,n_3=-\infty}^{\infty} \exp[-\tfrac{1}{2}\vartheta(p_1^2+p_2^2+p_3^2)/m] \quad (\vartheta>0). \tag{5·4·4}$$

A weight function ϖ has been inserted in the partition function to cover cases in which the translational energy levels are degenerate. This cannot occur for structureless particles, but a weight factor is necessary for atoms, such as alkali atoms, which have an unpaired electron. Since L_1, L_2 and L_3 are large, the energy levels can be made to lie as close together as we please by making the L's large enough. (If we wish to consider the effect of a small enclosure, the series (5·4·4) can be summed exactly, either by using transformations common in the theory of theta functions or by applying the Euler-Maclaurin sum formula. See §6·11.) We can therefore replace the sum over the integers n_1, n_2, n_3 by an integral over n_1, n_2, n_3 taken as continuous variables, but it is more convenient to change to the p's. We then have

$$Z(\vartheta)=\frac{\varpi V}{h^3}\int_{-\infty}^{\infty}\int_{-\infty}^{\infty}\int_{-\infty}^{\infty} \exp[-\tfrac{1}{2}\vartheta(p_1^2+p_2^2+p_3^2)/m]\,dp_1dp_2dp_3, \tag{5·4·5}$$

where $V=L_1L_2L_3$ is the volume of the enclosure. By using polar coordinates in the p space, we find that

$$Z(\vartheta)=\frac{4\pi\varpi V}{h^3}\int_0^{\infty} \exp(-\tfrac{1}{2}\vartheta p^2/m)\,p^2\,dp,$$

where $p^2=p_1^2+p_2^2+p_3^2$, which gives

$$Z(\vartheta)=BV\vartheta^{-\frac{3}{2}}, \tag{5·4·6}$$

with

$$B=\varpi(2\pi m)^{\frac{3}{2}}/h^3. \tag{5·4·7}$$

The pressure P is the generalized force corresponding to the generalized coordinate V, and so

$$P=\frac{N}{\vartheta}\frac{\partial \log Z}{\partial V}=\frac{N}{\vartheta V}=\frac{NkT}{V}, \tag{5·4·8}$$

which is the familiar equation of state. The constant k, Boltzmann's constant, is equal to $R/\mathscr{L}$, where R is the molar gas constant, and $\mathscr{L}$ is Avogadro's number, the number of molecules in a gram molecule, namely, $6{\cdot}0254\times10^{23}$ molecules/gram molecule. Hence $k=1{\cdot}3083\times10^{-16}$ erg/degree.

The internal energy U is given by

$$U=-N\,\partial\log Z/\partial\vartheta=\tfrac{3}{2}NkT, \tag{5·4·9}$$

so that the specific heat per gram molecule (at constant volume) is

$$C_V=\tfrac{3}{2}\mathscr{L}k=\tfrac{3}{2}R. \tag{5·4·10}$$

Also, the entropy is

$$S = U/T + k(N\log Z - N\log N + N) = \tfrac{5}{2}Nk + Nk\log(Z/N).$$

To put this in a more familiar form, we write

$$Z/N = B(kT)^{\frac{3}{2}}\,V/N = B(kT)^{\frac{5}{2}}/P,$$

and we then have

$$S = \tfrac{5}{2}Nk + Nk\log\left(\frac{V}{N}(kT)^{\frac{3}{2}}\frac{\varpi(2\pi m)^{\frac{3}{2}}}{h^3}\right) = \tfrac{5}{2}Nk + Nk\log\left(\frac{(kT)^{\frac{5}{2}}}{P}\,\frac{\varpi(2\pi m)^{\frac{3}{2}}}{h^3}\right). \tag{5·4·11}$$

The corresponding expressions for the free energies F and G and the thermodynamic potential μ are

$$F = -NkT - NkT\log\left(\frac{V}{N}(kT)^{\frac{3}{2}}\frac{\varpi(2\pi m)^{\frac{3}{2}}}{h^3}\right), \tag{5·4·12}$$

$$G = NkT\log\left(\frac{P}{(kT)^{\frac{5}{2}}}\,\frac{h^3}{\varpi(2\pi m)^{\frac{3}{2}}}\right) \tag{5·4·13}$$

and

$$\mu = kT\log\left(\frac{P}{(kT)^{\frac{5}{2}}}\,\frac{h^3}{\varpi(2\pi m)^{\frac{3}{2}}}\right). \tag{5·4·14}$$

These are the same as the formulae (3·51·12), (3·51·15) and (3·51·17) with $C_V = \tfrac{3}{2}R$, except that here an explicit expression has been obtained for the 'entropy constant'.

It will be shown in §§ 6·311 and 6·41 that the formulae (5·4·8)–(5·4·14) are the limiting forms of the corresponding quantal formulae for temperatures which are sufficiently high for the parameter λ to be large and negative. It will further be shown that, for fixed N and V, the entropy tends to zero as $T\to 0$. The difficulties associated with finding a convenient standard state from which to measure the entropy therefore arise from the use of classical statistics and disappear when quantal statistics are used.

The condition for the applicability of classical statistics is

$$P/(kT)^{\frac{5}{2}} \ll \varpi(2\pi m)^{\frac{3}{2}}/h^3, \tag{5·4·15}$$

and, if this is satisfied, S is positive, as it must be. In practice, this condition is always satisfied except for the electrons in a metal (see § 6·32), since condensation always occurs long before the condition breaks down. For example, for helium the right-hand side of (5·4·15) has the value 3×10^{33} c.g.s. units, whereas the left-hand side has the value $1·5\times 10^{26}$ c.g.s. units for $T=1$ and $P = 0·2$ mm. Hg, which is the approximate value of the vapour pressure of helium at 1° K.

The Transition to Classical Mechanics

5·5. *Remarks upon the derivation of statistical mechanics from classical mechanics.* Before deriving in § 5·52 the form which the statistical equations take when we pass to the limit $h \to 0$, it is desirable to comment briefly upon the historical development of the subject. The basic equations of statistical mechanics were derived before the fundamental principles of quantum mechanics were known, and from time to time there have been acute controversies as to how far it is possible to develop the theory using only classical mechanics. An outline of the classical procedure, which dates from the time of Boltzmann and Maxwell, is as follows. We consider a general conservative dynamical system with r degrees of freedom, and express the equations of motion in Hamiltonian form, using generalized coordinates q_i and generalized momenta p_i. The $2r$-dimensional space defined by the possible values of the q_i's and the p_i's is called the phase space, and to set up a statistical theory associated with the dynamical system the assumption is introduced that any element of phase space has an *a priori* probability proportional to its volume. That this assumption is consistent with the laws of classical mechanics is a consequence of Liouville's theorem.

We next consider a system composed of a large number N of practically independent particles or elements, and we divide the corresponding $2r$-dimensional phase space up into a large number of small cells, each having the same volume ω. To each cell is to be ascribed the *a priori* probability (or weight) $A\omega$, where A is a constant. Then if we specify a microscopic configuration of the N elements by the number of elements N_i in the ith cell, for all values of i, the number of complexions is

$$W(N_1, N_2, \ldots) = \frac{(A\omega)^{N_1+N_2+\ldots}\, N!}{N_1!\, N_2! \ldots}. \tag{5·5·1}$$

This expression replaces the expression (5·3·1) upon which the exposition of § 5·3 was based. Provided that we can choose the size of the cells to be such that the numbers N_i are sufficiently large, the above expression for $W(N_1, N_2, \ldots)$ will lead to a classical partition function which is correct except in two important respects. In the first place, the constant A is undefined, which leads to an arbitrariness in the dependence of the entropy constant of the system upon N, and in the second place the $N!$ in the numerator of (5·5·1) should be omitted when the system consists of indistinguishable unlocalized particles. Many arguments have been advanced to justify the omission of the $N!$, but none of them have stood up to hostile criticism. It is, in fact, difficult to see how a satisfactory argument can be constructed using solely the principles of classical mechanics, since the indistinguishability of particles is an idea which is quite foreign to the classical theory. It is for this reason that the exposition adopted in this book is based upon quantal and not classical ideas, but once the principles have been enunciated the basic formulae can be translated into classical terms either at an early or at a late stage.

It will be shown in § 5·52 that the constant A must be chosen to be h^{-r} in order to reproduce the correct value of the entropy constant, or, in other words, that for each pair of conjugate coordinates and momenta the weight factor for the phase space should be $dpdq/h$. Such a weight factor could obviously not be deduced from the classical theory alone. But this circumstance is not so serious an objection as the preceding one to basing the exposition on the classical theory, since the value of the weight factor could always be introduced as an extra hypothesis at a late stage in the theory, whereas the arguments leading to the omission or retention of the $N!$ are fundamental and must be settled (either explicitly or implicitly) at the outset.

It has been stated several times in the preceding sections that the $N!$ factor must be retained when the particles are localized, even when they are indistinguishable from one another. Before deriving the classical formulae as limits of the quantal formulae, we shall discuss, as an illustration, one problem in which the $N!$ factor must be retained.

5·51. The problem we shall discuss is the adsorption of atoms on a solid surface from a perfect gas in contact with it. We shall assume that there are Q possible fixed sites on the surface at which the atoms can be adsorbed (an adsorbed atom being incapable of free moment along the surface of the solid, though it will necessarily have a number of different vibrational states), and that the sites are all equivalent and sufficiently far apart for there to be no interactions between any of the adsorbed atoms.

Let the total number of atoms be N, of which N_G are in the gas and N_A are adsorbed. Also let the possible energy levels of a gaseous atom be $E_{G,r}$ $(r=1,2,\ldots)$ and let the energy levels of an adsorbed atom be $E_{A,s}$ $(s=1,2,\ldots)$. As in § 5·1, we group the various possible energy levels together so that there are $M_{G,r}$ energy levels and $N_{G,r}$ atoms with energy $E_{G,r}$, and $M_{A,s}$ levels and $N_{A,s}$ atoms with energy $E_{A,s}$. The number of complexions of the gaseous atoms is given by the analogue of (5·21·1). To find the number of complexions of the adsorbed atoms we proceed as follows. We can choose N_A sites for the adsorbed atoms in $Q!/\{N_A!\,(Q-N_A)!\}$ ways, and we then have to find the number of ways of distributing the N_A atoms in their various stationary states over the N_A sites. Now the sites are localized, i.e. they are distinguishable, and, if $\psi_1^{(i)}(\mathbf{r}_k)$ denotes the wave function of the kth atom in its ith excited state when it is on site 1, we must treat $\psi_1^{(i)}(\mathbf{r}_1)\,\psi_2^{(j)}(\mathbf{r}_2)$ $(i \neq j)$ as a different wave function from $\psi_1^{(j)}(\mathbf{r}_1)\,\psi_2^{(i)}(\mathbf{r}_2)$. On the other hand, $\psi_1^{(i)}(\mathbf{r}_1)\,\psi_2^{(j)}(\mathbf{r}_2)$ is to be considered the same as $\psi_1^{(i)}(\mathbf{r}_2)\psi_2^{(j)}(\mathbf{r}_1)$, since, if properly symmetrized, they would both be combined into the wave function $\psi_1^{(i)}(\mathbf{r}_1)\,\psi_2^{(j)}(\mathbf{r}_2) \pm \psi_1^{(i)}(\mathbf{r}_2)\,\psi_2^{(j)}(\mathbf{r}_1)$, the plus or minus sign being taken according as the wave functions belong to the symmetrical or to the antisymmetrical class. In other words, the sites are distinguishable but the atoms are not. The number of distinct complexions of the atoms on a given set of sites is therefore

$$\prod_s \frac{N_A!\,M_{A,s}^{N_{A,s}}}{N_{A,s}!}, \tag{5·51·1}$$

and the total number of complexions of all the atoms in the gas and on the solid is

$$W(N_G, N_A) = \prod_r \frac{M_{G,r}^{N_{G,r}}}{N_{G,r}!}\,\frac{Q!}{(Q-N_A)!\,N_A!}\prod_s \frac{N_A!\,M_{A,s}^{N_{A,s}}}{N_{A,s}!}. \tag{5·51·2}$$

If we maximize $\log W(N_{G,1}, N_{G,2}, \ldots; N_{A,1}, N_{A,2}, \ldots)$, subject to the usual conditions

$$\sum_r N_{G,r} = N_G, \quad \sum_s N_{A,s} = N_A, \quad N_G + N_A = N \tag{5·51·3}$$

and

$$\Sigma(N_{G,r}E_{G,r} + N_{A,s}E_{A,s}) = U, \tag{5·51·4}$$

which express the constancy of the number of atoms and of the energy, we obtain the equations

$$\overline{N}_{G,r} = M_{G,r}\exp(\lambda - \vartheta E_{G,r}), \quad \overline{N}_{A,r} = M_{A,s}(Q-\overline{N}_A)\exp(\lambda - \vartheta E_{A,s}). \tag{5·51·5}$$

It then follows that

$$N_G = e^\lambda Z_G, \quad N_A = \frac{Q\,e^\lambda Z_A}{1+e^\lambda Z_A}, \quad N = N_G + N_A, \tag{5·51·6}$$

and

$$U_G = -e^\lambda \frac{\partial Z_G}{\partial \vartheta}, \quad U_A = -\frac{Q\,e^\lambda}{1+e^\lambda Z_A}\frac{\partial Z_A}{\partial \vartheta}, \quad U = U_G + U_A, \tag{5·51·7}$$

where

$$Z_G = \Sigma\exp(-\vartheta E_{G,r}), \quad Z_A = \Sigma\exp(-\vartheta E_{A,s}). \tag{5·51·8}$$

In order to deduce the expression for the entropy we have to follow the procedure of § 5·34. We leave the details of this calculation to the reader. The result is that $S = S_G + S_A$, where

$$S_G = k\vartheta U_G - k\lambda N_G + kN_G, \quad S_A = k\vartheta U_A - k\lambda N_A + kQ\log\{Q/(Q-N_A)\}, \tag{5·51·9}$$

or, equivalently, $F = F_G + F_A$, where

$$F_G = -kT\log\frac{Z_G^{N_G}}{N_G!}, \quad F_A = -kT\log\left[\frac{Q!}{(Q-N_A)!\,N_A!}Z_A^{N_A}\right]. \tag{5·51·10}$$

It will be seen from the expressions for F_G and F_A how the different expressions for the numbers of complexions affect the structure of the thermodynamic functions. Further, a cursory examination of the expression

(5·51·10) for F_A, or of that part of (5·51·2) which refers to the adsorbed atoms, might suggest that the case in which $N_A = Q$ would be one in which the strict classical theory would apply. The expression given in (5·51·6) for N_A shows, however, that the case $N_A = Q$ is a singular one, and that it can only be treated properly by first considering the case $N_A < Q$ and then passing to the limit in which N_A approaches arbitrarily near to Q. Any problem in which F is of the form $-kT \log Z^N$, where Z is independent of N, will in fact give rise to difficulties. When a free energy of this form occurs, it means that the problem is an idealized limit of a more complex one, and a more searching investigation is required. (For example, see the discussion of the thermal properties of the vibrations of a solid given in § 6·21.)

5·511. *Langmuir's adsorption equation.* The discussion of the preceding problem was primarily undertaken in order to emphasize the importance of the presence or absence of the factor $N!$ in the number of complexions. Quite apart from this, the problem has a considerable physical interest of its own, and, although it lies off the main theme of the present discussion, we shall complete the physical interpretation of the results.

If we write $N_A/Q = \alpha$, so that α is the fraction of the available sites on the surface of the solid which are occupied by adsorbed atoms, and if we eliminate λ between the expressions given in (5·51·6) for N_G and N_A, we obtain the equation

$$N_G = \frac{\alpha}{1-\alpha} \frac{Z_G}{Z_A}. \qquad (5·511·1)$$

Now if the gas is a monatomic one, Z_G is equal to $BV_G(kT)^{\frac{3}{2}}$, where V_G is the volume of the gas, while Z_A is a function of T only. Therefore, since we have $PV_G = N_G kT$, (5·511·1) can be written in either of the forms

$$P = \frac{\alpha}{1-\alpha} \phi(T), \quad \alpha = \frac{P}{1 + P\phi(T)}, \qquad (5·511·2)$$

where

$$\phi(T) = B(kT)^{\frac{5}{2}}/Z_A(T). \qquad (5·511·3)$$

If the gas is perfect but not monatomic, the formulae (5·511·3) still hold, provided that the molecules do not dissociate or aggregate on adsorption and provided that (5·511·3) is replaced by

$$\phi(T) = B(kT)^{\frac{5}{2}} Z_{\text{int.}}(T)/Z_A(T), \qquad (5·511·4)$$

where $Z_{\text{int.}}$ is the 'internal partition function' of a molecule in the gaseous state.

The equations (5·511·2) are known as Langmuir's adsorption equations. They agree in a qualitative way with the experimental data, but it is necessary to consider less specialized models in order to obtain formulae of general applicability.

5·52. *The classical partition function.* In classical mechanics the energy levels are not discrete but form a continuum, and the partition function is represented by an integral instead of by a series. To derive the classical partition function we have to investigate the quantum formulae in the limit $h \to 0$. For a structureless particle moving under no forces it is clear from (5·4·2) that the limit $h \to 0$ gives rise to the same results as $V \to \infty$, and, by (5·4·5), that the classical partition function is

$$Z(\vartheta) = \frac{1}{h^3} \int \dots \int e^{-\vartheta E} \, dp_1 dp_2 dp_3 dx dy dz, \tag{5·52·1}$$

where the integration is to be taken over all the values of the momenta and of the coordinates. Correspondingly, if we denote the space coordinates by q_1, q_2, q_3, the partition function for a system of N such indistinguishable particles is given by

$$Z_{\text{system}} = \frac{1}{N!\,h^{3N}} \int \dots \int \exp\left[-\vartheta \mathscr{H}(p,q)\right] dp_1^{(1)} dp_2^{(1)} dp_3^{(1)} \dots dq_1^{(N)} dq_2^{(N)} dq_3^{(N)}, \tag{5·52·2}$$

where $\mathscr{H}(p,q)$ is the sum of the kinetic energies of the particles and where the integration is taken over all values of the $3N$ momenta and the $3N$ coordinates. If the particles are distinguishable we must omit the factor $1/N!$ in (5·52·2). We can generalize (5·52·2) to apply to complex systems in which the elements do not move independently of one another by taking $\mathscr{H}(p,q)$ to be the Hamiltonian function of the system, and in this form of the partition function we can take the p's and the q's to be generalized momenta $p_1, \dots, p_{3N}$ and generalized coordinates $q_1, \dots, q_{3N}$ of any type.

If we use a different set of momenta P and coordinates Q which are functions of the p's and q's, but which do not involve the time explicitly, then it follows from the transformation theory of Hamiltonian systems that both the Hamiltonian function $\mathscr{H}(p,q)$ and the 'extension in phase space' $dp_1 \dots dq_{3N}$ are invariants,‡ and it is therefore consistent to assume that (5·52·2) is the correct partition function for any system.

‡ Any canonical transformation which preserves the Hamiltonian equations of motion and which does not involve the time explicitly can be written in the form

$$p_i = \frac{\partial F(q_i, Q_j)}{\partial q_i}, \quad P_j = -\frac{\partial F(q_i, Q_j)}{\partial Q_j},$$

where F is an arbitrary function. By writing

$$\frac{\partial(P_1, \dots, P_{3N}; Q_1, \dots, Q_{3N})}{\partial(p_1, \dots, p_{3N}; q_1, \dots, q_{3N})} = \frac{\partial(P_1, \dots, P_{3N}; Q_1, \dots, Q_{3N})}{\partial(q_1, \dots, q_{3N}; Q_1, \dots, Q_{3N})} \Bigg/ \frac{\partial(p_1, \dots, p_{3N}; q_1, \dots, q_{3N})}{\partial(q_1, \dots, q_{3N}; Q_1, \dots, Q_{3N})}$$

$$= (-)^{3N} \frac{\partial(P_1, \dots, P_{3N})}{\partial(q_1, \dots, q_{3N})} \Bigg/ \frac{\partial(p_1, \dots, p_{3N})}{\partial(Q_1, \dots, Q_{3N})},$$

and inserting the above expressions for p_i and P_j, we see at once that the Jacobian has the value unity. Therefore, by the formula for the change of variable in a multiple integral, the extension in phase space is invariant.

5·521. In order to deduce (5·52·2) from the general expression (5·351·6) we introduce the density matrix $\Psi(q', q, \gamma)$, which, for a general quantal system, is defined as

$$\Psi(q', q, \gamma) = \sum_i \psi_i(q')^* \exp(-\gamma \mathscr{H}) \psi_i(q), \tag{5·521·1}$$

where $\mathscr{H}(q)$ is the Hamiltonian function of the system, and where $\psi_i(q)$ is the normalized eigenfunction of the ith stationary state. We use q to denote generically all the space coordinates of the system. The expression $\exp(-\gamma\mathscr{H})\psi$ is to be interpreted as

$$\exp(-\gamma\mathscr{H})\psi = \sum_{n=0}^{\infty} \frac{(-1)^n \gamma^n}{n!} \mathscr{H}^n \psi,$$

where $\mathscr{H}^n\psi = (\mathscr{H}(\mathscr{H} \ldots (\mathscr{H}(\mathscr{H}\psi)\ldots)))$. Therefore, if E_i is the energy of the ith stationary state, we have $\exp(-\gamma\mathscr{H})\psi_i(q) = \exp(-\gamma E_i)\psi_i(q)$, so that

$$\Psi(q', q, \gamma) = \sum_i \psi_i(q')^* \exp(-\gamma E_i) \psi_i(q), \tag{5·521·2}$$

and, since the wave functions are normalized, we have

$$Z_{\text{system}} = \sum_i \exp(-E_i/kT) = \int \Psi(q, q, \gamma)\, dq \quad (\gamma = 1/kT). \tag{5·521·3}$$

The density matrix $\Psi(q', q, \gamma)$ satisfies the Schrödinger equation

$$-\partial\Psi/\partial\gamma = \mathscr{H}(q)\Psi, \tag{5·521·4}$$

since $\mathscr{H}(q)$ only acts on $\psi_i(q)$. Further, $\Psi(q', q, \gamma)$ need not be defined in terms of the eigenfunctions of the stationary states, and the ψ_i's can be any complete set of normalized orthogonal functions.

To prove this last statement, let $\phi_i(q)$ and $\psi_i(q)$ be any two sets of normalized orthogonal functions, so that

$$\psi_i(q) = \sum_j \phi_j(q) \int \phi_j(q')^* \psi_i(q')\, dq', \tag{5·521·5}$$

$$\phi_k(q) = \sum_i \psi_i(q) \int \psi_i(q')^* \phi_k(q')\, dq'. \tag{5·521·6}$$

Then, if A is any operator, we have, using (5·521·5),

$$\begin{aligned}(q' \mid A \mid q) &\equiv \sum_i \psi_i(q')^* A \psi_i(q) \\ &= \sum_{i,k,l} \phi_k(q')^* A\phi_l(q) \int \phi_k(q'')\psi_i(q'')^*\, dq'' \int \psi_i(q''')\phi_l(q''')^*\, dq''' \\ &= \sum_{k,l} \phi_k(q')^* A\phi_l(q) \int \phi_k(q'')\phi_l(q'')^*\, dq'',\end{aligned}$$

by (5·521·6). Hence

$$(q' \mid A \mid q) = \sum_{k,l} \phi_k(q')^* A\phi_l(q)\,\delta_{kl} = \sum_k \phi_k(q')^* A\phi_k(q),$$

which proves that $(q' \mid A \mid q)$ is an invariant for orthogonal transformations of the ψ_i's. Also, for any arbitrary function $f(q)$, we have

$$f(q) = \sum_i \phi_i(q) \int \phi_i(q')^* f(q')\, dq',$$

which can be expressed as $(q' \mid 1 \mid q) = \delta(q-q')$. Hence $\Psi(q', q, \gamma)$ satisfies the Schrödinger equation (5·521·4) and also the 'initial condition'

$$\Psi(q', q, 0) = \delta(q-q'). \qquad (5·521·7)$$

5·522. The density matrix can be obtained in various exact or approximate forms. When the interactions are sufficiently small for the system to be nearly classical, it is convenient to take the ψ_i's to be the wave functions of free particles and to obtain Z as a series in ascending powers of h (Kirkwood, 1933). If we have only one particle present, we have

$$\Psi(\mathbf{r}', \mathbf{r}, \gamma) = \frac{1}{h^{\frac{3}{2}}} \int \exp(-i\mathbf{p}.\mathbf{r}'/\hbar) \exp(-\gamma\mathscr{H}) \exp(i\mathbf{p}.\mathbf{r}/\hbar)\, d\mathbf{p}, \qquad (5·522·1)$$

where
$$\mathscr{H} = -\frac{\hbar^2}{2m}\nabla^2 + \mathscr{V}(\mathbf{r}), \qquad (5·522·2)$$

and, if we write

$$\phi(\mathbf{r}, \gamma) = \exp(-\gamma\mathscr{H}) \exp(i\mathbf{p}.\mathbf{r}/\hbar) = w(\mathbf{r}, \gamma) \exp\left[-\left(\frac{\mathbf{p}^2}{2m} + \mathscr{V}(\mathbf{r})\right)\gamma + \frac{i\mathbf{p}.\mathbf{r}}{\hbar}\right], \qquad (5·522·3)$$

$\phi(\mathbf{r}, \gamma)$ satisfies the equation $\partial\phi/\partial\gamma = -\mathscr{H}\phi$. Hence $w(\mathbf{r}, \gamma)$ is the solution of the equation

$$\frac{\partial w}{\partial \gamma} = \frac{i\hbar}{m}(\mathbf{p}.\mathrm{grad}\, w - \gamma w \mathbf{p}.\mathrm{grad}\, \mathscr{V}) + \frac{\hbar^2}{2m}\{\nabla^2 w - 2\gamma\, \mathrm{grad}\, \mathscr{V}.\mathrm{grad}\, w - \gamma w \nabla^2 \mathscr{V} + \gamma^2 w (\mathrm{grad}\, \mathscr{V})^2\}, \qquad (5·522·4)$$

which can be solved by successive approximations.

The first approximation to w, which must be such as to satisfy (5·521·7), is $w = 1$. The second approximation is

$$w(\mathbf{r}, \gamma) = 1 - \tfrac{1}{2} i\hbar\gamma^2 \mathbf{p}.\mathrm{grad}\, \mathscr{V}/m, \qquad (5·522·5)$$

and the higher approximations can readily be determined. The first approximation gives the classical partition function (5·52·1), and the corrections to this can, in principle, be determined. But, unless these corrections are small, the series of successive approximations is only slowly convergent while the calculations are difficult to carry out so as to obtain a reliable numerical result. In general, therefore, special methods of evaluating the partition function must be found.

In the general case of a system containing N particles, it is necessary to use the correctly symmetrized wave functions for the N free particles.

We then obtain (5·52·2) as the first approximation to the partition function. It is also possible to carry out the calculations for free particles exactly, and in this way deduce the quantal formulae for a perfect gas (see, for example, Kahn, 1938). We shall not give details of these calculations since they are less direct and more difficult than those given later in §§ 5·61 and 5·62.

5·53. *The equipartition of energy.* One of the most important consequences of the classical theory is that, loosely speaking, every square term in the Hamiltonian function of a system contributes $\frac{1}{2}kT$ to the energy of the assembly.

To make this statement precise, we suppose that there are N degrees of freedom so that the kinetic energy is a homogeneous quadratic function of the N momenta $p_1, \ldots, p_N$, whose coefficients are functions of the q's. We suppose that the potential energy is a homogeneous quadratic function of $q_1, \ldots, q_M$ $(M \leqslant N)$, whose coefficients may be functions of the remaining $q_{M+1}, \ldots, q_N$, and we further suppose that the kinetic energy does not depend upon $q_1, \ldots, q_M$, but depends only on $p_1, \ldots, p_N$ and $q_{M+1}, \ldots, q_N$. In view of the invariance of Z against canonical transformations, we may further suppose that the coordinates and momenta have been chosen so as to reduce the quadratic forms to the sum of squares so that

$$\mathscr{H} = \alpha_1 p_1^2 + \ldots + \alpha_N p_N^2 + \beta_1 q_1^2 + \ldots + \beta_M q_M^2,$$

where $\alpha_1, \ldots, \beta_M$ are functions of $q_{M+1}, \ldots, q_N$ only. The range of integration of the variables $p_1, \ldots, q_M$ is $-\infty, \infty$, and we so obtain

$$Z(\vartheta) = \frac{1}{h^N}\left(\frac{\pi}{\vartheta}\right)^{\frac{1}{2}(N+M)} \frac{1}{(\alpha_1 \ldots \alpha_N)^{\frac{1}{2}}(\beta_1 \ldots \beta_M)^{\frac{1}{2}}} \int dq_{M+1} \ldots dq_N,$$

where the integral is clearly independent of ϑ. We then have

$$U = -\frac{\partial \log Z}{\partial \vartheta} = \frac{1}{2}\frac{N+M}{\vartheta} = \tfrac{1}{2}(N+M)\,kT, \tag{5·53·1}$$

which is the precise form of the statement in the first paragraph.

The statement of this relatively simple theorem may seem somewhat complicated, but it is required in this form if we are to apply it to cyclic coordinates which occur, for example, when we consider rotating bodies.

Quantal Statistical Mechanics

5·6. As mentioned in § 5·2, the use of unsymmetrized wave functions is logically unjustified, and any results obtained from such wave functions can only be assumed to have an approximate validity. We now investigate the typical quantum effects which appear when the influence of the symmetry properties of the wave functions upon the number of complexions is taken into account. As in § 5·21, we group the energy levels together into large sets containing M_r levels, and consider the distribution of N_r systems over these levels.

5·61. *Antisymmetrical wave functions.* The number of ways in which N_r elements can be distributed over M_r levels so that not more than one element is allocated to any particular level is the number of terms in the product

$$(1+x_1)(1+x_2)\dots(1+x_{M_r})$$

which are of degree N_r in the x's. This number is the coefficient of x^{N_r} in the expansion of $(1+x)^{M_r}$, i.e.

$$\frac{M_r!}{N_r!\,(M_r-N_r)!},$$

and the number of complexions for the whole system is

$$W=\prod_r \frac{M_r!}{N_r!\,(M_r-N_r)!}. \tag{5·61·1}$$

We may now repeat the argument of § 5·3 by considering two loosely coupled systems which we distinguish by the indices 1 and 2. The number of complexions for the composite system is the product of two expressions such as (5·61·1), and we determine the maximum value of $\log W$ subject to the conditions expressing the constancy of the numbers of elements and of the energy of the composite system, which take the form

$$N^{(1)}=\Sigma N_r^{(1)}, \quad N^{(2)}=\Sigma N_s^{(2)}, \tag{5·61·2}$$

$$U=\Sigma(N_r^{(1)}E_r^{(1)}+N_s^{(2)}E_s^{(2)}). \tag{5·61·3}$$

If we introduce undetermined multipliers λ_1, λ_2 and ϑ and maximize $\log W+\lambda_1 N^{(1)}+\lambda_2 N^{(2)}-\vartheta U$, we find that

$$\delta(\log W+\lambda_1 N^{(1)}+\lambda_2 N^{(2)}-\vartheta U)=\Sigma\delta N_r^{(1)}\left(\log\frac{M_r^{(1)}-N_r^{(1)}}{N_r^{(1)}}+\lambda_1-\vartheta E_r^{(1)}\right)$$
$$+\Sigma\delta N_s^{(2)}\left(\log\frac{M_s^{(2)}-N_s^{(2)}}{N_s^{(2)}}+\lambda_2-\vartheta E_s^{(2)}\right)=0.$$

Hence
$$\frac{\overline{N}_r^{(1)}}{M_r^{(1)}}=\frac{1}{\exp(\vartheta E_r^{(1)}-\lambda_1)+1}, \quad \frac{\overline{N}_s^{(2)}}{M_s^{(2)}}=\frac{1}{\exp(\vartheta E_s^{(2)}-\lambda_2)+1}. \tag{5·61·4}$$

We may now revert to the average number of elements $\overline{n}_r$ in the state r, where $\overline{n}_r=\overline{N}_r/M_r$, provided that in any sums we sum over all the energy levels and not merely over the different groups of levels. The parameters λ_1, λ_2 and ϑ are then determined by the conditions

$$N^{(1)}=\Sigma\overline{n}_r^{(1)}=\Sigma\frac{1}{\exp(\vartheta E_r^{(1)}-\lambda_1)+1}, \quad N^{(2)}=\Sigma\overline{n}_s^{(2)}=\Sigma\frac{1}{\exp(\vartheta E_s^{(2)}-\lambda_2)+1} \tag{5·61·5}$$

and
$$U=\Sigma\left(\frac{E_r^{(1)}}{\exp(\vartheta E_r^{(1)}-\lambda_1)+1}+\frac{E_s^{(2)}}{\exp(\vartheta E_s^{(2)}-\lambda_2)+1}\right). \tag{5·61·6}$$

When the formulae of the present section are applicable, the elements are said to obey Fermi-Dirac statistics, after Fermi (1926) and Dirac (1926), who first obtained the distribution function (5·61·4).

5·611. *The thermodynamics of systems obeying Fermi-Dirac statistics.* We consider a single system of elements and derive the thermodynamic relations analogous to those of §§ 5·32 and 5·33. The average value $\bar{A}_i$ of any generalized force taken over the whole assembly is the average value of $-\partial E_r/\partial a_i$. Hence the first law of thermodynamics, the conservation of energy, can be stated in the form

$$dQ=dU+\sum_i \bar{A}_i da_i = dU-\sum_i \frac{\partial}{\partial a_i}\left(\sum_r \bar{n}_r E_r\right) da_i. \tag{5·611·1}$$

To derive the second law we must show that ϑ is an integrating factor of this linear differential form, but it is convenient at this stage to allow the number of elements in the system to vary, and in this case it is necessary to show that we can write (5·611·1) in the form

$$dU=T\,dS-\Sigma\bar{A}_i da_i+\mu\, dN, \tag{5·611·2}$$

where μ is the thermodynamic potential.

Now from (5·61·4) and (5·611·1) we have

$$\begin{aligned}\vartheta(dU+\Sigma\bar{A}_i da_i) &= d(\vartheta U)-U\,d\vartheta+\Sigma\bar{A}_i da_i\\ &= d(\vartheta U)-\sum_r\frac{1}{\exp(\vartheta E_r-\lambda)+1}\left(E_r d\vartheta-d\lambda+\vartheta\sum_i\frac{\partial E_r}{\partial a_i}da_i\right)\\ &\quad -d\sum_r\frac{\lambda}{\exp(\vartheta E_r-\lambda)+1}+\lambda d\sum_r\frac{1}{\exp(\vartheta E_r-\lambda)+1}.\end{aligned} \tag{5·611·3}$$

Hence
$$k\vartheta(dU+\Sigma\bar{A}_i da_i)=dS+k\lambda\, dN, \tag{5·611·4}$$

where
$$S=k\vartheta U+k\Sigma\log[1+\exp(\lambda-\vartheta E_r)]-k\lambda N, \tag{5·611·5}$$

and where the relation between N and λ is

$$N=\sum_r\frac{1}{\exp(\vartheta E_r-\lambda)+1}. \tag{5·611·6}$$

Hence the absolute temperature T is equal to $1/k\vartheta$ and S is the entropy. Also (5·611·4) can be written as

$$dU=T\,dS-\Sigma\bar{A}_i da_i+kT\lambda\, dN, \tag{5·611·7}$$

and, by comparison with (5·611·2), we see that

$$\mu=kT\lambda. \tag{5·611·8}$$

Having identified the parameters λ and ϑ in terms of familiar thermodynamic quantities, we may write the expressions for the important thermodynamic functions in the following forms:

$$U = \Sigma \frac{E_r}{\exp[(E_r-\mu)/kT]+1}, \tag{5·611·9}$$

$$S = k\Sigma \log\{1+\exp[(\mu-E_r)/kT]\}+(U-N\mu)/kT, \tag{5·611·10}$$

$$F \equiv U-ST = N\mu - kT\Sigma\log\{1+\exp[(\mu-E_r)/kT]\}, \tag{5·611·11}$$

$$\Sigma \bar{A}_i a_i \equiv N\mu - F = kT\Sigma\log\{1+\exp[(\mu-E_r)/kT]\}. \tag{5·611·12}$$

If λ is large and negative, all the formulae given above reduce immediately to those derived in §§5·3ff. for systems with unsymmetrized wave functions.

The entropy S can be expressed directly in terms of the number of complexions W. For, from (5·61·1), we have

$$\log W = \Sigma\{M_r \log M_r - N_r \log N_r - (M_r-N_r)\log(M_r-N_r)\},$$

and, by (5·61·4),

$$\log W_{\text{max.}} = \Sigma M_r \log\{1+\exp[(\mu-E_r)/kT]\} + \Sigma \frac{M_r(E_r-\mu)/kT}{\exp[(E_r-\mu)/kT]+1}, \tag{5·611·13}$$

where the summation is over all the groups of levels. We may eliminate M_r from this formula by summing over all the levels instead of over the groups of levels, and we then see that

$$S = k \log W_{\text{max.}}, \tag{5·611·14}$$

which is the same relation as was found in §5·33 for systems with unsymmetrized wave functions.

To complete the above proof we ought to show that (5·611·4) is correct when N is not constant. The proof of this is exactly the same as that given in §5·34, and the details are left to the reader.

5·612. Instead of deriving the results *ab initio* for systems with antisymmetrical wave functions it is possible to start from the expression (5·351·6) for the free energy, which holds for any system. The detailed calculations are very similar to those given in §5·61, but it is instructive to see how the formula

$$F = -kT\log\{\sum_i W(U_i)\exp(-U_i/kT)\} = -kT\log Z_{\text{system}}, \tag{5·612·1}$$

derived by a consideration of classical statistics, leads to the correct quantal formulae.

As usual we consider the energy levels and elements to be divided up into groups of approximately the same energy, and the logarithm of the general term in the series for Z_{system} is

$$\log W(U_i) - U_i/kT = \Sigma \log \frac{M_r!}{N_r!\,(M_r - N_r)!} - \Sigma \frac{N_r E_r}{kT}, \qquad (5\cdot612\cdot2)$$

where
$$\Sigma N_r = N. \qquad (5\cdot612\cdot3)$$

(The expression for W is the same here as in § 5·61, but the energy U_i of the system is now given as $U_i = \Sigma N_r E_r$ for a given distribution of N's subject only to the condition (5·612·3).) We have to consider all possible values of U_i, but, as usual, there is a sharp maximum in the general term in the series for Z_{system}, and it is sufficient to determine this maximum term, which is given by maximizing $\log W_i - U_i/kT + \lambda N$. We find that

$$-\log N_r + \log (M_r - N_r) - E_r/kT + \lambda = 0,$$

i.e.
$$\frac{N_r}{M_r} = \frac{1}{\exp (E_r/kT - \lambda) + 1}, \qquad (5\cdot612\cdot4)$$

which is, of course, the same as (5·61·4). Inserting this into (5·612·3) we regain (5·611·6); if we replace Z_{system} by its maximum term, we have, by (5·611·13),

$$\log Z_{\text{system}} = \Sigma \log [1 + \exp (\lambda - E_r/kT)] - N\lambda, \qquad (5\cdot612\cdot5)$$

where the summation is over all the levels, which shows that F is given by (5·611·11). Replacing Z_{system} by its maximum term means introducing an error of order $\sqrt{N}$ into F, which, since the dominant part of F is proportional to N, is trivial.

Having shown that F is given by (5·611·11), we can now derive all the thermodynamic functions by forming the appropriate derivatives of $F(T, a_i, N)$. We first note that $\mu \equiv kT\lambda$ is a function of T, a_i and N given implicitly by (5·611·6), and that, by (5·611·6), differentiation of (5·611·11) gives

$$\partial F(T, a_i, N, \mu)/\partial \mu = 0. \qquad (5\cdot612\cdot6)$$

Hence in forming derivatives of F we do not need to include terms involving $\partial F/\partial \mu$, although μ itself is not constant.

The internal energy U is most simply found from the relation

$$U = -T^2 \frac{\partial}{\partial T} \frac{F}{T},$$

and we regain (5·611·9) at once. The entropy is then given by $S = (U - F)/T$, while the external force A_i is given by $-\partial F/\partial a_i$. Finally, we verify from the relation $\mu = \partial F/\partial N$ for the thermodynamic potential that we have correctly identified the parameter λ with μ/kT.

The difference between the present procedure and that followed in § 5·61 is largely a matter of the choice of the assumptions to be made. (The number of complexions of the system is necessarily the same, but it can be introduced at different stages of the argument.) In § 5·61 we began by considering two systems whose combined energy is fixed; we deduced the distribution function and the thermodynamic relations, and identified the parameters ϑ and λ. (To complete the argument it was necessary to consider three systems.) In the procedure starting from (5·612·1) the temperature is taken as given, and the thermodynamic relations are implicit in the form assumed for F. The internal energy U of the system, whether it is considered to be a simple or a composite system, is, however, not given at the outset, and its average value has to be deduced from the expression for F. We may therefore say that in § 5·61 we consider a set of systems with a given total energy and deduce the existence and value of the temperature, whereas in the present section we consider a system with a fixed temperature and deduce the energy of the system. The procedures are equivalent so long as we only consider mean values of U and T. They are not equivalent if we wish to consider fluctuations, since in one case U (for the composite system) is constant and there can only be fluctuations in the statistical parameter T and in the energy of a subsystem, while, in the other case, T is constant and there can be fluctuations in the total energy of the system. The former is the more general case since the temperature of a system can be made constant by coupling it to a very much larger system and keeping constant the total energy of the combined system. These matters are discussed further in § 5·7.

5·62. *Symmetrical wave functions.* If the wave functions are symmetrical in each pair of systems, the number of complexions for given M_r and N_r is the number of terms in the product

$$(1+x_1+x_1^2+\dots)(1+x_2+x_2^2+\dots)\dots(1+x_{M_r}+x_{M_r}^2+\dots)$$

which are of degree N_r in the x's. For, if we have any unsymmetrized wave function $\psi_1(1)\psi_2(2)\dots\psi_{N_r}(n_r)$, we can always make one and only one symmetrical wave function from it by permutation of the coordinates of the elements and addition of the resulting wave functions. In this case there is no restriction requiring the primitive wave functions $\psi_1, \psi_2, \dots$ to be all different, and any ψ can be repeated as many times as one wishes. The number of complexions is therefore the coefficient of x^{N_r} in the expansion of $(1-x)^{-M_r}$, i.e.

$$\frac{(M_r+N_r-1)!}{N_r!\,(M_r-1)!}. \tag{5·62·1}$$

The argument now follows closely that for antisymmetrical wave functions, and it is sufficient to quote the results.

The average number of systems $\bar{n}_r$ in the state r is given by

$$\bar{n}_r = \frac{1}{\exp[(E_r-\mu)/kT]-1}, \quad N = \Sigma \frac{1}{\exp[(E_r-\mu)/kT]-1}, \tag{5·62·2}$$

where μ, for prescribed T, is given in terms of N by the second of these relations. The thermodynamic functions U, S, F and $\Sigma \bar{A}_i a_i$ are given by

$$U = \Sigma \frac{E_r}{\exp[(E_r-\mu)/kT]-1}, \tag{5·62·3}$$

$$S = -k\Sigma \log\{1-\exp[(\mu-E_r)/kT]\} + (U-N\mu)/kT, \tag{5·62·4}$$

$$F \equiv U - ST = N\mu + kT\Sigma \log\{1-\exp[(\mu-E_r)/kT]\}, \tag{5·62·5}$$

$$\Sigma \bar{A}_i a_i \equiv N\mu - F = -kT\Sigma \log\{1-\exp[(\mu-E_r)/kT]\}. \tag{5·62·6}$$

Finally, the relations (5·311·13) and (5·311·14), connecting S and F with $W_{\text{max.}}$, hold unchanged for systems with symmetrical wave functions. Systems to which the formulae of this section are applicable are said to obey Einstein-Bose statistics (Einstein, 1924; Bose, 1924).

Fluctuations

5·7. We have so far assumed without proof that it is permissible to concern ourselves only with the average value of any physical quantity associated with a system. We now undertake a more searching investigation by calculating the mean-square deviations of the important physical quantities from their average values without assuming that the number of elements is effectively infinite.

We shall consider the composite system which has formed the basis for the derivation of the fundamental statistical formulae in §§ 5·3, 5·61 and 5·62, and we use the notation of those sections. If

$$W(N_1^{(1)}, N_2^{(1)}, \ldots; N_1^{(2)}, N_2^{(2)}, \ldots)$$

is the number of complexions of the composite system for given

$$N_1^{(1)}, N_2^{(1)}, \ldots, N_1^{(2)}, N_2^{(2)}, \ldots,$$

then the average value of $N_r^{(1)}$ is given by

$$C\overline{N_r^{(1)}} = \Sigma N_r^{(1)} W(N_1^{(1)}, N_2^{(1)}, \ldots; N_1^{(2)}, N_2^{(2)}, \ldots), \tag{5·7·1}$$

where the summation is over all sets of the N's which are compatible with the constraints on the system and with the constancy of the numbers of the elements and of the energy, and where C is the total number of the complexions, i.e. the sum of W over all the permissible sets of the N's. If W is given by (5·3·1) we can write the average value of $N_r^{(1)}$ as

$$C\overline{N_r^{(1)}} = M_r^{(1)} \partial C / \partial M_r^{(1)}, \tag{5·7·2}$$

provided that C is calculated subject to the restrictions (5·3·2) and (5·3·3). If, however, W is built up from expressions such as (5·61·1) or (5·62·1), such a definition is impossible. To avoid this difficulty we may, however, introduce factors $x_r^{N_r^{(1)}}$ and $y_s^{N_s^{(2)}}$ into the expressions for W. The purpose of these factors is to enable us to deduce identities by differentiation with respect to x_r and y_s, and after the differentiations we may put $x_r = y_s = 1$. With this formal generalization, the expressions for W are as follows. For unsymmetrized wave functions we have

$$W(N_r^{(1)}, x_r;\ N_s^{(2)}, y_s) = \prod_r \prod_s \frac{(x_r M_r^{(1)})^{N_r^{(1)}}}{N_r^{(1)}!} \frac{(y_s M_s^{(2)})^{N_s^{(2)}}}{N_s^{(2)}!}, \tag{5·7·3}$$

while $$W(N_r^{(1)}, x_r;\ N_s^{(2)}, y_s) = \prod_r \prod_s \frac{M_r^{(1)}!\, x_r^{N_r^{(1)}}}{N_r^{(1)}!\,(M_r^{(1)} - N_r^{(1)})!} \frac{M_s^{(2)}!\, y_s^{N_s^{(2)}}}{N_s^{(2)}!\,(M_s^{(2)} - N_s^{(2)})!} \tag{5·7·4}$$

for antisymmetrical wave functions, and

$$W(N_r^{(1)}, x_r;\ N_s^{(2)}, y_s) = \prod_r \prod_s \frac{(M_r^{(1)} + N_r^{(1)} - 1)!\, x_r^{N_r^{(1)}}}{N_r^{(1)}!\,(M_r^{(1)} - 1)!} \frac{(M_s^{(2)} + N_s^{(2)} - 1)!\, y_s^{N_s^{(2)}}}{N_s^{(2)}!\,(M_s^{(2)} - 1)!} \tag{5·7·5}$$

for symmetrical wave functions. It therefore follows that, in all cases,

$$\overline{N_r^{(1)}} = \frac{1}{C} x_r \frac{\partial C}{\partial x_r} \tag{5·7·6}$$

and $$\overline{N_r^{(1)2}} = \frac{1}{C} x_r \frac{\partial}{\partial x_r}\left(x_r \frac{\partial C}{\partial x_r}\right). \tag{5·7·7}$$

The mean value of any power of $N_r^{(1)}$ can be calculated by successive differentiations. Now (5·7·7) can be written as

$$\begin{aligned}\overline{N_r^{(1)2}} &= \frac{1}{C} x_r \frac{\partial}{\partial x_r}\left\{C\left(\frac{1}{C} x_r \frac{\partial C}{\partial x_r}\right)\right\} = \left(\frac{1}{C} x_r \frac{\partial C}{\partial x_r}\right)^2 + x_r \frac{\partial}{\partial x_r}\left(\frac{1}{C} x_r \frac{\partial C}{\partial x_r}\right) \\ &= (\overline{N_r^{(1)}})^2 + x_r \partial \overline{N_r^{(1)}} / \partial x_r. \end{aligned} \tag{5·7·8}$$

Hence the mean-square deviation of $N_r^{(1)}$ is determined by the fundamental relation

$$\text{Av.}\,(N_r^{(1)} - \overline{N_r^{(1)}})^2 = x_r \partial \overline{N_r^{(1)}} / \partial x_r \quad (x_r = 1), \tag{5·7·9}$$

where the differentiation must be carried out at constant $N^{(1)}$, $N^{(2)}$ and U.

It is obvious that $x_r \partial \overline{N_r^{(1)}} / \partial x_r$ must be of order $N^{(1)}$ at most. Hence the root-mean-square deviation of $N_r^{(1)}$ from its average value must be of order $\sqrt{N^{(1)}}$, and for large systems this is negligible compared with $\overline{N_r^{(1)}}$ itself, which is of order $N^{(1)}$. This completes the justification of the derivation of the equations of statistical mechanics. It is, however, of some interest to calculate (5·7·9) exactly, but the calculation is somewhat involved since the differentiation with respect to x_r has to be carried out at constant $N^{(1)}$, $N^{(2)}$ and U, and not at constant λ_1, λ_2 and ϑ. The reader who is not interested in the mathematical arguments may proceed at once to § 5·75.

5·71. To calculate $\overline{N_r^{(1)}}$ and $\overline{N_s^{(2)}}$ from first principles, the procedure is to maximize $\log W + \lambda_1 N^{(1)} + \lambda_2 N^{(2)} - \vartheta U$. Hence the effect of the introduction of the x and y factors into W is to replace $\lambda_1 N_r^{(1)}$ and $\lambda_2 N_s^{(2)}$ by $(\lambda_1 + \log x_r) N_r^{(1)}$ and $(\lambda_2 + \log y_s) N_s^{(2)}$ in the quantity to be maximized. Instead of $\overline{N_r^{(1)}}(\lambda_1, \vartheta)$, which is a function of λ_1 and ϑ only, we now have $\overline{N_r^{(1)}}$ which is a function of x_r, λ_1 and ϑ of the form $\overline{N_r^{(1)}}(x_r, \lambda_1, \vartheta) = \overline{N_r^{(1)}}(\lambda_1 + \log x_r, \vartheta)$, and similarly we have to replace $\overline{N_s^{(2)}}(\lambda_2, \vartheta)$ by $\overline{N_s^{(2)}}(y_s, \lambda_2, \vartheta) = \overline{N_s^{(2)}}(\lambda_2 + \log y_s, \vartheta)$. We therefore have the identities

$$x_r \frac{\partial \overline{N_r^{(1)}}}{\partial x_r} = \frac{\partial \overline{N_r^{(1)}}}{\partial \lambda_1}, \quad y_s \frac{\partial \overline{N_s^{(2)}}}{\partial y_s} = \frac{\partial \overline{N_s^{(2)}}}{\partial \lambda_2}, \tag{5·71·1}$$

and, since $\qquad N^{(1)} = \Sigma \overline{N_r^{(1)}}, \quad N^{(2)} = \Sigma \overline{N_s^{(2)}},$

we also have
$$x_r \frac{\partial N^{(1)}}{\partial x_r} = x_r \frac{\partial \overline{N_r^{(1)}}}{\partial x_r} = \frac{\partial \overline{N_r^{(1)}}}{\partial \lambda_1}, \quad y_s \frac{\partial N^{(2)}}{\partial y_s} = \frac{\partial \overline{N_s^{(2)}}}{\partial \lambda_2}. \tag{5·71·2}$$

Further, since $\overline{N_r^{(1)}}$ depends upon x_r, λ_1 and ϑ through the combination $\vartheta E_r - \lambda_1 - \log x_r$, and since $U = \Sigma(\overline{N_r^{(1)}} E_r^{(1)} + \overline{N_s^{(2)}} E_s^{(2)})$, we have

$$x_r \frac{\partial U}{\partial x_r} = -\frac{\partial \overline{N_r^{(1)}}}{\partial \vartheta}, \quad y_s \frac{\partial U}{\partial y_s} = -\frac{\partial \overline{N_s^{(2)}}}{\partial \vartheta}, \quad \frac{\partial U}{\partial \lambda_1} = -\frac{\partial N^{(1)}}{\partial \vartheta}, \quad \frac{\partial U}{\partial \lambda_2} = -\frac{\partial N^{(2)}}{\partial \vartheta}. \tag{5·71·3}$$

These relations enable us to replace differentiations with respect to x_r and y_s by differentiations with respect to λ_1, λ_2 and ϑ.

5·72. To calculate the quantity on the right-hand side of (5·7·9) we use the theorems on partial differentiation given in § 1·6. By (1·6·15) we have (omitting the bars over the N_r's)

$$x_r \left(\frac{\partial N_r^{(1)}}{\partial x_r} \right)_{N^{(1)}, N^{(2)}, U} = x_r \frac{\partial(N_r^{(1)}, N^{(1)}, N^{(2)}, U)}{\partial(x_r, \lambda_1, \lambda_2, \vartheta)} \bigg/ \frac{\partial(N^{(1)}, N^{(2)}, U)}{\partial(\lambda_1, \lambda_2, \vartheta)}. \tag{5·72·1}$$

We can eliminate the derivatives with respect to x_r by using the relations (5·71·1), (5·71·2) and (5·71·3), and the result can be obtained in a number of equivalent forms. The transformations involve the manipulation of Jacobians, and, although these are elementary, we give them in detail, since it is possible to arrive at complex rather than simple expressions if the correct sequence of steps is not followed. The numerator of (5·72·1), when expanded in terms of the elements of the first column, is

$$\frac{\partial N_r^{(1)}}{\partial \lambda_1}\frac{\partial(N^{(1)},N^{(2)},U)}{\partial(\lambda_1,\lambda_2,\vartheta)}-\left(\frac{\partial N_r^{(1)}}{\partial \lambda_1}\right)^2\frac{\partial(N^{(2)},U)}{\partial(\lambda_2,\vartheta)}$$
$$+\frac{\partial N_r^{(1)}}{\partial \lambda_1}\frac{\partial U}{\partial \lambda_1}\frac{\partial N^{(2)}}{\partial \lambda_2}\frac{\partial N_r^{(1)}}{\partial \vartheta}+\frac{\partial N_r^{(1)}}{\partial \vartheta}\frac{\partial(N_r^{(1)},N^{(1)},N^{(2)})}{\partial(\lambda_1,\lambda_2,\vartheta)},\quad (5·72·2)$$

since
$$\partial N_r^{(1)}/\partial\lambda_2=\partial N^{(1)}/\partial\lambda_2=\partial N^{(2)}/\partial\lambda_1=0.\quad (5·72·3)$$

Now multiply the second, third and fourth terms of (5·72·2) by $\partial N^{(1)}/\partial\lambda_1$. The second and third terms then give

$$-\left(\frac{\partial N_r^{(1)}}{\partial \lambda_1}\right)^2\left(\frac{\partial(N^{(1)},N^{(2)},U)}{\partial(\lambda_1,\lambda_2,\vartheta)}-\frac{\partial N^{(1)}}{\partial\vartheta}\frac{\partial N^{(2)}}{\partial\lambda_2}\frac{\partial N^{(1)}}{\partial\vartheta}\right)-\frac{\partial N_r^{(1)}}{\partial \lambda_1}\frac{\partial N_r^{(1)}}{\partial \vartheta}\frac{\partial N^{(1)}}{\partial \lambda_1}\frac{\partial N_1^{(1)}}{\partial \vartheta}\frac{\partial N^{(2)}}{\partial \lambda_2}$$

by using the relation $\partial U/\partial\lambda_1=-\partial N^{(1)}/\partial\vartheta.$ (5·72·4)

On rearranging the terms in (5·72·4) it becomes

$$-\left(\frac{\partial N_r^{(1)}}{\partial \lambda_1}\right)^2\frac{\partial(N^{(1)},N^{(2)},U)}{\partial(\lambda_1,\lambda_2,\vartheta)}+\frac{\partial N_r^{(1)}}{\partial \lambda_1}\frac{\partial N^{(1)}}{\partial\vartheta}\frac{\partial N^{(2)}}{\partial\lambda_2}\frac{\partial(N_r^{(1)},N^{(1)})}{\partial(\lambda_1,\vartheta)}.\quad (5·72·5)$$

Also
$$\frac{\partial(N_r^{(1)},N^{(1)},N^{(2)})}{\partial(\lambda_1,\lambda_2,\vartheta)}=-\frac{\partial N^{(2)}}{\partial\lambda_2}\frac{\partial(N_r^{(1)},N^{(1)})}{\partial(\lambda_1,\vartheta)},\quad (5·72·6)$$

by (5·72·3). Hence, collecting the terms together, the numerator of (5·72·1) is

$$\left(\frac{\partial N_r^{(1)}}{\partial\lambda_1}-\frac{(\partial N_r^{(1)}/\partial\lambda_1)^2}{\partial N^{(1)}/\partial\lambda_1}\right)\frac{\partial(N^{(1)},N^{(2)},U)}{\partial(\lambda_1,\lambda_2,\vartheta)}+\frac{\partial N^{(2)}/\partial\lambda_2}{\partial N^{(1)}/\partial\lambda_1}\left(\frac{\partial(N_r^{(1)},N^{(1)})}{\partial(\lambda_1,\vartheta)}\right)^2.$$

Now, by (1·6·15), we have

$$\left(\frac{\partial N_r^{(1)}}{\partial\vartheta}\right)_{N^{(1)},N^{(2)},x_r}=\frac{\partial(N_r^{(1)},N^{(1)},N^{(2)})}{\partial(\vartheta,\lambda_1,\lambda_2)}\Big/\frac{\partial(N^{(1)},N^{(2)})}{\partial(\lambda_1,\lambda_2)}=-\frac{\partial(N_r^{(1)},N^{(1)})}{\partial(\lambda_1,\vartheta)}\Big/\frac{\partial N^{(1)}}{\partial\lambda_1}$$

by (5·72·3), and (5·72·7)

$$\left(\frac{\partial U}{\partial\vartheta}\right)_{N^{(1)},N^{(2)},x_r}=\frac{\partial(U,N^{(1)},N^{(2)})}{\partial(\vartheta,\lambda_1,\lambda_2)}\Big/\frac{\partial(N^{(1)},N^{(2)})}{\partial(\lambda_1,\lambda_2)}=\frac{\partial(N^{(1)},N^{(2)},U)}{\partial(\lambda_1,\lambda_2,\vartheta)}\Big/\left(\frac{\partial N^{(1)}}{\partial\lambda_1}\frac{\partial N^{(2)}}{\partial\lambda_2}\right).$$
(5·72·8)

We therefore finally obtain, on reinserting the averaging bars in the appropriate places,

$$\text{Av.}\,(N_r^{(1)}-\overline{N_r^{(1)}})^2=\frac{\partial\overline{N_r^{(1)}}}{\partial\lambda_1}-\frac{(\partial\overline{N_r^{(1)}}/\partial\lambda_1)^2}{\partial\overline{N^{(1)}}/\partial\lambda_1}+\left(\frac{\partial\overline{N_r^{(1)}}}{\partial\vartheta}\right)^2_{N^{(1)}}\Big/\left(\frac{\partial U}{\partial\vartheta}\right)_{N^{(1)},N^{(2)}}\quad (5·72·9)$$

By exactly similar arguments it can be shown that

$$\text{Av.}\,\{(N_r^{(1)}-\overline{N_r^{(1)}})(N_s^{(1)}-\overline{N_s^{(1)}})\}=-\frac{(\partial\overline{N_r^{(1)}}/\partial\lambda_1)(\partial\overline{N_s^{(1)}}/\partial\lambda_1)}{\partial\overline{N^{(1)}}/\partial\lambda_1}$$
$$+\left(\frac{\partial\overline{N_r^{(1)}}}{\partial\vartheta}\right)_{N^{(1)}}\left(\frac{\partial\overline{N_s^{(1)}}}{\partial\vartheta}\right)_{N^{(1)}}\Big/\left(\frac{\partial U}{\partial\vartheta}\right)_{N^{(1)},N^{(2)}}\quad (5·72·10)$$

if $r \neq s$, and that

$$\text{Av.}\{(N_r^{(1)}-\overline{N_r^{(1)}})(N_s^{(2)}-\overline{N_s^{(2)}})\} = \left(\frac{\partial \overline{N_r^{(1)}}}{\partial \vartheta}\right)_{N^{(1)}} \left(\frac{\partial \overline{N_s^{(2)}}}{\partial \vartheta}\right)_{N^{(2)}} \Big/ \left(\frac{\partial U}{\partial \vartheta}\right)_{N^{(1)}, N^{(2)}}. \qquad (5·72·11)$$

We may, if we please, regard the above formulae as applying to the occupation numbers $n_r^{(1)}$ and $n_s^{(1)}$ of the individual states. For, if we write $N_r^{(1)} = \Sigma n_i^{(1)}$, where the summation is over the states in the rth group, and form the average of $(\Sigma n_i^{(1)} - \Sigma \overline{n_i^{(1)}})^2$, then, on using the formulae for the n's corresponding to (5·72·9) and (5·72·10), we recover (5·72·9). We may similarly deduce (5·72·10) and (5·72·11) from the formulae for the fluctuations of the corresponding n's.

5·73. To calculate the fluctuations in the energy of one of the systems we proceed as follows. We have

$$\overline{U_1} = \frac{1}{C}\Sigma \sum_r N_r^{(1)} E_r^{(1)} W(N_1^{(1)}, \ldots; N_1^{(2)}, \ldots) = \frac{1}{C}\sum_r E_r x_r \frac{\partial C}{\partial x_r}, \qquad (5·73·1)$$

where the first summation is over all the permissible complexions, and

$$\overline{U_1^2} = \frac{1}{C}\Sigma\left(\sum_r N_r^{(1)} E_r^{(1)}\right)^2 W(N_1^{(1)}, \ldots; N_1^{(2)}, \ldots) = \frac{1}{C}\sum_{r,s} E_r^{(1)} x_r \frac{\partial}{\partial x_r}\left(E_s^{(1)} x_s \frac{\partial C}{\partial x_s}\right). \qquad (5·73·2)$$

Hence

$$\begin{aligned}\overline{U_1^2} &= \frac{1}{C}\sum_r E_r^{(1)} x_r \frac{\partial}{\partial x_r}\left(C\frac{1}{C}\sum_s E_s^{(1)} x_s \frac{\partial C}{\partial x_s}\right) \\ &= \left(\frac{1}{C}\sum_r E_r^{(1)} x_r \frac{\partial C}{\partial x_r}\right)^2 + \sum_r E_r^{(1)} x_r \frac{\partial}{\partial x_r}\left(\frac{1}{C}\sum_s E_s^{(1)} x_s \frac{\partial C}{\partial x_s}\right) \\ &= \overline{U_1}^2 + \sum_r E_r^{(1)} x_r \, \partial \overline{U_1}/\partial x_r. \end{aligned} \qquad (5·73·3)$$

Now, by (1·6·15) (omitting the bar over U_1),

$$x_r\left(\frac{\partial U_1}{\partial x_r}\right)_{N^{(1)}, N^{(2)}, U} = x_r \frac{\partial(U_1, N^{(1)}, N^{(2)}, U)}{\partial(x_r, \lambda_1, \lambda_2, \vartheta)} \Big/ \frac{\partial(N^{(1)}, N^{(2)}, U)}{\partial(\lambda_1, \lambda_2, \vartheta)}, \qquad (5·73·4)$$

and $\Sigma E_r^{(1)} x_r \partial U_1/\partial x_r$ can be expressed as the ratio D_1/D_2 of two determinants. By using the relations (5·71·2) and (5·71·3) to eliminate differentiations with respect to x_r, we find that

$$D_1 = \begin{vmatrix} -\frac{\partial U_1}{\partial \vartheta} & \frac{\partial U_1}{\partial \lambda_1} & 0 & \frac{\partial U_1}{\partial \vartheta} \\ -\frac{\partial N^{(1)}}{\partial \vartheta} & \frac{\partial N^{(1)}}{\partial \lambda_1} & 0 & \frac{\partial N^{(1)}}{\partial \vartheta} \\ 0 & 0 & \frac{\partial N^{(2)}}{\partial \lambda_2} & \frac{\partial N^{(2)}}{\partial \vartheta} \\ -\frac{\partial U_1}{\partial \vartheta} & \frac{\partial U}{\partial \lambda_1} & \frac{\partial U}{\partial \lambda_2} & \frac{\partial U}{\partial \vartheta} \end{vmatrix} = - \begin{vmatrix} \frac{\partial U_1}{\partial \vartheta} & \frac{\partial U_1}{\partial \lambda_1} & 0 & 0 \\ \frac{\partial N^{(1)}}{\partial \vartheta} & \frac{\partial N^{(1)}}{\partial \lambda_1} & 0 & 0 \\ 0 & 0 & \frac{\partial N^{(2)}}{\partial \lambda_2} & \frac{\partial N^{(2)}}{\partial \vartheta} \\ \frac{\partial U_1}{\partial \vartheta} & \frac{\partial U}{\partial \lambda_1} & \frac{\partial U_2}{\partial \lambda_2} & \frac{\partial U_2}{\partial \vartheta} \end{vmatrix}$$

$$= -\frac{\partial(U_1, N^{(1)})}{\partial(\vartheta, \lambda_1)} \frac{\partial(N^{(2)}, U_2)}{\partial(\lambda_2, \vartheta)} = -\frac{\partial N^{(1)}}{\partial \lambda_1}\left(\frac{\partial U_1}{\partial \vartheta}\right)_{N^{(1)}} \frac{\partial N^{(2)}}{\partial \lambda_2}\left(\frac{\partial U_2}{\partial \vartheta}\right)_{N^{(2)}}, \qquad (5·73·5)$$

while, by (5·72·8),
$$D_2=\frac{\partial N^{(1)}}{\partial\lambda_1}\frac{\partial N^{(2)}}{\partial\lambda_2}\left(\frac{\partial U}{\partial\vartheta}\right)_{N^{(1)},N^{(2)}}. \tag{5·73·6}$$

Hence, reinserting the bars to denote averages,

$$\text{Av.}(U_1-\bar{U}_1)^2=-\left(\frac{\partial\bar{U}_1}{\partial\vartheta}\right)_{N^{(1)}}\left(\frac{\partial\bar{U}_2}{\partial\vartheta}\right)_{N^{(2)}}\Big/\left(\frac{\partial U}{\partial\vartheta}\right)_{N^{(1)},N^{(2)}}. \tag{5·73·7}$$

5·74. The fluctuations in the forces exerted by the system on its surroundings can be calculated by means of the formulae already obtained. If A is the generalized force corresponding to the generalized coordinate a, then

$$A^{(1)}=\sum_r N_r^{(1)}(-\partial E_r^{(1)}/\partial a). \tag{5·74·1}$$

Hence
$$\overline{A^{(1)}}=\sum_r \overline{N_r^{(1)}}(-\partial E_r^{(1)}/\partial a) \tag{5·74·2}$$

and

$$\text{Av.}(A^{(1)}-\overline{A^{(1)}})^2=\sum_r \text{Av.}(N_r^{(1)}-\overline{N_r^{(1)}})^2(\partial E_r^{(1)}/\partial a)^2$$
$$+2\sum_{r\neq s}\text{Av.}\{(N_r^{(1)}-\overline{N_r^{(1)}})(N_s^{(1)}-\overline{N_s^{(1)}})\}(\partial E_r^{(1)}/\partial a)(\partial E_s^{(1)}/\partial a). \tag{5·74·3}$$

By using equations (5·72·9) and (5·72·10), this can be reduced to

$$\text{Av.}(A^{(1)}-\overline{A^{(1)}})^2=\sum_r\frac{\partial\overline{N_r^{(1)}}}{\partial\lambda_1}\left(\frac{\partial E_r^{(1)}}{\partial a}\right)^2-\frac{(\partial\overline{A^{(1)}}/\partial\lambda_1)^2}{\partial N^{(1)}/\partial\lambda_1}+\left(\frac{\partial\overline{A^{(1)}}}{\partial\vartheta}\right)^2_{N^{(1)}}\Big/\left(\frac{\partial U}{\partial\vartheta}\right)_{N^{(1)},N^{(2)}}, \tag{5·74·4}$$

which may be transformed as follows. We have

$$\frac{\partial A^{(1)}}{\partial a}=-\sum_r N_r^{(1)}\frac{\partial^2E_r^{(1)}}{\partial a^2},\quad \left(\frac{\partial\overline{A^{(1)}}}{\partial a}\right)_{N^{(1)},N^{(2)}}=-\sum_r\overline{N_r^{(1)}}\frac{\partial^2E_r^{(1)}}{\partial a^2}-\sum_r\left(\frac{\partial\overline{N_r^{(1)}}}{\partial a}\right)_{N^{(1)},N^{(2)}}\frac{\partial E_r^{(1)}}{\partial a},$$

and, since $\overline{N_r^{(1)}}$ only depends upon ϑ and λ_1 through the combination $\vartheta E_r^{(1)}-\lambda_1$,

$$\left(\frac{\partial\overline{N_r^{(1)}}}{\partial a}\right)_{N^{(1)},N^{(2)}}=\left(\frac{\partial\overline{N_r^{(1)}}}{\partial a}\right)_{\lambda_1}+\left(\frac{\partial\overline{N_r^{(1)}}}{\partial\lambda_1}\right)_a\left(\frac{\partial\lambda_1}{\partial a}\right)_{N^{(1)}}=-\frac{\partial\overline{N_r^{(1)}}}{\partial\lambda_1}\left(\vartheta\frac{\partial E_r^{(1)}}{\partial a}+\frac{(\partial N^{(1)}/\partial a)_{\lambda_1}}{\partial N^{(1)}/\partial\lambda_1}\right).$$

Also
$$\left(\frac{\partial N^{(1)}}{\partial a}\right)_{\lambda_1}=-\sum_r\vartheta\frac{\partial\overline{N_r^{(1)}}}{\partial\lambda_1}\frac{\partial E_r^{(1)}}{\partial a}=\vartheta\frac{\partial\overline{A^{(1)}}}{\partial\lambda_1}.$$

Therefore

$$\left(\frac{\partial\overline{A^{(1)}}}{\partial a}-\overline{\frac{\partial A^{(1)}}{\partial a}}\right)_{N^{(1)},N^{(2)}}=-\sum_r\left(\frac{\partial\overline{N_r^{(1)}}}{\partial a}\right)_{N^{(1)},N^{(2)}}\frac{\partial E_r^{(1)}}{\partial a}$$
$$=\sum_r\frac{\partial E_r^{(1)}}{\partial a}\frac{\partial\overline{N_r^{(1)}}}{\partial\lambda_1}\left(\vartheta\frac{\partial E_r^{(1)}}{\partial a}+\vartheta\frac{\partial\overline{A^{(1)}}/\partial\lambda_1}{\partial N^{(1)}/\partial\lambda_1}\right)$$
$$=\vartheta\sum_r\frac{\partial\overline{N_r^{(1)}}}{\partial\lambda_1}\left(\frac{\partial E_r^{(1)}}{\partial a}\right)^2-\vartheta\frac{(\partial\overline{A^{(1)}}/\partial\lambda_1)^2}{\partial N^{(1)}/\partial\lambda_1} \tag{5·74·5}$$

and so

$$\text{Av.}\,(A^{(1)}-\overline{A^{(1)}})^2=\frac{1}{\vartheta}\left(\frac{\partial\overline{A^{(1)}}}{\partial a}-\frac{\overline{\partial A^{(1)}}}{\partial a}\right)_{N^{(1)},N^{(2)}}+\left(\frac{\partial\overline{A^{(1)}}}{\partial\vartheta}\right)^2_{N^{(1)}}\Bigg/\left(\frac{\partial U}{\partial\vartheta}\right)_{N^{(1)},N^{(2)}}. \quad (5·74·6)$$

5·75. It is impossible to give briefly an adequate account of all the possible fluctuation phenomena, and we therefore discuss only the most important points of interest in the formulae of the preceding sections. In the first place, if we have two loosely coupled systems, the fluctuations in the first system depend upon the properties of the second system except when the energy of the second system is very much larger than that of the first. This is equivalent to fixing the temperature of the first system by immersing it in an infinite heat bath, and we shall only consider this case. It is then possible to drop the index distinguishing the system under discussion.

For classical statistics, we have $\overline{N}_r=e^{\lambda-\vartheta E_r}$ and $\partial\overline{N}_r/\partial\lambda=\overline{N}_r$, $\partial N/\partial\lambda=N$. Hence (5·72·9) becomes

$$\text{Av.}\,(N_r-\overline{N}_r)^2=\overline{N}_r-\overline{N}_r^2/N, \quad (5·75·1)$$

which is more usually given in the forms

$$\frac{\overline{N_r^2}-\overline{N}_r{}^2}{\overline{N}_r{}^2}=\frac{1}{\overline{N}_r}-\frac{1}{N},\quad \frac{\overline{n_r^2}-\overline{n}_r{}^2}{\overline{n_r^2}}=\frac{1}{\overline{n}_r}-\frac{1}{N} \quad (5·75·2)$$

(compare equation (5·12·15)). Therefore, for large systems, the mean-square deviation of N_r from its average value $\overline{N}_r$ is negligible provided that $\overline{N}_r$ is itself large.

If we neglect the second term in (5·72·9), which is proportional to $1/N$. we have

$$\overline{n}_r=\frac{\overline{N}_r}{M_r}=\frac{1}{\exp(\vartheta E_r-\lambda)+1},\quad \overline{N_r^2}-\overline{N}_r^2=\overline{N}_r-\frac{\overline{N}_r^2}{M_r},\quad \overline{n_r^2}-\overline{n}_r{}^2=\overline{n}_r-\overline{n}_r{}^2 \quad (5·75·3)$$

for Fermi-Dirac statistics, and

$$\overline{n}_r=\frac{\overline{N}_r}{M_r}=\frac{1}{\exp(\vartheta E_r-\lambda)-1},\quad \overline{N_r^2}-\overline{N}_r^2=\overline{N}_r+\frac{\overline{N}_r^2}{M_r},\quad \overline{n_r^2}-\overline{n}_r{}^2=\overline{n}_r+\overline{n}_r{}^2 \quad (5·75·4)$$

for Einstein-Bose statistics. The fluctuations in a Fermi-Dirac system are less, and those in an Einstein-Bose system are greater, than the fluctuations in a classical system.

5·751. When a system is immersed in a heat bath, equation (5·73·7) shows that

$$\overline{U^2}=\overline{U}^2=-\partial\overline{U}/\partial\vartheta=kT^2\,\partial\overline{U}/\partial T. \quad (5·751·1)$$

Now $\partial\overline{U}/\partial T$ is of the order of Nk, so that

$$\overline{U^2}-\overline{U}^2=O(Nk^2T^2)=O(\overline{U}^2/N), \quad (5·751·2)$$

and the mean-square deviation of the energy is negligible for sufficiently large systems.

For classical statistics we have

$$\bar{U} = \Sigma E_r \exp(-\vartheta E_r)/\Sigma \exp(-\vartheta E_r) \tag{5·751·3}$$

and
$$-\left(\frac{\partial \bar{U}}{\partial \vartheta}\right)_N = \frac{\Sigma E_r^2 \exp(-\vartheta E_r)}{\Sigma \exp(-\vartheta E_r)} - \left(\frac{\Sigma E_r \exp(-\vartheta E_r)}{\Sigma \exp(-\vartheta E_r)}\right)^2. \tag{5·751·4}$$

Hence, if we define a new average of U^2 by the relation

$$[U^2] = \Sigma E_r^2 \exp(-\vartheta E_r)/\Sigma \exp(-\vartheta E_r), \tag{5·751·5}$$

then
$$[U^2] - \bar{U}^2 = -\partial \bar{U}/\partial \vartheta = kT^2 \partial \bar{U}/\partial T, \tag{5·751·6}$$

and $[U^2] = \overline{U^2}$. The equality of these different types of average does not hold for other statistics.

REFERENCES

Bose, S. N. (1924). Planck's law and the hypothesis of light quanta. *Z. Phys.* **26**, 178.

Dirac, P. A. M. (1926). The the quantum mechanics. *Proc. Roy. Soc.* A, **112**, 661.

Einstein, A. (1924). Quantum theory of the monatomic ideal gas. *S.B. preuss. Akad. Wiss.* 261.

Einstein, A. (1925). Quantum theory of the monatomic ideal gas. *S.B. preuss. Akad. Wiss.* 3, 18.

Fermi, E. (1926). The quantisation of the ideal monatomic gas. *Z. Phys.* **36**, 902.

Fowler, R. H. (1936). *Statistical mechanics*, 2nd ed. Cambridge.

Gibbs, J. W. (1902). Elementary principles in statistical mechanics (*Collected Works*, vol. **2**). New York.

Kahn, B. (1938). On the theory of the equation of state. Dissertation. Utrecht.

Kirkwood, J. G. (1933). Quantum statistics of almost classical assemblies. *Phys. Rev.* **44**, 31.

von Neumann, J. (1932). *Mathematische Grundlagen der Quantenmechanik.* Berlin.

Titchmarsh, E. C. (1939). *The theory of functions*, 2nd ed. Oxford.

Tolman, R. (1938). *The principles of statistical mechnics.* Oxford.

Valiron, G. (1923). *Lectures on the general theory of integral functions.* Paris.

Chapter 6

SOME APPLICATIONS OF STATISTICAL MECHANICS

The Specific Heat of Gases

6·1. The classical theory of the specific heat of gases is extremely simple, since, by the principle of the equipartition of energy, each square term in the Hamiltonian function contributes $\frac{1}{2}kT$ to the internal energy. A monatomic gas has therefore a specific heat (at constant volume) of $\frac{3}{2}k$ per particle. A diatomic gas, on the other hand, has rotational and vibrational energy as well as translational energy. The Hamiltonian contains two rotational square terms (arising from the kinetic energy of rotation) and two vibrational square terms (one arising from the kinetic and the other from the potential energy of the vibrational motion). The rotational and vibrational contributions to the specific heat are therefore each equal to k per molecule, and the total specific heat per molecule should be $\frac{7}{2}k$. A general polyatomic molecule containing n atoms has $3n$ degrees of freedom, so that there are three rotational square terms and $6n-12$ vibrational square terms in the Hamiltonian. The total specific heat per molecule should therefore be $3(n-1)k$ $(n \geqslant 3)$. At ordinary temperatures, however, the experimental values of C_V are considerably less than those predicted by the classical theory, and this is due to the rotational and vibrational energy levels being discrete and not forming a continuum, so that the principle of the equipartition of energy does not hold unless the temperature T is such that kT is much larger than the separation between the energy levels.

If ΔE_r is the average energy separation of the rotational levels, the quantity $T_r = \Delta E_r/k$ may vary from about 1° K. for a heavy molecule to about 20° K. for a light molecule (except for hydrogen, which is a special case), and at normal temperatures the rotational degrees of freedom are fully excited. This, however, is not true for the vibrational degrees of freedom, since, if ΔE_v is the average spacing of the vibrational levels, the quantity $T_v = \Delta E_v/k$ may range from 2000 to 5000° K. for diatomic molecules and from about 300 to 1000° K. for polyatomic molecules. (The latter have a considerable number of normal modes and have complex vibrational spectra.) Therefore, as a rough qualitative generalization we may expect the specific heat at constant volume of diatomic gases to be $\frac{5}{2}k$ per molecule at room temperature, while the specific heat of polyatomic molecules will in general be $3k$ per molecule, but may be substantially greater if there are any vibrations with comparatively low frequencies. Linear polyatomic molecules, on the other hand, are very similar to diatomic molecules, since the axial moment of inertia is very small and there are effectively only two rotational degrees of freedom at normal temperatures. At sufficiently high temperatures, however, the vibrations may be sufficiently pronounced to make the molecules non-linear, but in this case it is clearly impossible to treat the rotational and vibrational motions as independent. Some values of the specific heat at 18° C. are given in table 6·1.

Table 6·1. *Values of the specific heat at constant volume per molecule of various gases at* 291° K. *in units of k*

He	A	H_2	N_2	O_2	CO	NO
1·51	1·54	2·45	2·51	2·51	2·52	2·64

C_2H_2	CO_2	N_2O	NH_3	CH_4	C_2H_4
4·20	3·44	3·61	3·28	3·28	4·11

The molecules C_2H_2, CO_2 and N_2O are linear, and the 'normal' contribution of the translational and rotational motions is $2{\cdot}5k$. The molecules NH_3, CH_4 and C_2H_4 are non-linear, and the 'normal' contribution of the translational and rotational motion is $3k$. In all the polyatomic molecules there is a considerable vibrational contribution to the specific heat.

The band spectra of many of the simpler molecules have been analysed and the energy levels deduced. (See, for example, Herzberg (1945). This book is the source of the experimental data quoted later.) In such cases it is, in principle, a relatively straightforward but tedious matter to calculate numerically the partition function to any desired accuracy. It is, however, usually possible to avoid the direct summation by introducing suitable approximative methods, for a description of which the reader is referred to the review articles by Kassel (1936) and E. B. Wilson (1940) and to Chapter 5 of volume 2 of Herzberg's book. The account which follows is limited to the simplest cases, which are, nevertheless, sufficient to illustrate the principles involved.

The translational motion of a molecule is independent of the internal motion,‡ and the total partition function is therefore the product of a partition function relating to the translational motion and of one relating to the internal degrees of freedom. If, in addition, the rotational and vibrational motions are independent of one another, which is true as a first approximation, the partition function for the internal degrees of freedom

‡ For convenience in nomenclature we include the rotational motion in the internal motion, i.e. all motion which is not translational is termed internal. In rotation the molecule behaves, of course, as a rigid body.

again splits up into the product of two independent partition functions. We shall assume that all the motions are independent of one another and therefore that the logarithms of the partition functions are additive.

6·11. *The rotational motion of a diatomic molecule.* Let the orientation of the axis of the molecule be specified by the spherical polar coordinates θ and ϕ whose origin is at the centroid of the molecule, and let I be the moment of inertia of the molecule perpendicular to the axis. Then the classical kinetic energy is $\mathscr{T}=\frac{1}{2}I(\dot\theta^2+\dot\phi^2\sin^2\theta)$. Hence the generalized momenta are $p_\theta=\partial\mathscr{T}/\partial\dot\theta=I\dot\theta$, $p_\phi=\partial\mathscr{T}/\partial\dot\phi=I\dot\phi\sin^2\theta$, and the Hamiltonian function is

$$\mathscr{H}=\frac{1}{2I}\left(p_\theta^2+\frac{1}{\sin^2\theta}p_\phi^2\right).$$

Hence the Schrödinger equation is

$$\frac{h^2}{8\pi^2 I}\left(\frac{\partial^2\psi}{\partial\theta^2}+\frac{1}{\sin^2\theta}\frac{\partial^2\psi}{\partial\phi^2}\right)+E\psi=0. \tag{6·11·1}$$

The eigenfunctions of this equation are

$$\psi=P_j^{|m|}(\cos\theta)\,e^{im\phi}\quad(m=-j,\,-j+1,\,\ldots,\,j-1,\,j), \tag{6·11·2}$$

where $P_j^{|m|}(\cos\theta)$ is the associated Legendre function, while the energy levels are

$$E_j=j(j+1)\,h^2/(8\pi^2 I)\quad(j=0,1,\ldots). \tag{6·11·3}$$

The jth energy level possesses $2j+1$ independent wave functions and is therefore of weight $2j+1$. This gives the partition function as

$$Z(x)=\sum_{j=0}^{\infty}(2j+1)\,e^{-xj(j+1)}, \tag{6·11·4}$$

where

$$x=h^2/(8\pi^2 IkT). \tag{6·11·5}$$

The sum $Z(x)$ can only be evaluated over the whole range of x by numerical methods, but we can readily obtain the limiting forms at high and low temperatures. For high temperatures (small values of x) we may replace the sum by an integral and write

$$e^{\frac{1}{4}x}\sum_{j=0}^{\infty}(2j+1)\exp[-x(j+\tfrac{1}{2})^2]\sim e^{\frac{1}{4}x}\int_0^\infty(2t+1)\exp[-x(t+\tfrac{1}{2})^2]\,dt=1/x. \tag{6·11·6}$$

A more accurate result can be obtained by using the Euler-Maclaurin sum formula

$$\sum_{\nu=0}^{n}f(\nu)=\int_0^n f(t)\,dt+\tfrac{1}{2}\{f(n)+f(0)\}$$
$$+\sum_{k=1}^{\lambda}(-1)^{k-1}\frac{B_k}{(2k)!}[f^{(2k-1)}(n)-f^{(2k-1)}(0)]+R_\lambda, \tag{6·11·7}$$

where $B_1=\frac{1}{6}$, $B_2=\frac{1}{30}$, $B_3=\frac{1}{42}$, ..., and where R_λ is the remainder. It is readily found that

$$Z(x)=x^{-1}+\tfrac{1}{3}+\tfrac{1}{15}x+O(x^2). \tag{6·11·8}$$

Hence, with $\vartheta=1/kT$,

$$\frac{U}{N}=-\frac{\partial}{\partial\vartheta}\log Z=-\frac{h^2}{8\pi^2 I}\frac{\partial}{\partial x}\log Z, \tag{6·11·9}$$

and the specific heat per molecule is

$$C_V^{\text{rot.}}=\frac{\partial}{\partial T}\frac{U}{N}=-\frac{h^2}{8\pi^2 IkT^2}\frac{\partial}{\partial x}\frac{U}{N}=kx^2\frac{\partial^2}{\partial x^2}\log Z. \tag{6·11·10}$$

Hence
$$C_V^{\text{rot.}}=k\{1+\tfrac{1}{45}x^2+O(x^3)\}. \tag{6·11·11}$$

The first term gives the value k for $C_V^{\text{rot.}}$. This is the 'classical' value, which, as is clear from the preceding calculation, is obtained if we consider the spacing between the energy levels to be small compared with kT.

For low temperatures we have

$$Z=1+3e^{-2x}+\ldots,\quad \log Z=3e^{-2x}+\ldots \tag{6·11·12}$$

and
$$C_V^{\text{rot.}}=12kx^2e^{-2x}+\ldots.$$

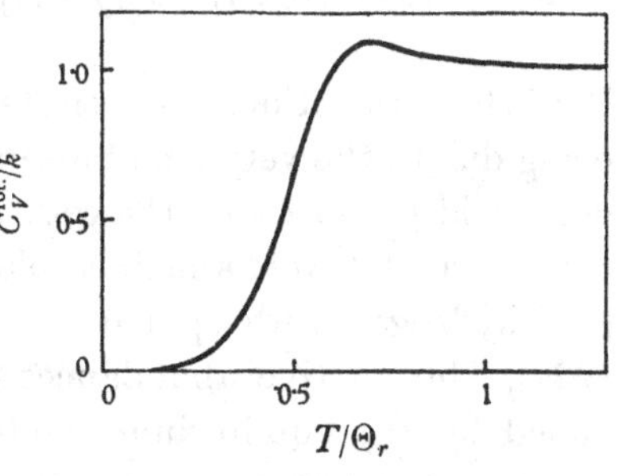

Fig. 6·1. The theoretical rotational specific heat of a heteronuclear diatomic molecule as a function of T/Θ_r.

The form of the rotational specific heat curve as a function of T/Θ_r, where

$$\Theta_r=h^2/(8\pi^2 kI), \tag{6·11·13}$$

is shown in fig. 6·1.

6·12. *The rotational specific heat of hydrogen.* The moments of inertia of most diatomic molecules are so large that in general the rotational specific heat per molecule has its classical value k for all temperatures at which the specific heat of the gas can be measured. Hydrogen is an exception, but in this case it is also necessary to take into account the influence of the spins of the protons upon the symmetry properties of the wave functions describing the motion of the molecule. The proton has spin $\frac{1}{2}$ (in units of $\hbar$), and the hydrogen molecule as a whole has a nuclear spin of either 0 (with weight 1) or 1 (with weight 3). Now the total wave function for the nuclear motion of a hydrogen molecule must be antisymmetrical in the nuclei, taking into account both the rotational and the spin coordinates. But, interchanging the two nuclei is equivalent to replacing θ by $\pi-\theta$ and ϕ by $\pi+\phi$, i.e. replacing $P_j^{|m|}(\cos\theta)\,e^{im\phi}$ by $(-1)^m P_j^{|m|}(-\cos\theta)\,e^{im\phi}$, which is equal to $(-1)^j P_j^{|m|}(\cos\theta)\,e^{im\phi}$, since $P_j^{|m|}(\cos\theta)$ is an odd or even function of $\cos\theta$ according as $j+|m|$ is odd or even. Hence the rotational wave function is symmetrical in the nuclei for even j and antisymmetrical for odd j, while the states with spin 0 have antisymmetrical and those with spin 1 have symmetrical spin wave functions. In order to obtain a total wave function which is antisymmetrical we must therefore associate the rotational wave functions having even j's with spin wave functions having spin 0, and rotational functions with odd j's with spin functions having spin 1. This means that all rotational states with even j's have weight 1, while those with odd j's have weight 3.

If hydrogen were an equilibrium mixture, the partition function would be given by

$$Z = \sum_{\text{even } j} (2j+1)\exp[-xj(j+1)] + 3\sum_{\text{odd } j}(2j+1)\exp[-xj(j+1)]. \quad (6\cdot12\cdot1)$$

But in fact transitions between states with different spins are extremely rare, being due to the very small interaction of the nuclear spins with the magnetic field produced by the rotation of the nuclei, and hydrogen normally behaves as if it were a metastable mixture of two entirely separate gases—para-hydrogen with spin 0 and ortho-hydrogen with spin 1 (Dennison, 1927). The relative abundances of the two types of molecule are determined by the equilibrium conditions at room temperatures. Since the approximation leading to (6·11·6) shows that at high temperatures

$$\sum_{\text{even } j}(2j+1)\exp[-xj(j+1)] = \sum_{\text{odd } j}(2j+1)\exp[-xj(j+1)]$$
$$= \tfrac{1}{2}\sum_{\text{all } j}(2j+1)\exp[-xj(j+1)] = \frac{1}{2x},$$

the relative abundances of para- and ortho-hydrogen are 1:3. If then we write

$$\left.\begin{aligned} Z_p = Z_{\text{even}} &= \sum_{\text{even } j}(2j+1)\exp[-xj(j+1)],\\ Z_o = Z_{\text{odd}} &= \sum_{\text{odd } j}(2j+1)\exp[-xj(j+1)],\end{aligned}\right\} \quad (6\cdot12\cdot2)$$

the specific heat per molecule is given by

$$C_V^{\text{rot.}} = \tfrac{1}{4}kx^2\frac{\partial^2}{\partial x^2}\log Z_p + \tfrac{3}{4}kx^2\frac{\partial^2}{\partial x^2}\log Z_o = \tfrac{1}{4}kx^2\frac{\partial^2}{\partial x^2}\log(Z_p Z_o^3), \quad (6\cdot12\cdot3)$$

instead of by
$$C_V^{\text{rot.}} = kx^2\frac{\partial^2}{\partial x^2}\log(Z_p + 3Z_o). \quad (6\cdot12\cdot4)$$

Similar considerations apply to all homonuclear molecules, i.e. those with identical nuclei, and we can generalize the above results to cases where the nuclear spins are different from $\tfrac{1}{2}\hbar$. If the spin of a nucleus is $\tau\hbar$ there are $\rho = 2\tau+1$ spin wave functions $\psi_r(a)$, $\psi_s(b)$ $(r, s = 1, \ldots, \rho)$ for each nucleus. From these we can form $\tfrac{1}{2}\rho(\rho-1)$ symmetrical and $\tfrac{1}{2}\rho(\rho-1)$

antisymmetrical combinations $\psi_r(a)\,\psi_s(b) \pm \psi_s(a)\,\psi_r(b)$ $(r \neq s)$, and ρ symmetrical products $\psi_r(a)\,\psi_r(b)$. Hence there are $\frac{1}{2}\rho(\rho-1)$ antisymmetrical spin wave functions and $\frac{1}{2}\rho(\rho+1)$ symmetrical ones, and these are the weight factors.

For deuterium we have $\tau = 1$ and $\rho = 3$, and since the deuterium nucleus contains two particles (a neutron and a proton) the complete nuclear wave function must be symmetrical. The rotational wave functions with even j's have therefore weight 6, while those with odd j's have weight 3, and so for the metastable mixture we have

$$C_V^{\text{rot.}} = \tfrac{6}{9}kx^2 \frac{\partial^2}{\partial x^2} \log Z_{\text{even}} + \tfrac{3}{9}kx^2 \frac{\partial^2}{\partial x^2} \log Z_{\text{odd}}. \tag{6·12·5}$$

In this case the para states have odd j's and the ortho states have even j's. On the other hand, for the mixed molecule HD there are no symmetry requirements, and the specific heat is determined by (6·11·4).

At high temperatures where $Z_{\text{even}} = Z_{\text{odd}} = \frac{1}{2}Z$, the partition function, including the nuclear weight factor, is $\rho_1\rho_2 Z$ for heteronuclear diatomic molecules and $\frac{1}{2}\rho^2 Z$ for homonuclear diatomic molecules, if the spin weights of the nuclei are ρ_1 and ρ_2 (or ρ for homonuclear molecules). At these temperatures it is convenient to define a symmetry number σ for homonuclear molecules of any type such that the complete partition function is

$$\frac{\Pi\rho}{\sigma} Z = \frac{\Pi\rho}{\sigma} \frac{8\pi^2 IkT}{h^2}, \tag{6·12·6}$$

where $\Pi\rho$ is the product of the spin weights of the nuclei. For diatomic homonuclear molecules we have $\sigma = 2$, but σ may take greater values for polyatomic molecules.

Although the transitions between the ortho and para states of hydrogen are so infrequent that the mixture maintains its high temperature composition (namely, a 3 : 1 ratio) during all normal measurements of specific heats it is possible to catalyse the ortho to para transitions at low temperatures by bringing the gas into contact with activated charcoal, thereby obtaining an equilibrium mixture. If the catalysis is carried out at sufficiently low temperatures, effectively pure para-hydrogen is obtained, and this will remain in metastable equilibrium, in the absence of a catalyst, if the temperature is raised. It is therefore possible in principle to measure the specific heats of the metastable mixture, of the equilibrium mixture and of pure para-hydrogen. The theoretical curves are shown in fig. 6·2, and similar results are shown for deuterium in fig. 6·3. These curves are in excellent agreement with the experimental results if we take $\Theta_r = 85{\cdot}4°$ K. for hydrogen and $\Theta_r = 42{\cdot}7°$ K. for deuterium. The experimental values of C_V for HD are in agreement with the curve given in fig. 6·1, with $\Theta_r = 64{\cdot}1°$ K.

It will be seen from figs. 6·1, 6·2 and 6·3 that the rotational specific heat differs appreciably from the classical value k only when T is less than about $4\Theta_r$. Some values of Θ_r for various gases, obtained from the band spectra of the molecules, are given in table 6·2 on p. 145, and it will be seen that it is

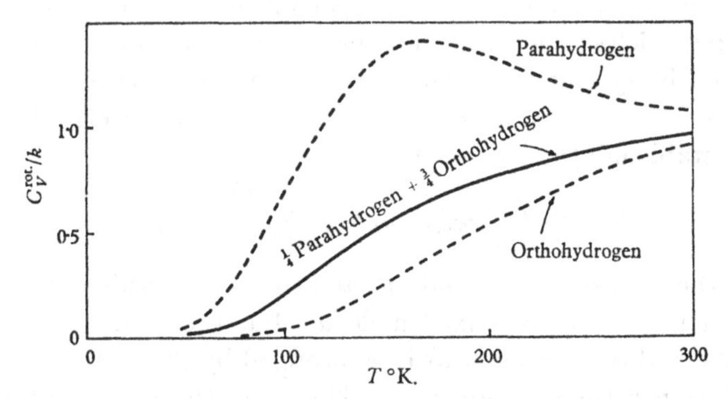

Fig. 6·2. The rotational specific heat of hydrogen. $\Theta_r = 85{\cdot}4°$ K.

only for the three hydrogen molecules that any temperature variation of the rotational specific heat is observable, since the other substances would be in the liquid state before the appropriate temperature range was reached.

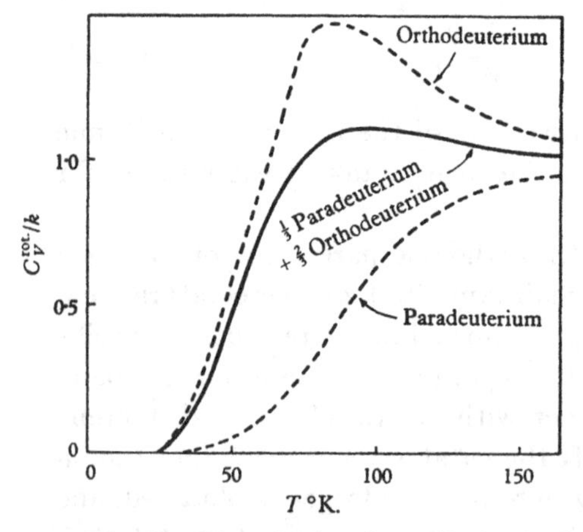

Fig. 6·3. The rotational specific heat of deuterium. $\Theta_r = 42{\cdot}7°$ K.

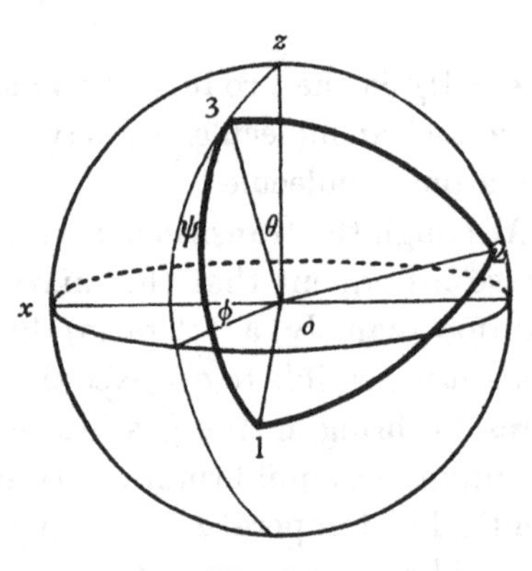

Fig. 6·4

6·13. *The rotational motion of polyatomic molecules.* The rotational motion of a linear polyatomic molecule is the same as that of a diatomic molecule, and the configuration can be described by the spherical polar coordinates θ and ϕ. To specify the configuration of a non-linear molecule we use moving axes O_1, O_2, O_3, the origin of which is at the centroid of the molecule, the axes being directed along the principal axes of inertia. Relative to axes Ox, Oy, Oz fixed in space, the position of these axes is defined by the spherical polar angles θ, ϕ, and the third Eulerian angle ψ. (See fig. 6·4 which portrays the spherical triangles defined by the points in which the axes cut the unit sphere.) The angular velocity $(\omega_1, \omega_2, \omega_3)$ of the molecule is then given by

$$\omega_1 = \dot{\theta}\sin\psi - \dot{\phi}\sin\theta\cos\psi, \quad \omega_2 = \dot{\theta}\cos\psi + \dot{\phi}\sin\theta\sin\psi, \quad \omega_3 = \dot{\psi} + \dot{\phi}\cos\theta, \tag{6·13·1}$$

while the kinetic energy $\mathscr{T}$ is given by

$$2\mathscr{T} = I_1\omega_1^2 + I_2\omega_2^2 + I_3\omega_3^2, \tag{6·13·2}$$

where I_1, I_2, I_3 are the principal moments of inertia. An explicit formula can be found for the energy levels if two of I_1, I_2, I_3 are equal, but this is not possible in the general case. The moments of inertia are, however, such that it is sufficiently accurate to use the classical partition function for all temperatures for which the gases exist, due allowance being made for the existence of identical configurations. We define the symmetry number σ as the number of distinct symmetry operations (rotations, reflexions and inversions) which keep invariant the configuration of a molecule. For H_2O (an isosceles triangle) we have $\sigma = 2$; for NH_3 (an equilateral triangular pyramid) we have $\sigma = 3$; and for CH_4 (a regular tetrahedron with the C atom at the centroid) we have $\sigma = 12$. When $\sigma \neq 1$, the various rotational terms form non-combining groups, as in the hydrogen molecule, and the gas behaves as a metastable mixture. In the classical limit, however, we have merely to introduce a factor $1/\sigma$ into the partition function. If we include the nuclear spin weight factor, the classical partition function is therefore

$$Z_{\text{rot.}} = \frac{\Pi\rho}{\sigma}\int\dots\int \exp(-\mathscr{T}/kT)\,d\theta\,d\phi\,d\psi\,dp_\theta\,dp_\phi\,dp_\psi, \tag{6·13·3}$$

where

$$\left.\begin{aligned} p_\theta &= \partial\mathscr{T}/\partial\dot{\theta} = I_1\omega_1\sin\psi + I_2\omega_2\cos\psi,\\ p_\phi &= \partial\mathscr{T}/\partial\dot{\phi} = -I_1\omega_1\sin\theta\cos\psi + I_2\omega_2\sin\theta\sin\psi + I_3\omega_3\cos\theta,\\ p_\psi &= \partial\mathscr{T}/\partial\dot{\psi} = I_3\omega_3. \end{aligned}\right\} \tag{6·13·4}$$

To evaluate (6·13·3) it is simplest to change the variables from p_θ, p_ϕ, p_ψ to $\omega_1, \omega_2, \omega_3$. The Jacobian of the transformation is readily found to be

$$\frac{\partial(p_\theta, p_\phi, p_\psi)}{\partial(\omega_1, \omega_2, \omega_3)} = I_1 I_2 I_3 \sin\theta, \tag{6·13·5}$$

and so

$$\begin{aligned} Z_{\text{rot.}} &= \frac{\Pi\rho}{\sigma} I_1 I_2 I_3 \int_0^\pi \sin\theta\,d\theta \int_0^{2\pi} d\phi \int_0^{2\pi} d\psi \int_{-\infty}^{\infty}\int_{-\infty}^{\infty}\int_{-\infty}^{\infty} e^{-\mathscr{T}/kT}\,d\omega_1\,d\omega_2\,d\omega_3 \\ &= \frac{\Pi\rho}{\sigma}\frac{8\pi^2}{h^3}(I_1 I_2 I_3)^{\frac{1}{2}}(2\pi kT)^{\frac{3}{2}}. \end{aligned} \tag{6·13·6}$$

Some values of $\Theta_r^i = h^2/(8\pi^2 k I_i)$, deduced from the band spectra, are given in table 6·3 below. The values of Θ_r^i do not affect the rotational specific heat, which necessarily has its classical value of $\frac{3}{2}k$ per molecule, but they appear explicitly in the expressions for the free energy and the entropy.

6·14. *The vibrational specific heat.* If the vibrations of the nuclei of a diatomic molecule are simple harmonic with frequency ν, the possible energy levels are given by $E_n = (n+\frac{1}{2})h\nu$. Hence

$$Z = \sum_{n=0}^{\infty} \exp[-(n+\tfrac{1}{2})h\nu/kT] = \frac{\exp(-\frac{1}{2}h\nu/kT)}{1-\exp(-h\nu/kT)}. \qquad (6\cdot14\cdot1)$$

If we put $x = h\nu/kT$, then

$$C_V^{\text{vib.}} = kx^2 \frac{\partial^2}{\partial x^2}\log Z = k\left(\frac{\Theta_v}{T}\right)^2 \frac{\exp(\Theta_v/T)}{[\exp(\Theta_v/T)-1]^2} \quad (\Theta_v = h\nu/k). \quad (6\cdot14\cdot2)$$

The general form of the vibrational specific heat as a function of T/Θ_v is shown in fig. 6·5, and it will be seen that its limiting value is k for large T

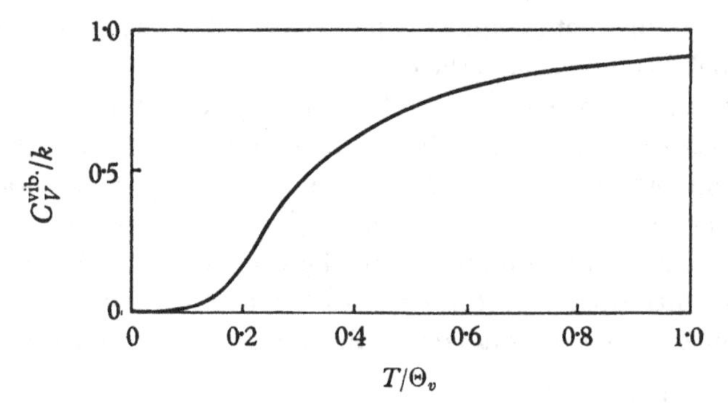

Fig. 6·5. The theoretical vibrational specific heat of a diatomic gas as a function of T/Θ_v.

but that it only becomes appreciable for temperatures greater than about $0\cdot1\Theta_v$. The experimental results, using the spectroscopic values of Θ_v given in table 6·2, are in good general agreement with the theoretical predictions. In general, however, there is a coupling between the vibrations and the rotations of a molecule, and it is not easy to disentangle the two contributions. In such cases empirical expressions for the energy levels have to be obtained by analysing the band spectra, and the partition function can then be calculated by numerical methods. Even when the coupling between the vibrations and rotation is negligible, it is often necessary to take into account the anharmonic nature of the vibrations at high temperatures. The expression for the energy levels is then of the form

$$E_n = (n+\tfrac{1}{2})h\nu - a(n+\tfrac{1}{2})^2 h\nu,$$

where a is a measure of the anharmonicity. The anharmonic contribution to the specific heat may amount to as much as $0{\cdot}25k$ per molecule at 1000° C.

Table 6·2. *Values of Θ_r and Θ_v for various diatomic molecules*

	H_2	HD	D_2	HCl	HBr	HI	CO	N_2	NO	O_2
Θ_r	85·4	64·1	42·7	15·2	12·1	9·0	2·77	2·86	2·42	2·07
$\Theta_v \times 10^{-3}$	6·1	5·3	4·3	4·14	3·7	3·2	3·07	3·34	2·69	2·23

A polyatomic molecule containing n atoms has in general $3n-6$ vibrational frequencies ν_i, some of which may coincide with one another. (A linear molecule, however, has $3n-5$ vibrational frequencies, since it only has two rotational degrees of freedom.) The vibrational partition function therefore consists of a number of terms such as (6·14·1). The values of $\Theta_v^i = h\nu_i/k$ for some of the simpler molecules, are given in table 6·3.

Table 6·3. *Values of σ, I and Θ_v for various polyatomic molecules*

The numbers in brackets give the number of coincident vibrational frequencies.

	CO_2	H_2O	NH_3	CH_4
σ	2	2	3	12
$I \times 10^{40}$ g.cm.^{2}	70·4	0·996	2·78	5·267
		1·908	2·78	5·267
		2·981	4·33	5·267
Θ_v	960(2)	2294	1366	1875(3)
	1900	5180	2340(2)	2185(2)
	3400	5400	4790	4190
			4990(2)	4340(3)

6·15. *Internal rotations in polyatomic molecules.* It is not always possible to consider the motion of a molecule as being that of an essentially rigid body, the constituent atoms of which execute small vibrations about their positions of equilibrium. In the paraffins, for example, the molecules can rotate round every carbon to carbon linkage. (See also § 14·1 for the corresponding phenomenon in rubber.) In ethane $CH_3.CH_3$, therefore, instead of there being 18 vibrational frequencies, there are only 17, the remaining vibration being replaced by a rotation of the two CH_3 groups relative to one another. If classical conditions apply, this rotational degree of freedom contributes $\frac{1}{2}k$ to the specific heat per molecule, but such internal rotations are far from being free. If ϕ denotes the angular displacement of one CH_3 group relative to the other, then, due to the interaction of the groups with one another, there is a potential energy resisting the rotation of the form $A_n(1-\cos n\phi)$ where n is an integer (or the sum of such terms), and for small energies the motion is a libration and not a complete rotation. The lowest energy levels are therefore essentially vibrational ones, whereas, if the total energy is greater than $2A_n$, there is free rotation. The intermediate energy levels are of a mixed character. The energy spectrum can be determined by solving the relevant Schrödinger equation, which in this case is the differential equation for the Mathieu functions (or its generalization the Hill equation when there are several terms in the potential energy). The calculations are complicated and have been considered by many authors (see, for example, Pitzer & Gwinn, 1942), and tables of the thermodynamic functions have been compiled. There is, however, considerable uncertainty in the value of A_n which determines the size of the potential hump, and it can only really be inferred by comparing the calculated and observed values of the thermodynamic functions, and assuming that the model used gives a correct description of the molecular motion. The most likely value of the potential hump in ethane seems to be about 2750 calories/gram molecule (i.e. 0·16 eV.).

6·16. *The electronic specific heat.* If the molecule is such that there is an excited electronic state whose energy is ΔE above the ground state, where ΔE is of the same order as kT, there is an appreciable electronic contribution to the specific heat. Let the ground state have weight ϖ_0 and the excited state weight ϖ_1. It is convenient to include the weight ϖ_0 in the translational partition function, and the electronic partition function $Z_{\text{el.}}$ is then given by

$$\varpi_0 Z_{\text{el.}} = \varpi_0 + \varpi_1 e^{-\Delta E/kT}. \tag{6·16·1}$$

Hence, if $x = \Delta E/kT$,

$$S_{\text{el.}} = -\frac{\partial}{\partial x}(x \log Z_{\text{el.}}) = k \log\left(1 + \frac{\varpi_1}{\varpi_0} e^{-\Delta E/kT}\right) + \frac{\Delta E/T}{1 + (\varpi_0/\varpi_1)\, e^{\Delta E/kT}}, \tag{6·16·2}$$

$$C_V^{\text{el.}} = kx^2 \frac{\partial^2}{\partial x^2} \log Z_{\text{el.}} = \frac{k(\Delta E/kT)^2}{\{1 + (\varpi_1/\varpi_0)\, e^{-\Delta E/kT}\}\{1 + (\varpi_0/\varpi_1)\, e^{\Delta E/kT}\}}. \tag{6·16·3}$$

The $C_V^{\text{el.}} - T$ curve has a sharp maximum near $T = \Delta E/k$, as shown in fig. 6·6. A specific heat with this temperature variation is often known as a specific heat of the Schottky type.

The molecule NO, which has an odd number of electrons, has two low-lying states, both of weight 2, separated by an energy of $178k$, so that $\varpi_0 Z_{\text{el.}} = 2(1 + e^{-178/T})$. The specific heat of NO over the range 130–180° K. is shown in fig. 6·7 (Eucken & d'Or, 1932). At these temperatures the electronic specific heat is adequately represented by (6·16·3), and it will be seen that its contribution is of the amount to be expected, but measurements over a wider range are lacking.

Apart from nitric oxide and free radicals, the only molecule which has a degenerate ground state is oxygen, for which $\varpi_0 = 3$. The excitation energy of oxygen is, however, considerable, and the electronic specific heat is r appreciable below 1500° K.

6·17. *The free energy.* The free energy of a gas of any type can be written down by combining the results of the preceding sections with those of § 5·4. Unlike the specific heat, the free energy, even under classical conditions, depends upon the moments of inertia and the vibrational frequencies of the molecules. If the temperature is high enough for the

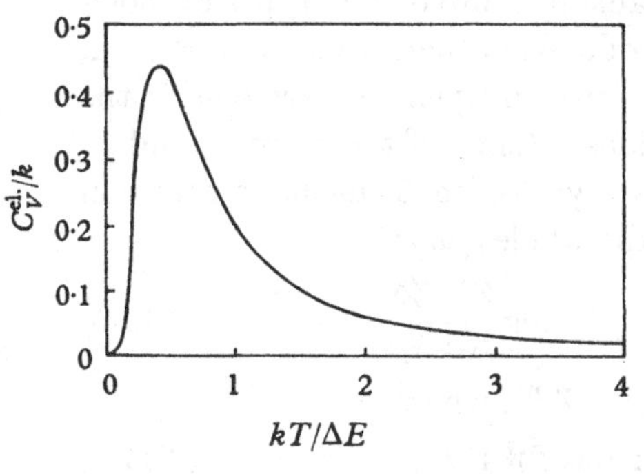

Fig. 6·6. The theoretical electronic specific heat as a function of $kT/\Delta E$ for $\varpi_0 = \varpi_1$.

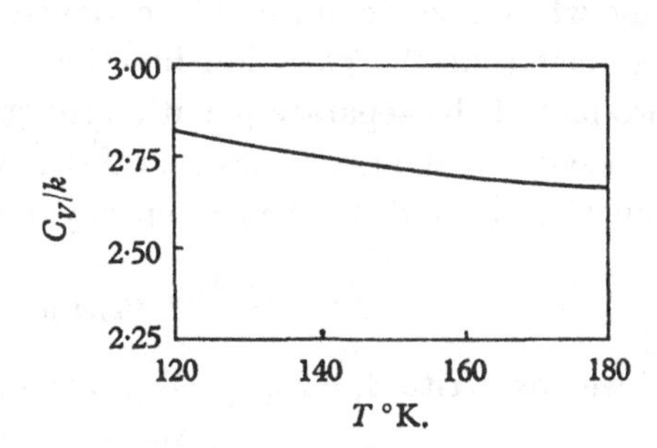

Fig. 6·7. The specific heat of nitric oxide.

rotations to be classical and if we include the nuclear spin weights, the free energy is given by the following formulae. For linear molecules we have

$$F = -NkT\{1+\log(V/N)\} - \tfrac{5}{2}NkT\log T + NkT\sum_i\{\log[1-\exp(-h\nu_i/kT)] - \tfrac{1}{2}h\nu_i/kT\} - NkT\log Z_{\text{el.}} - NkT\log\left(\frac{\varpi_0(2\pi mk)^{\frac{3}{2}}}{h^3}\,\frac{8\pi^2 Ik}{h^2}\,\frac{\Pi\rho}{\sigma}\right), \tag{6·17·1}$$

$$G = NkT\log P - \tfrac{7}{2}NkT\log T + NkT\sum_i\{\log[1-\exp(-h\nu_i/kT)] - \tfrac{1}{2}h\nu_i/kT\} - NkT\log Z_{\text{el.}} - NkT\log\left(\frac{\varpi_0(2\pi mk)^{\frac{3}{2}}}{h^3}\,\frac{8\pi^2 Ik}{h^2}\,\frac{\Pi\rho}{\sigma}\right), \tag{6·17·2}$$

while for non-linear molecules we have

$$F = -NkT\{1+\log(V/N)\} - 3NkT\log T + NkT\sum_i\{\log[1-\exp(-h\nu_i/kT)] - \tfrac{1}{2}h\nu_i/kT\} - NkT\log Z_{\text{el.}} - NkT\log\left(\frac{\varpi_0(2\pi mk)^{\frac{3}{2}}}{h^3}\,\frac{8\pi^2(2\pi k)^{\frac{3}{2}}(I_1I_2I_3)^{\frac{1}{2}}}{h^3}\,\frac{\Pi\rho}{\sigma}\right), \tag{6·17·3}$$

$$G = NkT\log P - 4NkT\log T + NkT\sum_i\{\log[1-\exp(-h\nu_i/kT)] - \tfrac{1}{2}h\nu_i/kT\} - NkT\log Z_{\text{el.}} - NkT\log\left(\frac{\varpi_0(2\pi mk)^{\frac{3}{2}}}{h^3}\,\frac{8\pi^2(2\pi k)^{\frac{3}{2}}(I_1I_2I_3)^{\frac{1}{2}}}{h^3}\,\frac{\Pi\rho}{\sigma}\right), \tag{6·17·4}$$

where m is the mass of the molecule. Any of the other thermodynamic functions can be obtained by performing the appropriate differentiations. These formulae are used in § 7·5 to calculate the 'spectroscopic entropies' of gases, and in § 11·7 to calculate the equilibrium constants of gas reactions.

To derive the expression for the free energy of a homonuclear gas such as hydrogen which consists of a metastable mixture of two types of molecule which behave quite independently of one another, we may start from the fact that the partition function for two independent systems is the product of the separate partition functions. Hence, if there are N_1 and N_2 molecules of the two types, and if the two types of molecule have partition functions Z_1 and Z_2, the free energy of the whole gas is

$$F = -kT \log Z_{\text{system}} = -kT \log \frac{Z_1^{N_1} Z_2^{N_2}}{N_1!\, N_2!}. \tag{6·17·5}$$

If we now write $Z_1 = Z^{(1)}_{\text{trans.}} Z^{(1)}_{\text{int.}}$, $Z_2 = Z^{(2)}_{\text{trans.}} Z^{(2)}_{\text{int.}}$, where

$$Z^{(1)}_{\text{trans.}} = Z^{(2)}_{\text{trans.}} = \varpi_0 (2\pi mkT)^{\frac{3}{2}}\, V/h^3, \tag{6·17·6}$$

and where $Z^{(1)}_{\text{int.}}$ and $Z^{(2)}_{\text{int.}}$ are independent of the volume, we have

$$\begin{aligned} F = kT[&N \log (N/V) - N - \tfrac{3}{2} N \log T + N_1 \log N_1 + N_2 \log N_2 - N \log N \\ &- N_1 \log Z^{(1)}_{\text{int.}} - N_2 \log Z^{(2)}_{\text{int.}} - N \log \{\varpi_0 (2\pi mk)^{\frac{3}{2}}/h^3\}], \end{aligned} \tag{6·17·7}$$

where $N = N_1 + N_2$.

In general, $Z_{\text{int.}} = Z_{\text{rot.}} Z_{\text{vib.}} Z_{\text{el.}}$, and, if the temperature is sufficiently low to invalidate (6·17·1) and (6·17·3), only $Z_{\text{rot.}}$ is different from unity. For hydrogen, and other diatomic molecules with half-integral nuclear spins, we have

$$\left.\begin{aligned} Z^{(1)}_{\text{rot.}} &= \tfrac{1}{2}\rho(\rho-1) \sum_{\text{even } j} (2j+1) \exp[-xj(j+1)], \\ Z^{(2)}_{\text{rot.}} &= \tfrac{1}{2}\rho(\rho+1) \sum_{\text{odd } j} (2j+1) \exp[-xj(j+1)], \end{aligned}\right\} \tag{6·17·8}$$

where ρ is the spin weight of a nucleus and where $x = h^2/(8\pi^2 IkT)$. For sufficiently small values of T we need only consider the terms $j = 0$ and $j = 1$, and this gives

$$\begin{aligned} F = kT\Bigg[&N \log \frac{N}{V} - N - \tfrac{3}{2} N \log T + N_1 \log \frac{N_1}{\frac{1}{2}\rho(\rho-1) N} \\ &+ N_2 \log \frac{N_2}{\frac{1}{2}\rho(\rho+1) N} - N_2 \log 3 - N \log \frac{\varpi_0 (2\pi mk)^{\frac{3}{2}}}{h^3}\Bigg] + N_2 \frac{h^2}{4\pi^2 I}, \end{aligned} \tag{6·17·9}$$

the last term being the zero-point energy of the state with $j = 1$. If the gas has the composition corresponding to equilibrium between the ortho and para states at high temperatures where the rotations are classical, we have

$$\frac{N_1}{\frac{1}{2}\rho(\rho-1) N} = \frac{N_2}{\frac{1}{2}\rho(\rho+1) N} = \frac{1}{\rho^2}, \tag{6·17·10}$$

and the fourth and fifth terms in (6·17·9) are equal to $-NkT \log \rho^2$.

For deuterium and other diatomic molecules with integral nuclear spins we have to interchange the spin factors in (6·17·8). The free energy is then given by (6·17·9) with the factors $\frac{1}{2}\rho(\rho-1)$, $\frac{1}{2}\rho(\rho+1)$ interchanged, and the normal mixture has

$$\frac{N_1}{\frac{1}{2}\rho(\rho+1)\,N}=\frac{N_2}{\frac{1}{2}\rho(\rho-1)\,N}=\frac{1}{\rho^2}. \qquad (6\cdot17\cdot11)$$

The fourth and fifth terms therefore give $-NkT\log\rho^2$ as before.

At high temperatures (6·17·8) becomes

$$Z^{(1)}_{\text{rot.}}=\tfrac{1}{2}\rho(\rho-1)\,4\pi^2IkT/h^2, \quad Z^{(2)}_{\text{rot.}}=\tfrac{1}{2}\rho(\rho+1)\,4\pi^2IkT/h^2,$$

and, if N_1 and N_2 are given by (6·17·10), we regain (6·17·2) with $\sigma=2$. The same result holds for deuterium.

Crystalline Solids

6·2. The atoms of an ideal crystalline solid, i.e. one without impurities and distortions and in thermodynamic equilibrium, are arranged upon a three-dimensional lattice, and the symmetry properties of the lattice give rise to the characteristic shapes of macroscopic crystals. Every crystal lattice possesses a fundamental or unit cell containing comparatively few atoms (in many cases one atom), and the crystal consists of the fundamental cell many times repeated, so that it is in effect a gigantic molecule.

One of the earliest empirical facts established concerning the specific heat of solids at ordinary temperatures was the law enunciated by Dulong & Petit in 1819, namely, that the specific heat at constant pressure of most elements is about 6·4 calories per gram atom per degree. When the correction to constant volume is made, as detailed in §3·212, the law states that the specific heat at constant volume is about 6 calories per gram atom per degree. An extension of Dulong & Petit's law, due to Neumann and to Regnault, states that the molecular heat of a solid chemical compound is equal to the sum of the atomic heats of the constituent elements. We may therefore say that each atom in a solid contributes about $3k$ to the specific heat at constant volume, which is just twice the specific heat per atom of a monatomic perfect gas.

This empirical result can be readily interpreted according to the classical theory as follows. Each atom in a crystal is associated with a definite lattice point, but can vibrate about its equilibrium position under the influence of the forces exerted upon it by the other atoms. The atoms cannot therefore be considered as independent systems, but, provided that the mean displacements from the equilibrium positions are not too large, the motion of the atoms can be considered to be simple harmonic and can be analysed into normal modes by the usual method of transforming the kinetic and the potential energies simultaneously into the sum of squares. If there are N atoms in the crystal, there are $3N$ independent normal modes, each of which can be treated as a simple harmonic oscillator with frequency ν. Since the Hamiltonian for a simple harmonic oscillator of mass M and frequency ν is $\mathscr{H} = \frac{1}{2}(p/M)^2 + 2\pi^2 M\nu^2 q^2$, there are $6N$ square terms in the total Hamiltonian of the vibrations of the crystal, and the internal energy is therefore $3NkT$, which is equivalent to Dulong & Petit's law.

It was early recognized that Dulong & Petit's law was only valid at high temperatures and that large deviations from it occur at low temperatures. Attempts to explain the divergencies from the classical theory had a considerable influence upon the early development of the quantum theory, the principal contributors in this field being Einstein (1907), Born & von Kármán (1912) and Debye (1912). A modern version of these theories is as follows.

To determine the thermodynamic properties of a crystal, we start from equation (5·351·5) for the free energy F of a system, namely,

$$F = -kT \log Z_{\text{system}}. \tag{6·2·1}$$

If $Z(\nu)$ is the partition function of a harmonic oscillator with frequency ν, then, since the normal modes are all independent, we have

$$Z_{\text{system}} = \prod_{\nu} Z(\nu). \tag{6·2·2}$$

Now the energy levels of the oscillator ν are given by

$$E_r = (r + \tfrac{1}{2})\, h\nu, \tag{6·2·3}$$

so that the corresponding partition function is

$$Z(\nu) = \sum_{r=0}^{\infty} \exp(-E_r/kT) = \frac{\exp(-\frac{1}{2}h\nu/kT)}{1 - \exp(-h\nu/kT)}. \tag{6·2·4}$$

The vibrational levels of a crystal, in the limit of an infinitely large number of atoms, form a continuum, and it is therefore convenient to write the number of normal modes with frequencies lying in the range $(\nu, \nu + d\nu)$ as $Ng(\nu)\, d\nu$, where the integral of $g(\nu)$ is equal to 3. It is also convenient for later purposes to choose the energy zero to refer to the state when all the atoms are at an infinite distance from each other, and we then have

$$F = U_0 + NkT \int_0^{\infty} \log[1 - \exp(-h\nu/kT)]\, g(\nu)\, d\nu, \tag{6·2·5}$$

where

$$U_0 = NE_0 + \tfrac{1}{2}N \int_0^{\infty} h\nu g(\nu)\, d\nu. \tag{6·2·6}$$

(The binding energy of the crystal at $T=0$ would be $-NE_0$ $(E_0<0)$ were it not for the residual energy $\frac{1}{2}N\int_0^\infty h\nu g(\nu)\,d\nu$ of the vibrations, which reduces the binding energy to the value $-U_0$.)

6·21. *The equation of state of a crystal.* The energy E_0 and the distribution function $g(\nu)$ (and hence also U_0) are determined by the nature of the atoms present and by the size and shape of the unit cell of the crystal, and the latter can be changed by external forces. If we confine our attention to the simplest case of crystals under hydrostatic pressures only, then U_0 and $g(\nu)$ are functions of the size of the unit cell, and hence are functions of V/N, where V is the total volume of the crystal. (More general stresses can be dealt with if we introduce changes in shape and not merely changes in volume of the unit cell. See, for example, Born & Huang (1954).) We may therefore write

$$F=-NkT\log Z(T,V/N)=N\Phi(T,V/N), \tag{6·21·1}$$

where

$$\log Z(T,V/N)=\frac{1}{N}\log Z_{\text{system}}$$
$$=-\frac{E_0}{kT}-\int_0^\infty\left\{\log\left[1-\exp\left(h\nu/kT\right)\right]+\frac{h\nu}{2kT}\right\}g(\nu)\,d\nu, \tag{6·21·2}$$

and we then have $$P=-\partial F/\partial V=-N(\partial\Phi/\partial V)_{T,N} \tag{6·21·3}$$

as the equation of state of the crystal.

We may also find an expression for μ, the thermodynamic potential. We have

$$\mu=(\partial F/\partial N)_{T,V}=\Phi+N(\partial\Phi/\partial N)_{T,V}. \tag{6·21·4}$$

Now $$\left(N\frac{\partial}{\partial N}\right)_V=\frac{N}{V}\frac{\partial}{\partial(N/V)}=\left(\frac{1}{V}\frac{\partial}{\partial(1/V)}\right)_N=-\left(V\frac{\partial}{\partial V}\right)_N, \tag{6·21·5}$$

and so $$\mu=\Phi-V(\partial\Phi/\partial V)_{T,N}=(F+PV)/N. \tag{6·21·6}$$

This could be derived directly by means of the definition $G=F+PV$ and the relation $G=N\mu$. For any condensed phase PV is small compared with the thermodynamic functions U and F, and the dependence of Φ upon V/N is small. If, however, Φ is treated as a function of T only, the pressure does not appear in the equations and we have $U=H$ and $F=G$.

The theory of the thermodynamics of crystals has now been reduced to the determination of the function $g(\nu)$ which characterizes the distribution of the frequencies. The calculation of $g(\nu)$ is a dynamical problem of great difficulty, and progress can only be made by considering especially simple models.

If F is of the form

$$F(T,V,N)=N\digamma_0\left(\frac{V}{N}\right)+NT\Psi\left\{\frac{f(V/N)}{T}\right\}, \tag{6·21·7}$$

which, as we shall see in § 6·23, is approximately true for certain models, we have

$$S = -N\Psi + Nf\Psi'/T, \qquad P = -F_0' - f'\Psi', \tag{6·21·8}$$

$$U = NF_0 + Nf\Psi', \qquad C_V = -f^2\Psi''/T^2. \tag{6·21·9}$$

Also, the isothermal compressibility κ_T is given by

$$\frac{1}{\kappa_T} = -V\left(\frac{\partial P}{\partial V}\right)_{T,N} = VF_0'' + Vf''\Psi' + V\frac{f'^2\Psi''}{T}, \tag{6·21·10}$$

while

$$(\partial P/\partial T)_{V,N} = ff'\Psi''/T^2 = -(f'/f)\,C_V. \tag{6·21·11}$$

6·22. *The dynamics of a linear chain.* A simple instructive example is provided by an infinite linear chain of equal atoms of mass m at a normal distance a apart, which are restricted to perform longitudinal vibrations. We suppose that the forces on the atoms are elastic forces obeying Hooke's law and that they are only exerted by each atom on its two neighbours. If x_g is the displacement of the atom g, the equations of motion are

$$md^2x_g/dt^2 = -\lambda(x_g - x_{g-1}) - \lambda(x_g - x_{g+1}), \tag{6·22·1}$$

where λ is the elasticity constant. To solve this we put $x_g = \alpha(t)\,e^{iqa}$. Then

$$md^2\alpha/dt^2 = -(4\lambda \sin^2 \tfrac{1}{2}qa)\,\alpha. \tag{6·22·2}$$

Therefore $\alpha \propto \cos(2\pi\nu_q t + \delta)$, where δ is the phase and where

$$2\pi\nu_q = 2(\lambda/m)^{\frac{1}{2}}\,|\sin \tfrac{1}{2}qa\,|. \tag{6·22·3}$$

The behaviour of ν_q as a function of qa is shown in fig. 6·8. The velocity of propagation $c = 2\pi\nu_q/q$ is a function of qa, and as qa tends to zero c increases to its maximum value of $a(\lambda/m)^{\frac{1}{2}}$.

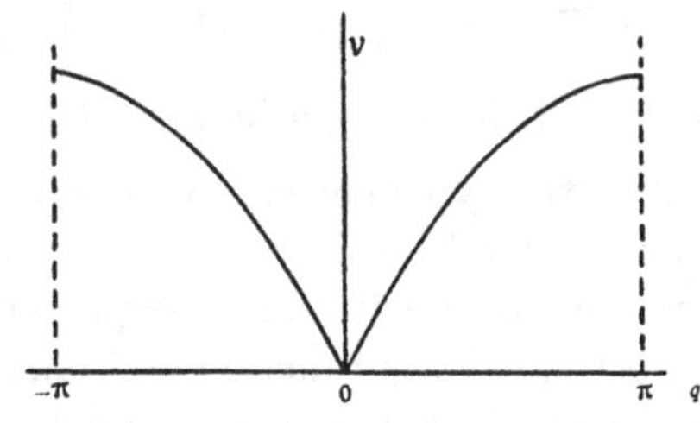

Fig. 6·8. The frequency spectrum for a simple linear chain.

The displacement x_g is necessarily real, and for each value of ν_q the general expression for x_g is

$$x_g = (A\cos qga + B\sin qga)\cos(2\pi\nu_q t + \delta), \tag{6·22·4}$$

where A and B are arbitrary constants. The q vibration is therefore the same as the $-q$ vibration, so that we obtain all frequencies by restricting qa to the range $0 \leqslant qa \leqslant \pi$, but there are two vibrations associated with each frequency. For small values of qa the frequency is proportional to q and is given by $2\pi\nu_q = (\lambda/m)^{\frac{1}{2}} qa$, while the maximum frequency $\nu_{\max.}$ is given by

$$2\pi\nu_{\max.} = 2(\lambda/m)^{\frac{1}{2}}. \qquad (6\cdot22\cdot5)$$

The more complicated problem of the vibrations of a three-dimensional lattice was considered by Born & von Kármán (1912). Almost simultaneously, a simpler theory was put forward by Debye, which for many purposes gives an adequate description of the phenomena.

6·23. *Debye's theory of the specific heat of a crystal.* For sufficiently low frequencies, the motion of the atoms of a crystal must resemble the vibrations of an elastic continuum, since the wavelengths of the elastic waves are then large compared with the dimensions of the unit cell. The elastic waves in a continuum are characterized by three velocity potentials ϕ_i $(i = 1, 2, 3)$ satisfying the wave equation $\nabla^2\phi_i = \partial^2\phi_i/\partial(c_i^2 t^2)$, where c_i is the velocity of propagation of the waves of type i (i.e. the velocity of sound). One type of wave $(i = 1)$ is longitudinal and compressional, while the other two are transverse. For the longitudinal waves ϕ_1 is the dilatation, while for the transverse waves ϕ_2 and ϕ_3 are any two independent components of the rotation, and we have $c_2 = c_3 < c_1$. It is convenient to consider a finite crystal, which we take to have the form of a parallelepiped with volume V and sides L_1, L_2, L_3, at the faces of which the displacement is taken to vanish. The wave equation then has solutions of the form

$$\phi_i = \sin q_1 x \sin q_2 y \sin q_3 z \cos 2\pi\nu_i t, \qquad (6\cdot23\cdot1)$$

where
$$\nu_i^2 = (q_1^2 + q_2^2 + q_3^2)\, c_i^2/(4\pi^2) = (qc_i/2\pi)^2 \qquad (6\cdot23\cdot2)$$

and where

$$q_1 = n_1\pi/L_1, \quad q_2 = n_2\pi/L_2, \quad q_3 = n_3\pi/L_3 \quad (n_1, n_2, n_3 = 0, 1, \ldots). \qquad (6\cdot23\cdot3)$$

The number of vibrations of each type with n's in the ranges

$$(n_1, n_1 + \Delta n_1), \quad (n_2, n_2 + \Delta n_2), \quad (n_3, n_3 + \Delta n_3)$$

is therefore
$$\Delta n_1 \Delta n_2 \Delta n_3 = (L_1 L_2 L_3/\pi^3)\, dq_1 dq_2 dq_3,$$

and, since the q's are positive and are confined to an octant of the $\mathbf{q}$ space, the number of vibrations per unit volume with $q = |\mathbf{q}|$ lying in the range $(q, q + dq)$ is

$$q^2 dq/(2\pi^2) = 4\pi c_i^{-3} \nu^2 d\nu. \qquad (6\cdot23\cdot4)$$

For a crystal we must therefore have, for sufficiently small values of ν,

$$g(\nu) = \sum_i g_i(\nu), \quad N g_i(\nu)\, d\nu = 4\pi V c_i^{-3} \nu^2 d\nu. \qquad (6\cdot23\cdot5)$$

Debye (1912) suggested that the formula (6·23·5), which is exact for a continuum, should be applied to a crystal even for high frequencies, and that the discrete nature of the crystal lattice should merely be taken into account by restricting the upper limit of the frequencies so as to obtain the correct number $3N$ of the possible modes of vibration of the N atoms. If we denote the maximum frequencies by $\nu_{i,\,\text{max.}}$, then, by (6·23·5), we have

$$\frac{N}{V}=\frac{4\pi}{c_i^3}\int_0^{\nu_{i,\,\text{max.}}}\nu^2\,d\nu=\frac{4\pi}{3c_i^3}\nu_{i,\,\text{max.}}^3 \tag{6·23·6}$$

and

$$g_i(\nu)\,d\nu=3\nu^2\,d\nu/\nu_{i,\,\text{max.}}^3. \tag{6·23·7}$$

We define characteristic temperatures Θ_i by the relation

$$h\nu_{i,\,\text{max.}}=k\Theta_i, \tag{6·23·8}$$

and then

$$\left.\begin{aligned} g_i(\nu)\,d\nu &= 3h^3\nu^2\,d\nu/(k\Theta_i)^3 \quad (\nu\leqslant k\Theta_i/h)\\ &=0 \qquad\qquad\qquad\qquad\; (\nu>k\Theta_i/h).\end{aligned}\right\} \tag{6·23·9}$$

With this approximation (6·2·5) becomes

$$F=U_0+3NkT^4\sum_{i=1}^{3}\frac{1}{\Theta_i^3}\int_0^{\Theta_i/T}\log\left(1-e^{-x}\right)x^2\,dx. \tag{6·23·10}$$

In view of the crude nature of the approximations made, it is scarcely justifiable to use three separate Θ's, and it is customary to introduce an average Θ defined by

$$\frac{3}{\Theta^3}=\sum_{i=1}^{3}\frac{1}{\Theta_i^3}. \tag{6·23·11}$$

Then

$$\left.\begin{aligned} g(\nu)\,d\nu &= 9h^3\nu^2\,d\nu/(k\Theta)^3 \quad (\nu\leqslant k\Theta/h)\\ &=0 \qquad\qquad\qquad\qquad (\nu>k\Theta/h),\end{aligned}\right\} \tag{6·23·12}$$

which gives

$$F=U_0+\frac{9NkT^4}{\Theta^3}\int_0^{\Theta/T}\log\left(1-e^{-x}\right)x^2\,dx \quad \left(U_0=NE_0+\tfrac{9}{8}Nk\Theta\right). \tag{6·23·13}$$

The characteristic temperature Θ can be expressed in terms of the atomic volume and the velocity of sound. For, by (6·23·6) and (6·23·8),

$$\Theta_i=\frac{hc_i}{k}\left(\frac{3N}{4\pi V}\right)^{\frac{1}{3}}, \tag{6·23·14}$$

and if we define the average velocity of sound c by

$$\frac{3}{c^3}=\frac{1}{c_1^3}+\frac{2}{c_2^3}, \tag{6·23·15}$$

we have, from (6·23·11),

$$\Theta=\frac{hc}{k}\left(\frac{3N}{4\pi V}\right)^{\frac{1}{3}}, \tag{6·23·16}$$

where V/N is the atomic volume.

6·231. We may now obtain any of the thermodynamic functions by differentiating the expression (6·23·13) for F. The internal energy U is given by

$$U = -T^2 \frac{\partial}{\partial T}\frac{F}{T} = U_0 - \frac{27NkT^4}{\Theta^3}\int_0^{\Theta/T} \log(1-e^{-x})\,x^2 dx + 9NkT\log(1-e^{-\Theta/T}) \tag{6·231·1}$$

$$= U_0 + \frac{9NkT^4}{\Theta^3}\int_0^{\Theta/T} \frac{x^3 dx}{e^x - 1} \tag{6·231·2}$$

by integrating by parts, while the specific heat C_V is given by

$$C_V = \frac{1}{N}\frac{\partial U}{\partial T} = \frac{36kT^3}{\Theta^3}\int_0^{\Theta/T} \frac{x^3 dx}{e^x - 1} - 9k\frac{\Theta}{T}\frac{1}{e^{\Theta/T}-1}$$

$$= 9k\left(\frac{T}{\Theta}\right)^3 \int_0^{\Theta/T} \frac{x^4 dx}{(e^x - 1)(1-e^{-x})}, \tag{6·231·3}$$

after an integration by parts.

The pressure P is given by

$$P = -\left(\frac{\partial F(T, V, \Theta/T)}{\partial V}\right)_{T,N} = -\left(\frac{\partial F}{\partial V}\right)_{T,N,\Theta} - \frac{\Theta'}{N}\left(\frac{\partial F}{\partial \Theta}\right)_{T,N,V}, \tag{6·231·4}$$

where we write

$$F = F_0(V, N) + F_1(\Theta, T) = U_0(V, N) + F_1(\Theta, T),$$

and where

$$\Theta' = N(\partial\Theta/\partial V)_N = \partial\Theta/\partial v. \tag{6·231·5}$$

Now F_1/T only contains T and Θ in the combination Θ/T, and so

$$\left(\frac{\partial F_1}{\partial \Theta}\right)_T = \left(\frac{\partial}{\partial(\Theta/T)}\frac{F_1}{T}\right)_T = \frac{1}{\Theta}\left(\frac{\partial}{\partial(1/T)}\frac{F_1}{T}\right)_\Theta = -\frac{T^2}{\Theta}\frac{\partial}{\partial T}\frac{F_1}{T}$$

$$= -\frac{T^2}{\Theta}\frac{\partial}{\partial T}\frac{F-U_0}{T} = \frac{U-U_0}{\Theta}.$$

Hence the equation of state can be written as

$$P = -\frac{\partial u_0}{\partial v} - \frac{\Theta'}{\Theta}(u - u_0). \tag{6·231·6}$$

The normal atomic volume $(V/N)_0 = v_0$ of the solid is given by the solution of (6·231·6) for $P=0$, while the isothermal compressibility κ is given by

$$\kappa = -\frac{1}{V}\left(\frac{\partial V}{\partial P}\right)_{T,N},$$

where $(\partial V/\partial P)_{T,N} = 1/(\partial P/\partial V)_{T,N}$ is determined by

$$\left(\frac{\partial P}{\partial V}\right)_{T,N} = -\frac{\partial^2 U_0}{\partial V^2} - \frac{\partial}{\partial V}\left(\frac{\Theta'}{\Theta}(u - u_0)\right)_{T,N}. \tag{6·231·7}$$

The volume coefficient of thermal expansion α is given by

$$\alpha = \frac{1}{V}\left(\frac{\partial V}{\partial T}\right)_P = -\frac{1}{V}\left(\frac{\partial V}{\partial P}\right)_T \left(\frac{\partial P}{\partial T}\right)_V. \tag{6·231·8}$$

But, from (6·231·6), $$(\partial P/\partial T)_{V,N} = -(\Theta'/\Theta)\, C_V, \tag{6·231·9}$$

and so $$\alpha = \gamma\kappa C_V/v, \tag{6·231·10}$$

where‡ $$\gamma = -v\Theta'/\Theta. \tag{6·231·11}$$

These equations are particular cases of those given in § 6·21.

6·24. *Survey of the experimental data.* Since the specific heat is a function of T/Θ only, it should be possible to make the specific heat curves for all substances (as functions of T) become identical by changing the scale of T suitably. This is true to a high degree of approximation, though, as is discussed later, the values of Θ deduced from the specific heat curves depend to a certain extent upon the temperature ranges over which the results are analysed. The most reliable values of Θ are given in table 6·4 below.

The high- and low-temperature forms of C_V are easily found by determining the behaviour of the Debye function $D(x)$, where

$$D(x) = \frac{3}{x^3}\int_0^x \frac{t^3\,dt}{e^t - 1}. \tag{6·24·1}$$

For small values of x, the denominator can be expanded in ascending powers of x and we find $$D(x) = 1 - \tfrac{3}{8}x + \tfrac{1}{20}x^2 + \dots. \tag{6·24·2}$$

For large values of x, the upper limit in the integral can be replaced by infinity with a negligible error. Then, since

$$\int_0^\infty \frac{t^3\,dt}{e^t - 1} = \sum_{n=1}^{\infty}\int_0^\infty t^3 e^{-nt}\,dt = 3!\sum_{n=1}^{\infty}\frac{1}{n^4} = \frac{\pi^4}{15}, \tag{6·24·3}$$

we have $$D(x) \sim \tfrac{1}{5}\pi^4 x^{-3} \tag{6·24·4}$$

for large x. Therefore $$C_V \sim 3k \quad (T \gg \Theta) \tag{6·24·5}$$

and $$C_V \sim \tfrac{12}{5}\pi^4 k (T/\Theta)^3 \quad (\Theta \gg T). \tag{6·24·6}$$

The general behaviour of C_V as a function of T/Θ is shown in fig. 6·9.

6·241. Since the characteristic temperature Θ can be expressed in terms of the velocities of the sound waves we should be able to determine Θ from the compressibility κ and Poisson's ratio σ. According to the theory of

‡ This quantity γ is usually known as Grüneisen's constant. It must not be confused with the symbol for the ratio of the specific heats of a gas.

elasticity of isotropic solids, the velocities c_1 and c_2 of longitudinal and transverse sound waves are given by

$$c_1^2 = \frac{3(1-\sigma)}{(1+\sigma)\kappa\rho}, \quad c_2^2 = \frac{3(1-2\sigma)}{2(1+\sigma)\kappa\rho}, \tag{6·241·1}$$

where ρ is the density, while (6·23·11) and (6·24·14) give

$$\frac{1}{\Theta^3} = \left(\frac{k}{h\nu_{\text{max.}}}\right)^3 = \frac{k^3}{h^3}\frac{4\pi V}{9N}\left(\frac{1}{c_1^3} + \frac{2}{c_2^3}\right). \tag{6·241·2}$$

Some values of Θ obtained from the elastic constants are shown in table 6·4, together with those calculated from the thermal data. (These have been collected together by Dr M. Blackman.) It will be seen that there is reasonable agreement between the two sets of constants.

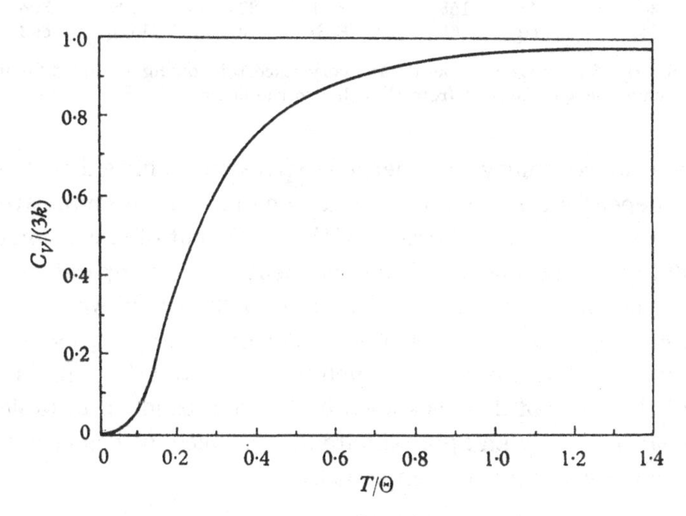

Fig. 6·9. The specific heat according to Debye's theory as a function of T/Θ.

6·242. *The compressibility and the thermal expansion.* The compressibility of a solid is determined primarily by the cohesive energy at $T = 0$, and the lattice vibrations, apart from their zero-point energy, have relatively little effect. As a first approximation it is therefore permissible to neglect the second term in (6·231·7), and the problem is then reduced to the calculation of the cohesive energy as a function of the volume. This is a difficult problem, and the reader is referred to the books by Born & Huang (1954), Seitz (1940) and Wilson (1953) for details of the calculations, which are different for metals and non-metals. Here we merely refer to the experimental results. The temperature variation of the compressibility is shown in table 6·5 for a few metals, and it will be seen that the variation is not large.

Table 6·4. *Characteristic temperatures* Θ

From specific heat data for the temperature range $\frac{1}{2}\Theta$ to $\frac{3}{2}\Theta$

Li	430	Cr	405	Sn	160	NaCl	275
Be	900	Fe	355	I	105	KCl	230
C	1850	Co	385	Ta	245	KBr	180
Na	160	Ni	320	W	315	KI	175
Mg	330	Cu	310	Ir	285	RbBr	132
Al	410	Zn	240	Pt	225	RbI	115
Si	580	Ge	380	Au	185	MgO	740
S	180	Mo	360	Hg	96	CaF_2	470
K	99	Ag	220	Pb	88	FeS_2	600
Ca	220	Cd	165	LiF	650		

From the elastic constants at room temperature

Al	394	Cd	189	LiF	715	KI	132
Cu	342	W	384	NaCl	302	CaF_2	499
Zn	306	Au	158	KCl	227	ZnS	306
Ag	212	Hg	69	KBr	179	FeS_2	682

The values for the ferromagnetic metals are very uncertain owing to the difficulty in separating the magnetic specific heat from that due to the lattice.

Unlike the compressibility, the thermal expansion coefficient is strongly temperature-dependent (see fig. 6·10), and an empirical law enunciated by Grüneisen in 1908 states that the ratio of the coefficient of expansion of an isotropic solid to its specific heat is independent of the temperature. This law is contained in equations (6·21·11) and (6·231·10) if we take κ as constant. According to Debye's theory, the quantity $\gamma = v\alpha/(\kappa C_V) = -v\Theta'/\Theta$ should be equal to $\frac{1}{3}$, but this value is much too small, and the experimental values are of the order of 2, as is shown by the figures given in table 6·6. (Since the relation is only an approximate one, no account has been taken of the variation of κ and v with temperature.)

Table 6·5. *The compressibility* $\kappa \times 10^7$ *in* $(kg./cm.^2)^{-1}$ *at various temperatures*

T°K. ...	0	83	290	404	438
Fe	(5·88)	5·93	6·22	6·52	6·62
Cu	(6·96)	7·04	7·58	7·98	8·12
Pt	(3·64)	3·67	3·84	3·93	3·96

It will be seen from the figures in table 6·6 that γ is reasonably constant at high temperatures. (Since iron is ferromagnetic the specific heat is abnormally large at high temperatures, and hence γ tends to decrease.) At low temperatures α tends to zero somewhat more rapidly than C_V, and γ decreases considerably when T/Θ falls below about 0·3 (see fig. 6·10), reaching values of the order of unity (Rubin, Altman & Johnston, 1954;

Bijl & Pullan, 1954, 1955). An analysis of this phenomenon, based upon a more exact theory of the vibrational spectrum than Debye's, has been given by Barron (1955).

Table 6·6. *Expansion coefficients and specific heats at various temperatures*

	Al $\kappa = 13 \times 10^{-13}$ dyne^{-1}cm.^{2} $V = 16{\cdot}5 \times 10^{-24}$ cm.^{3}/atom			Fe $\kappa = 6 \times 10^{-13}$ dyne^{-1}cm.^{2} $V = 11{\cdot}7 \times 10^{-24}$ cm.^{3}/atom			Cu $\kappa = 7{\cdot}5 \times 10^{-13}$ dyne^{-1}cm.^{2} $V = 11{\cdot}8 \times 10^{-24}$ cm.^{3}/atom		
T°K.	$\alpha \times 10^6$	$C_V/(3k)$	γ	$\alpha \times 10^6$	$C_V/(3k)$	γ	$\alpha \times 10^6$	$C_V/(3k)$	γ
73	21	0·3	2·0	9	0·25	1·9	20	0·39	2·1
173	57	0·75	2·3	30	0·72	2·1	42	0·8	2·1
273	67·5	0·9	2·3	35	0·97	1·8	48·5	0·93	2·1
373	73·5	0·94	2·4	38	1·1	1·7	50	0·96	2·1

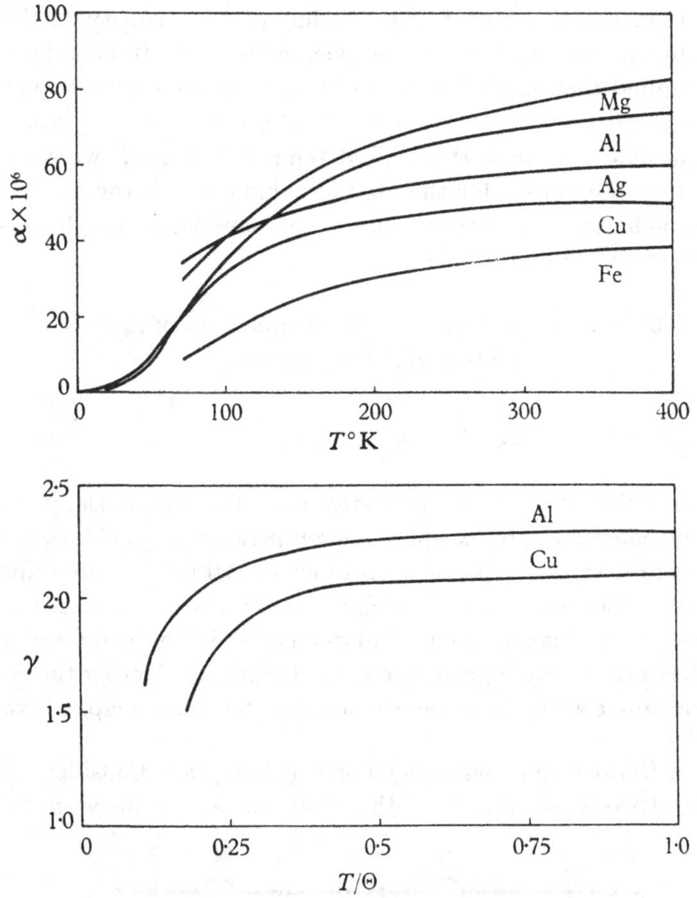

Fig. 6·10. The expansion coefficient of various metals as a function of the temperature, and Grüneisen's constant.

6·243. *The relation between the specific heats.* The relation

$$C_P - C_V = vT\alpha^2/\kappa, \tag{6·243·1}$$

which was derived in § 3·212, is used in order to deduce C_V from the measured values of the specific heat, which is necessarily C_P. (Note that in § 3·212 the specific heat is measured per gram molecule, whereas here it is per atom.) The right-hand side of (6·243·1) is proportional to T^7 at low temperatures, and so the difference between C_P and C_V is only significant at moderate and high temperatures where it is proportional to T.

By using the relation (6·231·10), equation (6·243·1) can be written in the more convenient form

$$\frac{C_P - C_V}{C_V T} = \gamma^2 \frac{\kappa C_V}{v}, \tag{6·243·2}$$

where γ is Grüneisen's constant. Some values of the quantity on the left, for room temperatures and above, are given in table 6·7. It must, however, be borne in mind that all that we can obtain from the measured value of C_P at atmospheric pressure and from (6·243·1) is the value of C_V relating to the actual atomic volume of the solid at temperature T. If we wished to determine the value of C_V for the atomic volume which the solid has at $T=0$, it would be necessary to carry out the measurements at high pressures of the order of 10,000 atmospheres.

Table 6·7. *The difference in the specific heats of various solids at high temperatures*

	Al	Cu	Ag	Pb	Pt
$\frac{C_P - C_V}{C_V T} \times 10^4$	1·4	0·8	1·6	1·8	0·6

6·25. *Extension of the theory to compounds.* The simple Debye theory cannot be expected to apply, except at low temperatures, to solids containing atoms of more than one type, or even to elemental solids whose unit cell contains more than one atom. In such solids there are, in addition to the sound waves, vibrations which are intermolecular in character and which therefore have relatively high frequencies. Before considering the general theory we examine a simple model which illustrates the principles involved.

6·251. *A linear chain containing atoms of two types.* Consider a linear chain with lattice constant $a+b$, with two atoms per cell unsymmetrically

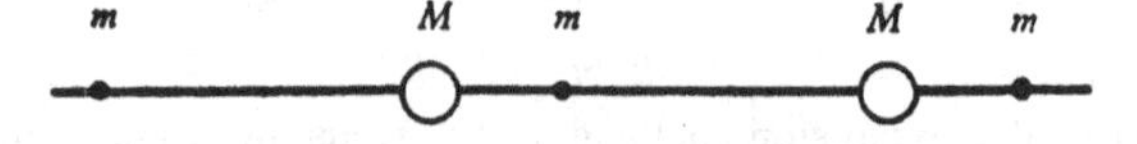

Fig. 6·11. A composite linear chain.

placed, the masses of the atoms being m and M (see fig. 6·11). We label the m atoms as the even atoms and the M atoms as the odd atoms. The equations of motion can then be written

$$\left.\begin{aligned} m\ddot{x}_{2g} &= -\lambda(x_{2g}-x_{2g+1})-\mu(x_{2g}-x_{2g-1}), \\ M\ddot{x}_{2g+1} &= -\lambda(x_{2g+1}-x_{2g})-\mu(x_{2g+1}-x_{2g+2}). \end{aligned}\right\} \quad (6\cdot251\cdot1)$$

We solve these equations by putting

$$x_{2g} = A\exp[igq(a+b)], \quad x_{2g+1} = B\exp[i(g+1)qa+igqb]$$

and
$$d^2A/dt^2 = -4\pi^2\nu_q^2 A, \quad d^2B/dt^2 = -4\pi^2\nu_q^2 B.$$

Then
$$(4\pi^2 m\nu_q^2-\lambda-\mu)A+(\lambda e^{iqa}+\mu e^{-iqb})B=0,$$
$$(\lambda e^{-iqa}+\mu e^{iqb})A+(4\pi^2 M\nu_q^2-\lambda-\mu)B=0,$$

and these are only compatible if

$$(4\pi^2 m\nu_q^2-\lambda-\mu)(4\pi^2 M\nu_q^2-\lambda-\mu)-\{\lambda^2+\mu^2+2\lambda\mu\cos q(a+b)\}=0.$$

The roots of this equation are

$$4\pi^2\nu_q^2 = [(m+M)(\lambda+\mu)\pm\sqrt{\{(\lambda+\mu)^2(m+M)^2 - 16\lambda\mu mM\sin^2\tfrac{1}{2}q(a+b)\}}]/(2mM), \quad (6\cdot251\cdot2)$$

and the ν_q, q curve consists of two separate branches, the general shape of which is shown in fig. 6·12. The branch relating to the minus sign before the square root gives the lower curve in the figure, and the corresponding vibrations are similar in character to those of a simple lattice. They are usually known as the acoustical vibrations, the velocity of propagation tending to $(a+b)[\lambda\mu/\{(\lambda+\mu)(m+M)\}]^{\frac{1}{2}}$ as $\nu_q\to 0$. The other branch is called the optical branch, and, as $q\to 0$, ν_q tends to the non-zero limit

$$[(m+M)(\lambda+\mu)/(4\pi^2 mM)]^{\frac{1}{2}}.$$

This division of the frequencies into acoustical and optical types is a general phenomenon and occurs in any crystal which cannot be described by a unit cell containing only one atom.

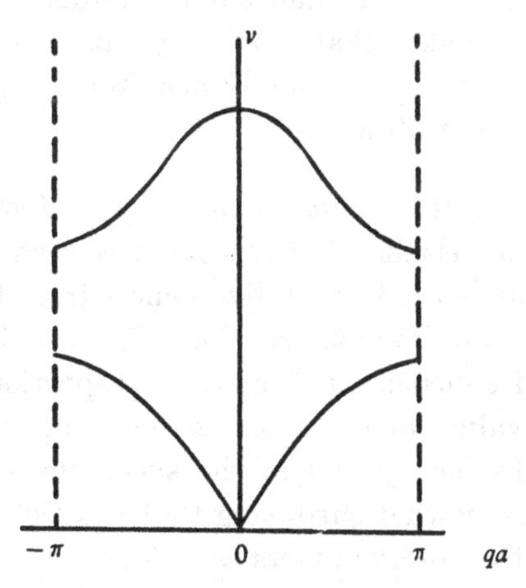

Fig. 6·12. The frequency spectrum of a composite lattice as a function of the wave number.

6·252. *Three-dimensional lattices.* The general theory of the vibrations of the atoms in a crystal lattice has been formulated by Born (1923). (See also Born & Huang (1954).) If there are s atoms in the unit cell and N unit cells in the crystal, there are $3N$ acoustical modes in which the atoms in any one unit cell move more or less as a whole, and the vibrations have the same general characteristics as those of a continuum except that there is a maximum frequency. The remaining $3(s-1)N$ vibrations are the optical ones and are essentially vibrations of the s particles in a unit cell relative to one another. As a rough approximation the frequencies of these vibrations can be taken to be independent of the wavelength and they then divide up into $3(s-1)$ sets of N vibrations with the same frequency ν_i^0 $(i=4, 5, \ldots, 3s)$.

If we apply Debye's approximate theory to the acoustical modes of vibration and assume that each optical branch can be characterized by a single frequency, we have

$$F = U_0 + \frac{9NkT^4}{\Theta^3}\int_0^{\Theta/T} \log(1-e^{-x})\,x^2\,dx + NkT\sum_{i=4}^{3s}\log\left[1-\exp\left(-h\nu_i^0/kT\right)\right] \tag{6·252·1}$$

and

$$U = U_0 + \frac{9NkT^4}{\Theta^3}\int_0^{\Theta/T}\frac{x^3\,dx}{e^x-1} + N\sum_{i=4}^{3s}\frac{h\nu_i^0}{\exp(h\nu_i^0/kT)-1}. \tag{6·252·2}$$

The terms in the summation are usually known as the Einstein terms. They are negligible at low temperatures, but can give a significant contribution at moderate and high temperatures.

In certain molecular crystals it is possible for the molecules to rotate provided that the temperature is not too low. The specific heat then behaves abnormally near the temperature at which the rotation changes into a libration.

6·26. *Criticism of Debye's theory.* By considering more complicated models than Debye's, it has been shown by Blackman (1935, 1941) that there are considerable divergences from Debye's theory even at temperatures of the order of 20° K. For sufficiently long waves the density of the vibrational frequencies $g(\nu)$ must be proportional to ν^2, but this is not true for larger values of ν. Since the specific heat does not depend merely on the low-frequency end of the spectrum, except at very low temperatures, it is somewhat surprising that the Debye theory gives as good a description of the observed facts as it does.

The usual way of analysing the departures from Debye's theory is to fit a Debye curve to the values of the specific heat in the neighbourhood of the temperature T, and to calculate the value of Θ from it. If Debye's formula were exact, all the values of Θ would be the same, but in practice Θ is a slowly varying function of T, and the variations are a measure of the inadequacy of the theory.

The calculation of $g(\nu)$ for real crystals is a very laborious and complicated matter. The most detailed calculations are those for sodium chloride (Kellermann, 1941), potassium chloride (Iona, 1941), diamond (Smith, 1948) and silver (Leighton, 1948). Kellermann's results for sodium chloride are

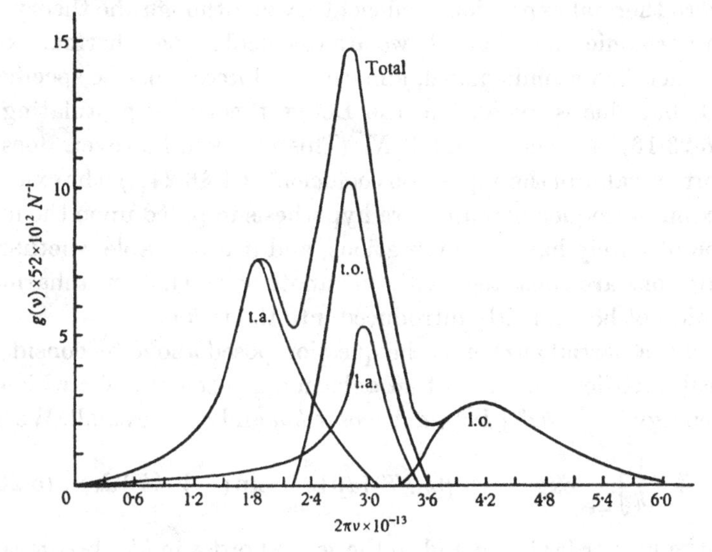

Fig. 6·13. The frequency spectrum for rock salt; total curve and four partial curves, transverse acoustic (t.a.), transverse optical (t.o.), longitudinal acoustic (l.a.) and longitudinal optical (l.o.).

shown in figs. 6·13 and 6·14. The frequency spectrum consists of six branches, the longitudinal acoustic, the transverse acoustic, the longitudinal optical and the transverse optical, the transverse waves being doubly degenerate.

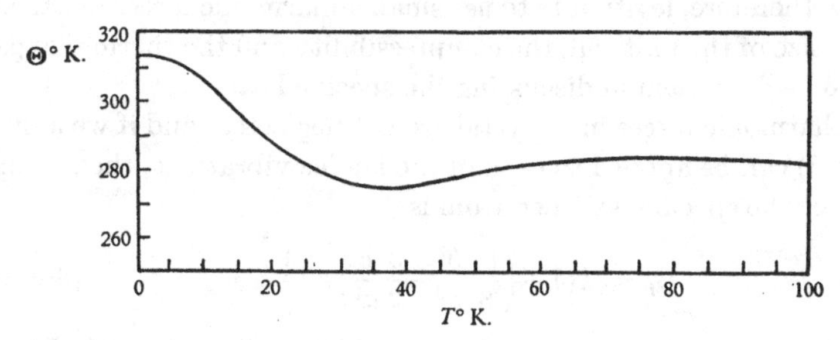

Fig. 6.14. The apparent Debye Θ as a function of T for rock salt.

The temperature dependence of Θ, given in fig. 6·14, shows that variations in Θ of up to 10 % are to be expected, and that the variations extend to very low temperatures. The observed variations in Θ at low temperatures are of the same order as those predicted by the calculations. For example,

Θ for sodium chloride varies from 320 at 0° K. to 300 at 20° K. and then falls to still lower values (Clusius, 1946; Martin, 1955).

6·261. It may seem strange at first sight that Debye's theory can lead to a non-zero thermal expansion coefficient, even although the theory only deals with harmonic vibrations. If we are to calculate the thermal expansion from a detailed atomic model, anharmonic forces must be specifically considered, but this is avoided in the Debye theory by postulating the relation (6·23·16) between Θ and V/N. (This relation, however, does not give the correct value of the expansion coefficient. Cf. § 6·242.) The existence of the maximum frequency is an extra hypothesis imposed upon the initial assumption of purely harmonic vibrations, and it is arguable whether the two assumptions are consistent with one another or whether anharmonic forces should not be explicitly introduced into the theory.

We can give a partial answer to the question posed above by considering the classical partition function of an anharmonic oscillator for which the potential energy is $V = ax^2 + bx^3 + cx^4$, where b/a and c/a are small. We have

$$Z = \frac{1}{h}\int_{-\infty}^{\infty} \exp\left[-\tfrac{1}{2}p^2/(mkT)\right] dp \int_{-\infty}^{\infty} \exp\left(-V/kT\right) dx, \qquad (6\cdot261\cdot1)$$

which, to the first order in c/a and to the second order in b/a, becomes

$$Z = \frac{2\pi}{h}\sqrt{\frac{m}{a}}\,kT\left(1 - \frac{3}{4}\frac{c}{a^2}kT + \frac{15}{16}\frac{b^2}{a^3}kT\right). \qquad (6\cdot261\cdot2)$$

This contains no term linear in b/a, so that small anharmonic forces do not affect the thermodynamic functions in the first-order approximation, and the calculations have to be taken to the second order to obtain a non-zero result. It is, therefore, legitimate to use small anharmonic forces in calculations of the size of the unit cell, the compressibility and the thermal expansion, and to neglect them in discussing the specific heat.

If the anharmonic forces in a crystal are not neglected, and if we assume that (6·261·2) can be applied to each of the lattice vibrations, then at high temperatures the specific heat per atom is

$$C_V = 3k\left\{1 + \left(\frac{15}{8}\frac{b^2}{a^3} - \frac{3}{2}\frac{c}{a^2}\right)kT\right\}. \qquad (6\cdot261\cdot3)$$

According to this formula the specific heat will vary linearly with T when the temperature is sufficiently high. Empirically the specific heat increases with temperature above the classical value, and we therefore conclude that the influence of the third-order terms in the potential energy predominates over that of the fourth-order terms. It must, however, be borne in mind that the accuracy of the measurements is not great. Further, the only reasonably reliable measurements refer to metals where the situation is complicated by the heat capacity of the free electrons (see § 6·32).

The specific heats of sodium and potassium above 250° K. are considerably larger than the limiting classical value (Carpenter & Steward, 1939; Ginnings, Douglas & Ball, 1950; Dauphinee, MacDonald & Preston-Thomas, 1954), and there is no temperature region in which the specific heat is constant (fig. 6·15). Owing to the low characteristic temperatures and the low melting-points of the alkali metals, this could be due either to the influence of the anharmonic forces or to a more profound modification of the crystal structure as the melting-point is approached, caused by the appearance of holes and other imperfections in the lattice. It is probable that the latter is the more important effect.

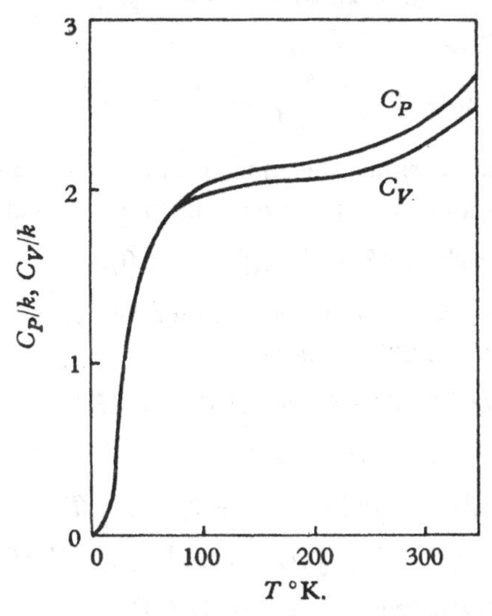

Fig. 6·15. The specific heat of potassium.

Systems Obeying Fermi-Dirac Statistics

6·3. The theory given in § 5·4 of a perfect gas obeying classical statistics can be extended at once to deal with a system of free particles when the wave functions of the assembly belong to the antisymmetrical system.‡ In this case the calculation is simplified by deriving the thermodynamic function PV as a function of T, V and ζ. If each state has weight ϖ, we have, by (5·611·11),

$$PV = \varpi kT \Sigma \log\{1 + \exp[(E - \zeta)/kT]\}. \tag{6·3·1}$$

‡ The main application of Fermi-Dirac statistics is to the electrons in a metal. In the theory of metals it is customary to use the symbol ζ for the thermodynamic potential instead of μ. We therefore use this notation in the present section.

Then, if we consider the particles to be in the enclosure $0 \leqslant x \leqslant L_1$, $0 \leqslant y \leqslant L_2$, $0 \leqslant z \leqslant L_3$, where L_1, L_2, L_3 are large, the energy levels are given by

$$E = \frac{h^2}{2m}\left(\frac{n_1^2}{L_1^2} + \frac{n_2^2}{L_2^2} + \frac{n_3^2}{L_3^2}\right),$$

where n_1, n_2, n_3 are integers. As usual we can replace the summation over n_1, n_2, n_3 in (6·3·1) by an integration, and the integration can be reduced to one with respect to the energy, so that we finally have

$$PV = \frac{4\sqrt{2}\,\pi\varpi}{h^3} m^{\frac{3}{2}} V \int_0^\infty \log\{1 + \exp[-(E-\zeta)/kT]\} E^{\frac{1}{2}} dE. \quad (6\cdot3\cdot2)$$

By integration by parts this can be transformed into

$$PV = \tfrac{4}{3}\pi\varpi(2mkT/h^2)^{\frac{3}{2}}\, VkT F_{\frac{3}{2}}(\xi), \quad (6\cdot3\cdot3)$$

where $\xi = \zeta/kT$ and where

$$F_n(\xi) = \int_0^\infty \frac{x^n}{e^{x-\xi}+1} dx \quad (n > -1). \quad (6\cdot3\cdot4)$$

The expressions for the various thermodynamic functions which can be obtained by differentiating (6·3·3) can be simplified by using the relation

$$dF_n(\xi)/d\xi = nF_{n-1}(\xi) \quad (n > 0). \quad (6\cdot3\cdot5)$$

(To prove this relation we differentiate under the integral sign and integrate by parts.) By (3·42·20) we have

$$N = \partial(PV)/\partial\zeta = 2\pi\varpi(2mkT/h^2)^{\frac{3}{2}}\, VF_{\frac{1}{2}}(\xi), \quad (6\cdot3\cdot6)$$

$$P = \tfrac{4}{3}\pi\varpi(2mkT/h^2)^{\frac{3}{2}}\, kTF_{\frac{3}{2}}(\xi), \quad (6\cdot3\cdot7)$$

$$S = \tfrac{10}{3}\pi\varpi(2mkT/h^2)^{\frac{3}{2}}\, VkF_{\frac{3}{2}}(\xi) - N\zeta/T. \quad (6\cdot3\cdot8)$$

Also

$$F \equiv N\zeta - PV = N\zeta - \tfrac{4}{3}\pi\varpi(2mkT/h^2)^{\frac{3}{2}}\, VkTF_{\frac{3}{2}}(\xi) \quad (6\cdot3\cdot9)$$

and

$$U = F + ST = 2\pi\varpi(2mkT/h^2)^{\frac{3}{2}}\, VkTF_{\frac{3}{2}}(\xi). \quad (6\cdot3\cdot10)$$

It will be seen that the reason why the PV thermodynamic function is the natural one to use is because of the difficulty of inverting the relation (6·3·6) to obtain ζ as an explicit function of N, which would be necessary if, for example, we wished to set up the $F(T, V, N)$ fundamental equation. It is therefore simpler to use a thermodynamic function which need not have N as one of the independent variables. It must, however, be borne in mind that it is necessary, at some stage or other, to determine ζ explicitly in terms of N.

The relation (6·3·6) can be used to express the various formulae in alternative forms which are sometimes convenient. For example, we can write

$$U = NkTF_{\frac{3}{2}}(\xi)/F_{\frac{1}{2}}(\xi), \quad F = N\zeta - \tfrac{2}{3}NkTF_{\frac{3}{2}}(\xi)/F_{\frac{1}{2}}(\xi), \quad (6\cdot3\cdot11)$$

$$PV = \tfrac{2}{3}NkTF_{\frac{3}{2}}(\xi)/F_{\frac{1}{2}}(\xi), \quad S = \tfrac{5}{3}NkF_{\frac{3}{2}}(\xi)/F_{\frac{1}{2}}(\xi) - N\zeta/T. \quad (6\cdot3\cdot12)$$

6·31. The properties of a Fermi-Dirac gas depend strongly upon the value of ξ, which can range from $-\infty$ to ∞. If ξ is large and negative, classical statistics hold, and even if ξ is small and negative the deviations from the classical theory are of quantitative rather than of qualitative significance. If ξ is positive, however, the behaviour is entirely different, as can be seen by inspection of the Fermi function

$$f_0(E)=\frac{1}{e^{(E-\zeta)/kT}+1}, \tag{6·31·1}$$

which gives the average occupation number of a single quantum state. For large values of $\xi=\zeta/kT$, its general behaviour is shown in fig. 6·16. For values of E less than ζ, $f_0(E)$ is effectively unity, and, as E approaches ζ from below, $f_0(E)$ decreases from unity, is equal to $\frac{1}{2}$ when $E=\zeta$ and falls exponentially to zero for larger values of E. The occupation numbers of the states for which $E<\zeta$ are therefore practically independent of the temperature, and it is only the small fraction of the occupied states for which $E>\zeta$ that contribute to any temperature-dependent effect. The fraction of the particles in these states is of the order of kT/ζ, and they have a Maxwell-Boltzmann distribution, the occupation number being effectively $e^{-(E-\zeta)/kT}$.

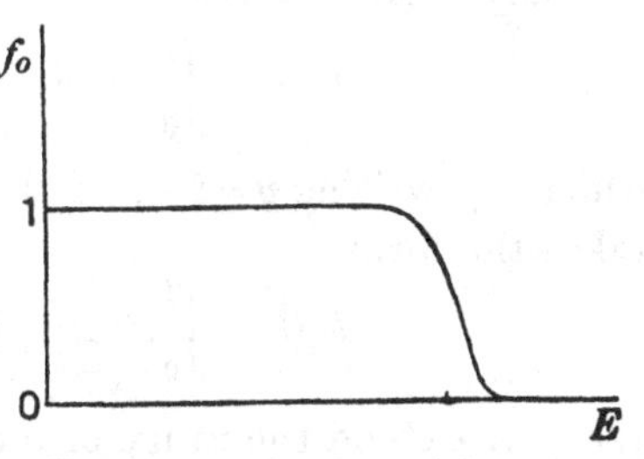

Fig. 6·16. The Fermi function.

To make these statements more precise we must first obtain ξ as a function of N by solving the implicit equation (6·3·6). If we write

$$T_0=\frac{h^2}{2mk}\left(\frac{3}{4\pi\varpi}\frac{N}{V}\right)^{\frac{2}{3}}, \tag{6·31·2}$$

equation (6·3·6) becomes $\quad F_{\frac{1}{2}}(\xi)=\frac{2}{3}(T_0/T)^{\frac{3}{2}}$. (6·31·3)

The solution of this equation can be carried out analytically for the two limiting cases of high and very low temperatures. For all negative values of ξ we may write

$$F_n(\xi)=\sum_{s=1}^{\infty}(-1)^{s+1}\int_0^{\infty}x^n\,e^{-s(x-\xi)}\,dx=\Gamma(n+1)\sum_{s=1}^{\infty}(-1)^{s+1}\frac{e^{s\xi}}{s^{n+1}}\quad(\xi<0), \tag{6·31·4}$$

and we can obtain ξ as a function of T by successive approximations. Since $\Gamma(\frac{3}{2})=\frac{1}{2}\sqrt{\pi}$, the first approximation is

$$e^{\xi}=\tfrac{4}{3}\pi^{-\frac{1}{2}}(T_0/T)^{\frac{3}{2}} \tag{6·31·5}$$

with $$f_0(E)=e^{-E/kT+\xi}=\frac{1}{\varpi}\left(\frac{h^2}{2\pi mkT}\right)^{\frac{3}{2}}\frac{N}{V}e^{-E/kT}, \tag{6·31·6}$$

and we regain all the formulae of § 5·4, including the value of the entropy constant given by (5·4·11). The approximations can be continued indefinitely, but the series is only convergent over a limited temperature range.

If ξ is large and positive the gas is said to be highly degenerate since its properties differ greatly from those of a classical gas. The temperature T_0 defined by (6·31·2) is called the degeneracy temperature. If $T_0 \gg T$ the gas is degenerate, whereas if $T \gg T_0$ it is non-degenerate. To obtain a first approximation to $F_n(\xi)$ for $\xi \gg 1$, we observe that in this case the function $1/(e^{x-\xi}+1)$ is effectively unity for $0 \leqslant x \leqslant \xi$ and effectively zero for $x > \xi$. We may therefore write

$$F_n(\xi) = \int_0^\xi x^n\,dx - \int_0^\xi \frac{x^n}{1+e^{\xi-x}}\,dx + \int_\xi^\infty \frac{x^n}{e^{x-\xi}+1}\,dx,$$

which, by writing $y = \xi - x$ in the second integral and $y = x - \xi$ in the third, takes the form

$$F_n(\xi) = \int_0^\xi x^n\,dx - \int_0^\xi \frac{(\xi-y)^n}{1+e^y}\,dy + \int_0^\infty \frac{(\xi+y)^n}{e^y+1}\,dy.$$

We may extend the range of integration in the second integral to infinity with an error $O(e^{-\xi})$, and we can expand the numerators in descending powers of ξ. The second approximation to $F_n(\xi)$ is therefore given by

$$F_n(\xi) = \frac{\xi^{n+1}}{n+1} + 2n\xi^{n-1}\int_0^\infty \frac{y}{e^y+1}\,dy.$$

To find the value of this integral we write

$$\int_0^\infty \frac{y}{e^y+1}\,dy = \sum_{s=1}^{\infty}\int_0^\infty (-1)^{s+1} y\,e^{-sy}\,dy = \sum_{s=1}^{\infty}\frac{(-1)^{s+1}}{s^2} = \frac{\pi^2}{12}. \tag{6·31·7}$$

Hence
$$F_n(\xi) = \frac{\xi^{n+1}}{n+1} + \frac{\pi^2}{6}n\xi^{n-1} + O(\xi^{n-3}) \quad (\xi \gg 1). \tag{6·31·8}$$

6·311. When $\xi \gg 1$, the equation (6·31·3) becomes

$$\tfrac{2}{3}\xi^{\frac{3}{2}} + \frac{\pi^2}{12}\xi^{\frac{1}{2}} + \ldots = \tfrac{2}{3}(T_0/T)^{\frac{3}{2}}. \tag{6·311·1}$$

If ζ_0 is the limiting value of ζ as $T \to 0$, we have

$$\zeta_0 = kT_0 = \frac{h^2}{2m}\left(\frac{3}{4\pi\varpi}\frac{N}{V}\right)^{\frac{2}{3}} \quad (T_0 \gg T), \tag{6·311·2}$$

while the second approximation to ζ is given by

$$\zeta = \zeta_0 - \pi^2k^2T^2/(12\zeta_0). \tag{6·311·3}$$

Hence, by (6·3·10) and (6·31·8), we have

$$U = \tfrac{3}{5}NkT\xi\left(1 + \frac{\pi^2}{2\xi^2} + \ldots\right) = \tfrac{3}{5}N\zeta_0\left(1 + \frac{5\pi^2k^2T^2}{12\zeta_0^2} + \ldots\right). \tag{6·311·4}$$

Similarly
$$F = \tfrac{3}{5}N\zeta_0\left(1 - \frac{5\pi^2k^2T^2}{12\zeta_0^2} + \ldots\right). \tag{6·311·5}$$
Therefore
$$P = -\left(\frac{\partial F}{\partial V}\right)_{T,N} = \frac{2}{5}\frac{N\zeta_0}{V} + \ldots = \left(\frac{3}{4\pi\varpi}\right)^{\frac{2}{3}}\frac{h^2}{5m}\left(\frac{N}{V}\right)^{\frac{5}{3}}\left\{1 + O\left(\frac{T^2}{T_0^2}\right)\right\}, \tag{6·311·6}$$
$$S = -\left(\frac{\partial F}{\partial T}\right)_{V,N} = \frac{\pi^2mk}{h^2}\left(\frac{4\pi\varpi}{3}\frac{V}{N}\right)^{\frac{2}{3}}NkT\left\{1 + O\left(\frac{T^2}{T_0^2}\right)\right\}, \tag{6·311·7}$$
and C_V the specific heat (at constant volume) per electron, is
$$C_V = \frac{1}{N}\left(\frac{\partial U}{\partial T}\right)_{V,N} = \frac{T}{N}\left(\frac{\partial S}{\partial T}\right)_{V,N} = \frac{\pi^2mk}{h^2}\left(\frac{4\pi\varpi}{3}\frac{V}{N}\right)^{\frac{2}{3}}kT\left\{1 + O\left(\frac{T^2}{T_0^2}\right)\right\}. \tag{6·311·8}$$

The above formulae show that, as $T \to 0$, the energy, free energy and pressure all tend to constant limits, while the entropy and the specific heat tend linearly to zero. We do not therefore encounter the difficulties which occur in the classical theory of perfect gases when we try to integrate thermodynamic relations down to $P = 0$ and $T = 0$. At temperatures which are neither very high nor very low, the calculations cannot be carried out analytically, but any of the thermodynamic functions can be obtained numerically by using the extensive tables of the functions $F_{\frac{1}{2}}(\xi)$ and $F_{\frac{3}{2}}(\xi)$ prepared by McDougall & Stoner (1938). Fig. 6·17 shows the specific heat C_V as a function of T/T_0.

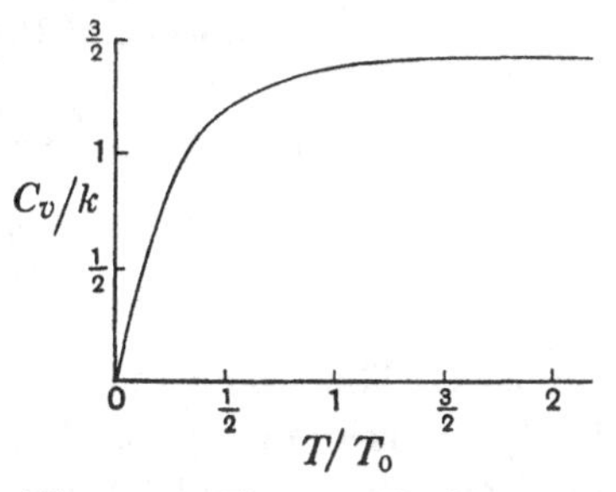

Fig. 6·17. The specific heat per electron of a gas of free electrons.

6·32. *The free electrons in a metal.* The high electrical and thermal conductivities of metals are due to the valency electrons of the metallic atoms being free to wander through the crystal. As a crude model we may suppose the conduction electrons to form an 'electron gas', moving under no forces in an enclosure bounded by the surface of the metal. The electrons obey Fermi-Dirac statistics, and, if we do not differentiate between the two directions of the spin, each translational state has weight 2. Also, the mass of the electron is so small that the electrons in a metal form a highly degenerate gas. For example, if we put $N/V = 5{\cdot}9 \times 10^{22}$, which is the number of valency electrons per unit volume for silver, we find that $T_0 = 6 \times 10^{4}$ ° K., so that the electron gas is degenerate for all ordinary temperatures.

The specific heat of the electrons per gram atom of the metal is given by
$$C_{V,A} = \mathscr{L}\frac{4\pi^3mk^2}{3h^2}\left(\frac{3z}{\pi n_a^2}\right)^{\frac{1}{3}}T \quad (T_0 \gg T), \tag{6·32·1}$$

where $\mathscr{L}$ is Avogadro's number, z is the valency and n_a is the number of atoms per cm.[3]. The specific heat of the electrons is normally masked by the specific heat of the lattice vibrations, and at ordinary temperatures the ratio of the specific heat of the electrons to that of the lattice vibrations is of the order of $10^{-4}T$, but the former must predominate at sufficiently low temperatures since the latter varies as T^3. The electronic specific heat was first measured by Keesom & Kok for silver in 1934, and since that time it has also been determined for a number of other metals. Some of these results are given in table 6·7 and are compared with the theoretical values, and it will be seen that there is reasonable agreement between the theoretical predictions and the experimental values.

Table 6·7. *Specific heats of metals at low temperatures in units of* $10^{-4}T$ *joule/degree/gram atom*

	$C_{V,A}$ observed	$n_a \times 10^{-22}$	$C_{V,A}$ calculated
Cu	7·4	8·5	5
Ag	6·7	5·9	6·7
Zn	6·3	6·7	7·4
Al	14·7	6·1	9·2

6·321. The assumption that the valency electrons in a metal are perfectly free is an unnecessarily crude one, and much more elaborate theories exist (see, for example, Wilson, 1953). The electrons in a metal move in a three-dimensional periodic field, and as a result the wave functions are of the form

$$\psi_{\mathbf{k}}(\mathbf{r}) = \exp(i\mathbf{k}\,.\,\mathbf{r})\,u_{\mathbf{k}}(\mathbf{r}), \tag{6·321·1}$$

where $\mathbf{k} = (k_1, k_2, k_3)$ is the wave vector and where $u_{\mathbf{k}}(\mathbf{r})$ is a periodic function with the unit cell of the crystal lattice as its period. Instead of the energy levels forming a continuum stretching from zero to infinity, they are grouped into an infinite number of bands, in each of which the energy is a continuous function of $\mathbf{k}$, separated by regions in which there are no energy levels. For a one-dimensional lattice in which the atoms are at a distance a apart, the energy discontinuities occur when $k_1 = \pm n\pi/a$, where n is an integer. The energy levels of such a lattice are shown schematically in fig. 6·18. An alternative description of the energy levels can be given by reducing the values of k_1 by the appropriate multiples of π/a so as to restrict k_1 to the range $\pm\pi/a$. The wave vector is then called the reduced wave vector. It will be seen from fig. 6·18 that the energy-level system can be

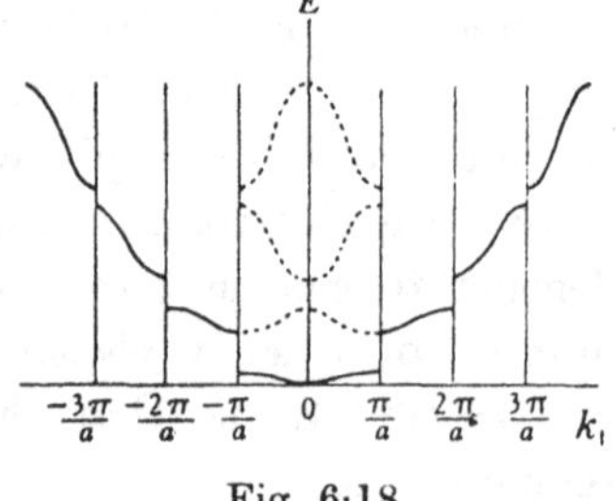

Fig. 6·18

considered to be derived from the 'free-electron' form ($E=\lambda\mathbf{k}^2$, where λ is a constant) by dissection and deformation of the continuous spectrum. Further, it can be shown that near the bottom of an energy band the energy is given by $E=\text{constant}+\lambda_1\mathbf{k}^2$, while near the top of a band we have $E=\text{constant}-\lambda_2\mathbf{k}^2$, where λ_1 and λ_2 are positive. If we write

$$\lambda_1=h^2/(8\pi^2m_1^*), \quad \lambda_2=h^2/(8\pi^2m_2^*),$$

m_1^* and m_2^* are called the effective masses of an electron in the given energy states.

6·322. A complete knowledge of the energy-level system is not required for the evaluation of the thermodynamic functions of the electrons in a metal. It is sufficient to know $\mathfrak{n}(E)$ the density of energy states, the definition of which is such that $\mathfrak{n}(E)\,dE$ is the number of energy levels per unit volume (for one direction of the electron spin) lying in the range E, $E+dE$. Since each state then has weight 2 we have

$$U=2V\int_0^\infty Ef_0(E)\,\mathfrak{n}(E)\,dE. \qquad (6\cdot322\cdot1)$$

We also have the relation

$$N=2V\int_0^\infty f_0(E)\,\mathfrak{n}(E)\,dE, \qquad (6\cdot322\cdot2)$$

which determines the parameter ξ.

These integrals are of the form

$$I=\int_0^\infty \frac{\phi(x)}{e^{x-\xi}+1}\,dx,$$

which can be evaluated approximately by the same method as was used in § 6·31 to evaluate $F_n(\xi)$. We write

$$I=\int_0^\xi \phi(x)\,dx-\int_0^\xi \frac{\phi(x)}{1+e^{-x+\xi}}\,dx+\int_x^\infty \frac{\phi(x)}{e^{x-\xi}+1}\,dx.$$

We then substitute $y=\xi-x$ in the second integral and $y=x-\xi$ in the third, and extend the integration with respect to y in the second integral to infinity, with an error $O(e^{-\xi})$. Expansion of the numerators in powers of y/ξ gives, to the second order,

$$\int_0^\infty \frac{\phi(x)}{e^{x-\xi}+1}\,dx=\int_0^\xi \phi(x)\,dx+2\phi'(\xi)\int_0^\infty \frac{y}{e^y+1}\,dy=\int_0^\xi \phi(x)\,dx+\frac{\pi^2}{6}\phi'(\xi). \qquad (6\cdot322\cdot3)$$

To this approximation (6·322·1) and (6·322·2) become

$$\frac{U}{V}=2\int_0^\zeta E\mathfrak{n}(E)\,dE+\frac{\pi^2}{3}k^2T^2\frac{d}{d\zeta}\{\zeta\mathfrak{n}(\zeta)\}, \qquad (6\cdot322\cdot4)$$

$$\frac{N}{V}=2\int_0^\zeta \mathfrak{n}(E)\,dE+\frac{\pi^2}{3}k^2T^2\frac{d\mathfrak{n}(\zeta)}{d\zeta}. \qquad (6\cdot322\cdot5)$$

Now ζ_0, the value of ζ when $T=0$, is given by the relation

$$\frac{N}{V}=2\int_0^{\zeta_0}\mathfrak{n}(E)\,dE,$$

and if we write
$$\int_0^{\zeta}\mathfrak{n}(E)\,dE=\int_0^{\zeta_0}\mathfrak{n}(E)\,dE+(\zeta-\zeta_0)\,\mathfrak{n}(\zeta_0),$$

the second approximation to ζ is

$$\zeta=\zeta_0-\frac{\pi^2}{6}k^2T^2\frac{1}{\mathfrak{n}(\zeta)}\frac{d\mathfrak{n}(\zeta)}{d\zeta}. \tag{6·322·6}$$

On substituting this into (6·322·4) we find

$$\frac{U}{V}=2\int_0^{\zeta_0}E\mathfrak{n}(E)\,dE+\tfrac{1}{3}(\pi kT)^2\,\mathfrak{n}(\zeta), \tag{6·322·7}$$

so that
$$C_{\Gamma}=\tfrac{2}{3}\pi^2k^2T\mathfrak{n}(\zeta). \tag{6·322·8}$$

This important formula allows us to determine the density of states of the electrons from measurements of the specific heat.

The electronic specific heats of the transition metals are especially interesting since they are extremely high and are of the order of ten times those of normal metals (see table 6·8). This means that the density of states in a transition metal is some ten times as large as it would be if the electrons were perfectly free.

Table 6·8. *Specific heats of transition metals at low temperatures in units of* $10^{-4}T$ *joule/degree/gram atom*

	W	Ta	Mn	Pd	Pt
$C_{V,A}$	220	60	178	130	67

According to the detailed theory of metals, it can be shown that a large density of states is to be expected for metals whose atoms in the free state contain incomplete *d*- or *f*-shells, but the calculated densities of states, though large, are not as large as those observed. For further details and for a discussion of the electronic specific heat at high temperatures, the reader is referred to books on the theory of metals (e.g. Wilson, 1953).

Systems obeying Einstein-Bose Statistics

6·4. Gases obeying Einstein-Bose statistics do not play such an important role in the theory of the properties of matter as do assemblies obeying Fermi-Dirac statistics. In the first place the departure from the classical theory is by no means so extreme as for Fermi-Dirac assemblies. Secondly, at absolute zero temperature all the particles in an Einstein-Bose assembly are in the zero-energy state, and the assembly then has none of the interesting features which characterize a completely degenerate Fermi-Dirac gas. Thirdly, the departures from the classical behaviour are only significant for the lightest particles, and, whereas for electrons these anomalies occur at ordinary temperatures, any real gas whose wave functions belong to the symmetrical type is far from perfect, and is normally in the liquid or solid state, at those temperatures at which quantum effects would occur which are due to the symmetry properties of the wave functions. Nevertheless, an Einstein-Bose gas shows some interesting special features, and we therefore give a short account of its properties. The most important result is that the specific heat rises above the classical value as the temperature decreases and reaches a maximum value of $1{\cdot}925k$ per atom at a certain characteristic temperature. Below this temperature a type of condensation occurs, and the specific heat falls rapidly to zero as the temperature approaches $T=0$.

A special case of an Einstein-Bose assembly is that of radiation in equilibrium with matter. This is treated in § 6·5.

6·41. If we follow the same procedure as that set out in § 6·3 all the formulae apply to a perfect gas obeying Einstein-Bose statistics if we replace $F_n(\xi)$ by

$$F_n^-(\xi)=\int_0^\infty \frac{x^n}{e^{x-\xi}-1}\,dx \quad (n>-1). \tag{6·41·1}$$

The integral is convergent only if ξ is negative, so that, unlike the Fermi-Dirac case, we do not have to consider positive values of ξ. We therefore calculate $F_n^-(\xi)$ from the expansion

$$F_n^-(\xi)=\Gamma(n+1)\sum_{s=1}^{\infty}\frac{e^{s\xi}}{s^{n+1}} \quad (\xi\leqslant 0). \tag{6·41·2}$$

(Note that $F_n^-(\xi)$ is an increasing function of ξ and that its greatest value is $F_n^-(0)$.) The relation

$$F_{\frac{1}{2}}^-(\xi)=\frac{1}{2\pi\varpi}\left(\frac{h^2}{2mkT}\right)^{\frac{3}{2}}\frac{N}{V} \tag{6·41·3}$$

can then be inverted to give ξ as a function of $N/(VT^{\frac{3}{2}})$. It is convenient to define T_1 as the solution of (6·41·3), with given N/V, for which $\xi=0$. Then‡

$$\frac{1}{\varpi}\left(\frac{h^2}{2\pi mkT_1}\right)^{\frac{3}{2}}\frac{N}{V}=\frac{2}{\sqrt{\pi}}F_{\frac{1}{2}}^-(0)=\sum_{s=1}^{\infty}\frac{1}{s^{\frac{3}{2}}}=2{\cdot}612, \tag{6·41·4}$$

and (6·41·3) can be written as

$$\sum_{s=1}^{\infty}\frac{e^{s\xi}}{s^{\frac{3}{2}}}=2{\cdot}612\left(\frac{T_1}{T}\right)^{\frac{3}{2}}, \tag{6·41·5}$$

‡ Note that, if $n>0$, $F_n(0)=\Gamma(n+1)\,\zeta(n+1)$, where $\zeta(z)$ is the Riemann ζ-function.

of which the solution is

$$e^{\xi} = 2{\cdot}612\left(\frac{T_1}{T}\right)^{\frac{3}{2}} - \frac{(2{\cdot}612)^2}{2^{\frac{3}{2}}}\left(\frac{T_1}{T}\right)^3 + \left(\frac{1}{4} - \frac{1}{3^{\frac{3}{2}}}\right)(2{\cdot}612)^3\left(\frac{T_1}{T}\right)^{\frac{9}{2}} + \dots \qquad (6{\cdot}41{\cdot}6)$$

$$= 2{\cdot}612\left(\frac{T_1}{T}\right)^{\frac{3}{2}}\left\{1 - 0{\cdot}92347\left(\frac{T_1}{T}\right)^{\frac{3}{2}} + 0{\cdot}3926\left(\frac{T_1}{T}\right)^3 - 0{\cdot}1027\left(\frac{T_1}{T}\right)^{\frac{9}{2}} + \dots\right\}. \qquad (6{\cdot}41{\cdot}7)$$

The thermodynamic functions can be most easily calculated from the formulae corresponding to (6·3·11) and (6·3·12). Thus

$$\tfrac{3}{2}PV = U = \tfrac{3}{2}NkT\left(1 - \frac{1}{2^{\frac{5}{2}}}e^{\xi} + \dots\right)$$

$$= \tfrac{3}{2}NkT\left\{1 - 0{\cdot}4618\left(\frac{T_1}{T}\right)^{\frac{3}{2}} - 0{\cdot}0225\left(\frac{T_1}{T}\right)^3 - 0{\cdot}00196\left(\frac{T_1}{T}\right)^{\frac{9}{2}} + \dots\right\}, \qquad (6{\cdot}41{\cdot}8)$$

$$S = Nk\{\tfrac{5}{2} - \log(2{\cdot}612T_1^{\frac{3}{2}})\} + \tfrac{3}{2}Nk\log T$$

$$- Nk\left\{0{\cdot}2309\left(\frac{T_1}{T}\right)^{\frac{3}{2}} + 0{\cdot}0225\left(\frac{T_1}{T}\right)^3 + 0{\cdot}0023\left(\frac{T_1}{T}\right)^{\frac{9}{2}} + \dots\right\}. \qquad (6{\cdot}41{\cdot}9)$$

The free energy F can be written down at once as $F = U - ST$, but as we shall not have occasion to use it we shall not give it explicitly. The first two terms in S can be written as

$$\tfrac{5}{2}Nk + Nk\log\left(\frac{V}{N}(kT)^{\frac{3}{2}}\frac{\varpi(2\pi m)^{\frac{3}{2}}}{h^3}\right),$$

which shows that for high temperatures (6·41·9) passes over into the correct classical value (5·4·11).

6·42. *The condensation phenomenon.* Since $F_n^-(\xi) \leqslant F_n^-(0)$, the equation (6·41·5) has no solution if $T < T_1$. This does not mean that the theory breaks down, but that (6·41·5) must be replaced by the exact equation

$$N = \varpi\sum_i \frac{1}{\exp[(E_i - \mu)/kT] - 1} \quad (E_0 = 0), \qquad (6{\cdot}42{\cdot}1)$$

from which (6·41·5) was derived by replacing the summation by an integration. When $T = 0$, all the particles have zero energy and we have

$$N = \varpi/(e^{-\xi} - 1),$$

so that

$$e^{-\xi} = 1 + \varpi/N. \qquad (6{\cdot}42{\cdot}2)$$

This suggests that, for $T < T_1$, $e^{-\xi} = 1 + O(1/N)$, and that it is necessary to treat the lowest energy level differently from the others. It is clear that a concentration of all the particles in one level cannot be described by an integral (unless we allow the integrand to contain a δ-function), since this assumes that, in the limit of large N, only an infinitesimal number of particles are located in any one level. We therefore write (6·42·1) in the form

$$N = \varpi \frac{1}{e^{-\xi} - 1} + \varpi \sum_{i \geq 1} \frac{1}{\exp[-\xi + E_i/kT] - 1}, \tag{6·42·3}$$

and we then have to consider the double limit $N \to \infty$, $V \to \infty$ (London, 1938). If we keep N fixed and make V sufficiently large, the number of particles in any of the levels E_i ($i \geq 1$) is infinitesimal, and, since the energy levels can be made to lie as close as we please, we may replace the summation in the second term by an integration. Then

$$1 = \frac{\varpi/N}{e^{-\xi} - 1} + \frac{2\pi\varpi V}{N} \left(\frac{2mkT}{h^2}\right)^{\frac{3}{2}} F_{\frac{1}{2}}^{-}(\xi), \tag{6·42·4}$$

i.e.
$$e^{-\xi} = 1 + \frac{\varpi}{N}\left[1 - \frac{2\pi\varpi V}{N}\left(\frac{2mkT}{h^2}\right)^{\frac{3}{2}} F_{\frac{1}{2}}^{-}(\xi)\right]^{-1}. \tag{6·42·5}$$

If we now let N and V both tend to infinity in such a way as to make V/N finite, we regain (6·41·3). If, however, we consider V and N to be large but not infinite we have to solve the more general equation (6·42·5). Since $e^{-\xi} = 1 + O(1/N)$, we may obtain the solution by successive approximations, the first approximation being given by writing $F_{\frac{1}{2}}^{-}(\xi) = F_{\frac{1}{2}}^{-}(0)$. Then

$$e^{-\xi} = 1 + \frac{\varpi}{N} \frac{1}{1 - (T/T_1)^{\frac{3}{2}}} + O(N^{-2}) \quad (T < T_1). \tag{6·42·6}$$

This is valid for any fixed $T < T_1$, and T can be taken to lie as close as we please to T_1 if we choose N large enough. We therefore see that, in the limit of infinitely large N, ξ is given by (6·41·5) for $T > T_1$ and by $\xi = 0$ for $T \leq T_1$.

Throughout the whole temperature range the thermodynamic functions are given by the analogues of (6·3·3), (6·3·9) and (6·3·10), but not by the analogues of (6·3·11) and (6·3·12), which are not valid if $T < T_1$. The internal energy is given by

$$\frac{U}{NkT} = 2\pi\varpi \left(\frac{2mkT}{h^2}\right)^{\frac{3}{2}} \frac{V}{N} F_{\frac{3}{2}}^{-}(\xi),$$

which, for $T \leq T_1$, becomes

$$\frac{U}{NkT} = 2\pi\varpi \left(\frac{2mkT}{h^2}\right)^{\frac{3}{2}} \frac{V}{N} F_{\frac{3}{2}}^{-}(0) = \frac{F_{\frac{3}{2}}^{-}(0)}{F_{\frac{1}{2}}^{-}(0)} \left(\frac{T}{T_1}\right)^{\frac{3}{2}} \quad (T \leq T_1) \tag{6·42·7}$$

on substituting for V/N from (6·41·4). Corresponding expressions can be obtained for PV, S and F, and since $\zeta(\frac{5}{2}) = 1·341$ we have

$$\tfrac{3}{2}PV = U = 0·7701 NkT(T/T_1)^{\frac{3}{2}} \quad (T \leq T_1), \tag{6·42·8}$$

$$S = 1·2835 Nk(T/T_1)^{\frac{3}{2}}, \quad F = -0·5134 NkT(T/T_1)^{\frac{3}{2}} \quad (T \leq T_1). \tag{6·42·9}$$

It can be shown by summing the series (6·41·8) and (6·41·9) numerically that we obtain the same values for $T = T_1$ as are given by (6·42·8) and

(6·42·9), so that U and S are continuous at $T=T_1$. Further, the specific heat per atom at constant volume is

$$C_V=\tfrac{3}{2}k\left\{1+0\cdot2309\left(\frac{T_1}{T}\right)^{\frac{3}{2}}+0\cdot04504\left(\frac{T_1}{T}\right)^{3}+0\cdot00686\left(\frac{T_1}{T}\right)^{\frac{9}{2}}+\ldots\right\}\quad(T\geqslant T_1),\tag{6·42·10}$$

$$C_V=1\cdot925k(T/T_1)^{\frac{3}{2}}\quad(T\leqslant T_1),\tag{6·42·11}$$

so that C_V is continuous at $T=T_1$. The temperature coefficient of C_V, however, is discontinuous at $T=T_1$, and

$$dC_V/dT=-0\cdot722k\quad(T=T_1+0),\quad dC_V/dT=2\cdot888k\quad(T=T_1-0).\tag{6·42·12}$$

The behaviour of C_V as a function of T is shown over the whole temperature range in fig. 6·19.

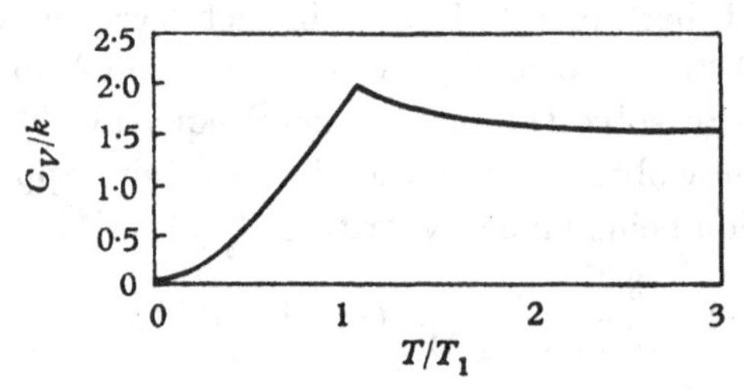

Fig. 6·19. The specific heat per atom as a function of T/T_1 for a perfect gas obeying Einstein-Bose statistics.

6·43. *Alternative investigation of the condensation phenomenon.* In view of the delicate nature of the arguments involved, we give a more rigorous investigation of the condensation phenomenon connected with the Einstein-Bose statistics. (Numerous accounts have been published. Those which bear most resemblance to that given here are the ones by Fowler & Jones (1938) and by de Groot, Hooyman & Ten Seldam (1950).)

We expand the right-hand side of the exact equation (6·42·1) in ascending powers of $\exp[(\mu-E_i)/kT]$. We take $\varpi=1$ and we consider N atoms in a finite enclosure, which we take to be a cube of side L. Then

$$N=\sum_i\sum_{j=1}^{\infty}\lambda^j\exp(-jE_i/kT),$$

where $\lambda=e^{\mu/kT}$. (Note that in the present section we use λ to denote $e^{\mu/kT}$ and not μ/kT.) If we make the wave functions vanish on the boundary they are given by $\psi=\sin q_1x\sin q_2y\sin q_3z$, where $q_1=\pi s_1/L$, $q_2=\pi s_2/L$, $q_3=\pi s_3/L$, the s's being positive integers including zero, while the energy levels are $h^2(s_1^2+s_2^2+s_3^2)/(8mL^2)$. The relation between N and λ is therefore

$$N=\sum_{j=1}^{\infty}\lambda^j \sum_{s_1,s_2,s_3} \exp\left[-j\gamma(s_1^2+s_2^2+s_3^2)\right],$$

where
$$\gamma=h^2/(8mkTL^2).$$

There is a corresponding expression for U. The problem is to examine the behaviour of these expressions when N and L tend to infinity, with the number of atoms N/L^3 per unit volume remaining finite.

We discuss a slightly more general problem than that of the thermodynamic functions of free particles, and we investigate the properties of the functions defined by the series

$$N=\sum_{j=1}^{\infty}\lambda^j \sum_{s_1,\dots,s_r} \exp\left[-j\gamma(s_1^\alpha+\dots+s_r^\alpha)\right], \tag{6·43·1}$$

$$Nv=U=\gamma kT\sum_{j=1}^{\infty}\lambda^j \sum_{s_1,\dots,s_r}(s_1^\alpha+\dots+s_r^\alpha)\exp\left[-j\gamma(s_1^\alpha+\dots+s_r^\alpha)\right], \tag{6·43·2}$$

where $0<\lambda\leqslant 1, \gamma>0, 1\leqslant\alpha\leqslant 2$, and where $s_i=0, 1, 2, \dots$. We are particularly interested in the behaviour of λ and v in the limit $\gamma\to 0$, $N\to\infty$ with $N\gamma^{r/\alpha}$ remaining finite, and we shall show that λ is a steadily increasing function of $N\gamma^{r/\alpha}$. We shall also prove that, if $r/\alpha>1$, we have $\lambda\equiv 1$ for all $N\gamma^{r/\alpha}$ greater than a certain critical value.

Let ϵ be a positive number such that ϵ/γ is a large integer. We split the summation with respect to j into the sum from 1 to ϵ/γ and the sum from $1+\epsilon/\gamma$ to infinity. In the first of these the maximum value of $j\gamma$ is ϵ, and we require an expression for the summand which is appropriate for small values of $j\gamma$. This is provided by the Euler-Maclaurin sum formula (6·11·7), which gives

$$\sum_{s=0}^{\infty}\exp(-j\gamma s^\alpha)=\frac{\Gamma(1+1/\alpha)}{(j\gamma)^{1/\alpha}}\left[1+\frac{(j\gamma)^{1/\alpha}}{2\Gamma(1+1/\alpha)}+O\{(j\gamma)^{1+1/\alpha}\}\right]. \tag{6·43·3}$$

In the second part of the summation, we write $j=j'+\epsilon/\gamma$ $(j'=1,2,\dots)$ and

$$\sum_{s=0}^{\infty}\exp(-j\gamma s^\alpha)=\sum_{s=0}^{\infty}\exp\left[-(j'\gamma+\epsilon)s^\alpha\right]<\sum_{s=0}^{\infty}\exp(-\epsilon s^\alpha). \tag{6·43·4}$$

By a second application of the Euler-Maclaurin sum formula we have

$$\sum_{s=0}^{\infty}\exp(-\epsilon s^\alpha)=\frac{\Gamma(1+1/\alpha)}{\epsilon^{1/\alpha}}\left[1+\frac{\epsilon^{1/\alpha}}{2\Gamma(1+1/\alpha)}+O(\epsilon^{1+1/\alpha})\right]. \tag{6·43·5}$$

We may therefore write (6·43·1) as

$$\frac{N\gamma^q}{[\Gamma(1+1/\alpha)]^r}=\sum_{j=1}^{\epsilon/\gamma}\lambda^j\gamma^q\sum_{s_1,\dots,s_r}\frac{\exp\left[-j\gamma(s_1^\alpha+\dots+s_r^\alpha)\right]}{[\Gamma(1+1/\alpha)]^r}$$
$$+\sum_{j'=1}^{\infty}\lambda^{j'+\epsilon/\gamma}\gamma^q\sum_{s_1,\dots,s_r}\frac{\exp\left[-(j'\gamma+\epsilon)(s_1^\alpha+\dots+s_r^\alpha)\right]}{[\Gamma(1+1/\alpha)]^r}, \tag{6·43·6}$$

where $q=r/\alpha$. Since the quantity ϵ is at our disposal and since we are only interested in the limit $\gamma\to 0$, we may choose ϵ small and ϵ/γ large. Then, provided that $\lambda<1$, the second sum can be made as small as we please, and the first sum converges to $\sum_{j=1}^{\infty}\lambda^j j^{-q}$. Hence, if we put $N\gamma^q=n(h^2/8mkT)^q$, where $n=N/V$, (6·43·1) becomes

$$\lim\frac{N\gamma^q}{[\Gamma(1+1/\alpha)]^r}\equiv\frac{n}{[\Gamma(1+1/\alpha)]^r}\left(\frac{h^2}{8mkT}\right)^q=\sum_{j=1}^{\infty}\frac{\lambda^j}{j^q}\equiv R_q(\lambda)\quad(0<\lambda<1). \tag{6·43·7}$$

This, however, breaks down if the expression on the left is greater than the maximum value of $R_q(\lambda)$, namely, $R_q(1)$, which can only occur if $R_q(\lambda)$ is convergent for $\lambda=1$, i.e. if $q>1$. We cannot then neglect the second summation in (6·43·6).

If $R_q(1)$ is finite we define T_0 by

$$R_q(1)=\sum_{j=1}^{\infty}\frac{1}{j^q}=\frac{n}{[\Gamma(1+1/\alpha)]^r}\left(\frac{h^2}{8mkT_0}\right)^q. \tag{6·43·8}$$

Now, from (6·43·6) we have

$$\frac{N\gamma^q}{[\Gamma(1+1/\alpha)]^r}<\sum_{j=1}^{\epsilon/\gamma}\frac{\lambda^j}{j^q}[1+O(\epsilon)]^r+\sum_{j'=1}^{\infty}\lambda^{j'+\epsilon/\gamma}\left(\frac{\gamma}{\epsilon}\right)^q,$$

that is,
$$\frac{N}{[\Gamma(1+1/\alpha)]^r}<\frac{1}{\gamma^q}\sum_{j=1}^{\infty}\frac{1}{j^q}[1+O(\epsilon)]^r+\frac{\lambda^{\epsilon/\gamma}}{\epsilon^q}\frac{\lambda}{1-\lambda}. \tag{6·43·9}$$

If $T<T_0$, the first term on the right is less than the term on the left, and, by (6·43·8), the inequality can be written as

$$1-\lambda<\frac{\lambda^{1+\epsilon/\gamma}}{\epsilon^q}[\Gamma(1+1/\alpha)]^r\frac{1}{N}\frac{1}{1-(T/T_0)^q}\quad(T<T_0). \tag{6·43·10}$$

Hence λ differs from unity by a fraction of the order of $1/N$.

6·431. We have now shown that, in the limit of infinite N, λ is determined by (6·43·7) if $T>T_0$ and that $\lambda\equiv 1$ if $T<T_0$, where T_0 is defined by (6·43·8). The parameter λ or one of its derivatives with respect to T may therefore be discontinuous at $T=T_0$, and we proceed to investigate these discontinuities. We write (6·43·7) as

$$x=R_q(1)\left(\frac{T_0}{T}\right)^q\equiv\frac{n}{[\Gamma(1+1/\alpha)]^r}\left(\frac{h^2}{8mkT}\right)^q=\sum_{j=1}^{\infty}\frac{\lambda^j}{j^q}\equiv R_q(\lambda), \tag{6·431·1}$$

and consider λ as a function of x. We must first obtain some properties of the function $R_q(\lambda)$.

The series $R_q(\lambda)$ converges for all values of q if $0\leqslant\lambda<1$, and the derivative $dR_q/d\lambda$ is given by

$$dR_q/d\lambda=R_{q-1}/\lambda\quad(\text{all}\,q). \tag{6·431·2}$$

Hence $R_q = 0$ and $dR_q/d\lambda = 1$ for $\lambda = 0$. We also have

$$R_0(\lambda) = \lambda/(1-\lambda), \quad R_1(\lambda) = -\log(1-\lambda), \tag{6·431·3}$$

so that by repeated applications of (6·431·2) we can obtain $R_q(\lambda)$ if q is an integer. If $q > 1$, the series converges for $\lambda = 1$ and $R_q(1)$ is finite. If, however, $q < 1$, then, as $\lambda \to 1$,

$$R_q(\lambda) \sim \Gamma(1-q)(1-\lambda)^{q-1} \quad (q<1, \lambda \to 1). \tag{6·431·4}$$

To prove this we write $\lambda = e^{-t}$ $(t > 0)$, so that $t \to 0$ as $\lambda \to 1$. Then

$$(1-\lambda)^{-q+1} R_q(\lambda) = \left(\frac{1-e^{-t}}{t}\right)^{-q+1} \sum_{j=1}^{\infty} (tj)^{-q} e^{-tj} t,$$

and, if we put $z = tj$, $dz = t\Delta j = t$, we have

$$(1-\lambda)^{-q+1} R_q(\lambda) \to \int_0^\infty z^{-q} e^{-z} dz = \Gamma(1-q),$$

which proves (6·431·4). It then follows from (6·431·2) that

$$dR_q/d\lambda \sim \Gamma(2-q)(1-\lambda)^{q-2} \quad (q<2, \lambda \to 1). \tag{6·431·5}$$

Differentiating (6·431·1) with respect to x, we have

$$\frac{d\lambda}{dx} = \frac{1}{dR_q/d\lambda} = \frac{\lambda}{R_{q-1}}, \tag{6·431·6}$$

by (6·431·2), and a second differentiation with respect to x gives

$$\frac{d^2\lambda}{dx^2} = \frac{\lambda}{R_{q-1}^2} - \frac{\lambda R_{q-2}}{R_{q-1}^3}. \tag{6·431·7}$$

With the help of these relations and of (6·431·4) we can now discuss the behaviour of $d\lambda/dx$ and $d^2\lambda/dx^2$. There are various cases to be considered.

(*a*) $0 \leqslant q \leqslant 1$. In this case $R_q(1)$ is infinite, and λ increases from 0 to 1, while $d\lambda/dx$ decreases steadily from 1 to 0, as x increases from 0 to infinity.

(*b*) $q > 1$. In this case $R_q(1)$ is finite and λ increases steadily from 0 to 1 as x increases from 0 to $R_q(1)$. For $x \geqslant R_q(1)$, λ is constant and equal to 1. There are now four sub-cases to consider, which differ in the behaviour of $d\lambda/dx$ and $d^2\lambda/dx^2$ at $x = R_q(1) - 0$.

(i) $1 < q < \frac{3}{2}$. In this case

$$d\lambda/dx = 0, \quad d^2\lambda/dx^2 = 0 \quad \text{at} \quad x = R_q(1) - 0. \tag{6·431·8}$$

There is, however, a discontinuity in the higher derivatives of λ.

(ii) $q = \frac{3}{2}$. In this case

$$d\lambda/dx = 0, \quad d^2\lambda/dx^2 = -1/(2\pi) \quad \text{at} \quad x = R_q(1) - 0. \tag{6·431·9}$$

(iii) $\frac{3}{2} < q \leqslant 2$. In this case

$$d\lambda/dx = 0, \quad d^2\lambda/dx^2 = -\infty \quad \text{at} \quad x = R_q(1) - 0. \tag{6·431·10}$$

(iv) $q>2$. Since $q-1>1$, $R_{q-1}(1)$ is finite, and we now have

$$d\lambda/dx = 1/R_{q-1}(1) \quad \text{at} \quad x = R_q(1)-0. \tag{6·431·11}$$

In the first three of the above sub-cases $d\lambda/dx$ is continuous at $x=R_q(1)$, while the second or higher derivatives of λ are discontinuous. In the fourth sub-case, $d\lambda/dx$ itself is discontinuous. To calculate the derivatives of λ with respect to T we use the formulae

$$\frac{d\lambda}{dT} = -qR_q(1)\frac{T_0^q}{T^{q+1}}\frac{d\lambda}{dx}, \tag{6·431·12}$$

$$\frac{d^2\lambda}{dT^2} = q(q+1)\,R_q(1)\frac{T_0^q}{T^{q+2}}\frac{d\lambda}{dx} + q^2\{R_q(1)\}^2\frac{T_0^{2q}}{T^{2q+2}}\frac{d^2\lambda}{dx^2}. \tag{6·431·13}$$

6·432. *The internal energy.* To calculate the internal energy we write (6·43·2) as

$$U = -kT\sum_{j=1}^{\infty}\lambda^j\frac{\partial}{\partial j}\prod_{i=1}^{r}\sum_{s_i=0}^{\infty}\exp(-j\gamma s_i^{\alpha}), \tag{6·432·1}$$

which is true for all values of λ including $\lambda=1$. By definition, the sum with respect to j in (6·432·1) is $\lim_{J\to\infty}\sum_{j=1}^{J}$. To determine the limit of U as $\gamma\to 0$, we therefore choose γ so that γJ is small and apply (6·43·3). We then have

$$\lim_{N\to\infty,\,\gamma\to 0} U = -NkT\lim_{J\to\infty}\sum_{j=1}^{J}\lambda^j\frac{\partial}{\partial j}\lim_{\substack{\gamma\to 0\\ N\to\infty}}\frac{[\Gamma(1+1/\alpha)]^r}{N(j\gamma)^q} = \frac{q}{R_q(1)}NkT\left(\frac{T}{T_0}\right)^q\sum_{j=1}^{\infty}\frac{\lambda^j}{j^{q+1}} \tag{6·432·2}$$

$$= \frac{q}{R_q(1)}NkT\left(\frac{T}{T_0}\right)^q R_{q+1}(\lambda). \tag{6·432·3}$$

Since λ is continuous at $T=T_0$, U is continuous there. The specific heat C_V is given by

$$C_V = \frac{1}{N}\left(\frac{\partial U}{\partial T}\right)_{V,N} = \frac{1}{N}\frac{\partial U(T,\lambda)}{\partial T} + \frac{1}{N}\frac{\partial U(T,\lambda)}{\partial\lambda}\frac{d\lambda}{dT}. \tag{6·432·4}$$

When $T<T_0$, the second term disappears since $\partial\lambda/\partial T\equiv 0$.

6·433. If we insert the values $r=3$, $\alpha=2$, $q=\frac{3}{2}$ into the preceding formulae we obtain all the formulae derived in § 6·42, the specific heat being continuous at $T=T_0$, but $\partial C_V/\partial T$ having a finite discontinuity there of an amount given by equation (6·42·12). If, however, the energy levels are such that $\alpha<\frac{3}{2}$ (in which case the particles cannot be moving under no forces), the specific heat is discontinuous at $T=T_0$. We then have

$$C_V(T_0-0) = kq(q+1)\,R_{q+1}(1)/R_q(1), \tag{6·433·1}$$

$$C_V(T_0+0) = C_V(T_0-0) - kq^2R_q(1)/R_{q-1}(1), \tag{6·433·2}$$

so that the specific heat decreases discontinuously as T increases from T_0-0 to T_0+0. To find the numerical values when say $q=3$, we have

$$R_2(1)=1{\cdot}645, \quad R_3(1)=1{\cdot}202 \quad \text{and} \quad R_4(1)=1{\cdot}0823.$$

These give

$$C_V(T_0-0)=10{\cdot}8k, \quad C_V(T_0+0)=6{\cdot}6k. \tag{6·433·3}$$

6·434. *The isothermal curves.* The preceding equations are expressed in a form which is suitable for calculating the variation of the thermodynamic functions at constant volume. If, however, we wish to discuss the isothermal curves, we recast the equations by writing $PV=\frac{2}{3}U$ in (6·432·3) and by substituting for T_0 from (6·43·8). We obtain

$$P=\tfrac{2}{3}q[\Gamma(1+1/\alpha)]^r\,(8mk/h^2)^q\,kT^{q+1}R_{q+1}(\lambda) \quad (\text{all } \lambda), \tag{6·434·1}$$

which shows that P is independent of v when $\lambda \equiv 1$, where $v=V/N=1/n$. For fixed T, we have $\lambda<1$ for $v>v_0(T)$ and $\lambda=1$ for $v<v_0(T)$, where

$$v_0(T)=\frac{1}{[\Gamma(1+1/\alpha)]^r}\left(\frac{h^2}{8mkT}\right)^q\frac{1}{R_q(1)}. \tag{6·434·2}$$

The isothermal curves may have continuous or discontinuous gradients at $v=v_0(T)$. To calculate the gradient we have

$$\frac{dR_{q+1}(\lambda)}{dv}=\frac{dR_{q+1}(\lambda)}{d\lambda}\left(\frac{\partial\lambda}{\partial v}\right)_{T,N}=-\frac{1}{v^2}\frac{1}{[\Gamma(1+1/\alpha)]^r}\left(\frac{h^2}{8mkT}\right)^q\frac{R_q(\lambda)}{R_{q-1}(\lambda)}, \tag{6·434·3}$$

by (6·431·1) and (6·431·2). Hence

$$\left(\frac{\partial P}{\partial v}\right)_T=-\tfrac{2}{3}q\frac{kT}{\{v_0(T)\}^2}\frac{R_q(1)}{R_{q-1}(1)} \tag{6·434·4}$$

for $v=v_0(T)+0$, while $\partial P/\partial v=0$ for $v=v_0(T)-0$. The various cases are as follows:

(i) If $0\leqslant q\leqslant 1$, we have $v_0(T)=0$, and $(\partial P/\partial v)_T$ is continuous everywhere.
(ii) If $1<q\leqslant 2$, we have $v_0(T)\neq 0$, but $(\partial P/\partial v)_T$ is continuous at $v_0(T)$.
(iii) If $q>2$, $v_0(T)\neq 0$, but $(\partial P/\partial v)_T$ is discontinuous at $v_0(T)$.

6·435. As will be discussed in § 9·41, liquid helium undergoes a transition at 2·2° K., for which no completely satisfactory explanation has yet been given. Helium consisting of the normal isotope He^4 obeys the Einstein-Bose statistics, and the transition phenomenon may be due either to the effect of the interatomic forces, or to the condensation phenomenon discussed in the preceding section, or to a combination of both factors. The characteristic temperature T_1 of a perfect gas with the density 0·15 gram/cm.^{3} which is about that of liquid helium, is 3·1° K., so that the influence of the peculiarities of the Einstein-Bose statistics cannot be ignored. It is, however, impossible to treat liquid helium as a perfect gas, and it has so far not proved possible to extend the theory to include the effect of the intermolecular forces. Also, in the liquid helium transition it is C_V and not $\partial C_V/\partial T$ which is discontinuous. If, therefore, the liquid helium transition is to be explained by the foregoing theory it is necessary for the parameter q to be greater than 2. As a speculative hypothesis it might be hoped that the effect of the interatomic forces might be such that the energy levels were given by

$$E = h^2(s_1^\alpha + s_2^\alpha + s_3^\alpha)/(8mL^2) \quad (\alpha < \tfrac{3}{2}), \tag{6·435·1}$$

but no model with this property has so far been proposed. In spite of this it is widely held that there must be some connexion between the liquid helium transition and the Einstein-Bose condensation.

Radiation

6·5. *Elementary concepts.* When two bodies A and B at temperatures T_A and T_B ($T_A > T_B$) are placed in a vacuum enclosed by adiabatic walls, the temperature of A falls and that of B rises. Both bodies radiate and absorb energy, but A emits more than it absorbs, while B absorbs more than it emits, and only the net transfer is observable.

If we consider radiation in a vacuum which is enclosed by walls maintained at a uniform temperature T, then, as a consequence of the laws of thermodynamics, the energy of the radiation per unit volume can only depend upon T and not upon the nature of the enclosing walls. (Radiation which is in equilibrium with its enclosure is known as complete or black-body radiation.) For, consider the radiation whose wavelength lies between λ and $\lambda + d\lambda$, and set up a second enclosure at the same temperature but with different walls. Also, let us establish communication between the two enclosures by making small windows in the walls transparent only to radiation with wavelength between λ and $\lambda + d\lambda$. Then, if the energy density of the radiation with the wavelength λ is greater in one enclosure than in the other, the first enclosure will lose energy and its temperature will fall, while the second enclosure will gain energy, and its temperature will rise, which is impossible.

The existence of radiation as a form of energy implies the existence of radiation pressure. For, let R be a perfectly reflecting tube and let A and B be two radiating bodies inside R, with $T_A > T_B$, behind perfectly reflecting screens C and D (fig. 6·20). If D is opened, the radiation from B will fill the tube. Then if D is closed and C is opened, the radiation in R will not be absorbed by A because the radiation from A is of greater intensity than that

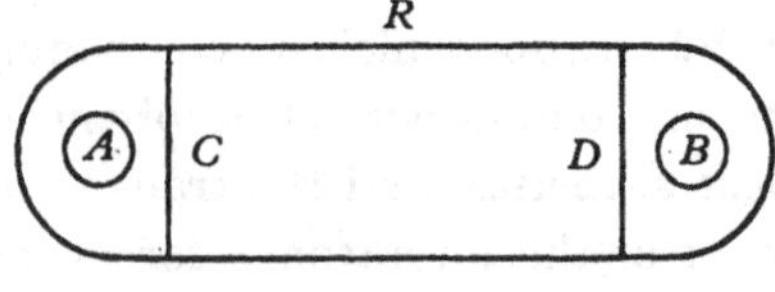

Fig. 6·20

emitted by B. But if the screen D is moved up to the position of C the whole of the radiation must be absorbed by A, and heat will have been transferred from the colder to the hotter body. An equivalent amount of external work must have been supplied to the system and therefore the radiation must have exerted a pressure on the perfectly reflecting screen D. According to the electromagnetic theory of light, the radiation pressure P of isotropic complete radiation is given by

$$P = \tfrac{1}{3}U/V, \tag{6·5·1}$$

where U is the energy and V is the volume.

6·51. *The thermodynamics of radiation.* If we consider radiation in an enclosure as a thermodynamic system it must be characterized by the equation

$$T\,dS = dU + P\,dV,$$

where P is given by (6·5·1) and where, by the arguments of the preceding section, U/V and P are functions of T only. Hence

$$dS = \frac{3V}{T}\frac{dP}{dT}dT + \frac{4P}{T}dV, \tag{6·51·1}$$

and, since dS is a perfect differential,

$$\frac{\partial}{\partial V}\left(\frac{3V}{T}\frac{dP}{dT}\right) = 4\frac{d}{dT}\frac{P}{T}.$$

This gives

$$\frac{dP}{dT} = \frac{4}{T}P$$

and

$$P = \tfrac{1}{3}aT^4, \quad U/V = aT^4, \tag{6·51·2}$$

where a is an absolute constant. The relation $U/V = aT^4$ is known as Stefan's law, and $\sigma = \tfrac{1}{4}ac$ is called Stefan's constant, c being the velocity of light. Further, (6·51·1) becomes

$$dS = \tfrac{4}{3}a(3VT^2dT + T^3dV) = \tfrac{4}{3}ad(T^3V),$$

and so

$$S = \tfrac{4}{3}aVT^3. \tag{6·51·3}$$

The value of σ can be determined directly by measuring the energy emitted per second from a black body, or indirectly from the theoretical expression (6·52·5) given later. Its value is $5{\cdot}735 \times 10^{-5}$ erg cm.$^{-2}$ sec.$^{-1}$ degree^{-4}.

6·52. *The statistical mechanics of radiation.* The thermodynamics of the radiation field can be elaborated much further, and such considerations were of great importance at the turn of the century when they led Planck to the beginnings of the quantum theory. They are of less importance today, since more direct methods are available for establishing all the important results. We therefore only give the most direct proof of Planck's formula for the spectral distribution of the energy density of black-body radiation.

The electric field $\mathbf{E}$ of electromagnetic waves in free space satisfies the wave equation $\nabla^2\mathbf{E} - \partial^2\mathbf{E}/c^2\partial t^2 = 0$ subject to the condition $\operatorname{div}\mathbf{E} = 0$. There is therefore a formal analogy between the electromagnetic oscillations in a vacuous enclosure and the vibrations of an elastic continuum of the same size, the only difference being that the condition $\operatorname{div}\mathbf{E} = 0$ means that there are only two (transverse) vibrations for a given set of wave numbers instead of three. We may therefore write down at once the free energy of the radiation from equations (6·2·5) and (6·23·5) as

$$F = \frac{8\pi}{c^3} kTV \int_0^\infty \log\left[1 - \exp\left(-h\nu/kT\right)\right] \nu^2 d\nu. \tag{6·52·1}$$

It should be noted that we have here omitted the residual energy $\frac{1}{2}h\nu$ of the various oscillations, since this would add an infinite term to F, and that the upper limit of the integral must be taken to be infinity since there is no reason for restricting the frequency in any way. To find the energy we differentiate F/T, and we obtain

$$U = \frac{8\pi V}{c^3} \int_0^\infty \frac{h\nu^3 d\nu}{\exp\left(h\nu/kT\right) - 1}. \tag{6·52·2}$$

If $E_\nu d\nu$ is the energy density of the radiation with frequencies lying between ν and $\nu + d\nu$, we have

$$E_\nu = \frac{8\pi h\nu^3}{c^3} \frac{1}{\exp\left(h\nu/kT\right) - 1}, \tag{6·52·3}$$

which is Planck's law. Stefan's law can be deduced immediately from (6·52·3). For, putting $x = h\nu/kT$, we have

$$U = \frac{8\pi k^4}{h^3c^3} VT^4 \int_0^\infty \frac{x^3 dx}{e^x - 1} = \frac{8\pi^5 k^4}{15h^3c^3} VT^4, \tag{6·52·4}$$

which gives the explicit value of Stefan's constant as

$$\sigma = \frac{2\pi^5 k^4}{15h^3c^2}. \tag{6·52·5}$$

If for any given temperature the position of the maximum of the spectral distribution curve is measured, we can obtain a relation for Planck's constant h. We may choose either E_ν or E_λ, which are connected by the relation $E_\nu d\nu = E_\lambda d\lambda$. If we choose E_ν, the maximum occurs when

$$\exp\left(-h\nu/kT\right) = 1 - \tfrac{1}{3}h\nu/kT,$$

and since the equation $e^{-x} = 1 - \frac{1}{3}x$ has the root $x = 2·8$, we have the relation

$$h\nu_m/kT = 2·8$$

for the value of the frequency ν_m at which the maximum in E_ν occurs. The measurement of both ν_m and σ enables us to determine both h and k, while if k is considered to be known from other phenomena, the measurement of ν_m will give us h and therefore σ.

6·53. *Radiation treated as being composed of photons.* It is possible to give an extreme particle description of the radiation field in quantum theory, and to consider the radiation to be composed of light quanta or photons which have zero rest-mass and for which the relation $h\nu \equiv E = c\,|\,\mathbf{p}\,|$ holds between the energy and the momentum. For every $\mathbf{p}$ there are two photons with different directions of polarization. Also there is no upper limit to the number of photons which may be present in any state. The proof that this description is a correct one involves the theory of the quantization of the radiation field. (See, for example, Heitler, 1936, Chapter 2.)

If we take these results as our starting point, we can calculate the energy of a system of photons from (5·62·3) with the introduction of a factor 2 to take care of the two possible states of polarization. The parameter μ must be omitted (or put equal to zero), since in deriving the distribution function we no longer have to impose any condition on the number of particles present. Just as in § 5·4, the wave function of a photon is

$$\exp\{2\pi i(p_1x + p_2y + p_3z)/h\},$$

and if the enclosure is bounded by the planes $0 \leqslant x \leqslant L_1$, $0 \leqslant y \leqslant L_2$, $0 \leqslant z \leqslant L_3$, the possible values of p_1, p_2, p_3 are once again hn_1/L_1, hn_2/L_2, hn_3/L_3. We therefore have

$$U = 2\sum_r \frac{E_r}{\exp(E_r/kT) - 1} = \frac{2V}{h^3}\int_{-\infty}^{\infty}\int_{-\infty}^{\infty}\int_{-\infty}^{\infty} \frac{E}{\exp(E/kT) - 1}\,dp_1\,dp_2\,dp_3.$$

Since $|\,\mathbf{p}\,| = h\nu/c$, this is the same result as that obtained by the more elementary method given in the preceding section.

REFERENCES

Barron, T. H. K. (1955). The thermal expansion of solids at low temperatures. *Phil. Mag.* (7), **46**, 720.

Bijl, D. & Pullan, H. (1954). Thermal expansion of simple solids at low temperatures and Grüneisen's law. *Phil. Mag.* (7), **45**, 290.

Bijl, D. & Pullan, H. (1955). A new method ot measuring the thermal expansion of solids at low temperatures. The thermal expansion of copper and aluminium and the Grüneisen rule. *Physica*, **21**, 285.

Blackman, M. (1935). Contributions to the theory of the specific heat of crystals. *Proc. Roy. Soc.* A, **148**, 365, 384.

Blackman, M. (1941). The theory of the specific heat of solids. *Rep. Progr. Phys.* **8**.

Born, M. (1923). *Atomtheorie des festen zustandes.* Leipzig.

Born, M. & Huang, K. (1954). *Dynamical theory of crystal lattices.* Oxford.

Born, M. & von Kármán, T. (1912). Vibrations in space lattices. *Phys. Z.* **13**, 297.

Carpenter, L. G. & Steward, C. J. (1939). The atomic heat of potassium. *Phil. Mag.* (7), **27**, 551.

Clusius, K. (1946). The molecular heat of lithium fluoride between 18° and 273·2° K. *Z. Naturf.* **1**, 79.

Dauphinee, T. M., MacDonald, D. K. C. & Preston-Thomas, H. (1954). The specific heat of sodium between 55° and 315° K. *Proc. Roy. Soc.* A, **221**, 267.

Debye, P. (1912). The theory of specific heats. *Ann. Phys., Lpz.,* (4), **39**, 789.

Dennison, D. M. (1927). A note on the specific heat of the hydrogen molecule. *Proc. Roy. Soc.* A, **115**, 483.

Einstein, A. (1907). Planck's theory of radiation and the theory of specific heat. *Ann. Phys., Lpz.,* (4), **22**, 180.

Eucken, A. & d'Or, L. (1932). The molecular heat of gaseous nitric oxide at low temperatures. *Nachr. Ges. Wiss. Göttingen,* p. 107.

Fowler, R. H. & Jones, H. (1938). The properties of a perfect Einstein-Bose gas at low temperatures. *Proc. Camb. Phil. Soc.* **34**, 523.

Ginnings, D. C., Douglas, T. B. & Ball, A. F. (1950). The heat capacity of sodium between 0° and 900° C. *J. Res. Nat. Bur. Stand.* **45**, 23.

de Groot, S. R., Hooyman, G. J. & Ten Seldam, C. A. (1950). On the Bose-Einstein condensation. *Proc. Roy. Soc.* A, **203**, 266.

Grüneisen, E. (1908). The thermal expansion and the specific heat of metals. *Ann. Phys., Lpz.,* (4), **26**, 211.

Heitler, W. (1936). *The quantum theory of radiation.* Oxford.

Herzberg, G. (1945). *Infra-red and Raman spectra.* New York.

Iona, M. (1941). The distribution of lattice vibrations of the potassium chloride crystal. *Phys. Rev.* **60**, 822.

Kassel, L. S. (1936). The calculation of the rmodynamic functions from spectroscopic data. *Chem. Rev.* **18**, 277.

Kellerman, E. W. (1941). The specific heat of the sodium chloride lattice. *Proc. Roy. Soc.* A, **178**, 17.

Leighton, R. B. (1948). The vibrational spectrum and specific heat of a face-centred cubic crystal. *Rev. Mod. Phys.* **20**, 165.

London, F. (1938). The Bose-Einstein condensation. *Phys. Rev.* **54**. 947.

McDougall, J. & Stoner, E. C. (1938). The computation of Fermi-Dirac functions. *Phil. Trans.* A, **237**, 67.

Martin, D. L. (1955). The specific heats of lithium fluoride, sodium chloride and zinc sulphide at low temperatures. *Phil. Mag.* (7), **46**, 751.

Pitzer, K. S. & Gwinn, W. D. (1942). Energy levels and thermodynamic functions for molecules with internal rotations. *J. Chem. Phys.* **10**, 428.

Rubin, T., Altman, W. H. & Johnston, H. L. (1954). Coefficients of thermal expansion of solids at low temperatures. *J. Amer. Chem. Soc.* **76**, 5289.

Seitz, F. (1940). *The modern theory of solids.* New York.

Smith, H. J. M. (1948). The theory of the vibrations and the Raman spectrum of the diamond lattice. *Phil. Trans.* A, **241**, 105.

Wilson, A. H. (1953). *The theory of metals,* 2nd ed. Cambridge.

Wilson, E. B. (1940). The present status of the statistical method of calculating thermodynamic functions. *Chem. Rev.* **27**, 17.

Chapter 7

THE THIRD LAW OF THERMODYNAMICS

INTRODUCTION

7·1. It has been pointed out on numerous occasions in the preceding chapters that the expressions for the entropy and the free energy of any homogeneous phase of a substance contain one and two arbitrary constants respectively. For many purposes the values of these constants are immaterial, but for phenomena involving a change of phase or a chemical reaction, the energy and entropy constants appear explicitly in the relevant formulae, and the question arises as to how the constants can be determined, either by measurement or by calculation.

For a classical perfect gas, the entropy depends logarithmically upon the temperature and the pressure, and we can therefore make no progress by trying to relate the entropy constant to the value of the entropy at the absolute zero, at least if we restrict ourselves to the strict classical theory. For solids, on the other hand, the specific heat tends to zero as $T \to 0$ (C_V is of order T for metals and of order T^3 for insulators), and the same difficulties do not occur. We may then write $H(T, P) = U_0(P) + O(T^2)$ for sufficiently small values of T, and the Gibbs-Helmholtz equation

$$\frac{\partial}{\partial T}\frac{G}{T} = -\frac{H}{T^2} \tag{7·1·1}$$

can be integrated to give

$$G(T, P) = U_0(P) + B(P)\,T + O(T^2). \tag{7·1·2}$$

A similar formula applies to more general systems characterized by s deformation coordinates a_i (and s generalized forces A_i). The hypothesis was put forward by Nernst in 1906 that the arbitrary function $B(P)$ of the pressure (and the corresponding functions $B(A_i)$ for more general systems), should be put equal to zero for all solid substances, which is equivalent to postulating that the derivative with respect to the temperature of the free energy, like that of the heat function or of the internal energy, vanishes at $T = 0$.

It should be noted that putting $B(P) = 0$ for one particular value of P would be a trivial hypothesis, since this could be attained merely by a change in the definition of the entropy. Similarly, Nernst's hypothesis entails more than taking $B(P) = 0$, for, say, each elemental solid, since it implies that we can take $B(P) = 0$ for all compounds as well.

Nernst later extended his hypothesis to include gases as well as solids. At that time this extension led to considerable difficulties, since the entropy of a classical perfect gas is infinite at $T=0$, and it was necessary to introduce the *ad hoc* assumption that a gas would undergo a degeneration process of some kind or other at low temperatures, which would result in its entropy tending to zero instead of minus infinity at $T=0$. As discussed in Chapters 5 and 6 we now know that such a degeneration is a consequence of the laws of quantum mechanics. The conditions for the validity of the classical theory have been derived in §§ 6·3 and 6·4, and expressions have been given for the entropy and the free energy when quantum effects must be taken into account. For many purposes, however, a knowledge of the full quantum theory is not necessary.

Nernst's own viewpoint is given in his book published in 1924. He believed that his hypothesis could be derived from the single condition that the specific heat should vanish at $T=0$, but this is not so, and the history of the development of the third law of thermodynamics is remarkable for the number of controversies that it has evoked.

Much of the early work to test the validity of Nernst's hypothesis was carried out by G. N. Lewis and his collaborators. As a result, he came to the conclusion that the third law should be stated in the following terms (Lewis & Randall, 1923, p. 448): 'Every substance has a finite positive entropy, but at the absolute zero of temperature the entropy may become zero, and does so become in the case of perfect crystalline substances.' In 1930 Simon published an extensive and critical survey of Nernst's principle, and it is largely due to this article and Simon's later extensions of it that it is now recognized that Lewis & Randall's formulation of the third law is unduly restrictive. The status of the third law is more or less the same today as it was in 1930, and the reader is referred to Simon's article for an account of the historical development of the subject and for a full list of references up to 1930. The following account is based largely upon Simon's exposition.

7·11. An insight into the various possibilities that could conceivably occur can be obtained by considering the expression for the entropy of a classical perfect gas, namely,

$$s(T,P)=C_P^0\log\frac{T}{T^\dagger}+\int_0^T\frac{C_P^1(T')}{T'}dT'-R\log\frac{P}{P^\dagger}+s_0, \qquad (7\cdot11\cdot1)$$

where C_P^0 and $C_P^1(T)$ are as usual the constant and variable parts of the specific heat. This expression for s contains one arbitrary constant, namely, $s_0-C_P^0\log T^\dagger+R\log P^\dagger$, and its behaviour as a function of T for various values of P is as shown in fig. 7·1. If, however, we were to assume that $C_P^0=0$

(which is impossible in the classical theory), the entropy would not tend to minus infinity at $T=0$ for any non-zero value of P, but it would behave as shown in fig. 7·2. According to Nernst's hypothesis neither of these cases can occur for any real substance which is in complete thermodynamic equilibrium, and the only remaining possibility is for the entropy curves to be concurrent at $T=0$, as shown in fig. 7·3. Whether the common end-point of the curves is taken to be $s=0$ or some positive value of s, is a comparatively minor matter. But, if for two substances C and D we take the values

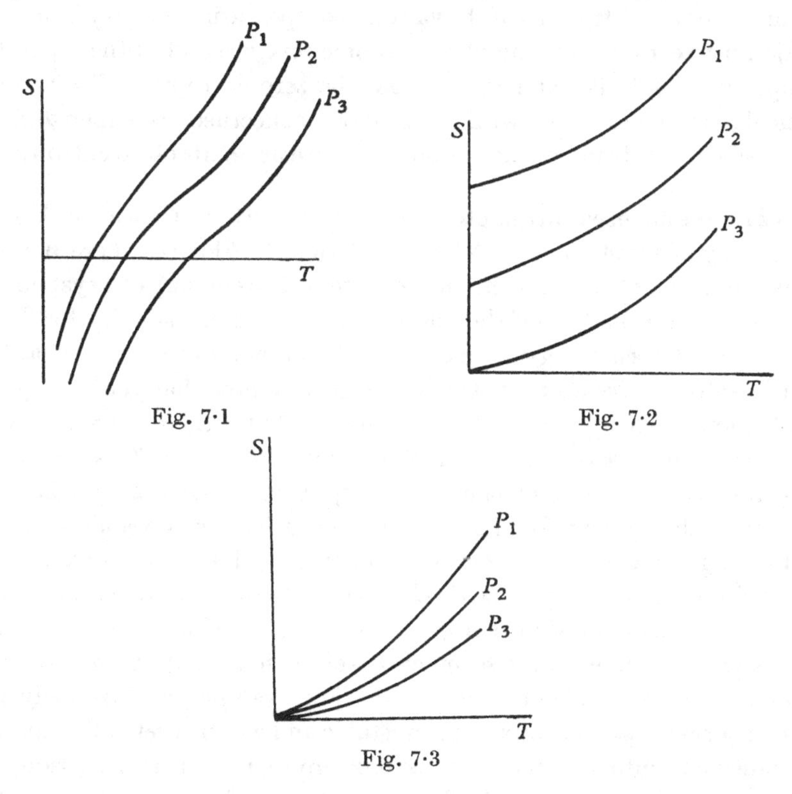

Fig. 7·1

Fig. 7·2

Fig. 7·3

of s at $T=0$ to be s_C^0 and s_D^0 respectively per atom, then for a compound with the composition $C_m D_n$ we must take the value of s at $T=0$ to be $ms_C^0+ns_D^0$ per molecule.

7·12. *Absolute entropy.* It is usually convenient but never necessary to take the limiting value of s as $T \to 0$ to be zero. The term absolute entropy is often used to denote the entropy of a substance when the above limit is assumed to be zero, but in practice the term is not well defined. For example, most elements consist of a mixture of isotopes, and as shown in § 11·142, a random mixture of a number of distinguishable substances has a greater entropy than its pure separated constituents at the same temperature and pressure. In principle, therefore, it is incorrect to ascribe zero value to the entropy at $T=0$ to the naturally occurring isotopic mixture of a given element, since the mixture could be separated into its constituent isotopes, and to each pure isotope we would then have to ascribe a negative entropy at $T=0$ in order to be consistent. It is therefore more correct to give each isotope zero entropy at $T=0$ and the natural mixture a corresponding positive entropy. But, unless we are dealing with a problem in which the concentrations of the various isotopes do in fact vary, the exact relation between the entropy of the isotopic mixture and of the separated isotopes is immaterial, and we may ascribe zero entropy at $T=0$ to the particular isotopic mixture with which we are concerned. It is then difficult to describe the entropy of the isotopic mixture as an absolute entropy.

7·121. We have seen in Chapter 5 that, according to statistical mechanics, the entropy of a system is given by the relation $S=k\log W$, where W is the number of relevant complexions of the system. The entropy of a system will therefore be zero at $T=0$ if the number of complexions is unity for $T=0$, that is, if the lowest energy state of the system is non-degenerate. That this is universally true seems probable, but no general proof has yet been given.

If the lowest energy state is non-degenerate, but if there are ϖ low-lying states whose total separation is much less than kT_0, where T_0 is the lowest temperature that can be reached in an experiment, the apparent limiting value of S will be $k\log\varpi$ and not zero. But in such a case we could still take the limiting value of S to be zero without running into any contradictions by redefining the entropy to be $S-k\log\varpi$. We should, however, be unable to maintain this definition if it subsequently proved possible to observe the system at temperatures so much lower than T_0 that the ϖ low-lying states had to be considered separately. For example, at extremely low temperatures an isotopic mixture is unstable and will theoretically separate spontaneously into a heterogeneous mixture of its constituent isotopes. At such temperatures it is logically incorrect to ascribe zero entropy to the isotopic mixture. In practice, however, we should probably still do so, since, unless some catalytic process could be found which would influence the equilibrium, the separation would not physically take place. We must therefore recognize that the experimental limiting value of the entropy of a system may depend upon the lowest temperature which is attainable, and the means available for making a system pass from a state of apparent equilibrium into a lower state of equilibrium, which may itself be either apparently or truly stable.

NERNST'S PRINCIPLE AND THE UNATTAINABILITY OF THE ABSOLUTE ZERO

7·2. In its strictest form, Nernst's hypothesis only applies to systems which are in thermodynamic equilibrium. It is then equivalent as a hypothesis in thermodynamics to the unproved theorem in statistical mechanics that the lowest energy state of a dynamical system is non-degenerate. However, as we have seen in the preceding section, such an interpretation is unduly restrictive owing to the existence, particularly at low temperatures, of states which are only in apparent equilibrium. A formulation of Nernst's principle which is sufficiently wide to include such states is the following (Simon, 1937):

GENERALIZATION E. *The contribution of the entropy due to each factor which is in internal equilibrium within a system becomes zero at* $T=0$.

From time to time, apparent exceptions to this principle have been discovered, but in all such cases it has been found either by direct experiment or by the consideration of a plausible theoretical model that the discrepancy can be attributed to the system being in a state of apparent equilibrium and undergoing an irreversible transition.

A generalization which is equivalent to Nernst's principle is the unattainability of the absolute zero (Nernst, 1912). The unattainability of the absolute zero is trivial for any classical system, but, as already mentioned in § 7·1, the specific heat of every real substance tends to zero as $T\to 0$, and in this case the principle of the unattainability of the absolute zero is not a consequence of the first and second laws of thermodynamics alone. In precise language, the principle can be stated as follows:

GENERALIZATION E′. *It is impossible to reduce the temperature of any assembly to the absolute zero in a finite number of operations.*

It is clear from fig. 7·3 that generalizations E and E′ are equivalent. A formal proof is as follows.

Consider any reversible adiabatic process connecting two states 1 and 2 with temperatures T and T', and entropies $S_1(T)$ and $S_2(T')$. We have $S_1(T)=S_2(T')$, and we can attain $T'=0$ if the equation $S_1(T)=S_2(0)$ has a non-zero solution for T. Since dS/dT is always positive there will in fact be a solution unless $S_1(0)\geqslant S_2(0)$. Similarly, the equation $S_1(0)=S_2(T)$ will always have a non-zero solution for T unless $S_2(0)\geqslant S_1(0)$. We must therefore have $S_1(0)=S_2(0)$, which is generalization E. If the process is irreversible it can only proceed in one direction, and if it is in the direction $1\to 2$ the temperature T'' of the final state must be such that $S_2(T'')>S_2(T')$. The temperature attained in this irreversible process will therefore be higher than that attained in a reversible process, and, as before, $T=0$ is unattainable. We have therefore proved the equivalence of generalizations E and E′. Generalization E′, namely, the unattainability of the absolute zero of temperature, is generally known as the third law of thermodynamics. In the present exposition it is taken as a general principle, and generalization E, which is usually known as Nernst's theorem, is deducible from it.

Elementary Consequences of Nernst's Principle

7·3. Since $\Delta S \to 0$ as $T \to 0$ for any isothermal reversible process, it follows that both $(\partial S/\partial V)_T$ and $(\partial S/\partial P)_T$ tend to zero as $T \to 0$ for any solid which does not show a strain hysteresis. By the Maxwell equations, these relations are equivalent to

$$\lim_{T\to 0}\left(\frac{\partial P}{\partial T}\right)_V = 0, \quad \lim_{T\to 0}\left(\frac{\partial V}{\partial T}\right)_P = 0. \qquad (7\cdot 3\cdot 1)$$

The coefficient of expansion therefore vanishes at $T=0$, a prediction which, as we have already seen in § 6·242, is in accordance with the experimental facts for crystalline solids. A special and unusual case is the vanishing at $T=0$ of the coefficient of expansion of liquid helium (see fig. 9·9, p. 279).

7·31. There are many systems which are describable by other variables than the pressure P and the volume V. If we have a system with s degrees of freedom, for which the generalized coordinates are a_i and the generalized forces are A_i, then according to Nernst's principle we must have

$$\lim_{T\to 0}\left(\frac{\partial S}{\partial A_i}\right)_T = -\lim_{T\to 0}\left(\frac{\partial a_i}{\partial T}\right)_{Ai} = 0, \quad \lim_{T\to 0}\left(\frac{\partial S}{\partial a_i}\right)_T = \lim_{T\to 0}\left(\frac{\partial A_i}{\partial T}\right)_{ai} = 0. \quad (7\cdot 31\cdot 1)$$

There are $s-1$ variables not given explicitly which can be chosen by picking one and only one variable out of each of the sets of conjugate variables $(a_1, A_1), \ldots, (a_r, A_r)$ excluding (a_i, A_i). The equality of say $(\partial S/\partial A_i)_T$ and $-(\partial a_i/\partial T)_{Ai}$ follows from the analogues of the Maxwell relations given in § 3·2. Some examples of the relations (7·31·1) are the following.

The surface tension σ and the surface area A of a liquid are conjugate variables, and in this case (7·31·1) becomes

$$\lim_{T\to 0}\left(\frac{\partial \sigma}{\partial T}\right)_A = 0. \qquad (7\cdot 31\cdot 2)$$

The only liquid which is stable down to $T=0$ is liquid helium, and, as shown by fig. 7·4, the relation (7·31·2) holds for helium.

It is shown in Chapter 10 that the intensity of magnetization I and the magnetic field H in a magnetic body are conjugate variables, and it therefore follows that

$$\lim_{T\to 0}\left(\frac{\partial I}{\partial T}\right)_H = 0. \tag{7·31·3}$$

This relation is obeyed by ferromagnetic bodies, which are bodies possessing an intrinsic magnetization (see § 10·5), but not by paramagnetic bodies if we assume that the paramagnetic susceptibility obeys Curie's law, according to which $I \propto H/T$. The relation (7·31·3) shows that although Curie's law is

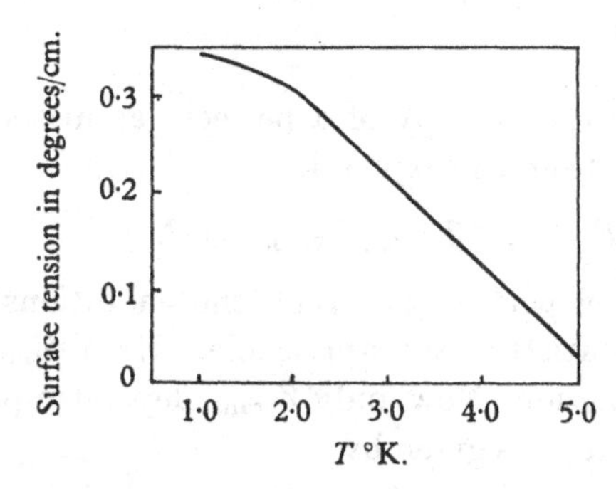

Fig. 7·4. The surface tension of liquid helium.

known to hold down to about 1° K. for many paramagnetic substances (see § 10·41), it must break down at still lower temperatures and be replaced by a law of the form $I = I_0 + \alpha T^r$, where I_0 may or may not be zero and where $r > 1$. Similar considerations hold for dielectrics.

More complicated phenomena are dealt with in later chapters.

The Spectroscopic and the Calorimetric Entropy of a Gas

7·4. It has already been pointed out in § 7·1 that Nernst's principle cannot be directly applied to a real gas without going beyond the bounds of the classical theory. For a permanent gas, the entropy at normal temperatures is given entirely satisfactorily by the expression (7·11·1) for the entropy of a perfect gas, but the value of the entropy constant is determined by the behaviour of the gas near $T = 0$, where (7·11·1) is no longer true. To obviate this difficulty we can proceed in two ways.

7·41. In principle, the straightforward procedure is to replace (7·11·1) by the correct quantal formula (either (6·3·8) and (6·311·7) or (6·41·9) and (6·42·9) according as the gas obeys Fermi-Dirac or Einstein-Bose statistics) since these reduce to (7·11·1) when $(kT)^{\frac{5}{2}}/P \gg h^3/(2\pi m)^{\frac{3}{2}}$, and they also give the correct behaviour near $T=0$. We are, however, never interested in regions in which the influence of quantum statistics is apparent, and we can avoid these unnecessarily complicated calculations by using the formulae of § 5·4, which relate to a gas whose energy levels are calculated according to quantum mechanics, but whose partition function is derived using the Maxwell-Boltzmann distribution. As was pointed out in §§ 6·3 and 6·4 this theory gives the correct high-temperature form of the thermodynamic functions, and the value of the entropy constant is that required to make $S=0$ for $T=0$ in the quantal expressions of which (5·4·11) is the high-temperature limit.

7·411. To calculate the entropy of a perfect gas in the wide sense in this way, we may start from the expression

$$F=-NkT\log(Z_{\text{trans.}}Z_{\text{int.}}/N!) \tag{7·411·1}$$

for the free energy of a perfect gas when the conditions are such that Maxwell-Boltzmann statistics are applicable. Here $Z_{\text{int.}}$ includes the rotational partition function. Now, only $Z_{\text{trans.}}$ depends upon the volume, and by (5·4·11) the entropy is given by

$$S=-\left(\frac{\partial F}{\partial T}\right)_V=\tfrac{5}{2}Nk+Nk\log\left(\frac{V}{N}\frac{\varpi_G(2\pi mkT)^{\frac{3}{2}}}{h^3}\right)+Nk\frac{\partial}{\partial T}(T\log Z_{\text{int.}}) \tag{7·411·2}$$

$$=\tfrac{5}{2}Nk+Nk\log\left(\frac{(kT)^{\frac{5}{2}}}{P}\frac{\varpi_G(2\pi m)^{\frac{3}{2}}}{h^3}\right)+Nk\frac{\partial}{\partial T}(T\log Z_{\text{int.}}), \tag{7·411·3}$$

where ϖ_G is the weight of the lowest electronic state. The internal partition function $Z_{\text{int.}}$ and the weight factor ϖ_G can both be obtained from spectroscopic data, and we can therefore calculate the entropy of the gas. The entropy obtained in this way is known as the 'spectroscopic entropy'. If the gas is not perfect, a correction can be introduced by using the formulae given later in § 8·5.

7·42. The second method of obtaining the entropy of a classical gas with the correct entropy constant is to determine the entropy of the corresponding solid up to some appropriate temperature T_1, and then to add on the entropy of sublimation at the temperature T_1. (In many cases it will be more convenient to determine the entropy of the solid up to the melting-point, the entropy of fusion, the entropy of the liquid from the melting-point to T_1 and the entropy of vaporization at T_1.) This will give us the correct entropy of the gas at the temperature T_1 and at the corresponding vapour pressure. Provided that T_1 is sufficiently high for classical statistics to be applicable, we may then determine the entropy at any higher temperature, and at arbitrary pressures, by using (7·11·1) (or its generalization

(8·5·5) for an imperfect gas), since the difference in the entropy of a perfect gas between two states does not depend upon the entropy constant. The entropy obtained in this way is known as the 'calorimetric entropy', and the applicability of the method depends upon the specific heat of a solid vanishing at the absolute zero.

7·43. If both the gas and the solid are in perfect thermodynamic equilibrium, the spectroscopic and calorimetric entropies must necessarily be the same if we ascribe zero entropy to the solid at $T=0$, but if either the gas or the solid can exist in metastable states there may be discrepancies. For the entropy constant of the gas has been chosen so as to make the entropy tend to zero at $T=0$ when quantum effects are taken into account, and by Nernst's principle we must then take the limiting value of the entropy of the solid to be zero. If, however, the gas is in a state of metastable equilibrium, as occurs, for example, with hydrogen, we shall not obtain the correct spectroscopic entropy by using (7·411·3), and we have to use the more elaborate theory for homonuclear gases given in § 6·17, which assumes that the metastable equilibrium is maintained down to the lowest temperatures attainable. At low temperatures the full quantum theory must be used for hydrogen, but for medium and high temperatures, and for all other substances, it is sufficient to use the formulae which apply when classical conditions hold.

7·431. Metastable states are more common in solids than in gases, since if a particular non-equilibrium state is attained at a low temperature the relaxation time may be effectively infinite and the non-equilibrium state is then 'frozen in'. If the determination of the calorimetric entropy of a gas involves a solid in a metastable state, the spectroscopic and calorimetric entropies will not agree unless the same metastable states occur in both the gas and the solid, or, in other words, unless the metastable equilibrium is unaffected by the (ideal) transformation of the solid into the gas at $T=0$.

In general, a solid will not contain fewer frozen-in metastable states than the corresponding gas, and, since the calorimetric entropy is calculated on the assumption that the entropy of the solid is zero at $T=0$ and that the solid is in perfect thermodynamic equilibrium, the calorimetric entropy of a gas will in general be less than or equal to the spectroscopic entropy. If the calorimetric entropy is less than the spectroscopic entropy, the discrepancy can be removed by assigning a non-zero value to the entropy of the solid at $T=0$, and this non-zero value of the entropy will be a measure of the degree of randomness of the structure due to the frozen-in states over and above any factors, such as the existence of isotopes, which are the same for the solid and for the gas.

NUMERICAL RESULTS

7·5. We now discuss the numerical results obtained for various gases, dealing in the present section with those gases for which the spectroscopic and calorimetric entropies agree, and in § 7·51 with those gases for which there is a discrepancy. On account of the typical quantum effects which occur, a separate discussion is given of hydrogen in § 7·52.

The spectroscopic entropy of a monatomic gas is given by (7·411·3). To obtain corresponding expressions for diatomic and polyatomic gases at temperatures where the rotations are classical, we differentiate (6·17·2) and (6·17·4). If we omit the nuclear spin weights we have

$$S = \tfrac{7}{2}Nk + Nk\log\left(\frac{(kT)^{\frac{7}{2}}}{P}\frac{\varpi_G(2\pi m)^{\frac{3}{2}}}{h^3}\right)$$
$$+ Nk\sum_i\left\{\frac{h\nu_i}{kT}\frac{1}{\exp(h\nu_i/kT)-1} - \log[1-\exp(-h\nu_i/kT)]\right\}$$
$$+ Nk\frac{\partial}{\partial T}(T\log Z_{\mathrm{el.}}) + Nk\log\frac{8\pi^2 Ik}{h^2\sigma} \qquad (7·5·1)$$

for linear molecules, and

$$S = 4Nk + Nk\log\left(\frac{(kT)^4}{P}\frac{\varpi_G(2\pi m)^{\frac{3}{2}}}{h^3}\right)$$
$$+ Nk\sum_i\left\{\frac{h\nu_i}{kT}\frac{1}{\exp(h\nu_i/kT)-1} - \log[1-\exp(-h\nu_i/kT)]\right\}$$
$$+ Nk\frac{\partial}{\partial T}(T\log Z_{\mathrm{el.}}) + Nk\log\frac{8\pi^2(2\pi k)^{\frac{3}{2}}(I_1 I_2 I_3)^{\frac{1}{2}}}{h^3\sigma} \qquad (7·5·2)$$

for non-linear molecules. The numerical values for a number of gases at their boiling-points are given in tables 7·1 and 7·2, most of the calculations being due to Giauque and his collaborators (1929, 1930, 1932), based upon the data given in tables 6·2 and 6·3. The calorimetric entropy is determined from the heat capacity of the solid and the liquid and from the latent heat of fusion and of vaporization. In addition, the calorimetric entropy is corrected for the departure of the vapour from the perfect gas laws, by using equation (8·5·5) on p. 230 and assuming that the vapour obeys Berthelot's equation. The calorimetric entropy therefore refers to an ideal and not to a real gas. (Alternatively, we could correct the spectroscopic entropy to make it refer to a real and not to an ideal gas.) The correction introduced is small, of the order of 0·3 calorie/degree/gram molecule.

It will be seen that the agreement between the calculated spectroscopic entropy and the measured calorimetric entropy is excellent. This means that if we neglect nuclear weights and isotopic factors the residual entropies of the solids are correctly taken to be zero. This expected result calls for no further comment, except for oxygen, since it means that in solid oxygen the electronic weight factor is 1 and not 3, as in the gas. We must therefore assume that two oxygen molecules become coupled together in the solid state in such a way that the combined spin of this binary complex is zero.

Table 7·1. *Spectroscopic and calorimetric entropies in calories/degree/gram molecule of some monatomic and diatomic gases at their boiling points at atmospheric pressure*

	Boiling-point in °K.	Ground state of the molecule	Weight of the ground state	Spectroscopic entropy	Calorimetric entropy
A	87·3	^1S	1	30·87	30·85
Kr	120·2	^1S	1	34·65	34·63
O_2	90·2	$^3\Sigma$	3	40·68	40·70
N_2	77·4	$^1\Sigma$	1	36·42	36·53
Cl_2	238·6	$^1\Sigma$	1	51·55	51·56
HCl	188·2	$^1\Sigma$	1	41·45	41·3
HBr	204·5	$^1\Sigma$	1	44·92	44·9
HI	237·5	$^1\Sigma$	1	47·8	47·8

Table 7·2. *Spectroscopic and calorimetric entropies in calories/degree/gram molecule of some polyatomic gases at their boiling points at atmospheric pressure*

	Boiling-point in °K.	Symmetry number	Spectroscopic entropy	Calorimetric entropy
H_2S	212·8	2	46·42	46·33
NH_3	239·7	3	44·10	44·13
CH_4	111·5	12	36·61	36·53
CO_2	194·7	2	47·55	47·59
CS_2	318·4	2	57·60	57·48

7·51. *Apparent exceptions to Nernst's principle.* As shown in table 7·3 there are a number of gases for which the calorimetric entropy is significantly less than the spectroscopic entropy, and the values of ϖ_S required to account for the residual entropies of the solid are given in the last column of the table. In all cases plausible reasons can be given for assuming that at sufficiently low temperatures the solids exist in disordered states.

The CO molecule has a very small electric moment ($0{\cdot}12 \times 10^{-18}$ e.s.u.) and is nearly symmetrical about its centroid. We should therefore expect there to be only a small difference in the energies of two adjacent molecules in a solid whether they point in the same or opposite directions. In the most stable state at $T=0$ the axes of all the molecules in a solid mass of carbon monoxide must be arranged in a regular pattern and there can be only one orientational configuration of the whole crystal. At ordinary temperatures, however, this will not be so, and, if we assume that the axis of each molecule can arrange itself randomly in any one of two directions in the crystal, the number of different orientational configurations will be 2^N, where N is the number of molecules. If this random orientation persists down to temperatures which are so low that the molecules possess insufficient energy to reorientate themselves into a regular array, there will be a 'frozen-in' or residual entropy amounting to $Nk \log 2$, and this would account for the smaller value of the calorimetric entropy. The same argument applies to the linear molecule N_2O, which has a high degree of symmetry. Another molecule in the same category is CH_3D, for which it is plausible that there are four nearly equivalent equilibrium configurations in a crystal.

Table 7·3. *Spectroscopic and anomalous calorimetric entropies in calories/degree/gram molecule, for certain gases at their boiling points at atmospheric pressure*

	Boiling-point in °K.	Symmetry number	Weight of ground state	Spectroscopic entropy	Calorimetric entropy	ϖ_S deduced
CO	83	1	1	38·32	37·2	2
NO	121·4	1	2	43·75	43·03	$2^{\frac{1}{2}}$
NNO	184·6	1	1	48·50	47·36	2
H_2O	—	2	1	45·10	44·29	$\frac{3}{2}$
D_2O	—	2	1	46·66	45·89	$\frac{3}{2}$
CH_3D	99·7	3	1	39·49	36·73	4

(The values for H_2O and D_2O refer to 298·1° K.)

The value of the calorimetric entropy of NO can be explained (Johnston & Giauque, 1929) if it is assumed that the solid consists of N_2O_2 molecules, each of which can exist in two isomeric forms having only slightly different energies. There is then an entropy of mixing of the two isomers (see § 11·142) of an amount $Nk \log 2$ per N_2O_2 molecule, i.e. $Nk \log 2^{\frac{1}{2}}$ per NO molecule.

In a gaseous water molecule the HOH angle is 105° and the two hydrogen nuclei are at a distance of 0·95 Å. from the oxygen atom. In ice the various molecules are bonded together into a loose structure by hydrogen bonds, and each oxygen atom is surrounded by four other oxygen atoms, tetrahedrally placed, at a distance of 2·76 Å. The residual entropy of ice can be explained by making the following four assumptions (Pauling, 1935): (i) The water molecules retain their individuality to a large extent, each oxygen atom in ice having two hydrogen atoms attached to it at a distance of approximately 0·95 Å. and the HOH angle being approximately 105°. (ii) The two hydrogen atoms in any particular water molecule are directed approximately along two of the four tetrahedral O—O directions. (iii) Only one hydrogen atom lies approximately along each O—O direction. (iv) All the configurations satisfying the foregoing conditions have approximately the same energy.

According to the first and second assumptions, any particular molecule can orientate itself in six ways. But each molecule has two tetrahedral directions occupied by hydrogen atoms and two unoccupied, so that the chance of any particular O—O direction being unoccupied is $\frac{1}{2}$, and the chance of two being simultaneously unoccupied is $\frac{1}{4}$. Thus assumption (iii) reduces the average number of configurations per molecule to $\frac{3}{2}$, in agreement with the experimental value of the residual entropy. A similar argument applies to D_2O.

7·52. *The entropy of hydrogen and deuterium.* The moment of inertia of the hydrogen molecule is so small that the expression (7·5·1) for the entropy is only valid for temperatures above about 300° K. For lower temperatures we must base the calculations upon the expression (6·17·5) for the free energy, with the rotational partition functions given by (6·12·2). It is then convenient to include the nuclear weights in the expression for the entropy. A comparison of equations (5·4·12) and (6·17·9) then shows that the residual entropy of the gas is not zero but

$$S_G^0 = -N_1 k \log \frac{N_1}{\frac{1}{2}\rho(\rho-1)N} - N_2 k \log \frac{N_2}{\frac{1}{2}\rho(\rho+1)N} + N_2 k \log 3, \quad (7·52·1)$$

where N_1 and N_2 are the numbers of molecules in para and ortho states respectively, and where $N = N_1 + N_2$. For the ordinary 1 : 3 mixture this gives $S_G^0 = Nk \log 4 + \frac{3}{4}Nk \log 3$. The first term is due to the nuclear spin weight ρ^2, while the second term is due to the lowest ortho state having weight 3 excluding the nuclear spin weight. Alternatively, we can divide the contribution $Nk \log 4$ into two parts in the following way. The nuclear spin weight of an ortho state is normally 3, but at temperatures of the order of 10^{-3} ° K. these spin states cannot be considered to be degenerate, and only the lowest spin state will be effectively present. We may therefore say that at higher temperatures than 10^{-3} ° K. the ortho nuclear spin weight contributes $\frac{3}{4}Nk \log 3$ (or more generally $N_2 k \log 3$) to S_G^0, while the remainder, namely, $Nk \log 4 - \frac{3}{4}Nk \log 3$, is due to the 'entropy of mixing' of the ortho and para molecules. This latter contribution can be written as $Nk \log N - \frac{1}{4}Nk \log(\frac{1}{4}N) - \frac{3}{4}Nk \log(\frac{3}{4}N)$, or more generally

$$Nk \log N - N_1 k \log N_1 - N_2 k \log N_2,$$

which agrees with the general expression given in § 11·142 for the entropy of mixing. The residual entropy $N_2 k \log 3$ due to the ortho nuclear spin weight is removable by going to sufficiently low temperatures, whereas the entropy of mixing is not, unless we can catalyse the ortho-para transitions.

For most molecules the period of rotation at low temperatures is so large that the molecules have no chance of rotating in the solid state, and we can calculate the heat capacities of the solids according to the theory of § 6·252, in which it is assumed that all the atoms in the solid perform simple harmonic vibrations about their equilibrium positions. The masses of the hydrogen isotopes are, however, so small, as are likewise the force fields in the crystals, that the molecules in solid H_2, D_2 and HD continue rotating down to very low temperatures of the order of 10° K. The rotational contribution to the heat capacity of the solid is, of course, automatically taken into account in measuring the calorimetric entropy, but we have the choice as to whether to include or exclude the various contributions to the residual entropy. If we wish the spectroscopic and calorimetric entropies to lead to the same results, it is necessary to include the contributions to the residual entropies from corresponding factors in the partition functions of both the gas and the solid (or to exclude them from both). This conclusion is made clear by an inspection of equation (6·17·3) for the free energy, which applies equally to the gas and to the solid. In the solid we have $Z_1 = Z^{(1)}_{\text{vib.}} Z^{(1)}_{\text{int.}}$ and $Z_2 = Z^{(2)}_{\text{vib.}} Z^{(2)}_{\text{int.}}$, where $Z_{\text{vib.}}$ is the partition function of the lattice vibrations and which gives zero contribution to the residual entropy. $Z_{\text{vib.}}$ for the solid corresponds to $Z_{\text{trans.}}$ for the gas, and the latter gives the correct value zero for the entropy of the gas in its degenerate state at $T = 0$. Now the spectroscopic entropy includes the residual entropy (if any) arising from the partition function $Z_{\text{int.}}$ of the gas, whereas the calorimetric entropy does not include the residual entropy of the solid; and to obtain the same results we must therefore add the corresponding residual entropy to the calorimetric entropy. If we exclude the nuclear spin weights, i.e. if we calculate the spectroscopic entropy from (7·5·1) (or from its analogue with non-classical rotations if the temperature is low), the residual entropy of normal H_2 will be $\frac{3}{4}Nk \log 3$ and that of normal D_2 will be $\frac{1}{3}Nk \log 3$. These assumed residual entropies are in agreement with the calculations and measurements of Giauque (1930) and of Clusius & Bartholmé (1935), whose results are given in table 7·4.

Table 7·4. *The spectroscopic and calorimetric entropies, excluding nuclear weights, in calories/degree/gram molecule of hydrogen and deuterium at* 298·1° K. *and at atmospheric pressure*

	Spectroscopic entropy	Calorimetric entropy	ϖ_S deduced
H_2	31·23	29·7	$3^{\frac{3}{4}}$
D_2	34·62	33·9	$3^{\frac{1}{3}}$

7·521. *The specific heat and entropy of solid hydrogen at very low temperatures.* The solids discussed in § 7·51 are in 'frozen' non-equilibrium states, and, once the various molecules have become locked in random orientations, further cooling will not result in the removal of the residual entropy. In hydrogen (and deuterium), however, the molecules are relatively free to rotate, and even at very low temperatures it is possible for them to take up ordered positions of absolute stability. The lowest ortho state is normally considered to be triply degenerate, but owing to the crystalline fields the three states in the solid are in fact split, although the energy separation is small. It is found (Mendelssohn, Ruhemann & Simon, 1931) that at temperatures below 12° K., where the splitting becomes important, the molecules still have sufficient freedom to redistribute themselves over the three separate ortho states. This redistribution will result in a reduction in the entropy and a corresponding hump in the

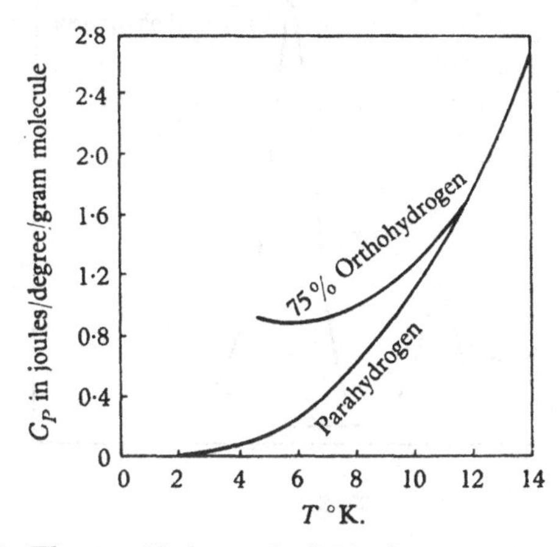

Fig. 7·5. The specific heat of solid hydrogen between 2 and 14° K.

specific heat curve. The total reduction possible in the entropy due to this cause is $k \log 3$ per ortho molecule, and the specific heat curve will depend upon the proportion of ortho molecules present, there being no anomaly for pure para-hydrogen. The measurements have been carried down to lower temperatures by Hill & Ricketson (1954). The results are shown in figs. 7·5 and 7·6, and it appears as if most of the entropy change (and the corresponding variation in the specific heat) takes place over a very narrow temperature range. The change from rotation to libration of the molecules may therefore be a cooperative phenomena of the type discussed later in connexion with ferromagnetism (§ 10·5) and the ordering of alloys (§ 14·2). On account of the specific heat curve resembling the shape of the Greek letter Λ such anomalous specific heats are often called λ-anomalies. The temperature at which the λ-anomaly occurs depends considerably upon the amount of ortho-hydrogen present, but in all cases the total decrease in entropy is very slightly less than $k \log 3$ per molecule of ortho-hydrogen.

As already mentioned above there is still a residual entropy of $k \log 3$ per ortho-molecule, due to the nuclear spin states, which could be removed by lowering the temperature to about 10^{-3} °K., but the entropy of mixing would still remain. The entropy per molecule of normal hydrogen will therefore be $s_0 = k \log 4 - \frac{3}{4} k \log 3$ at $T = 0$. As the temperature is increased the entropy s will remain constant until T reaches about 10^{-3} °K., when s will increase rapidly to $k \log 4$. It will then stay constant again until a temperature of about 1·5° K. is reached, when s will increase rapidly to $k \log 4 + \frac{3}{4} k \log 3$. After this, s will increase slowly, due to the excitation of the thermal vibrations and rotations.

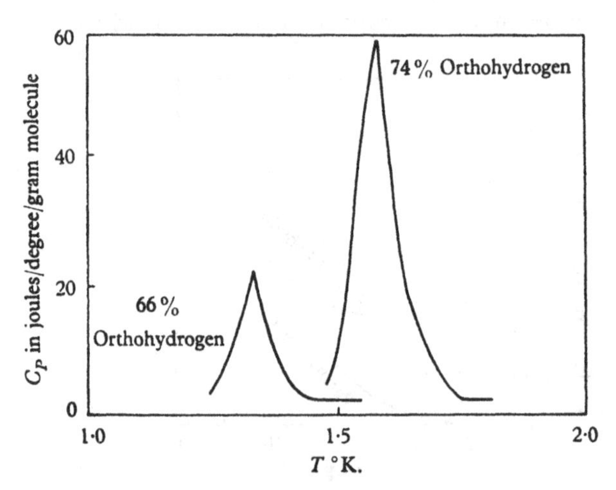

Fig. 7·6. The specific heat of hydrogen below 2° K.

The Entropy of a Solid which can Exist in Allotropic Forms

7·6. Allotropic transformations are exceedingly common, especially in metals. The majority of metals have either body-centred cubic, face-centred cubic or hexagonal close-packed structures, and, since the energies of these structures at $T = 0$ lie very close together, transformations from one to the other can take place readily at high temperatures provided that the metal has a sufficiently high melting-point. In each allotropic form the substance is in a state of stable (though not absolutely stable) equilibrium, and the entropy of each of the allotropic forms will therefore tend to zero as $T \to 0$. We consider here two well-known cases. The most interesting case, the graphite-diamond transition, is treated in detail in § 9·15.

7·61. *Tin.* Tin usually exists as white tin, which is a more or less normal metal with a tetragonal structure. But below room temperature the stable modification is grey tin which, in its pure state, is an insulator and has a crystal structure like that of diamond. In accordance with the third law, both modifications have zero entropy at $T=0$. Hence if T_0 is the temperature of the transition point, and if $C_P^{(g)}(T)$, $C_P^{(w)}(T)$ and $H(T_0)$ are the specific heat of grey tin, the specific heat of white tin and the heat of transformation of grey tin to white tin at $T=T_0$, the entropy of white tin at $T=T_0$ must be the same whether we calculate it from the specific heat of white tin or from the specific heat of grey tin and the entropy of transformation. That is, we must have

$$\int_0^{T_0} \frac{C_P^{(w)}(T)}{T}\,dT = \int_0^{T_0} \frac{C_P^{(g)}(T)}{T}\,dT + \frac{H(T_0)}{T_0}. \tag{7·61·1}$$

It is not difficult to keep white tin in a metastable equilibrium state below the transition point $T_0 = 291°$ K., and detailed measurements of the specific heats of both modifications have been carried out by Lange (1924). He found that the entropies of grey tin and white tin at 291° K. were 44 and 52 joules/degree/gram atom respectively, while $H_0(T_0)/T_0$ was 7·5 joules/degree/gram atom, which is in excellent agreement with equation (7·61·1).

7·611. *Sulphur.* The stable form of sulphur at low temperatures has a rhombic crystal structure, but above 368·6° K. a monoclinic structure is more stable. The monoclinic form can be obtained in a metastable state below the transition temperature T_0, and it behaves as a normal crystalline solid down to very low temperatures. We can therefore calculate the entropies at the transition point from the equation corresponding to (7·61·1). According to the measurements of Eastman & McGavock (1937), the entropies of the rhombic and monoclinic forms at $T_0 = 368·6°$ K. are 36·9 and 37·8 joules/degree/gram atom, so that the entropy of transition should be 0·9 joule/degree/gram atom. The measured value is 1·1 joules/degree/gram atom, which agrees with the previous figure within the accuracy of the measurements.

Supercooled Liquids and Glasses

7·7. A supercooled liquid is a thermodynamic system in metastable, or even in unstable, equilibrium, but if a true liquid could be obtained at $T=0$ its entropy, like that of the corresponding solid, would be zero there. The only true liquid which is known to exist in a state of thermodynamic equilibrium at $T=0$ is liquid helium, and its entropy at $T=0$ is zero (see § 9·41). Other liquids either pass over spontaneously into the solid state or become glass-like in character at sufficiently low temperatures.

The best known example of a supercooled liquid is provided by glycerine (Simon, 1922; Gibson & Giauque, 1923; Simon & Lange, 1926). The normal freezing-point of glycerine is 291° K., and the specific heat of the liquid at that point is about 200 joules/degree/gram molecule, while that of the solid is about 120 joules/degree/gram molecule. If the glycerine is kept in the supercooled liquid state its specific heat varies more or less linearly with the temperature, remaining about 80 joules/degree/gram molecule above that of the solid, until the temperature reaches 180° K. At this point the properties of the supercooled glycerine change considerably, and it then has most of the properties of an elastic solid. Its specific heat, however, drops rapidly and becomes more or less equal to that of the crystalline solid (fig. 7·7). If the excess of the entropy of the liquid over that of the solid is determined as a function of the temperature, it is found to be 58 joules/degree/gram molecule at the melting-point and it decreases as the temperature decreases, but it tends to the limiting value of 19 joules/degree/gram molecule as $T \to 0$. We therefore have an apparent exception to Nernst's principle.

The explanation of the discrepancy is as follows (see, for example, Simon, 1931; Jones & Simon, 1949). Although the molecules in a liquid are in a state of irregular motion, they possess a high degree of short-distance order, and the molecules which are nearest neighbours of a given molecule arrange themselves so as to give the most ordered structure possible at a given temperature. When a supercooled liquid passes into the glassy state, the viscosity becomes so high that any further rearrangement of the molecules to produce a greater degree of order as the temperature is reduced becomes very difficult, and the specific heat falls. The substance is then in a supercooled disordered state. If, however, the time taken to measure the specific heat at any particular temperature is made sufficiently long, it is possible for the molecules to take up their true equilibrium positions, and the measurements would reveal a larger specific heat than that normally obtained. Such experiments, with equilibrium times of the order of one week, were carried out on glycerine in the glassy state by Oblad & Newton (1937), and they were able to determine the specific heat curve of the true liquid down to about 165° K. The excess entropy of the true liquid over that of the solid is shown by the dotted curve in fig. 7·8, and, if it could be measured to sufficiently low temperatures, it would become zero at $T = 0$.

7·71. Although a glass is not a system in thermodynamic equilibrium, we can apply a restricted form of Nernst's principle to those phenomena in which the disorder remains unchanged. For example, the compressibility and the expansion coefficient of a given specimen of glass, as normally measured, depend only upon the displacements of the atoms from their frozen-in positions of apparent equilibrium. We should therefore expect that the coefficient of expansion of a glass, like that of a crystalline solid, should tend to zero as $T \to 0$. It is probable that this is true (Keesom & Doborzyński, 1933), but measurements are very scanty.

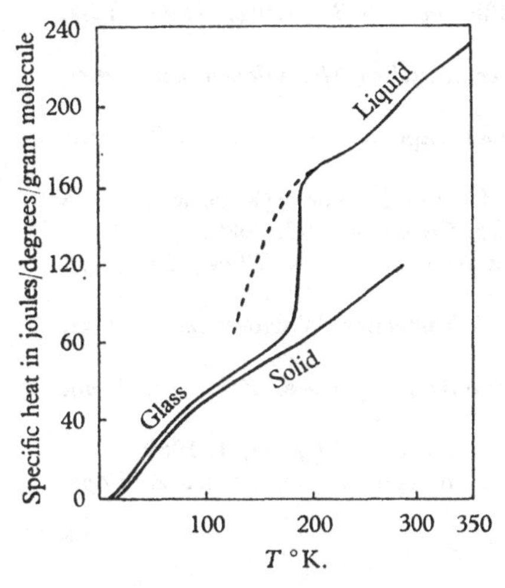

Fig. 7·7. The specific heat of glycerine.

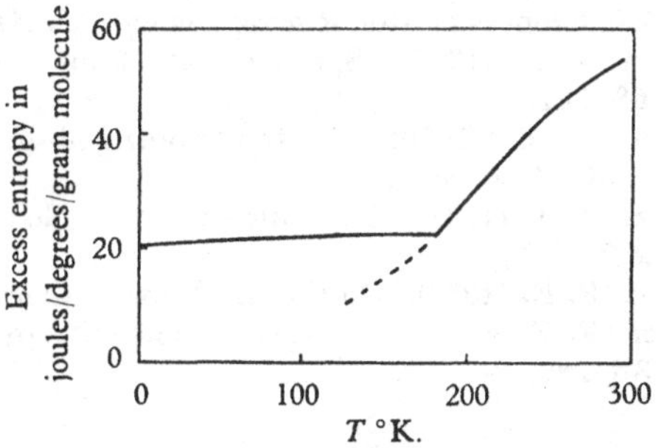

Fig. 7·8. The excess of the entropy of glycerine in the supercooled liquid state and in the glassy state.

REFERENCES

Clayton, J. O. & Giauque, W. F. (1932). The heat capacity and entropy of carbon monoxide. *J. Amer. Chem. Soc.* **54**, 2610.

Clusius, K. & Bartholmé, E. (1935). Entropy of heavy hydrogen. *Z. phys. Chem.* B, **30**, 258.

Eastman, E. D. & McGavock, W. C. (1937). The heat capacity and entropy of rhombic and monoclinic sulphur. *J. Amer. Chem. Soc.* **59**, 145.

Giauque, W. F. (1930). The entropy of hydrogen and the third law of thermodynamics. The free energy and dissociation of hydrogen. *J. Amer. Chem. Soc.* **52**, 4816.

Gibson, G. E. & Giauque, W. F. (1923). The third law of thermodynamics. Evidence from the specific heats of glycerol that the entropy of a glass exceeds that of a crystal at the absolute zero. *J. Amer. Chem. Soc.* **45**, 93.

Hill, R. W. & Ricketson, B. W. A. (1954). A λ-anomaly in the specific heat of solid hydrogen. *Phil. Mag.* (7), **45**, 277.

Johnston, H. L. & Giauque, W. F. (1929). The heat capacity of nitric oxide. *J. Amer. Chem. Soc.* **51**, 3194.

Jones, G. O. & Simon, F. E. (1949). What is a glass? *Endeavour*, **8**, 175.

Keesom, W. H. & Doborzyński, D. W. (1933). Measurements by the interferometric method on the thermal expansion of Jena glass down to 4° K. *Physica*, **1**, 1089.

Lange, F. (1924). Investigation of the specific heat at low temperatures. *Z. phys. Chem.* **110**, 343.

Lewis, G. N. & Randall, H. M. (1923). *Thermodynamics*. New York.

Mendelssohn, K., Ruhemann, W. & Simon, F. E. (1931). Specific heats of solid hydrogen at liquid helium temperatures. *Z. phys. Chem.* B, **15**, 121.

Nernst, W. (1906). The calculation of chemical equilibria from thermal measurements. *Nachr. Ges. Wiss. Göttingen*, p. 1.

Nernst, W. (1912). Thermodynamics and specific heat. *S.B. preuss. Akad. Wiss.* p. 134.

Nernst, W. (1924). *Die theoretischen und experimentellen Grundlagen des neuen Wärmesatzes*, 2nd ed. Berlin.

Oblad, A. G. & Newton, R. F. (1937). The heat capacity of supercooled liquid glycerol. *J. Amer. Chem. Soc.* **59**, 2495.

Pauling, L. (1935). The structure and entropy of ice and other crystals with some randomness of atomic arrangement. *J. Amer. Chem. Soc.* **57**, 2680.

Simon, F. E. (1922). Specific heat at low temperatures. *Ann. Phys., Lpz.*, (4), **68**, 241.

Simon, F. E. (1930). Fünfundzwanzig Jahre Nernstscher Wärmesatz. *Ergebn. exakt. Naturw.* **9**.

Simon, F. E. (1931). The state of supercooled liquids and glasses. *Z. anorg. Chem.* **203**, 219.

Simon, F. E. (1937). On the third law of thermodynamics. *Physica*, **4**, 1089.

Simon, F. E. & Lange, F. (1926). The entropy of amorphous substances. *Z. Phys.* **38**, 227.

Chapter 8

IMPERFECT GASES

Introduction

8·1. Real gases differ from perfect gases in two ways. First, the equation of state is not given by $PV = nRT$ and, secondly, real gases can be liquefied (and solidified) by reducing the temperature sufficiently. The present chapter deals mainly with the properties of the gaseous state, while the phenomena which are concerned with the simultaneous existence of the gas and the liquid (or the solid) are discussed in Chapter 9.

The experimental facts are briefly as follows. When a given mass of gas at a sufficiently high temperature is compressed isothermally, its volume diminishes steadily, the isothermal curve and its first derivative being continuous everywhere. For very high temperatures the isothermal curves are very nearly hyperbolas, as for a perfect gas, but as the temperature is lowered the isothermals deviate more and more from those of a perfect gas. For sufficiently low temperatures there is a portion of each isothermal which is parallel to the V axis, and there are two discontinuities in the slope of such an isothermal. The general behaviour of the isothermals is shown schematically in fig. 8·1.

Let us consider in more detail what happens when a gas is compressed so as to pass along an isothermal such as $ABCD$. Along the portion AB of the isothermal, increasing pressure produces a decrease in volume, but when the point B is reached part of the gas liquefies and the volume diminishes while the pressure remains unaltered. Further compression produces no increase in pressure until the point C is reached, when all the gas has been converted into liquid. It is now possible to increase the pressure once more, but since liquids are highly incompressible the portion CD of the isothermal is very nearly parallel to the P axis. With increasing temperature, the flat portion of the isothermals becomes less and less until at a certain temperature, called the critical temperature, the flat portion just vanishes. For temperatures higher than the critical temperature no isothermal possesses a flat portion, and the gas cannot be liquefied however much it is compressed. A gas is often referred to as a vapour when its temperature is below the critical temperature.

The horizontal portions of the isothermals correspond to heterogeneous states in which the substance is partly liquid and partly gaseous, and so the part of the P, V plane corresponding to coexistent liquid and gaseous states is the portion bounded by the curve $ECFBG$ in fig. 8·1. This region is bounded by two curves, called boundary curves, the left-hand one ECF referring to liquid just about to vaporize and the right-hand one GBF to gas just about to liquefy. These curves meet at a point F which lies on the isothermal belonging to the critical temperature. When the state of the substance corresponds to this point the substance is said to be at the critical point, and the pressure, volume and temperature are called the critical pressure, volume and temperature. At the critical point the distinction between the liquid and the gas vanishes, and all their properties become the same.

By careful handling it is possible to extend the curved portions of the isothermals such as $ABCD$ in fig. 8·1 beyond the points at which they join

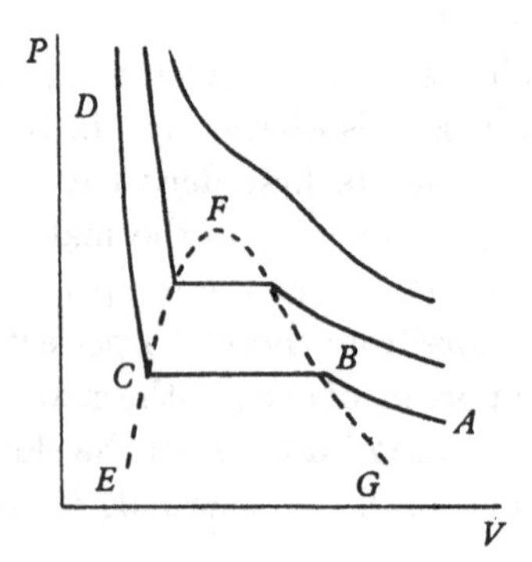

Fig. 8·1. The general behaviour of the isothermal curves of a real gas.

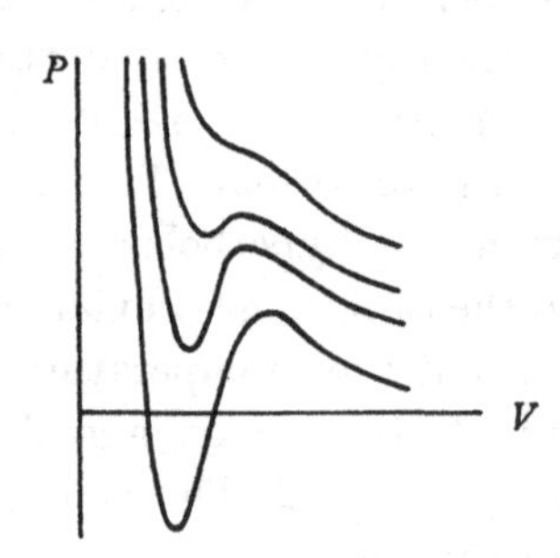

Fig. 8·2. The 'ideal' isothermal curves of a real gas including metastable and unstable states.

the flat portions. If we purify a vapour very thoroughly, removing all dust and impurities, and compress it isothermally, the vapour can be compressed beyond the point B without condensation taking place. The vapour is then said to be in a supersaturated state, and is in stable equilibrium for small disturbances. If a large disturbance is created in a supersaturated system, for example, by mechanical shock or by the introduction of a drop of liquid or a dust particle, the equilibrium is disturbed and a portion of the vapour will immediately condense. The supersaturated state may therefore be called a metastable state, stable for small disturbances but unstable for large ones.

Similarly, it is possible to obtain the liquid under a pressure less than that corresponding to the point C and yet to avoid vaporization. The liquid is then said to be superheated and is in a metastable state. By taking suitable precautions it is possible to extend the metastable portions of the isothermals quite a long way, and this suggests that if all disturbances could be avoided the 'ideal' isothermals would be as shown in fig. 8·2, and that ideally it ought to be possible to make the vapour pass over continuously into the liquid. However, this would entail making the substance pass through states for which $(\partial P/\partial V)_T > 0$. Such states, though equilibrium states, would be unstable, and any disturbance, however small, would destroy the homogeneity of the system, so that it is impossible to make a substance pass along the whole of the 'ideal' isothermal.

Equations of State

8·2. A very large number of equations of state have been proposed, some of which purport to describe the behaviour of imperfect gases for certain restricted ranges of pressure, temperature and volume, while others have been derived in order to account for the properties of fluids extending over the whole range of gas, vapour and liquid. No attempt will be made here to deal exhaustively with the subject, and the equations of state discussed are introduced mainly to illustrate the thermodynamic methods rather than to obtain results applicable to particular substances.

(i) The most general type of equation of state is that used by Kamerlingh Onnes in either of the forms

$$Pv = A + BP + CP^2 + \dots \tag{8·2·1}$$

and

$$Pv = A' + \frac{B'}{v} + \frac{C'}{v^2} + \dots, \tag{8·2·2}$$

where v, as usual, is the volume of one gram molecule of gas and where the coefficients $A, B, C, \dots$ (and $A' = A$, $B' = AB$, $C' = AB^2 + A^2C$, ...) are functions of the temperature, and are usually called the first, second, third, ... virial coefficients. Since, for any given temperature, every gas becomes perfect when the pressure is sufficiently small and the volume sufficiently large, we have $A = A' = RT$. The other coefficients must be obtained either from the empirical data or from calculations based upon assumed intermolecular forces.

It is instructive to plot PV against P for various temperatures, as is done in fig. 8·3. For one particular temperature, called the Boyle temperature, the PV, P curve has a horizontal tangent at $P = 0$. For higher temperatures the curves increase monotonically, while for lower temperatures each curve has a minimum, at a point called the Boyle point. Near its Boyle point, an isothermal is flat and Boyle's law is a good approximation, while the isothermal corresponding to the Boyle temperature is flat up to very considerable pressures so that, for this isothermal, Boyle's law has a very extended range of validity.

Below the Boyle temperature, the second virial coefficient B is negative. It is zero at the Boyle temperature and positive above the Boyle temperature.

(ii) Van der Waals's equation of state

$$(P+a/v^2)(v-b)=RT, \tag{8·2·3}$$

where a and b are constants, is the best known of the equations of state which have a simple closed form. It was originally derived by van der Waals by a consideration of the forces acting upon the molecules of an

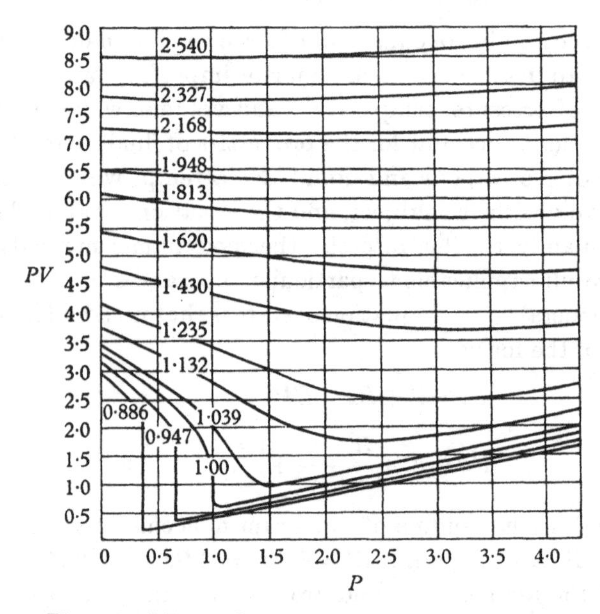

Fig. 8·3. PV as a function of P for various temperatures.

imperfect gas, and it can be justified theoretically for a certain type of intermolecular force when the departure from the equation of state of a perfect gas is small. The term a/v^2 represents a diminution of the pressure in an imperfect gas owing to the attractions between the molecules, while the occurrence of the term b is due to the non-zero size of the molecules, owing to which the volume of the gas cannot be reduced below the value b, however large the pressure may be.

Although van der Waals's equation can only be justified theoretically when the departures from the perfect gas laws are small, it nevertheless represents qualitatively the equation of state over the whole range of gas, vapour and liquid in a very remarkable way. It does not give the numerical details correctly, but it is a relatively simple equation and is widely used to illustrate the general behaviour of imperfect gases. We shall therefore make considerable use of it in discussing phenomena which do not occur with perfect gases, but the reader must bear in mind that in order to obtain reliable quantitative results a complicated equation of the type of (8·2·1), with a large number of adjustable constants, must be used.

(iii) Dieterici's equation

$$P(v-b) = RT\exp\left[-a/(RTv)\right], \tag{8·2·4}$$

where a and b are constants, has the same theoretical validity as van der Waals's equation, but represents some of the data better and some worse. We shall illustrate the results of the present chapter by means of both equations.

(iv) Berthelot's equation is a variation on van der Waals's equation and is

$$\{P + a'/(Tv^2)\}(v-b) = RT, \tag{8·2·5}$$

where a' and b are constants. It is of considerable practical importance in estimating corrections to the perfect gas laws when these are small. If we write it in the form

$$Pv = RT\left(1 + \frac{b}{v-b} - \frac{a'}{RT^2 v}\right),$$

and substitute for v the approximate value $v = RT/P$ on the right-hand side, we obtain

$$Pv = RT\left\{1 - \frac{P}{RT}\left(\frac{a'}{RT^2} - b\right)\right\}, \tag{8·2·6}$$

which is sufficiently accurate when the departures from the perfect gas laws are not too large. It has the advantage of giving v explicitly in terms of P and T. It cannot, of course, be used in the neighbourhood of the critical point or in the heterogeneous region since it gives v as a single-valued function of P.

(v) Callendar's equation is

$$P(v-b) = RT - aP/T^r, \tag{8·2·7}$$

where a, b and r are constants. It is a purely empirical equation, and is used to represent the properties of steam.

8·21. *Properties of van der Waals's equation.* The isothermals given by van der Waals's equation are of the general type shown in fig. 8·2, and they intersect any line $P =$ constant in three (real or imaginary) points. The limiting isothermal, given by $T = 0$, consists of portions of the two curves $P = -a/v^2$ and $v = b$. Some of the isothermals therefore have portions for which P is negative, but there is no reason to exclude these portions since liquids (but not gases) can exist under tension in a metastable state.

The isothermals cut the line $P=0$ in the two points given by

$$RTv^2-av+ab=0,$$

which are real if $4RT<a/b$. If $4RT=a/b$ the isothermal touches the line $P=0$ at $v=2b$, while for larger values of T there are no real intersections with $P=0$, and the liquid states are never under tension.

For sufficiently small values of T it is clear that the isothermals have two real maxima and minima, $P_{\text{max.}}$ and $P_{\text{min.}}$, and they cut any line $P=$ constant, where $P_{\text{max.}} \geqslant P \geqslant P_{\text{min.}}$, in three real points. For the critical isothermal the maximum and the minimum coincide. The critical isothermal therefore separates the curves with three real points of intersection with some line $P=$ constant from those with only one real point of intersection with all lines $P=$ constant, and it is such that the line $P=P_c$, where P_c is the critical pressure, touches it at three coincident points. Hence the critical isothermal has a point of inflexion at which the tangent is parallel to the v axis, and the equations defining the critical point P_c, v_c, T_c are

$$(\partial P/\partial v)_T=0, \quad (\partial^2 P/\partial v^2)_T=0. \tag{8·21·1}$$

For van der Waals's equation we have

$$\frac{\partial P}{\partial v}=-\frac{RT}{(v-b)^2}+2\frac{a}{v^3}, \quad \frac{\partial^2 P}{\partial v^2}=\frac{2RT}{(v-b)^3}-\frac{6a}{v^4}.$$

The equations (8·21·1) then give

$$v_c=3b, \quad RT_c=8a/(27b), \quad P_c=a/(27b^2). \tag{8·21·2}$$

The measured values of the critical constants can be used to determine the two constants a and b. These in turn can be calculated from the measured values of the second virial coefficient. For van der Waals's equation is equivalent to the equation

$$Pv=RT+(RTb-a)/v+O(v^{-2}), \tag{8·21·3}$$

and hence a and b can be found if the second virial coefficient is known as a function of T. It is found that the values of a and b deduced from the critical constants are not in agreement with those found from the second virial coefficient. That this should be so is not surprising, since van der Waals's equation is only valid according to the theoretical calculations when the deviations from the perfect gas laws are small, and we cannot expect it to give quantitatively correct results near the critical point.

The numerical discrepancies are perhaps best shown by considering the quantity $RT_c/(P_c v_c)$, which, according to the equations (8·21·2), should have the value $\frac{8}{3}$ for all gases. The observed values are of the same order of magnitude as this, but are consistently larger, as is shown by the figures given in table 8·1.

Table 8·1. *The critical constants of various gases*

	SO_2	CO_2	C_2H_4	CH_4	O_2	A
T_c in °K.	432	304	283	190	154·3	150·7
v_c in cm.^{3}/gram molecule	124	98	127	99	75	75
P_c in atmospheres	78	73	51	46	50	48
$RT_c/(P_c v_c)$	3·6	3·48	3·46	3·46	3·42	3·43

	CO	N_2	Ne	H_2	He
T_c in °K.	133	126	44·5	33·2	5·2
v_c in cm.^{3}/gram molecule	93	90	42	67	57·8
P_c in atmospheres	35	34	26	12·8	2·26
$RT_c/(P_c v_c)$	3·42	3·42	3·37	3·29	3·33

8·211. *The reduced equation.* Any equation of state such as that of van der Waals, which contains only two arbitrary constants, can be put into a dimensionless form by expressing the pressure, volume and temperature as ratios of the critical pressure, volume and temperature. Put

$$\pi = P/P_c, \quad \phi = v/v_c, \quad \vartheta = T/T_c. \tag{8·211·1}$$

Then substituting for P, v and T into van der Waals's equation and using the values (8·21·2) for the critical constants we obtain the equation

$$(\pi + 3/\phi^2)(3\phi - 1) = 8\vartheta. \tag{8·211·2}$$

Such an equation, in which all the adjustable constants have disappeared, is called a reduced equation of state, and π, ϕ and ϑ are called the reduced pressure, the reduced volume and the reduced temperature. Gases which have the same values of π, ϕ and ϑ are said to be in corresponding states.

8·212. *The Boyle point.* To investigate the Boyle point we use the reduced equation of state and put $y = \pi\phi$ and $x = \pi$, so that

$$(y^2 + 3x)(3y - x) = 8\vartheta y^2 \tag{8·212·1}$$

and

$$\left(\frac{\partial y}{\partial x}\right)_\vartheta = \frac{y^2 - 9y + 6x}{9y^2 - 2xy - 16\vartheta y + 9x}.$$

The Boyle points are given by $(\partial y/\partial x)_\vartheta = 0$, that is, by

$$y^2 - 9y + 6x = 0. \tag{8·212·2}$$

This represents a parabola and is the locus of the Boyle points.

The Boyle temperature is the temperature at which the Boyle point occurs at $x = 0$, $y \neq 0$. Equation (8·212·2) gives $y = 9$, and then equation (8·212·1) gives the Boyle temperature as $\frac{27}{8}T_c$. This is of the right order of magnitude, but is by no means quantitatively correct, as is shown by the figures given in table 8·2.

Table 8·2. *The Boyle temperatures of certain gases*

	He	H_2	Ne	N_2	A	O_2
Boyle temperature T_B in °K.	19	106	134	323	410	423
Critical temperature T_c in °K.	5·2	33·2	44·5	126	150·7	154·3
T_B/T_c	3·65	3·21	2·98	2·5	2·73	2·72

8·213. *The heterogeneous region.* The isothermals given by van der Waals's equation have the same analytical form for the gaseous and the liquid states, and the equation of state does not therefore lead directly to an explanation of the phenomenon of condensation. A complete discussion of why certain portions of the ideal isothermals correspond to metastable states involves the general theory of stability, which is dealt with in § 3·7. But, if we accept the implications of van der Waals's equation and the fact that the stable states have isothermals which have flat portions, corresponding to a heterogeneous mixture of vapour and liquid, we can obtain the position of the flat portions by the following argument. Suppose that the substance is taken round the reversible isothermal cycle *abcdeca* shown in fig. 8·4, where *abcde* represents the metastable and unstable portion of an ideal isothermal and *ae* is the flat portion of the real isothermal. Then $\oint du = T\oint ds - \oint P\,dv$. But, since u and s are both functions of the state, $\oint du = 0$ and $\oint ds = 0$. Hence the real isothermal must be such that $\oint P\,dv = 0$. This means that the real isothermal must be drawn so that the areas *abc* and *ced* are equal, a result that is known as the rule of equal areas.

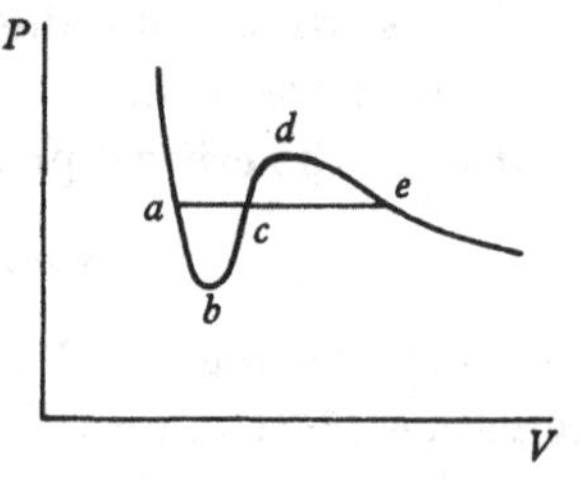

Fig. 8·4

We can obtain a set of implicit equations giving the position of the real isothermal as follows. Let the flat portion of the real isothermal extend from the reduced volume ϕ_1 to the reduced volume $\phi_2 > \phi_1$. The area enclosed by the ideal isothermal and the two ordinates at ϕ_1 and ϕ_2 is

$$\int_{\phi_1}^{\phi_2} \pi\,d\phi = \int_{\phi_1}^{\phi_2} \frac{8\vartheta}{3\phi-1}\,d\phi - \int_{\phi_1}^{\phi_2} \frac{3}{\phi^2}\,d\phi = \tfrac{8}{3}\vartheta \log\frac{3\phi_2-1}{3\phi_1-1} + 3\left(\frac{1}{\phi_2}-\frac{1}{\phi_1}\right),$$

and this must be equal to $\pi(\phi_2-\phi_1)$. Hence

$$\tfrac{8}{3}\vartheta \log\frac{3\phi_2-1}{3\phi_1-1} + 3\left(\frac{1}{\phi_2}-\frac{1}{\phi_1}\right) = \pi(\phi_2-\phi_1). \qquad (8\cdot213\cdot1)$$

Also the points π, ϕ_1 and π, ϕ_2 lie on the ideal isothermal, and so

$$\pi = \frac{8\vartheta}{3\phi_1 - 1} - \frac{3}{\phi_1^2} = \frac{8\vartheta}{3\phi_2 - 1} - \frac{3}{\phi_2^2}. \tag{8·213·2}$$

The three equations (8·213·1) and (8·213·2) determine π, ϕ_1 and ϕ_2 as functions of ϑ, but they cannot be solved explicitly.

We can also find the limits of the metastable states since they occur where $(\partial\pi/\partial\phi)_\vartheta = 0$. We have

$$\left(\frac{\partial\pi}{\partial\phi}\right)_\vartheta = -\frac{24\vartheta}{(3\phi-1)^2} + \frac{6}{\phi^3},$$

and this is zero if $4\vartheta = (3\phi-1)^2/\phi^3$. Eliminating ϑ between this equation and (8·211·2) we obtain

$$\pi = (3\phi - 2)/\phi^3 \tag{8·213·3}$$

as the equation of the curve separating the metastable states from the unstable ones. The curve (8·213·3) touches the critical isothermal at the critical point.

8·22. *Berthelot's equation.* The critical constants given by equation (8·2·5) are

$$v_c = 3b, \quad RT_c^2 = 8a'/(27b), \quad P_c = \tfrac{1}{2}(RT_c/b) - a'/(9T_c b^2), \tag{8·22·1}$$

and the reduced form of Berthelot's equation is

$$\{\pi + 3/(\vartheta\phi^2)\}(3\phi - 1) = 8\vartheta. \tag{8·22·2}$$

This form of Berthelot's equation is very similar to van der Waals's equation, and it has similar properties. Neither equation gives more than a qualitative picture of the properties of fluids for all values of the variables.

For practical purposes a modified form of Berthelot's equation is of much greater importance than (8·22·2). We denote by π, ϕ and ϑ the reduced variables calculated from the measured critical constants P_c, v_c and T_c, and not from the values given by (8·22·1). Then Berthelot's modified equation is an empirical equation given by

$$\{\pi + 16/(3\vartheta\phi^2)\}(4\phi - 1) = 128\vartheta/9. \tag{8·22·3}$$

The critical point 1, 1, 1 does not satisfy this equation, since the critical constants have not been taken from (8·22·1). The form of this equation corresponding to (8·2·6) is

$$\pi\phi = \tfrac{32}{9}\vartheta\left\{1 + \frac{9\pi}{128\vartheta}\left(1 - \frac{6}{\vartheta^2}\right)\right\}, \tag{8·22·4}$$

or, explicitly,

$$Pv = RT\left\{1 + \frac{9}{128}\frac{T_c}{T}\frac{P}{P_c}\left(1 - 6\frac{T_c^2}{T^2}\right)\right\}, \tag{8·22·5}$$

where P_c and T_c are the measured values of the critical pressure and temperature. This empirical equation represents the facts extremely well for moderate pressures up to a few atmospheres.

8·23. *Properties of Dieterici's equation.* The properties of Dieterici's equation are qualitatively similar to those of van der Waals's equation except that the isothermal $T=0$ consists of the lines $P=0$, $v=b$, but the numerical results are different. The quantities a and b that occur in Dieterici's equation are the same as those which occur in van der Waals's equation if they are calculated from the second virial coefficient, since Dieterici's equation reduces to (8·21·3) if we expand P in powers of $1/v$. However, the relation of a and b to the critical constants is different. At the critical point we have

$$\left(\frac{\partial P}{\partial v}\right)_T = \frac{RT\exp\left[-a/(RTv)\right]}{v-b}\left(\frac{a}{RTv^2}-\frac{1}{v-b}\right)=0,$$

$$\left(\frac{\partial^2 P}{\partial v^2}\right)_T = \frac{RT\exp\left[-a/(RTv)\right]}{v-b} \times\left[\left(\frac{a}{RTv^2}\right)^2-\frac{2a}{RTv^3}-\frac{2a}{RTv^2(v-b)}+\frac{2}{(v-b)^3}\right]=0.$$

These give $$v_c=2b, \quad RT_c=\tfrac{1}{4}a/b, \quad P_c=\tfrac{1}{4}e^{-2}a/b^2, \tag{8·23·1}$$

which differ considerably from the values given by (8·21·2). The quantity $RT_c/(P_c v_c)$ has the value $\frac{1}{2}e^2=3·69$, which is in somewhat better agreement with the experimental values than the value $\frac{8}{3}$ given by van der Waals's equation. On the other hand, the values of the constants a and b separately calculated from the critical data do not agree very well with those calculated from the second virial coefficient.

The reduced from of Dieterici's equation is easily shown to be

$$\pi(2\phi-1)=\vartheta\exp\left[2-2/(\vartheta\phi)\right], \tag{8·23·2}$$

and the Boyle point at temperature ϑ is given by

$$y=2\exp\left[2-x/(\vartheta y)\right], \tag{8·23·3}$$

where $y=\pi\phi$ and $x=\pi$. The locus of the Boyle points is obtained by eliminating ϑ between the equations (8·23·2) and (8·23·3), but the resulting expression is complicated. The Boyle temperature is obtained by putting $x=0$ in (8·23·3), which gives $y=2e^2$. Also equation (8·23·2) gives $2y=\vartheta e^2$ when $x=0$, and hence we have $\vartheta_B=4$, which is even further from the truth than the value $\frac{27}{8}$ given by van der Waals's equation.

The positions of the real isothermals in the heterogeneous region are determined by the rule of equal areas, but since the integral cannot be evaluated in finite terms the positions of the flat portions of the isothermals can only be found by numerical methods. On the other hand, the equation of the curve separating the metastable from the unstable states is easily found to be

$$\pi\phi^2 = \exp[1 - 1/(2\phi - 1)]. \tag{8·23·4}$$

On the whole, Dieterici's equation gives no better agreement with the experimental facts than van der Waals's equation, and its more complicated analytical nature renders it less suitable for purposes of illustration. The agreement with experiment can, however, be improved by introducing another parameter into Dieterici's equation and writing it in the form

$$P(v - b) = RT \exp[-a/(RT^r v)]. \tag{8·23·5}$$

The value $r = \frac{3}{2}$ seems to be the most suitable one.

8·24. *The isobars and the isochores.* Instead of considering the isothermal curves it is often more convenient to plot v as a function of T for various values of P, or P as a function of T for various values of v. These two families of curves are called the isobars and the isochores respectively.

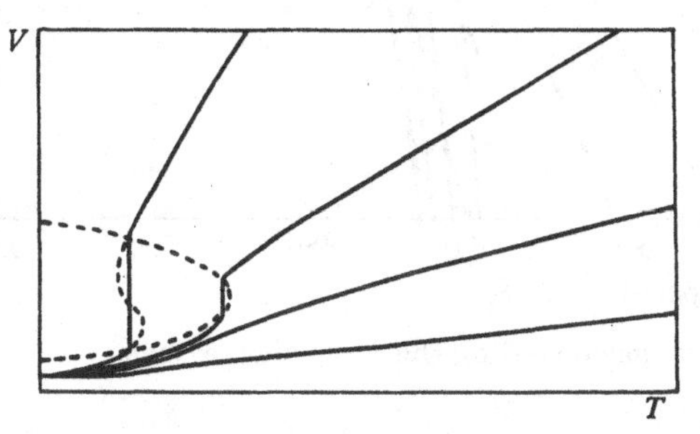

Fig. 8·5. The isobars of a real gas.

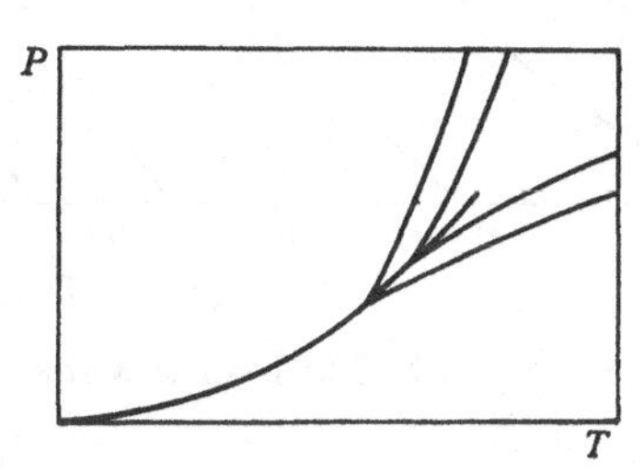

Fig. 8·6. The isochores of a real gas.

For large values of T the isobars are given by $v = RT/P$ and are straight lines which, if produced, would pass through the origin. For medium values of T the isobars are convex upwards, and in the heterogeneous region there are three values of v for each value of T for the ideal isobars, while the real isobars are parallel to the v axis (fig. 8·5).

In the P, T diagram, the vapour-liquid line is a single curve ending at the critical point. For large values of T the isochores are straight lines, but for moderate values of T they have a small curvature and they ultimately intersect the liquid-vapour line from one side or the other at a non-zero angle (fig. 8·6). If we include the metastable states, the isochores cross the liquid-vapour line, and the metastable states form a curvilinear triangular region, one of whose apexes is at the critical point. In this region three isochores pass through every point. This behaviour is illustrated in fig. 8·7, which gives some results obtained by Ramsay & Young (1887) for ethyl ether, extending into the region of negative pressures.

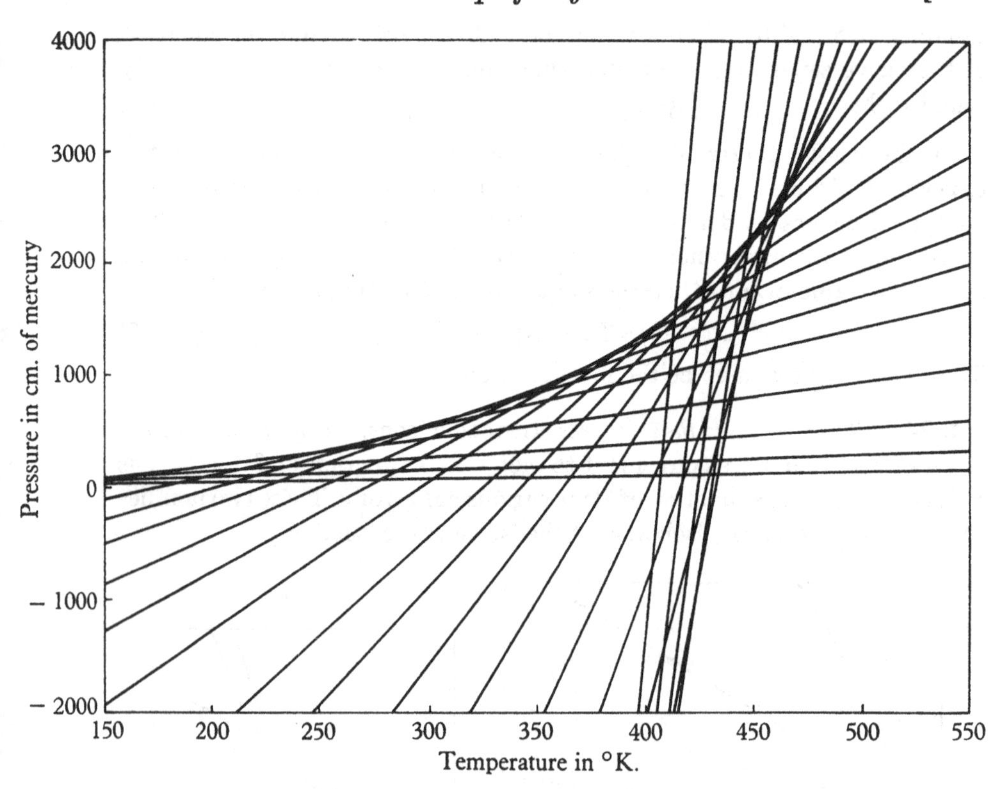

Fig. 8·7. The isochores of a real gas including the unstable portions.

The Critical Point

8·3. The critical point is defined by the conditions

$$(\partial P/\partial v)_T = 0, \quad (\partial^2 P/\partial v^2)_T = 0. \qquad (8\cdot3\cdot1)$$

This means that the isothermal compressibility is infinite at the critical point, and the properties of a substance are therefore highly singular in the critical region. We now give a more detailed analysis of the behaviour of a fluid near the critical point.

For any simple substance the pressure increases steadily with the temperature at constant volume, and so

$$(\partial P/\partial T)_V > 0 \quad (v = v_c, T = T_c). \qquad (8\cdot3\cdot2)$$

Now, by (1·6·5), $$\left(\frac{\partial v}{\partial T}\right)_P = -\left(\frac{\partial P}{\partial T}\right)_V \Big/ \left(\frac{\partial P}{\partial v}\right)_T,$$

and hence we have $(\partial v/\partial T)_P = \infty \quad (T = T_c, P = P_c).$ (8·3·3)

The coefficient of expansion is therefore infinite at the critical point. This in turn means that the specific heat C_P at constant pressure is infinite at the critical point since, by (3·212·1),

$$C_P = C_V + T\left(\frac{\partial P}{\partial T}\right)_V \left(\frac{\partial v}{\partial T}\right)_P.$$

Since for the ideal isothermals $(\partial P/\partial v)_{T, v=v_c}$ is positive for $T < T_c$ and negative for $T > T_c$, we have

$$\partial^2 P/\partial v\,\partial T < 0 \quad (v = v_c,\ T = T_c). \tag{8·3·4}$$

Also, the critical isothermal crosses the tangent at the critical point and lies above the tangent for $v < v_c$ and below the tangent for $v > v_c$. Hence

$$(\partial^3 P/\partial v^3)_T < 0 \quad (v = v_c,\ T = T_c). \tag{8·3·5}$$

If we now expand P near the critical point in powers of $v - v_c$ and $T - T_c$ and use the relations (8·3·1), (8·3·2), (8·3·4) and (8·3·5) we find, neglecting terms of all but the lowest orders,

$$P - P_c = l(T - T_c) - m(T - T_c)(v - v_c) - n(v - v_c)^3, \tag{8·3·6}$$

where l, m and n are positive constants given by

$$l = (\partial P/\partial T)_V, \quad m = -\partial^2 P/\partial v\,\partial T, \quad n = -\tfrac{1}{6}(\partial^3 P/\partial v^3)_T \quad (v = v_c,\ T = T_c). \tag{8·3·7}$$

The equation (8·3·6) defines the isothermals both in the gaseous and in the heterogeneous regions sufficiently near the critical point. The real isothermals, as opposed to the metastable ones, in the heterogeneous region can be found by the rule of equal areas, and when (8·3·6) is valid they can be found explicitly. Let v_1 and $v_2 > v_1$ be the volumes corresponding to the end-points of the flat portion of the isothermal with pressure $P < P_c$ and temperature $T < T_c$. Then v_1 and v_2 both satisfy (8·3·6), so that by addition we obtain

$$\begin{aligned} P - P_c = {} & l(T - T_c) - \tfrac{1}{2}(v_1 + v_2 - 2v_c) \\ & \times [m(T - T_c) + n\{(v_1 - v_c)^2 - (v_1 - v_c)(v_2 - v_c) + (v_2 - v_c)^2\}], \end{aligned} \tag{8·3·8}$$

and, by subtraction,

$$m(T_c - T) = n\{(v_2 - v_c)^2 + (v_1 - v_c)(v_2 - v_c) + (v_2 - v_c)^2\}. \tag{8·3·9}$$

Also, the condition that $P(v_2 - v_1) = \int_{v_1}^{v_2} P\,dv$ gives, on dividing by $v_2 - v_1$,

$$P - P_c = l(T - T_c) - \tfrac{1}{2}(v_1 + v_2 - 2v_c)[m(T - T_c) + \tfrac{1}{2}n\{(v_1 - v_c)^2 + (v_2 - v_c)^2\}]. \tag{8·3·10}$$

Comparing (8·3·8) and (8·3·10) we see that

$$v_1 + v_2 - 2v_c = 0, \tag{8·3·11}$$

and then (8·3·9) gives

$$v_2 - v_c = v_c - v_1 = (m/n)^{\frac{1}{2}}(T_c - T)^{\frac{1}{2}}. \qquad (8·3·12)$$

The condensation and vaporization curves therefore together form a parabola in the v, T plane with its vertex at v_c, T_c and with $v = v_c$ as its axis. In the P, v plane the equation of the parabola is

$$v_2 - v_c = v_c - v_1 = (m/ln)^{\frac{1}{2}}(P_c - P)^{\frac{1}{2}}. \qquad (8·3·13)$$

If ρ_1 and ρ_2 are the densities of the liquid and of the vapour, then since v_1 and v_2 are nearly equal to v_c we have to the first order

$$\rho_1 = 1/v_1 = 1/\{v_c + (v_1 - v_c)\} = \rho_c - \rho_c^2(v_1 - v_c)$$

and similarly $$\rho_2 = \rho_c - \rho_c^2(v_2 - v_c).$$

Thus $$\rho_1 + \rho_2 - 2\rho_c = 0 \qquad (8·3·14)$$

and $$\rho_1 - \rho_c = \rho_c - \rho = (m/n)^{\frac{1}{2}}\rho_c^2(T_c - T)^{\frac{1}{2}}. \qquad (8·3·15)$$

The relation expressed by (8·3·14) is of considerable importance, and is a particular case of the 'law of rectilinear diameters'. Although it has only been derived for states near the critical state, in practice its range of validity is quite large. It was discovered empirically by Cailletet & Mathias that if the densities of a liquid and its saturated vapour are plotted against the temperature, the resulting curves are nearly parabolic in shape and join smoothly at the critical point. The means of the densities of the liquid and the vapour lie very nearly as a straight line when plotted against the temperature. This line passes through the point ρ_c, T_c, but instead of being parallel to the axis of T as predicted by (8·3·14) it has a small negative gradient. Since the direct determination of the critical volume is a matter of some difficulty, while the critical pressure and temperature can be measured much more easily, the most accurate method of determining the critical volume is to determine the densities of the liquid and vapour near the critical point and extrapolate them to the critical temperature, using the law of rectilinear diameters.

The Second Virial Coefficient

8·4. The calculation of the equation of state of an imperfect gas requires the evaluation of the partition function for the whole gas, and, on account of the interactions between the molecules, the partition function cannot be split up into the product of the partition functions of the N individual molecules. We must therefore deal with the partition function Z_{system} which, according to § 5·52, is given by

$$Z_{\text{system}} = \frac{1}{N!\,h^{3N}} \int \cdots \int \exp[-\mathscr{H}(p, q)/kT]\, dp_1^{(1)} \ldots dq_3^{(N)}, \qquad (8·4·1)$$

if the conditions are such that classical mechanics is applicable to the system. The Hamiltonian function $\mathscr{H}(p,q)$ of the system is given by

$$\mathscr{H}(p,q)=\frac{1}{2m}\sum_{i=1}^{N}(p_1^{(i)2}+p_2^{(i)2}+p_3^{(i)2})+\tfrac{1}{2}\sum_{i\neq j}\mathscr{V}(r_{ij}), \tag{8·4·2}$$

where $\mathscr{V}(r_{ij})$ is the mutual potential energy of the molecules i and j, and which we assume to depend only on r_{ij} the distance apart of the molecules. The integration with respect to the momenta can be carried out at once and we obtain

$$Z_{\text{system}}=\left(\frac{2\pi mkT}{h^2}\right)^{\frac{3}{2}N}\frac{Q_N}{N!}, \tag{8·4·3}$$

with

$$Q_N=\int\ldots\int\exp\left\{-\frac{1}{2}\frac{\sum\limits_{i\neq j}\mathscr{V}(r_{ij})}{kT}\right\}d\tau_1\ldots d\tau_N, \tag{8·4·4}$$

where $d\tau_i$ denotes an integration over the space coordinates of the ith molecule.

The complete evaluation of the 'configurational' partition function Q_N presents formidable mathematical difficulties, and we postpone a discussion of these until § 8·8. If, however, we merely wish to derive a theoretical expression for the second virial coefficient it is possible to avoid the complexities of the complete theory by the use of Clausius's virial theorem.

8·41. *The virial theorem.* If a particle of mass m is acted upon by a force $\mathbf{F}$, its equation of motion is

$$m\,d^2\mathbf{r}/dt^2=\mathbf{F}, \tag{8·41·1}$$

where $\mathbf{r}$ is the position vector of the particle. If we form the scalar product of this with $\mathbf{r}$, we obtain the relation

$$\frac{1}{4}\frac{d^2}{dt^2}(m\mathbf{r}^2)-\tfrac{1}{2}m\left(\frac{d\mathbf{r}}{dt}\right)^2=\tfrac{1}{2}\mathbf{F}\,.\,\mathbf{r}. \tag{8·41·2}$$

If we sum this over all the particles in a system and form the time average over a long time t_0, we find that

$$-\frac{1}{4t_0}\left[\frac{d}{dt}\Sigma m\mathbf{r}^2\right]_0^{t_0}+\tfrac{1}{2}\overline{\Sigma m\mathbf{v}^2}=-\tfrac{1}{2}\overline{\Sigma\mathbf{F}\,.\,\mathbf{r}}, \tag{8·41·3}$$

where the bars denote time averages. Now if the system is in a steady state, $d(\Sigma m\mathbf{r}^2)/dt$ must be constant, at least as measured by macroscopic methods, and hence the first term in (8·41·3) is effectively zero for large t_0. We therefore have

$$\tfrac{1}{2}\overline{\Sigma m\mathbf{v}^2}=-\tfrac{1}{2}\overline{\Sigma\mathbf{F}\,.\,\mathbf{r}}. \tag{8·41·4}$$

The expression on the right is called the virial, and the equation is known as Clausius's virial theorem.

8·411. We now apply the virial theorem to an imperfect gas bounded by a surface S. Both the external forces, which balance the pressure P of the gas, and the internal forces between the molecules contribute to the virial. Let dS be an element of the surface and let the unit vector $\mathbf{n}$ define its outward normal. Then there is an external force $-P\mathbf{n}dS$ on the surface which makes a contribution to the virial of

$$\frac{1}{2}\int P\mathbf{n}\,dS = \tfrac{3}{2}P\int d\tau,$$

the surface integral being transformed into a volume integral by Green's theorem, where $d\tau$ is an element of volume. The external forces therefore contribute $\frac{3}{2}PV$ to the virial, where V is the total volume.

To calculate the effect of the intermolecular forces, we suppose that they are radial forces only and that the force between two molecules with position vectors $\mathbf{r}$ and $\mathbf{r}'$ is given by

$$\mathbf{F} = \frac{\mathbf{r}-\mathbf{r}'}{|\mathbf{r}-\mathbf{r}'|}\phi(|\mathbf{r}-\mathbf{r}'|), \quad \mathbf{F}' = \frac{\mathbf{r}'-\mathbf{r}}{|\mathbf{r}-\mathbf{r}'|}\phi(|\mathbf{r}-\mathbf{r}'|). \qquad (8·411·1)$$

The pair of molecules therefore contribute

$$-\tfrac{1}{2}\overline{\mathbf{r}.\mathbf{F}} - \tfrac{1}{2}\overline{\mathbf{r}'.\mathbf{F}} = -\tfrac{1}{2}\overline{r\phi(r)}$$

to the virial, where $r = |\mathbf{r}-\mathbf{r}'|$. The virial theorem therefore becomes

$$PV = \tfrac{1}{3}\Sigma\overline{m\mathbf{v}^2} - \tfrac{1}{3}\Sigma\Sigma\overline{r\phi(r)}, \qquad (8·411·2)$$

where the second term is to be summed over all pairs of molecules.

If there are N molecules present and if the temperature is T, we have $\Sigma m\mathbf{v}^2 = \frac{3}{2}NkT$. Also, if the molecules were uniformly distributed in space, the chance of two molecules being at a distance from each other lying between r and $r+dr$ would be $4\pi r^2 dr/V$, and, since there are $\frac{1}{2}N(N-1)$ pairs of molecules in all, the number of pairs with centres at distances between r and $r+dr$ would be $2\pi N^2 r^2 dr/V$, if we neglect the difference between N and $N-1$. The molecules are, however, not uniformly distributed in space, since the intermolecular forces introduce a Boltzmann factor $\exp[-\mathscr{V}(r)/kT]$, where $\mathscr{V}(r)$ is the mutual potential energy of two molecules. In addition, a multiplying factor should be introduced to make the total probability of finding two molecules at any distance apart equal to unity; but, since this factor does not affect the second virial coefficient, we omit it. We therefore have

$$\Sigma\Sigma\overline{r\phi(r)} = 2\pi\frac{N^2}{V}\int_0^\infty r^3\phi(r)\exp[-\mathscr{V}(r)/kT]\,dr \quad (\phi(r) = -d\mathscr{V}/dr) \qquad (8·411·3)$$

$$= 6\pi\frac{N^2kT}{V}\int_0^\infty r^2\{\exp[-\mathscr{V}(r)/kT]-1\}\,dr \qquad (8·411·4)$$

on integrating by parts, provided that $\mathscr{V}(r)$ tends to zero sufficiently rapidly as $r \to \infty$. Hence equation (8·411·2) becomes

$$PV = NkT\left[1 - 2\pi\frac{N}{V}\int_0^\infty r^2\{\exp[-\mathscr{V}(r)/kT] - 1\}\,dr\right], \qquad (8\cdot411\cdot5)$$

which gives

$$\left.\begin{aligned} B(T) &= 2\pi\mathscr{L}\int_0^\infty r^2\{1 - \exp[-\mathscr{V}(r)/kT]\}\,dr \\ &= -\frac{2}{3}\frac{\pi\mathscr{L}}{kT}\int_0^\infty r^3\frac{d\mathscr{V}}{dr}\exp[-\mathscr{V}(r)/kT]\,dr \end{aligned}\right\} \qquad (8\cdot411\cdot6)$$

as the expression for the second virial coefficient, where $\mathscr{L}$ is Avogadro's number.

8·42. *Special cases.* There are various special cases which suggest themselves.

(i) If the molecules repel each other according to the law $\phi(r) = m\lambda r^{-m-1}$ $(m > 3)$ we have $\mathscr{V}(r) = \lambda r^{-m}$ and

$$\begin{aligned} \frac{B(T)}{\mathscr{L}} &= \frac{2\pi\lambda m}{3kT}\int_0^\infty r^{2-m}\exp\left(-\frac{\lambda}{kTr^m}\right)dr \\ &= \frac{2\pi}{3}\left(\frac{\lambda}{kT}\right)^{3/m}\Gamma(1 - 3/m). \end{aligned} \qquad (8\cdot42\cdot1)$$

(ii) If the molecules behave like rigid spheres of diameter σ surrounded by an attractive field $\phi(r) = -n\mu r^{-n-1}$ $(n > 3)$, we have $\mathscr{V}(r) = \infty$ $(r < \sigma)$, $\mathscr{V}(r) = -\mu r^{-n}$ $(r \geqslant \sigma)$. We therefore write (Keesom, 1912)

$$\frac{B(T)}{\mathscr{L}} = 2\pi\int_0^\infty \{1 - \exp[-\mathscr{V}(r)/kT]\}\,r^2\,dr$$

$$= 2\pi\int_0^\sigma r^2\,dr + 2\pi\int_\sigma^\infty\left[1 - \exp\frac{\mu}{kTr^n}\right]r^2\,dr \qquad (8\cdot42\cdot2)$$

$$= \tfrac{2}{3}\pi\sigma^3\left[1 - \sum_{s=1}^\infty \frac{3(E_0/kT)^s}{s!\,(sn-3)}\right], \qquad (8\cdot42\cdot3)$$

where

$$E_0 = \mu/\sigma^n. \qquad (8\cdot42\cdot4)$$

(iii) The expression for $B(T)$ for the more general case where

$$\mathscr{V}(r) = \lambda r^{-m} - \mu r^{-n} \quad (m > n > 3)$$

can also be obtained (Lennard-Jones, 1924). We may, if we please, express λ and μ in terms of σ and E_0, where σ is the value of r where $\mathscr{V}(r) = 0$ (σ may be considered to be the 'diameter' of the molecule), and where $-E_0$ is the minimum value of $\mathscr{V}(r)$. Then

$$\frac{\lambda}{\sigma^m} = \frac{\mu}{\sigma^n} = \frac{m}{m-n}\left(\frac{m}{n}\right)^{n/(m-n)}E_0. \qquad (8\cdot42\cdot5)$$

To calculate $B(T)$ we have

$$\frac{B(T)}{\mathscr{L}}=\frac{2\pi}{3kT}\int_0^\infty\left(\frac{m\lambda}{r^{m+1}}-\frac{n\mu}{r^{n+1}}\right)\exp\left(-\frac{\lambda r^{-m}-\mu r^{-n}}{kT}\right)r^3dr$$

$$=\frac{2\pi}{3kT}\int_0^\infty\left(\lambda x^{-3/m}-\frac{n\mu}{m}x^{(n-m-3)/m}\right)\exp\left(-\frac{\lambda x-\mu x^{n/m}}{kT}\right)dx.$$

We now expand $\exp(x^{n/m}\mu/kT)$ in powers of $x^{n/m}$ and integrate term by term, obtaining $B(T)$ in the form $\lambda\Sigma C_s-(n\mu/m)\,\Sigma D_s$. This can be simplified by grouping together C_{s+1} and D_s, the result being

$$\frac{B(T)}{\mathscr{L}}=\tfrac{2}{3}\pi\left(\frac{kT}{\lambda}\right)^{-3/m}\Gamma\left(\frac{m-3}{m}\right)-\frac{2\pi}{m}\sum_{s=1}^{\infty}\left(\frac{kT}{\lambda}\right)^{(sn-3)/m}\left(\frac{\mu}{kT}\right)^s\frac{1}{s!}\Gamma\left(\frac{sn-3}{m}\right).\quad(8\cdot42\cdot6)$$

8·421. *Experimental results.* The experimental data for the second virial coefficient can be fitted by the formula (8·42·6) with a considerable

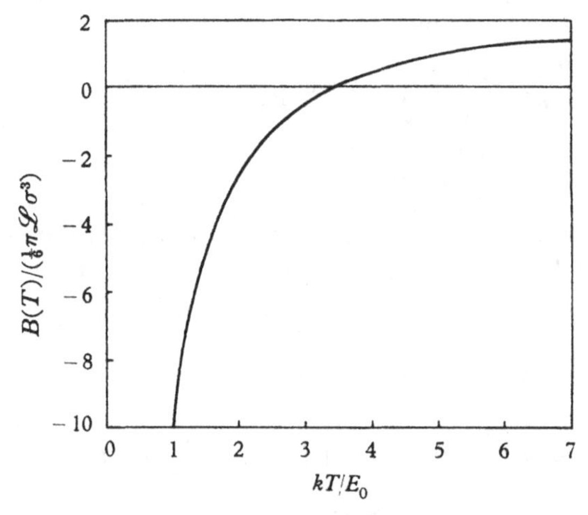

Fig. 8·8. The second virial coefficient in reduced units.

degree of latitude in the choice of the parameters. Calculations based upon the quantum theory of the interaction of two molecules suggest that n should be taken to be 6, but even with this restriction any value of m between about 8 and 14 will fit the data for gases with a simple molecular structure. If we put $kT^*=E_0$ and $\vartheta=T/T^*$, then with $m=12$, $n=6$, we have

$$\mathscr{V}(r)=4\left[\left(\frac{\sigma}{r}\right)^{12}-\left(\frac{\sigma}{r}\right)^{6}\right]E_0,\quad(8\cdot421\cdot1)$$

and (8·42·6) can be written

$$\frac{B(T)}{\frac{1}{6}\pi\mathscr{L}\sigma^3}=-\sum_{s=0}^{\infty}\frac{1}{s!}\Gamma(\tfrac{1}{2}s-\tfrac{1}{4})\left(\frac{4}{\vartheta}\right)^{\frac{1}{2}s+\frac{1}{4}}\quad(8\cdot421\cdot2)$$

Fig. 8·8 shows $B(T)$ calculated from (8·421·2). Some values of σ and T^* found by analysing the equations of state and using equation (8·421·2) are given in table 8·3 (Lennard-Jones, 1931; Buckingham, 1938; Hirschfelder, Curtiss & Bird, 1954).

Table 8·3. *Values of the potential energy constants σ and T^* for various gases*

	H_2	He	Ne	A	Kr	Xe
σ in 10^{-8} cm.	2·87	2·63	2·75	3·4	3·6	4·1
$\frac{2}{3}\pi\mathscr{L}\sigma^3$ in cm.^{3}/gram molecule	7·44	5·75	6·55	12·45	14·7	21·7
$v_c/(\frac{2}{3}\pi\mathscr{L}\sigma^3)$	8·7	10·1	6·4	6·0	6·4	5·3
T^* in °K.	29·2	6·03	35·6	120	171	221
T_c/T^*	1·14	1·15	1·24	1·23	1·23	1·31

	N_2	O_2	CO	CO_2	NO
σ in 10^{-8} cm.	3·7	3·58	3·76	4·5	3·17
$\frac{2}{3}\pi\mathscr{L}\sigma^3$ in cm.^{3}/gram molecule	16	14·5	16·8	28·5	10
$v_c/(\frac{2}{3}\pi\mathscr{L}\sigma^3)$	5·6	5·2	5·6	5·8	5·8
T^* in °K.	95	118	100	189	131
T_c/T^*	1·32	1·30	1·33	1·6	1·36

	N_2O	CH_4	C_2H_6	C_3H_8	$n-C_4H_{10}$
σ in 10^{-8} cm.	4·6	3·82	3·95	5·64	4·97
$\frac{2}{3}\pi\mathscr{L}\sigma^3$ in cm.^{3}/gram molecule	30	17·5	19·5	56·5	39
$v_c/(\frac{2}{3}\pi\mathscr{L}\sigma^3)$	3·2	5·6	7·6	3·5	6·6
T^* in °K.	189	148	243	242	297
T_c/T^*	1·64	1·3	1·25	1·53	1·4

8·422. *The compressibility factor.* One of the most useful ways of depicting the P, v, T data for gases is to plot the compressibility factor, defined as Pv/RT, as a function of P. For all gases which satisfy a reduced equation of state, $\pi\phi/\vartheta$ can be expressed as a universal function of π for a given value of ϑ. Some graphs of $\pi\phi/\vartheta$ are shown in fig. 8·9 (Dodge, 1932), where the reduced variables are the ratios of the actual pressure volume and temperature to their values at the critical point.

8·423. *The reduced equation of state.* If the law of force between the molecules contains only two constants which depend upon the nature of the molecules, the equation of state can be put into a dimensionless form by using reduced variables. If $\mathscr{V}(r_{ik})$ is of the form $E_0 f(r_{ik}/\sigma)$, where E_0 is an energy and σ is a length, each characteristic of a particular type of molecule, we may write (8·4·1) in the form

$$Z_{\text{system}} = \frac{(2\pi mkT)^{\frac{3}{2}N}\sigma^{3N}}{N!\,h^{3N}}\int\cdots\int \exp\left\{-\frac{1}{2\vartheta}\sum_{i\neq j} f(\rho_{ij})\right\} d\boldsymbol{\rho}_1\ldots d\boldsymbol{\rho}_N, \quad (8\cdot423\cdot1)$$

where

$$\vartheta = kT/E_0 = T/T^*, \quad \boldsymbol{\rho} = \mathbf{r}/\sigma. \quad (8\cdot423\cdot2)$$

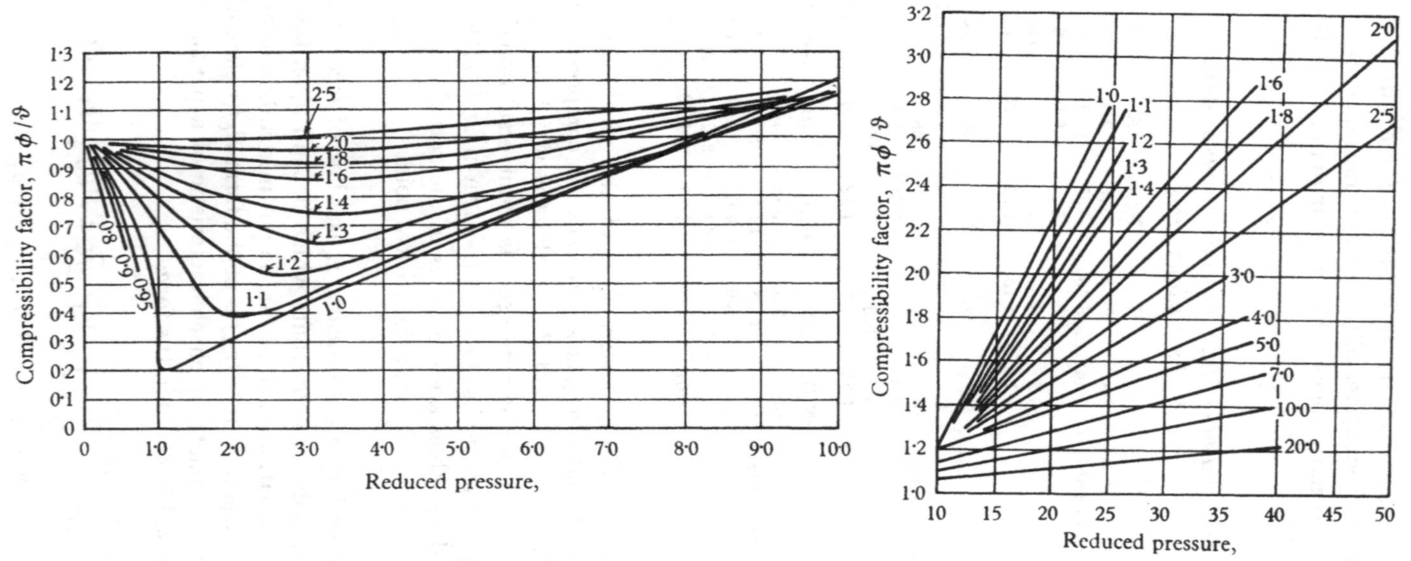

Fig. 8·9. The compressibility factor in reduced units.

Thus $T^* = E_0/k$ is a characteristic temperature. We may also define a characteristic volume v^* per gram molecule and a reduced volume ϕ by the relations

$$v^* = \tfrac{1}{6}\pi\mathscr{L}\sigma^3, \quad \phi = v/v^*, \tag{8·423·3}$$

and the partition function is then of the form

$$Z_{\text{system}} = (m\sigma^2/E_0)^{\frac{3}{2}N}\,\Phi(\phi, \vartheta). \tag{8·423·4}$$

Now
$$P = -kT\frac{\partial}{\partial v}\log Z_{\text{system}} = -\frac{6E_0}{\pi\mathscr{L}\sigma^3}\vartheta\frac{\partial \log \Phi}{\partial \phi},$$

and so if we define the reduced pressure π by the relation

$$\pi = \sigma^3 P/E_0, \tag{8·423·5}$$

the equation of state can be written in the dimensionless form

$$\pi \propto -\vartheta\partial \log \Phi/\partial\phi. \tag{8·423·6}$$

The characterization of the properties of a fluid by a set of reduced (dimensionless) variables is often referred to as the principle of corresponding states, and, if it is true, the properties of a fluid can be expressed in terms of a universal equation of state or partition function if the thermodynamic variables are expressed as ratios relative to the parameters of any state which is characteristic of the fluid. The principle of corresponding states has been used by Pitzer (1939) and by Guggenheim (1945) to correlate many of the properties of the simpler gases. If the validity of the reduced equation extends as far as the critical point, it is often convenient to express the reduced variables as ratios relative to P_c, v_c and T_c. The ratio $RT_c/(P_c v_c)$ should then be the same for all the gases which conform to a given reduced equation. It will be seen by reference to table 8·1 that, for a number of gases with simple structures, $RT_c/(P_c v_c)$ lies between 3·4 and 3·5, but that there are large divergences for highly polar gases and smaller divergences for gases with low molecular weights (in particular, hydrogen and helium). A more complex theory than that given here is required to deal with highly polar gases, for which the mutual potential energy of two molecules is not a function of their distance apart only. For the lighter gases, quantum effects play a part even although classical statistics is always applicable.

8·43. *Quantum effects.* The quantum analogue of (8·411·6) is (Slater, 1931)

$$B(T) = 2\pi\mathscr{L}\int_0^\infty r^2[1 - S(r)]\,dr, \tag{8·43·1}$$

where
$$S(r) = \sum_i \exp(-E_i/kT)\,\psi_i^*\psi_i. \tag{8·43·2}$$

The ψ_i's and the E_i's are the eigenfunctions and the energy levels of the relative motion of two molecules, and the normalization factor is such that

$S(r) \to 1$ as $r \to \infty$. In general the ψ_i's belong to the continuous spectrum, but in extreme cases there may be one or more discrete levels to be included in the sum (8·43·2).

For hydrogen and helium at low temperatures, it is necessary to use the exact quantum theory, since the differences from the classical theory, though small, are significant. The calculations are very complex, and for a survey of this part of the theory the reader is referred to the review article by de Boer (1949).

8·431. For many purposes it is possible to apply the classical principle of corresponding states to hydrogen and helium by defining the reduced temperature and volume by $\vartheta = T/T^*$ and $\phi = v/v^*$, where T^* and v^* differ considerably from the critical temperature and volume. The best choice of the parameters is $T^* = 43{\cdot}4°$ K., $v^* = 50$ cm.^{3}/gram molecule for hydrogen ($T_c = 33{\cdot}2°$ K., $v_c = 67$ cm.^{3}/gram molecule), and $T^* = 7{\cdot}66°$ K., $v^* = 33{\cdot}7$ cm.^{3}/gram molecule for helium ($T_c = 5{\cdot}2°$ K., $v_c = 57{\cdot}8$ cm.^{3}/gram molecule). Such a procedure can only apply far from the critical point.

In general, if quantum effects are important it is necessary to introduce a further dimensionless parameter into the equation of state, and this can be taken to be

$$\Lambda = \frac{h}{(mkT_c)^{\frac{1}{2}}} \left(\frac{\mathscr{L}}{v_c}\right)^{\frac{1}{3}}, \tag{8·431·1}$$

where m is the mass of the molecule. Alternatively, we may define it in terms of the parameters of the law of force by the relation

$$\Lambda^* = h/(\sigma m^{\frac{1}{2}} E_0^{\frac{1}{2}}). \tag{8·431·2}$$

Calculations based upon the Lennard-Jones law of force (8·421·1) have been carried out for the helium and hydrogen isotopes by de Boer & Lunbeck (1948) and by Hammel (1950), and the values of σ and E_0 have been deduced by comparing the calculated and experimental values of $B(T)$ at sufficiently low temperatures for the quantum effects to be appreciable. The values of Λ^* can then be determined from (8·431·2). If we plot the reduced critical constants as functions of Λ^* we obtain the curves shown in fig. 8·10, the limiting values for $\Lambda^* = 0$ being $\vartheta_c = 1{\cdot}26$, $\pi_c = 0{\cdot}117$, $\phi_c = 6{\cdot}15$, where

$$\pi = \frac{P\sigma^3}{E_0}, \quad \phi = \frac{v}{\frac{1}{6}\pi \mathscr{L}\sigma^3}, \quad \vartheta = \frac{kT}{E_0}. \tag{8·431·3}$$

By means of these curves the critical constants of He^3 were predicted in 1948 by de Boer & Lunbeck before they had been determined experimentally.

Since He^3 and He^4 only differ in their mass and their spin (and the latter plays no part in the present discussion), the values of σ and E_0 must be the same for the two isotopes, while the value of Λ^* for He^3 must be $\sqrt{\frac{4}{3}}$ times the value of Λ^* for He^4. The critical constants for He^3 can therefore be determined from those for He^4 by extrapolating the curves in fig. 8·10. They are in close agreement with the experimentally determined constants (Sydoriak, Grilly & Hammel, 1949), as shown in table 8·4. Similarly, if we assume that all the molecules of the various hydrogen isotopes satisfy the same reduced equation, a knowledge of the critical constants of the H_2

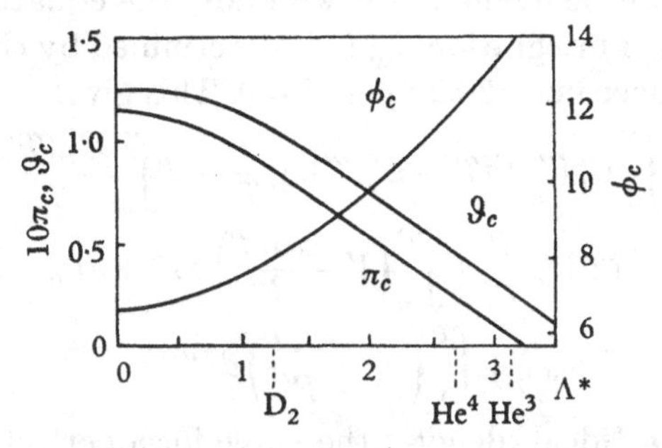

Fig. 8·10. The critical constants as functions of the parameter Λ^*.

molecule is sufficient to determine the constants of the other molecules. The constants calculated in this way are compared with the experimental values (Friedman, White & Johnston, 1951) in table 8·4.

Table 8·4. *The critical constants of the various helium and hydrogen isotopes*

	He^3	He^4	H_2	HD	D_2; *HT*	*DT*	T_2
Λ^*	3·08	2·67	1·73	1·41	1·22	1·10	1·00
ϑ_c	0·33	0·51	0·90	1·03	1·10	1·14	1·17
$100\pi_c$	1·35	2·71	6·46	7·8	8·6	9·2	9·6
ϕ_c	13·6	11·2	8·55	7·95	7·55	7·4	7·2
T_c °K.:							
Calculated	3·37	5·20	33·18	38	41	42	43
Observed	3·34			36	38	—	—
P_c in atmospheres:							
Calculated	1·12	2·26	12·98	15·5	17·5	18·5	19
Observed	1·15			14·6	16	—	—
v_c in cm.³/gram molecule:							
Calculated	70	57·76	66·95	62	59	57	56
Observed	72			63	60	—	—

The Thermodynamic Functions of an Imperfect Gas

8·5. There are numerous methods of deriving a general expression for the free energy of an imperfect gas, but the following is probably the simplest. We use the relation $(\partial G/\partial P)_T = V$ and substitute for V from the equation of state. At constant temperature, any gas becomes perfect as

$P \to 0$, but we cannot integrate the above relation directly down to $P=0$ on account of the resulting logarithmic singularity. Instead we write

$$G(T,P) = G_1(T) + \int^P V(T,P')\,dP'$$

$$= G_0(T) + nRT\log\frac{P}{P^\dagger} + \int_0^P \left(V - \frac{nRT}{P'}\right) dP', \qquad (8\cdot5\cdot1)$$

where $P^\dagger$ is as usual a standard pressure. The integral now converges at the lower limit, and can be evaluated if we know the equation of state. The arbitrary function of integration $G_0(T)$ is determined by the condition that (8·5·1) must pass over into (3·52·4) as $P \to 0$. This gives

$$G(T,P) = n\left(\int_0^T C_P(T')\,dT' - C_P^0 T\log\frac{T}{T^\dagger} - T\int_0^T \frac{C_P^1(T')}{T'}dT'\right)_{\text{ideal}}$$

$$+ nRT\log\frac{P}{P^\dagger} + \int_0^P \left(V - \frac{nRT}{P'}\right) dP' + nv_0 - Ts_0, \qquad (8\cdot5\cdot2)$$

$$= G(T,P)_{\text{ideal}} + \int_0^P \left(V - \frac{nRT}{P'}\right) dP', \qquad (8\cdot5\cdot3)$$

where the subscript 'ideal' denotes the value for a perfect gas.

All the thermodynamic functions can be obtained from (8·5·2) and (8·5·3) by the appropriate differentiations. In particular we have

$$H(T,P) = -T^2\frac{\partial}{\partial T}\frac{G}{T} = H(T)_{\text{ideal}} - \int_0^P T^2\frac{\partial}{\partial T}\frac{V}{T}dP', \qquad (8\cdot5\cdot4)$$

$$S(T,P) = -\frac{\partial G}{\partial T} = S(T,P)_{\text{ideal}} + \int_0^P \left(\frac{nR}{P'} - \frac{\partial V}{\partial T}\right) dP', \qquad (8\cdot5\cdot5)$$

$$C_P(T,P) = \frac{1}{n}\frac{\partial H}{\partial T} = \frac{T}{n}\frac{\partial S}{\partial T} = C_P(T)_{\text{ideal}} - \int_0^P T\frac{\partial^2 v}{\partial T^2}dP'. \qquad (8\cdot5\cdot6)$$

There are corresponding formulae for $F(T,V)$, $U(T,V)$, $S(T,V)$ and $C_V(T,V)$, which are derived from the relation $(\partial F/\partial V)_T = -P$. The explicit formulae are as follows:

$$F(T,V) = n\left(\int_0^T C_V(T')\,dT' - C_V^0 T\log\frac{T}{T^\dagger} - T\int_0^T \frac{C_V^1(T')}{T'}dT'\right)_{\text{ideal}}$$

$$- nRT\log\frac{V}{nV^\dagger} + \int_V^\infty \left(P - \frac{nRT}{V'}\right) dV' + nv_0 - Ts_0, \qquad (8\cdot5\cdot7)$$

$$= F(T,V)_{\text{ideal}} + \int_V^\infty \left(P - \frac{nRT}{V'}\right) dV', \qquad (8\cdot5\cdot8)$$

$$U(T,V) = U(T)_{\text{ideal}} - \int_V^\infty T^2\frac{\partial}{\partial T}\frac{P}{T}dV', \qquad (8\cdot5\cdot9)$$

$$S(T,V) = S(T,V)_{\text{ideal}} + \int_V^\infty \left(\frac{nR}{V'} - \frac{\partial P}{\partial T}\right) dV', \qquad (8\cdot5\cdot10)$$

$$C_V(T,V) = C_V(T)_{\text{ideal}} - \int_V^\infty T\frac{\partial^2 P}{\partial T^2}dv'. \qquad (8\cdot5\cdot11)$$

All these formulae are particular cases of those given in §§ 3·21 and 3·211.

8·51. *The fugacity.* For a perfect gas it is possible to obtain the thermodynamic potential μ in a number of equivalent forms, of which the most useful is given by equation (3·52·4) expressing μ as a function of T and P. But for practical purposes μ is not a very convenient function, since it tends to minus infinity as the pressure tends to zero, and it has been found desirable to introduce a new function called the fugacity (Lewis, 1901), which is a kind of corrected pressure. The fugacity has no theoretical significance, but it is of such great practical importance that some account of it is necessary.

The fugacity $f(T, P)$ is defined by the relation

$$\mu(T,P) = \mu^*(T) + RT\log f, \tag{8·51·1}$$

where the function $\mu^*(T)$ is such that

$$f(T,P) \to P \quad \text{as} \quad P \to 0. \tag{8·51·2}$$

Since $G(T,P) = n\mu(T,P)$, the above relations, in conjunction with equation (8·5·2), give an explicit expression for f, namely,

$$RT\log f = RT\log P + \int_0^P \left(v - \frac{RT}{P'}\right) dP'. \tag{8·51·3}$$

As already mentioned in § 8·422, the data concerning the equation of state are often given in terms of the compressibility factor Pv/RT. It is then convenient to rewite (8·51·3) as

$$\log f = \log P + \int_0^P \left(\frac{P'v}{RT} - 1\right)\frac{dP'}{P'}. \tag{8·51·4}$$

When Pv/RT can be expressed in terms of the reduced temperature and pressure, f/P can also be expressed as a universal function of the reduced variables. The necessary calculations have been carried out by Newton (1935), and his results are shown in fig. 8·11.

Any of the thermodynamic functions can be expressed in terms of the compressibility factor by appropriate substitutions, and the results can be expresed in reduced form with a variable degree of accuracy. Some results for $H - H_{\text{ideal}}$, obtained by Watson & Smith (1936), are given in fig. 8·12.

8·511. *Particular equations of state.* If the equation of state is given in the form

$$Pv = RT + \sum_{r=1}^{\infty} B_r(T)\,P^r, \tag{8·511·1}$$

the fugacity is given by

$$f = P\exp\frac{\Sigma B_r(T)\,P^r/r}{RT}. \tag{8·511·2}$$

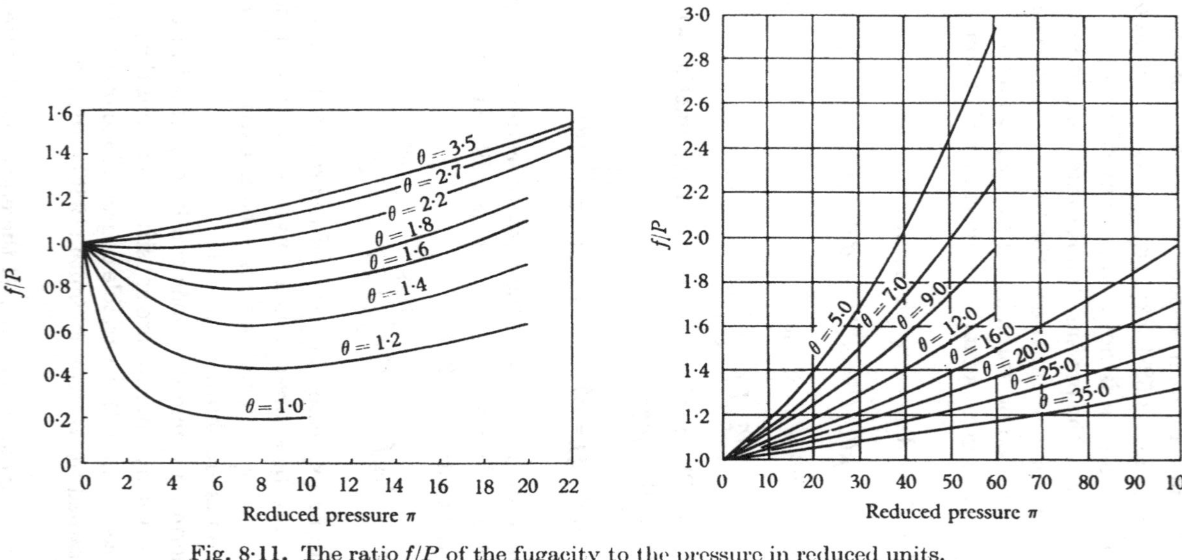

Fig. 8·11. The ratio f/P of the fugacity to the pressure in reduced units.

If the pressures are such that only the second virial coefficient $B(T)$ need be considered we have

$$f = P\,e^{B(T)\,P/RT}. \tag{8·511·3}$$

A special case of this which is much used for moderate pressures is obtained from Berthelot's equation (8·22·5). We then have

$$f = P \exp\left\{\frac{9}{128}\frac{T_c}{T}\frac{P}{P_c}\left(1 - 6\frac{T_c^2}{T^2}\right)\right\}. \tag{8·511·4}$$

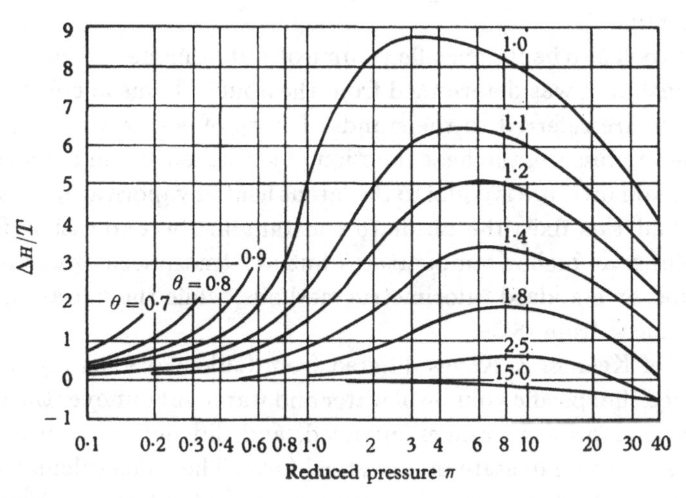

Fig. 8·12. The heat function $\Delta H = H - H_{\text{ideal}}$ as a function of the reduced variables. ΔH is in calories per gram molecule and T is in degrees K.

If we use van der Waals's equation it is simplest to write (8·51·3) in the form

$$RT\log f = \int_0^P v\,dP' + \lim_{P\to 0} RT\log P$$

$$= [vP']_0^P + \int_v^\infty P(T, v')\,dv' + \lim_{P\to 0} RT\log P. \tag{8·511·5}$$

Now $Pv \to RT$ as $P \to 0$, and so, after some simplification, we find that

$$\log f = -\log(v-b) - \frac{2a}{RTv} + \frac{b}{v-b} + \lim_{v\to\infty}\left\{\log(v-b) + \frac{a}{RTv}\right\} + \lim_{P\to 0}\log P.$$

But $\lim P(v-b) = RT$, and so

$$\log f = \log\frac{RT}{v-b} - \frac{2a}{RTv} + \frac{b}{v-b}. \tag{8·511·6}$$

8·52. *The practical determination of the thermodynamic functions of a real gas.* A complete description of the determination of any of the thermodynamic functions of a gas from directly measurable quantities involves many details of experimental methods and computational details, and only a brief reference can be made here to the methods used and to the results obtained. (For further details see, for example, Callendar (1920) and Keenan & Keyes (1936).)

The substance most thoroughly investigated is steam, and much work has been based upon Callendar's empirical equation (8·2·7) with $r=\frac{10}{3}$, which is of the form (8·2·1) with $B=b-a/T^r$. All the formulae of the preceding section can therefore be applied, provided that the various constants can be determined.

Callendar took b to be the specific volume of water, namely, 1 cm.³/gram, while the constant a was determined from the Joule-Thomson effect. If all measurements are referred to the standard state of water at 100° C. and at atmospheric pressure, the heat function of saturated steam at the same temperature and pressure is equal to the latent heat of vaporization of water at 100° C., and this fixes the absolute constant in the expression for H. The values of H at higher temperatures but at atmospheric pressure can then be found by standard calorimetric methods, while the variation with P can be deduced from (8·5·4).

The work of Keenan & Keyes differed from Callendar's mainly in that they measured the specific volume of water and water vapour over the whole temperature and pressure ranges involved, and did not make use of an approximate equation of state in analytical form. They also calculated the specific heat of water vapour (at zero pressure) from the known vibrational frequencies of the molecule.

8·521. For pressures below the critical pressure, the specific heat C_P shows some anomalies which can best be understood by a consideration of the formula (3·212·1), namely,

$$C_P=C_V+T\left(\frac{\partial v}{\partial T}\right)_P\left(\frac{\partial P}{\partial T}\right)_V,$$

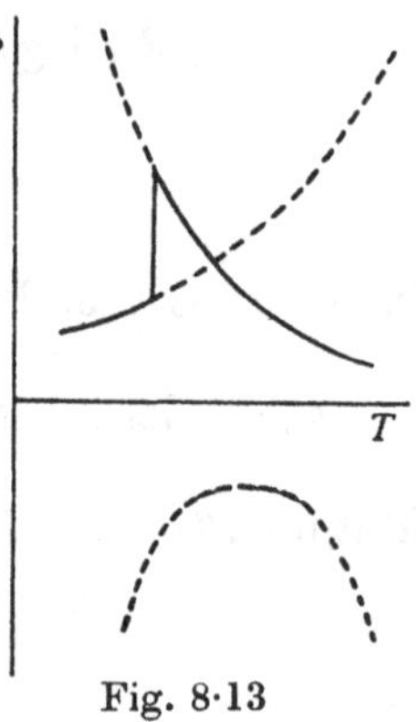

Fig. 8·13

and by an inspection of the curves in figs. 8·6 and 8·7. It was shown in § 3·72 that, at the limits of absolute stability of the liquid and vapour phases, $C_P=\infty$, while C_V remains finite, and we may therefore conclude that any anomalies in C_P are largely derived from anomalies in $(\partial v/\partial T)_P$ rather than in C_V. We then see that the general behaviour of C_P must be as depicted in fig. 8·13, there being two branches of the C_P, T curve referring to the liquid and the vapour respectively (including the metastable states), while there is a third branch, for which $C_P<0$, referring to the essentially unstable states. At the critical pressure the liquid and vapour branches both become infinite at the same value of T, while for pressures above the critical pressure, the C_P, T curve consists of one branch only. These anomalies are well illustrated by the work of Keenan & Keyes on steam, some of whose results are shown in fig. 8·14.

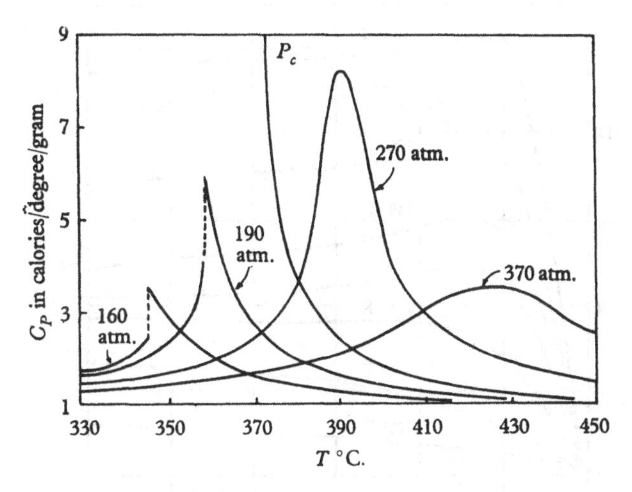

Fig. 8·14. The specific heat at constant pressure of water and steam.

8·522. *Results at high pressures.* If the equation of state is given numerically in the form of a table of measured values of v as a function of T for various values of P, the determination of the thermodynamic functions is not as simple as might be expected owing to the approximate linearity of the $v(T)$ curve. To avoid large errors in the numerical calculations, Deming & Shupe (1931) based their computations upon the quantities α and Δ defined by

$$\alpha = -v + RT/P, \quad \Delta = v(-1 + Pv/RT). \tag{8·522·1}$$

The curves giving the variations of α and Δ with T have large curvatures, and the first and second derivatives can be obtained without difficulty, which facilitates the calculations. The results obtained by Deming & Shupe for C_P and $C_P - C_V$ for nitrogen are shown in figs. 8·15 and 8·16. These curves are typical of the thermal behaviour of gases at high pressures and at temperatures well above the critical temperature. As would be expected, the effect of pressure is greatest when the departure of the gas from ideality is greatest. Therefore, although the specific heat increases with increasing temperature at low pressures, it increases with decreasing temperature at even quite moderate pressures.

8·53. *Thermodynamic diagrams.* Extensive tables exist of the various thermodynamic functions for substances of industrial importance, but it is often convenient to represent some of them in diagrammatic form. Two important graphical representations are provided by the $s(T, P)$ and the $H(s, P)$ diagrams.

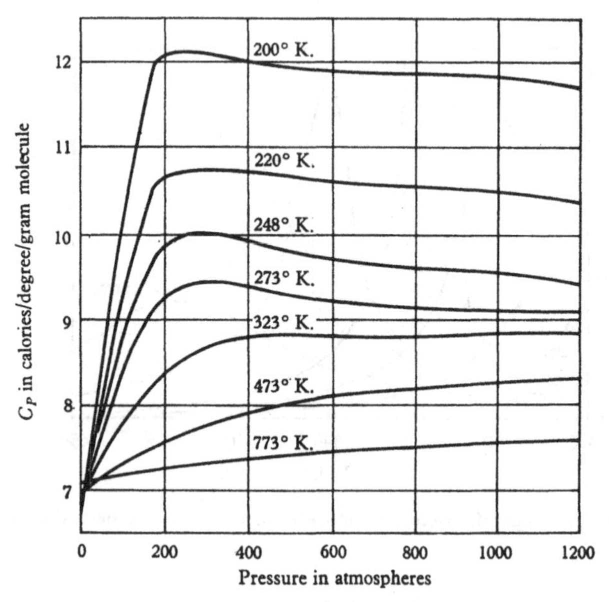

Fig. 8·15. The specific heat at constant pressure of nitrogen.

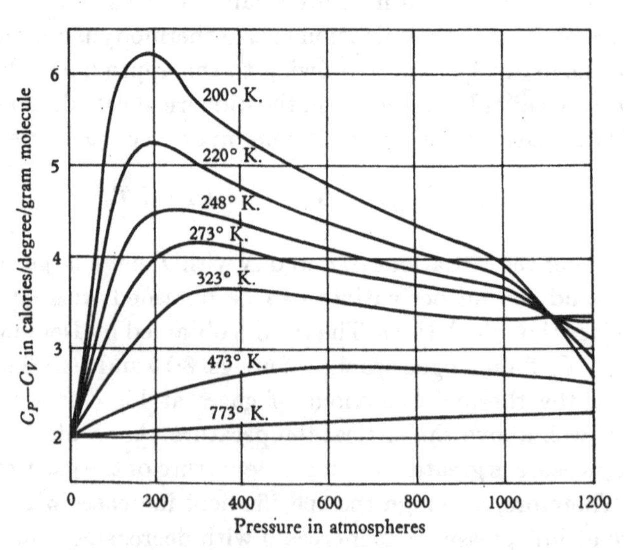

Fig. 8·16. The difference $C_P - C_V$ of the specific heats of nitrogen.

8·531. The $s(T, P)$ data are usually presented as a set of temperature-entropy curves for different values of the pressure. A schematic diagram for water and water vapour is shown in fig. 8·17, the parts of the diagram involving the presence of ice being omitted. Curves of constant H are also shown, i.e. the family of curves $T(s, H)$ for different values of H.

The portion of the T, s diagram lying below the curve ACB refers to heterogeneous mixtures of water and steam. The portions AC and BC of the boundary curve are known as the liquid and the vapour lines respectively, and they meet at the critical point C. The portion of the T, s diagram above the curve ACB refers to homogeneous liquid or gaseous states. Since it is possible to pass continuously from the liquid to the vapour state without encountering heterogeneous states, there is no sharp distinction in the

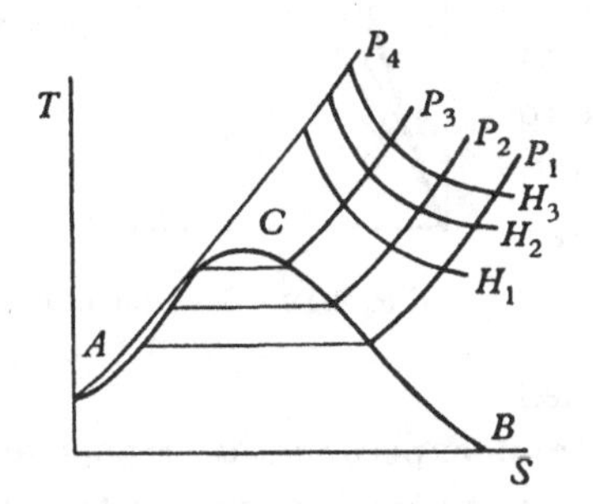

Fig. 8·17. Schematic T, s diagram for water.

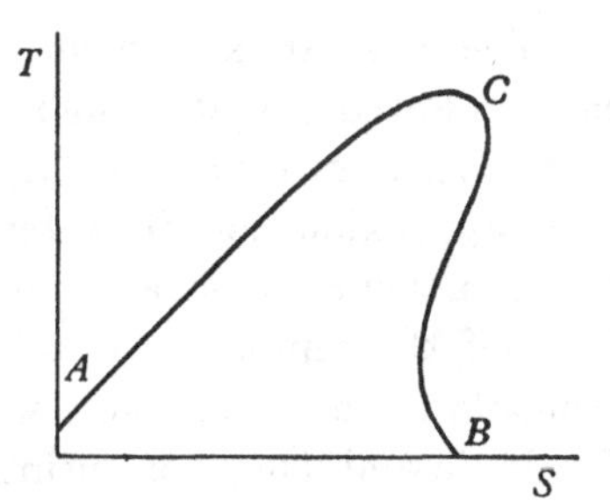

Fig. 8·18. Schematic T, s diagram for benzene.

diagram between the regions referring to the gas and to the liquid. The critical isobar, i.e. the curve with $P = P_c$, is, however, often taken as arbitrarily dividing the homogeneous liquid states from the homogeneous gaseous states.

Although the boundary curve is usually as shown in fig. 8·17, it is possible for it to have the shape given in fig. 8·18, an example of a substance having a boundary curve of this type being benzene. The shape of the boundary curve has a considerable influence on the value of the specific heat of the saturated vapour (see § 9·21, p. 266).

8·532. The $H(s, P)$ diagram is usually known as the Mollier diagram, and H is plotted against s for a series of values of P, as shown in fig. 8·19. The curve ACB is the boundary curve separating the homogeneous and heterogeneous states, AC being the liquid line and BC being the vapour line, which meet at the critical point C. The portion of the diagram below and to the right of the boundary curve refers to the heterogeneous states, and the remainder to the homogeneous states, there being no sharp distinction between the liquid and gaseous states.

The Mollier diagram has the great advantage that $H(s, P)$ is a fundamental thermodynamic function, and hence that all the other thermodynamic functions and variables can be obtained from it. For example, since $\partial H(s,P)/\partial s = T$, the gradient of the isobars gives the absolute temperature.

The thermodynamic diagrams are much used in engineering applications, and they normally contain a variety of curves relating to properties of particular technical importance.

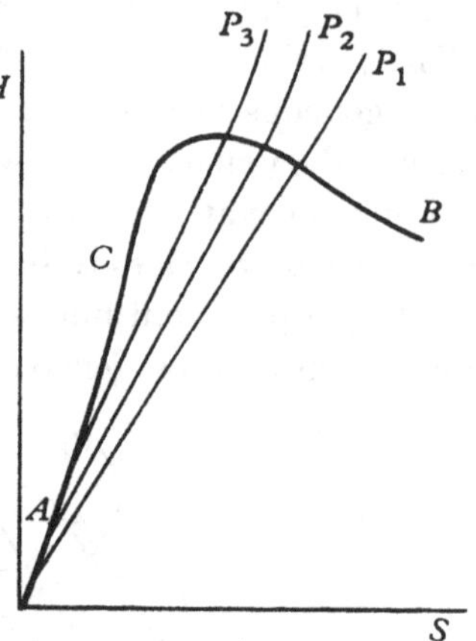

Fig. 8·19. Schematic H, S diagram.

THE JOULE-THOMSON EFFECT

8·6. The original experiment of Joule, in which he allowed gas at a high pressure to rush into a vacuum, established the fact that, to a first approximation, the internal energy of the permanent gases per unit mass is a function of the temperature only. Further experiments to test this result were carried out by Joule and Thomson, using a different experimental arrangement. They allowed gas at a pressure of 4·7 atmospheres to expand to atmospheric pressure in a throttling process by forcing it through a porous plug of cotton-wool. A slight cooling effect was observed for all the gases with which they experimented, with the exception of hydrogen, which showed a slight heating effect.

The theory of the experiment is as follows. The gas passing through the porous plug undergoes a throttling process, and the experimental arrangement is such that the kinetic energy generated is negligible. The heat function H is therefore constant in the process (§ 3·1) and hence, if the initial and final temperatures and pressures are T_1, T_2 and P_1, P_2 respectively, we have

$$H(T_1, P_1) = H(T_2, P_2). \tag{8·6·1}$$

This is the general relation, but if the pressure drop is small the cooling effect is determined by the quantity $(\partial T/\partial P)_H$. To find this differential coefficient we use the general result given in equation (3·211·5), namely,

$$dH = C_P dT - T^2\left(\frac{\partial}{\partial T}\frac{V}{T}\right)_P dP, \tag{8·6·2}$$

which, on putting $dH = 0$, gives at once

$$\left(\frac{\partial T}{\partial P}\right)_H = -\left(\frac{\partial H}{\partial P}\right)_T \bigg/ \left(\frac{\partial H}{\partial T}\right)_P = \frac{T^2}{C_P}\left(\frac{\partial}{\partial T}\frac{V}{T}\right)_P. \tag{8·6·3}$$

If we write‡ μ for the right-hand side of (8·6·3), then for not too large pressure drops we have $\Delta T = \mu \Delta P$. Since ΔP is negative, there is a cooling effect if μ is positive and a heating effect if μ is negative.

8·61. The cooling effect is entirely determined by the specific heat and the equation of state, and it is zero for a perfect gas since H is then a function of T only. We now use van der Waals's equation to discuss the general behaviour of μ as a function of the temperature. If we use the reduced variables, we have

$$\mu C_\pi = \vartheta \left(\frac{\partial \phi}{\partial \vartheta}\right)_\pi - \phi = \frac{\vartheta}{(\partial \vartheta / \partial \phi)_\pi} - \phi. \tag{8·61·1}$$

Now

$$8\left(\frac{\partial \vartheta}{\partial \phi}\right)_\pi = 3\pi - \frac{9}{\phi^2} + \frac{6}{\phi^3} \tag{8·61·2}$$

and

$$\mu C_\pi = \left(-\frac{\pi}{3} + \frac{6}{\phi} - \frac{3}{\phi^2}\right) \bigg/ \left(\pi - \frac{3}{\phi^2} + \frac{2}{\phi^3}\right). \tag{8·61·3}$$

But the denominator is $\frac{8}{3}(\partial \vartheta / \partial \phi)_\pi$, and this must be positive for any stable substance. Therefore μ has the sign of the numerator, and if $\pi < 9(2\phi - 1)/\phi^2$ there is a cooling effect, while if $\pi > 9(2\phi - 1)/\phi^2$ there is a heating effect.

The curve $\pi = 9(2\phi - 1)/\phi^2$, on which $\mu = 0$, is called the inversion curve. If we use the variables $y = \pi\phi$, $x = \pi$, the inversion curve becomes

$$9x = 18y - y^2, \tag{8·61·4}$$

and there is a cooling effect if $9x < 18y - y^2$, that is, if the point (x, y) lies to the left of the inversion curve. It is also possible to obtain the equation of the inversion curve explicitly in terms of ϑ and π by eliminating ϕ from the equations $\pi = 9(2\phi - 1)/\phi^2$ and $(\pi + 3/\phi^2)(3\phi - 1) = 8\vartheta$. The simplest way to do this is first to eliminate π, obtaining $4\vartheta = 3(3\phi - 1)^2/\phi^2$ and hence $1/\phi = 3 - 2\vartheta^{\frac{1}{2}}/\sqrt{3}$. Substituting this into the first of the above equations, we obtain the inversion curve in the form

$$(\pi + 12\vartheta + 27)^2 = 1728\vartheta. \tag{8·61·5}$$

This parabola is shown as curve I in fig. 8·20.

8·611. Dieterici's equation can be discussed similarly. We have from (8·23·2),

$$\left(\frac{\partial \vartheta}{\partial \phi}\right)_\pi = \frac{2\vartheta(\vartheta\phi^2 - 2\phi + 1)}{\phi(\phi\vartheta + 2)(2\phi - 1)} \tag{8·611·1}$$

and

$$\mu C_\pi = -\frac{\phi\vartheta + 8\phi - 4}{2(\phi^2\vartheta - 2\phi + 1)}. \tag{8·611·2}$$

‡ The symbol μ is always used for the Joule-Thomson coefficient. It should not be confused with the thermodynamic potential.

The denominator is always positive, and the sign of μ is determined by that of the numerator. The equation of the inversion curve is

$$\phi(8-\vartheta)=4, \tag{8·611·3}$$

and it can also be put into either of the forms

$$y=4\exp[2-x/(4y-2x)], \tag{8·611·4}$$

where $y=\pi\phi$ and $x=\pi$, and

$$\pi=(8-\vartheta)\exp(\tfrac{5}{2}-4/\vartheta). \tag{8·611·5}$$

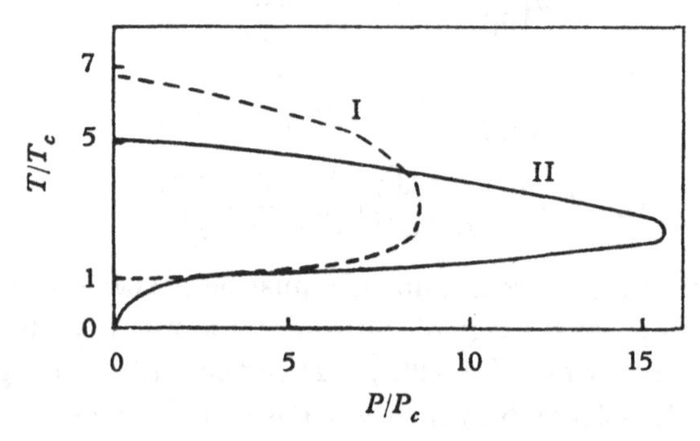

Fig. 8·20. The inversion curve for the Joule-Thomson effect.

The numerical results given by the generalized form (8·23·5) of Dieterici's equation are somewhat better. In the reduced variables, (8·23·5) becomes

$$\pi(2\phi-1)=\vartheta\exp[2-2/(\vartheta^r\phi)], \tag{8·611·6}$$

and the equation of the inversion curve can easily be found and put into either of the forms

$$\phi\{4(r+1)-\vartheta^r\}=2(r+1), \tag{8·611·7}$$

$$\pi\vartheta^{r-1}=\{4(r+1)-\vartheta^r\}\exp\left(\frac{2r+3}{r+1}-\frac{4}{\vartheta^r}\right). \tag{8·611·8}$$

The $\pi\phi, \pi$ form can also be obtained but it is distinctly cumbersome. Equation (8·611·8) with $r=\frac{3}{2}$ is shown as curve II in fig. 8·20.

8·612. The measured inversion curve for nitrogen (Roebuck & Osterberg, 1935) is shown in fig. 8·21, and it will be seen that the maximum and minimum inversion temperatures occur at about $\vartheta=5$ and $\vartheta=0·85$ respectively. Van der Waals's equation gives much too large a maximum inversion temperature, whereas Dieterici's equation is much better in this respect.

There is a cooling effect at room temperature for all gases with critical temperatures greater than about 55° K. The only gases which show a heating effect at room temperature are therefore neon, hydrogen and helium. If we wish to cool these gases by the Joule-Thomson effect, they must first be precooled below their inversion temperatures.

8·613. The Joule-Thomson effect may be used to investigate the virial coefficients which occur in the equation of state. If we write

$$C_P(T,P)=C_{P,0}(T)+C_{P,1}(T)\,P+\ldots,\quad \mu(T,P)=\mu_0(T)+\mu_1(T)\,P+\ldots, \tag{8·613·1}$$

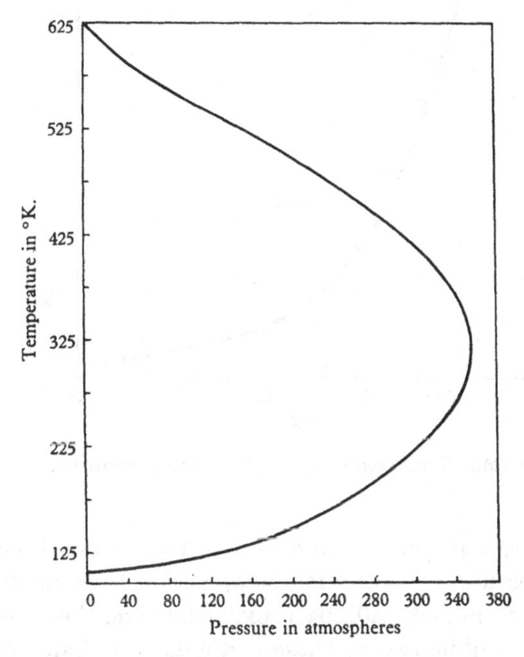

Fig. 8·21. The inversion curve for nitrogen.

and if the equation of state is given in the form (8·2·1), we have

$$\mu_0(T)=\frac{T^2}{C_{P,0}(T)}\frac{d}{dT}\frac{B(T)}{T}, \tag{8·613·2}$$

which provides a convenient means of comparing theoretical expressions for the second virial coefficient with experiment. The values of $\mu_0(T)$ deduced from equation (8·613·2) are shown in fig. 8·22 and are compared with the experimental results for nitrogen obtained by Roebuck & Osterberg (1935). The agreement is excellent, but it must be borne in mind that measurements of the Joule-Thomson effect merely provide an alternative but not an independent method of analysing the experimental results to a direct investigation of the P, V, T data.

8·614. The magnitude of the Joule-Thomson coefficient is not large. The values for carbon dioxide (Burnett, 1923) are shown in fig. 8·23 as a function of the pressure for different temperatures. Above the critical temperature, the $\mu(T, P)$ curve behaves normally, but when $T < T_c$ the $\mu(T, P)$ curve is discontinuous and consists of two branches, which end and begin on the vapour limit curve and the liquid limit curve respectively.

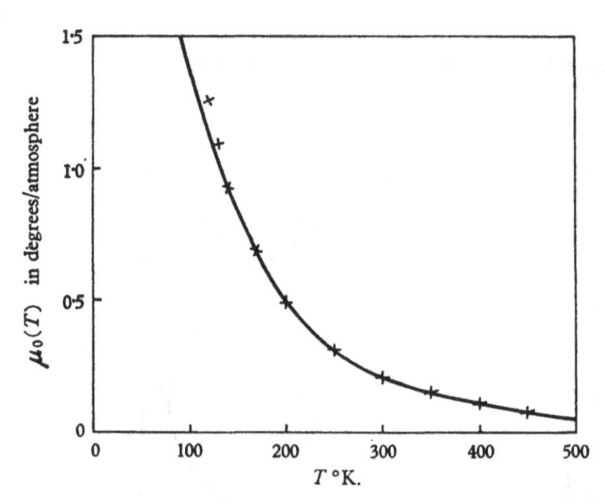

Fig. 8·22. The Joule-Thomson coefficient at zero pressure for nitrogen.

8·62. *The cooling and liquefaction of gases.* Apart from direct refrigeration at constant pressure and isothermal compression, there are two different ways of cooling gases, namely, adiabatic expansion and the Joule-Thomson expansion, or any combination of these methods. We shall consider these methods very briefly.

In an adiabatic expansion from the temperature T_1 and the pressure P_1 to the temperature T_2 and the pressure P_2, the entropy remains constant. We therefore have

$$S(T_1, P_1) = S(T_2, P_2), \tag{8·62·1}$$

which determines T_2 if T_1, P_1 and P_2 are given. For a perfect gas this relation is

$$T_2/T_1 = (P_2/P_1)^{1-1/\gamma}, \tag{8·62·2}$$

where γ is the ratio C_P/C_V of the specific heats. Since γ ranges from $\frac{5}{3}$ for monatomic gases to a limit of 1 for polyatomic gases with an infinite molecular weight, the cooling by adiabatic expansion is most effective for monatomic gases and decreases as the complexity of the molecule increases. For $P_1 = 50$ atmospheres, $P_2 = 1$ atmosphere and $T_1 = 300°$ K., the value of T_2 would be 63° K. for $\gamma = \frac{5}{3}$ and 98° K. for $\gamma = \frac{7}{5}$.

For an imperfect gas no such simple theory is available, and calculations made must be based upon equation (8·5·5) and an accurate equation of state. If a T, S diagram is available, the relation between T_2 and P_2 can be obtained at once. With the values given above for T_1, P_1 and P_2 a gas such as ethylene with a high critical temperature is cooled to its condensation point. In this case T_2 is 167° K., and about 20 % of the ethylene is condensed.

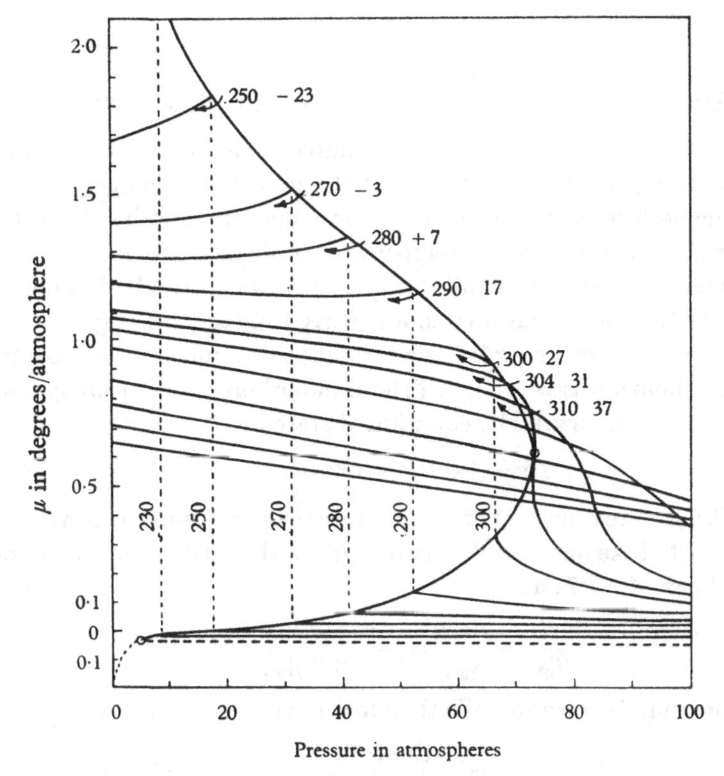

Fig. 8·23. The Joule-Thomson coefficient for carbon dioxide.

The Joule-Thomson effect is the basis of the Linde method of cooling and liquefying gases. It is a thermodynamically inefficient process, as is seen from the fact that it would not cool a perfect gas at all, but it avoids the difficult mechanical problems associated with the adiabatic expansion of gases at low temperatures. Until comparatively recent times the Linde method was the only practicable method of liquefying the permanent gases, on account of the difficulty in lubricating moving parts at low temperatures, but efficient expansion machines have now been developed, which enable the adiabatic expansion method to be used when desired.

If the pressure drop is small, the cooling effect can be obtained from equation (8·6·3) and the equation of state. If the pressure drop is large, it is simplest to apply the general equation (8·6·1) with H given by (8·5·4), and when a Mollier diagram is available the cooling effect can be read off it directly. If ethylene at 300° K. and 50 atmospheres pressure expands irreversibly to 1 atmosphere pressure, it is cooled by the Joule-Thomson effect to about 240° K., which is a very much smaller cooling effect than is obtained by the corresponding adiabatic expansion.

THE DETERMINATION OF THE ABSOLUTE TEMPERATURE

8·7. As pointed out in § 2·721, the simplest method of determining the absolute temperature is by means of the perfect gas temperature. To establish the perfect gas temperature scale we may start with the constant pressure or the constant volume temperature scale of a real gas and apply certain corrections. The most suitable gas to use as a standard is helium, since it obeys the perfect gas laws more nearly than any other gas.

The method of determining the corrections to be applied to the constant pressure thermometer is as follows. (The standard pressure is usually taken to be 100 cm. of mercury.) The equation of state is

$$Pv = A + BP + CP^2 + \ldots, \tag{8·7·1}$$

where $A, B, C, \ldots$ are functions of the absolute temperature T, and in general it is sufficient to take only the first two terms in the series. Now for 1 gram molecule of gas, $A = RT$, and so

$$\frac{T}{T_{0^\circ\,\text{C.}}} = \frac{A(T)}{A_{0^\circ\,\text{C.}}} = \frac{v - B(T)}{\{v - B(T)\}_{0^\circ\,\text{C.}}}, \tag{8·7·2}$$

since the pressure is constant. We therefore have

$$T = \frac{100A(T)}{A_{100^\circ\,\text{C.}} - A_{0^\circ\,\text{C.}}},$$

since $T_{100^\circ\,\text{C.}} - T_{0^\circ\,\text{C.}} = 100$, and by definition the temperature ϑ of the constant pressure thermometer is given by

$$\vartheta = 100v/(v_{100^\circ\,\text{C.}} - v_{0^\circ\,\text{C.}}).$$

Hence
$$T - \vartheta = 100\left(\frac{A(T)}{A_{100^\circ\,\text{C.}} - A_{0^\circ\,\text{C.}}} - \frac{v}{v_{100^\circ\,\text{C.}} - v_{0^\circ\,\text{C.}}}\right). \tag{8·7·3}$$

Now by measurements at varying pressures we can find A and B (and, if necessary, C) as functions of the temperature ϑ, that is, of the volume v, but it is $A(T)$ and not $A(\vartheta)$ which occurs on the right of (8·7·3). However, T and ϑ are nearly the same and we are only calculating a small correction.

It is therefore usually sufficient to write $A(\vartheta)$ instead of $A(T)$. If greater accuracy is required, equation (8·7·3) can be solved by successive approximations. In practice the corrections which have to be applied to helium thermometers are of the order of 0·05°.

By working at very low pressures it is possible to extend the determination of the absolute temperature by means of helium thermometers down to about 1° K. Below this temperature different methods are required, based upon the magnetic equation of state of paramagnetic salts (see § 10·422).

8·71. A method of calibrating any thermometer is provided by the results of the porous plug experiment. This is only one of the many ways of determining the absolute temperature, but it is of considerable historical interest.

Let ϑ be the temperature measured by any thermometer and let $(\partial\vartheta/\partial P)_H = \mu(\vartheta, P)$ be the differential cooling effect as determined by the thermometer. Now the thermodynamic formula for the cooling effect is

$$\left(\frac{\partial T}{\partial P}\right)_H = \frac{1}{C_P}\left\{T\left(\frac{\partial v}{\partial T}\right)_P - v\right\},$$

and if C'_P is the specific heat at constant pressure as measured by the thermometer in question we have

$$C'_P = T\left(\frac{\partial s}{\partial \vartheta}\right)_P = T\frac{dT}{d\vartheta}\left(\frac{\partial s}{\partial T}\right)_P = \frac{dT}{d\vartheta}C_P.$$

Hence
$$\mu(\vartheta, P) = \left(\frac{\partial \vartheta}{\partial P}\right)_H = \frac{d\vartheta}{dT}\left(\frac{\partial T}{\partial P}\right)_H = \frac{1}{C'_P}\left\{T\left(\frac{\partial v}{\partial \vartheta}\right)_P\frac{d\vartheta}{dT} - v\right\}. \qquad (8·71·1)$$

On rearranging this we find

$$\frac{1}{T}\frac{dT}{d\vartheta} = \left(\frac{\partial v}{\partial \vartheta}\right)_P \frac{1}{\mu C'_P + v}.$$

Now the right-hand side contains only quantities which can be measured by the thermometer, namely, the coefficient of expansion and the specific heat at constant pressure of the gas, and the differential cooling coefficient. Hence by integrating the equation between any two temperatures we obtain

$$\log\frac{T_2}{T_1} = \int_{\vartheta_1}^{\vartheta_2}\left(\frac{\partial v}{\partial \vartheta}\right)_P \frac{d\vartheta}{\mu C'_P + v}. \qquad (8·71·2)$$

This result takes an especially simple form if we use as the thermometer a constant pressure gas thermometer containing the same gas as is used in the porous plug experiment, the pressure of the gas in the thermometer being the mean of the pressures in the porous plug experiment. We then have $v = \vartheta$ and $(\partial v/\partial\vartheta)_P = 1$. Hence equation (8·71·3) becomes

$$\log\frac{T_2}{T_1} = \int_{\vartheta_1}^{\vartheta_2} \frac{d\vartheta}{\mu C'_P + \vartheta}. \qquad (8\cdot71\cdot3)$$

The most recent determination of the absolute temperature by means of the Joule-Thomson effect is by Roebuck (1936). His result is that the ice-point is $273{\cdot}16 \pm 0{\cdot}02°$ K.

The General Theory of the Equation of State

8·8. In order to be able to discuss the equation of state when the divergence from the perfect gas laws is large, it is necessary to evaluate the configurational partition function Q_N exactly. If we write

$$\exp[-\mathscr{V}(r_{ij})/kT] = 1 + f(r_{ij}), \qquad (8\cdot8\cdot1)$$

then the expression (8·4·4) for Q_N can be written as

$$Q_N = \int \dots \int \prod_{i>j} \{1 + f(r_{ij})\}\, d\tau_1 \dots d\tau_N. \qquad (8\cdot8\cdot2)$$

Now the function $\mathscr{V}(r_{ij})$ is large and positive for small values of r_{ij}, since the molecules repel each other at sufficiently small distances apart, while $\mathscr{V}(r_{ij}) = 0$ outside the range of molecular interaction, i.e. when r_{ij} is greater than about 10^{-8} to 10^{-7} cm. Hence $f(r_{ij})$ tends to -1 as $r_{ij} \to 0$, while it only differs appreciably from zero for very small values of r_{ij}. The first approximation to Q_N is therefore given by putting $f(r_{ij}) = 0$ in (8·8·2), while the second approximation is

$$Q_N = V^N + \sum_{i>j} \int \dots \int f(r_{ij})\, d\tau_1 \dots d\tau_N, \qquad (8\cdot8\cdot3)$$

where V is the total volume available to the gas. If we disregard the effect of configurations in which the molecules i and j are both close to the boundary of the containing vessels, i.e. if we ignore surface as opposed to volume effects, the integral $\iint f(r_{ij})\, d\tau_i d\tau_j$ is equal to $V \int f(r_{ij})\, d\tau_i$, and this latter integral is independent of i and j. We may therefore write (8·8·3) as

$$Q_N = (VB_1)^N + \tfrac{1}{2}N(N-1)(VB_1)^{N-1}\, VB_2, \qquad (8\cdot8\cdot4)$$

since there are $\frac{1}{2}N(N-1)$ terms in the summation on the right-hand side of (8·8·3). Here

$$B_1 = 1, \quad VB_2 = \iint f(r_{ij})\, d\tau_i d\tau_j = 4\pi V \int_0^\infty f(r)\, r^2 dr. \qquad (8\cdot8\cdot5)$$

If we express the integrand of (8·8·2) as sums of products of the f's, there are various ways of ordering the terms. It might be thought that the simplest method of ordering would correspond to binary, ternary, quaternary, ... encounters between the molecules, but this turns out not to be a useful expansion. Instead, we introduce the concept of a (complex) chain or cluster of molecules, which is such that each member of the cluster is so close to at least one other member (but not necessarily to all members) of the cluster that the corresponding $f(r_{ij})$ is effectively different from zero (Ursell, 1927). We may picture the various molecules in the cluster as being joined together by 'bonds', and the proper method of ordering the various terms in the expansion for the partition function is to group together the terms corresponding to all the 'bonds' that can be formed between the molecules in a given cluster, and then to consider various configurations of clusters. For clusters of 2, 3, 4 molecules, the various types of 'bonds' are shown schematically in fig. 8·24.

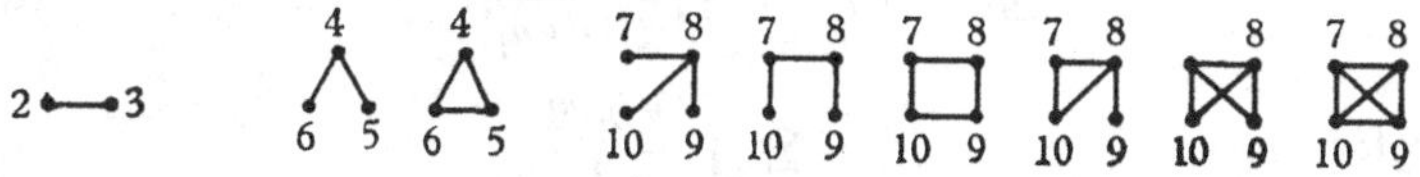

Fig. 8·24. Clusters of 2, 3 and 4 molecules.

A general configuration of clusters consists of n_1 isolated molecules, n_2 clusters each containing two molecules, n_3 clusters each containing three molecules, etc., where the total number of molecules is N. All the terms in the partition function will be obtained if we consider all possible configurations of clusters and all possible 'bonds' in each cluster.

An isolated molecule corresponds to a factor unity in the integrand of (8·8·2), a cluster of two molecules corresponds to a factor $f(r_{23})$, a cluster of three molecules to a factor

$$\sum_{\text{triple cluster}}\prod f(r_{ij}) = f(r_{45})\,f(r_{46}) + f(r_{46})\,f(r_{56}) + f(r_{45})\,f(r_{56}) + f(r_{45})\,f(r_{56})\,f(r_{64}), \tag{8·8·6}$$

and so on. (The molecules must be considered as distinguishable in making this enumeration, since the configurational partition function has been written as $Q_N/N!$, the factor $1/N!$ being introduced to correct for the permutations of the molecules among themselves.)

An exact expression for the configurational partition function can be obtained as follows (Mayer, 1937; Mayer & Ackerman, 1937; Mayer & Harrison, 1938; see also Mayer & Mayer, 1940). If we write

$$VB_l = \int \cdots \int \sum_{l\text{-fold cluster}} \prod f(r_{ij})\, d\tau_1 \ldots d\tau_l, \tag{8·8·7}$$

the configuration considered above will contribute

$$(VB_1)^{n_1}(VB_2)^{n_2}\dots(VB_l)^{n_l} \tag{8·8·8}$$

to the partition function, and, hence, if $g(N, n_l)$ is the number of ways in which the N molecules can be divided up into n_1 single molecules, n_2 double clusters, n_3 triple clusters, etc., we have

$$Q_N = \Sigma g(N, n_l) \prod_l (VB_l)^{n_l}, \tag{8·8·9}$$

where the summation is to be taken over all the sets of numbers $n_1, \dots, n_l, \dots$ which satisfy the condition

$$\sum_l l n_l = N. \tag{8·8·10}$$

Now the total number of permutations of the N molecules amongst themselves is $N!$, but the $l!$ permutations of the l molecules in one of the l-fold clusters and the $n_l!$ permutations of the n_l l-fold clusters amongst each other do not give rise to different configurations, and hence

$$g(N, n_l) = N! \prod_l \frac{1}{(l!)^{n_l} n_l!}. \tag{8·8·11}$$

We therefore have

$$\frac{Q_N}{N!} = \sum_{n_l} \prod_l \frac{(Vb_l)^{n_l}}{n_l!}, \tag{8·8·12}$$

where the summation is subject to the condition (8·8·10), and where

$$b_l = B_l / l!. \tag{8·8·13}$$

If V is sufficiently large, the integrals B_l (and hence b_l) will be independent of V. We shall assume in what follows that this condition is always satisfied. The theory will therefore certainly apply when the departures from the perfect gas laws are small, but it may have an extended range of validity and may even apply to the heterogeneous states consisting of both liquid and vapour, provided that the amount of liquid present is sufficiently small for the largest cluster (in this case probably the whole of the liquid) to occupy only a small fraction of the total volume. Without considering special models it is impossible to lay down conditions under which the b_l's are independent of the volume, but we shall assume that they are so and see how far the theory leads us.

8·81. *Approximate evaluation of the partition function.* If we assume that every term in (8·8·12) is positive we can obtain an expression for Q_N valid for large N by determining the maximum term in the series. The logarithm of the general term is

$$\sum_l \{n_l \log(Vb_l) - n_l \log n_l + n_l\}, \tag{8·81·1}$$

and, if we make this a maximum for varying n_l subject to (8·8·10), we have

$$\log(Vb_l) - \log n_l + \lambda l = 0 \quad (l = 1, 2, \dots), \tag{8·81·2}$$

where λ is an undetermined multiplier. The maximum is therefore given by

$$n_l = V b_l \xi^l, \tag{8·81·3}$$

where $\xi = e^\lambda$. By (8·8·10), ξ is the root of the equation

$$V \sum_{l=1}^{N} l b_l \xi^l = N. \tag{8·81·4}$$

If we replace the series for $Q_N/N!$ by its maximum term we find that

$$\log(Q_N/N!) = V \sum_{l=1}^{N} b_l \xi^l - N \log \xi. \tag{8·81·5}$$

We have now merely to invert the series (8·81·4) to find ξ as a function of V, and when this expression for ξ is inserted into (8·81·5) we obtain $Q_N/N!$, and hence the free energy F, as a function of T, V and N. Since N must in any case be large, we may extend the summation in (8·81·5) to infinity with a negligible error.

We may now obtain all the thermodynamic functions by the appropriate differentiations. We have

$$F = -kT \log Z = -kT \log\{(2\pi mkT/h^2)^{\frac{3}{2}N} Q_N/N!\}, \tag{8·81·6}$$

and
$$U = -T^2 \frac{\partial}{\partial T}\frac{F}{T} = \tfrac{3}{2}NkT + kT^2 V \sum_{l=1}^{\infty} \frac{db_l}{dT}\xi^l + k\left(\frac{\partial \xi}{\partial T}\right)_{V,N}\left(V\sum_{l=1}^{\infty} l b_l \xi^l - N\right)$$

$$= \tfrac{3}{2}NkT + kT^2 V \sum_{l=1}^{\infty} \frac{db_l}{dT}\xi^l, \tag{8·81·7}$$

since the coefficient of $(\partial\xi/\partial T)_{V,N}$ vanishes by (8·81·4). Similarly, the pressure is given by

$$P = -(\partial F/\partial V)_{T,N} = kT \sum_{l=1}^{\infty} b_l \xi^l, \tag{8·81·8}$$

while the thermodynamic potential is given by

$$\mu = (\partial F/\partial N)_{T,V} = -\tfrac{3}{2}kT \log(2\pi mkT/h^2) - kT \log \xi, \tag{8·81·9}$$

which shows the physical significance of the parameter ξ.

Equation (8·81·4) can be solved by successive approximations if V/N is large enough. Since $b_1 = 1$, the first approximation is $\xi = N/V$, and the second approximation is

$$\xi = \frac{N}{V} - 2b_2\left(\frac{N}{V}\right)^2 + \dots. \tag{8·81·10}$$

Hence, to this approximation,

$$\log(Q_N/N!) = -N\log(N/V) + N + N^2 b_2/V \tag{8·81·11}$$

and

$$F(T,V,N) = -NkT\log(2\pi mkT/h^2)^{\frac{3}{2}} + NkT\{\log(N/V) + 1 + Nb_2/V\}. \tag{8·81·12}$$

Now b_2 is independent of V, and so the pressure is given by

$$P=-\left(\frac{\partial F}{\partial V}\right)_{T,N}=\frac{NkT}{V}\left(1-\frac{Nb_2}{V}\right), \qquad (8\cdot81\cdot13)$$

which, by (8·8·5) and (8·8·13), is the same expression as was obtained in § 8·411 by means of the virial theorem.

8·811. *The condensation phenomenon.* It will be noted that the equations (8·81·4)–(8·81·9) bear a strong formal resemblance to those describing a perfect gas which obeys the Einstein-Bose statistics. It was shown in § 6·42, and more rigorously in § 6·43, that at a certain critical temperature a type of condensation sets in, and that the anomalies in the behaviour of an Einstein-Bose gas are connected with the singularities of certain series. In the present case of a classical imperfect gas, it is clear that the arguments used to derive the partition function and the equation of state may break down if the series $\Sigma b_l \xi^l$ (or its derivatives) is not convergent. If, for example, the series $\Sigma l b_l \xi^l$ has a branch point at $\xi=\xi_0$, where ξ_0 is real and positive, we cannot apply equation (8·81·4) for $\xi>\xi_0$, and so the present theory is not valid for values of V/N less than $1/\Sigma l b_l \xi_0^l$. It was suggested by Mayer in 1937 that for smaller values of V/N the pressure should remain constant and equal to $kT\Sigma b_l \xi_0^l$, and therefore that (8·81·4) and (8·81·8) will reproduce the flat portions of the isothermals in the P, V diagram as well as the isothermals of the vapour phase.

If this suggestion of Mayer's is correct, the b_l's must satisfy certain conditions. For large values of T, there must be no condensation, and (8·81·4) must apply to all values of V/N. Sufficient conditions for this are either that the series should have an infinite radius of convergence, or, if it has a singularity on the positive real axis at ξ_0, that $\Sigma l b_l \xi_0^l$ should be infinite. For small values of T, on the other hand, the b_l's must be such that the series (8·81·4) has a branch point, not a pole, at ξ_0 on the positive real axis, $\Sigma l b_l \xi_0^l$ being finite. (Note that it is not sufficient for the series to diverge at $\xi=\xi_0$. The series $\Sigma(-1)^l x^{2l}$ diverges at $x=1$, but the function which it represents, namely, $1/(1+x^2)$, is finite along the whole of the real axis.) If the series (8·81·4) has a branch point at ξ_0, the various thermodynamic functions will, in the limit of infinitely large N, have singularities of various types at the corresponding value $v_0(T)$ of V/N, and they will have different analytical expressions for $v>v_0(T)$ and for $v<v_0(T)$. It is, however, impossible to deduce rigorously by the arguments given in § 8·81 what the form of the functions should be for $v<v_0(T)$, and we cannot assume without proof that there is more than a formal analogy between the present theory and that of § 6·43. A more searching investigation is therefore required.

8·82. *Rigorous evaluation of the partition function.* The calculation given in the preceding subsection can only be justified if all the b_l's are positive, and if the series have no branch points. A more general method, which is always applicable, is as follows (Born & Fuchs, 1938; Kahn & Uhlenbeck, 1938).

The expression (8·8·12) for $Q_N/N!$ is the coefficient of ζ^N in $\exp\left(V\sum_{l=1}^{\infty} b_l \zeta^l\right)$, and if we treat ζ as a complex variable we have, by Cauchy's theorem,

$$\frac{Q_N}{N!} = \frac{1}{2\pi i}\int_C \exp\{V\phi(\zeta)\}\frac{d\zeta}{\zeta^{N+1}}, \tag{8·82·1}$$

where
$$\phi(\zeta) = \sum_{l=1}^{\infty} b_l \zeta^l, \tag{8·82·2}$$

and where C is any contour encircling the origin once in the positive direction. Since V and N are both large in any physical system, we are faced with a double-limit problem, and the quantity which determines the free energy per unit volume or per unit mass is not $Q_N/N!$ itself but

$$\lim_{N\to\infty}\frac{1}{N}\log\frac{Q_N}{N!} \tag{8·82·3}$$

for fixed T and fixed $v = V/N$. We therefore write $V = Nv$ in (8·82·1) and form the function

$$F(z, b_l) = \sum_{N=1}^{\infty}\frac{Q_N}{N!}z^N. \tag{8·82·4}$$

If we choose the contour C so that

$$|\zeta| > |z e^{v\phi(\zeta)}|, \tag{8·82·5}$$

which can always be done if $|z|$ is sufficiently small, we can carry out the summation over N and obtain

$$F(z, b_l) = \frac{1}{2\pi i}\int_C \left(\frac{1}{\zeta - z\exp[v\phi(\zeta)]} - \frac{1}{\zeta}\right)d\zeta. \tag{8·82·6}$$

Since the condition (8·82·5) must hold, there are two poles of the integrand inside the contour C, namely, the pole at $\zeta = 0$ and the pole at $\zeta = \zeta_0$, where ζ_0 is the smallest root of the equation

$$\zeta_0 = z\exp[v\phi(\zeta_0)] = z\exp[v\Sigma b_l \zeta_0^l]. \tag{8·82·7}$$

(It is possible for this equation to have two roots, but in order to satisfy the condition (8·82·5) the values of $|\zeta|$ on the contour must be greater than the modulus of the smaller root for ζ_0 and less than that of the larger root.) If, therefore, we evaluate (8·82·6) by the theory of residues, we obtain

$$F(z, b_l) = \frac{1}{1 - zv\phi'(\zeta_0)\exp[v\phi(\zeta_0)]} - 1 = \frac{1}{1 - v\zeta_0\phi'(\zeta_0)} - 1. \tag{8·82·8}$$

Now, by Cauchy's test, the radius of convergence r of the power series (8·82·4) is given by

$$\lim_{N\to\infty}\left(\frac{Q_N}{N!}\right)^{1/N} r=1, \tag{8·82·9}$$

i.e.

$$\lim_{N\to\infty}\frac{1}{N}\log\frac{Q_N}{N!}=\log\frac{1}{r}. \tag{8·82·10}$$

The radius of convergence of the series (8·82·4) is determined by the singularity of the function $F(z, b_l)$ which is nearest to the origin, and, since Q_N is necessarily positive, this singularity must be situated on the positive real axis (see, for example, Titchmarsh, 1939, p. 214).

The singularities of $F(z, b_l)$ are either the zeros of the denominator or the branch points of $\phi'(\zeta_0)$, provided that these are also branch points of the inverse function. (Note that a pole of $\phi'(\zeta_0)$ will give rise to a zero and not a singularity of $F(z, b_l)$.) Now the branch points of $\phi'(\zeta_0)$ only depend upon the b_l's and are independent of v. We denote by $\bar{z}$ the position of the branch point of $F(z, b_l)$ nearest to the origin, and ξ^* the corresponding value of ζ_0. If all the b_l's are positive, $\bar{z}$ is real and positive; but this condition is not necessarily satisfied, and $\bar{z}$, if it exists at all, may be real or complex, while $\phi'(\xi^*)$ may be finite or infinite (Titchmarsh, 1939, pp. 217 ff.). The position of the smallest zero of the denominator of $F(z, b_l)$, on the other hand, is a function of v, and for large values of v this zero z_0 determines the radius of convergence. Hence for sufficiently large values of v we have

$$1/z_0=v\phi'(\zeta_0)\exp[v\phi(\zeta_0)], \tag{8·82·11}$$

where z_0 and ζ_0 are also connected by the relation (8·82·7). If we denote this particular value of ζ_0 by ξ, then on eliminating z_0 between (8·82·7) and (8·82·11), we see that it is determined by the relation

$$1=v\xi\phi'(\xi)=v\sum_{l=1}^{\infty}lb_l\xi^l. \tag{8·82·12}$$

We then determine the partition function from (8·82·10) with $r=z_0$, which gives

$$\log(Q_N/N!)\sim V\phi(\xi)+N\log\{V\phi'(\xi)/N\} \quad (N,\ V \text{ large}) \tag{8·82·13}$$

$$=V\sum_{l=1}^{\infty}b_l\xi^l-N\log\xi. \tag{8·82·14}$$

These equations for ξ and Q_N are the same as the equations (8·81·4) and (8·81·5), and the simplified method given in § 8·81 therefore leads to the correct result when v is sufficiently large.

8·821. *The condensation phenomenon.* The preceding argument breaks down if $|\bar{z}|<z_0$, which can only happen if v is sufficiently small. (We may note that although $\bar{z}$ may be complex, we cannot have $|\bar{z}|<z_0$ unless $\bar{z}$ is

real and positive, since the smallest singularity of $F(z, b_l)$ necessarily lies on the real axis.) If we assume that there are conditions in which, for certain temperature ranges, $\bar{z}$ is the smallest singularity of $F(z, b_l)$, there is a critical specific volume $v^*(T)$ for which $\bar{z}$ and z_0 coincide. If for a given temperature we have $v > v^*(T)$, equations (8·82·12) and (8·82·14) apply. But if $v < v^*(T)$ the radius of convergence of the series (8·82·3) is $\bar{z}$ which is independent of T, and in this case we have

$$\log(Q_N/N!) = V \sum_{l=1}^{\infty} b_l \xi^{*l} - N \log \xi^* \quad (v < v^*(T)), \qquad (8\cdot821\cdot1)$$

where ξ^* and $\bar{z}$ are connected by the relation (8·82·7). According to the assumptions made above, ξ^* is not given by (8·82·12) but by the condition that

$$\phi'(\xi) = \sum_{l=1}^{\infty} l b_l \xi^l \qquad (8\cdot821\cdot2)$$

is singular at $\xi = \xi^*$ and that $\phi'(\xi^*)$ is finite. The critical value $v^*(T)$ for which (8·82·14) and (8·821·1) hold simultaneously is determined by the relation

$$1 = v^*(T) \sum_{l=1}^{\infty} l b_l \xi^{*l}. \qquad (8\cdot821\cdot3)$$

When (8·821·1) applies, the equations corresponding to (8·81·7), (8·81·8) and (8·81·9) are as follows:

$$U = \tfrac{3}{2}NkT + kT^2 V \sum_{l=1}^{\infty} \frac{db_l}{dT} \xi^{*l}, \qquad (8\cdot821\cdot4)$$

$$P = kT \sum_{l=1}^{\infty} b_l \xi^{*l}, \qquad (8\cdot821\cdot5)$$

$$\mu = -\tfrac{3}{2}kT \log(2\pi mkT/h^2) - kT \log \xi^*. \qquad (8\cdot821\cdot6)$$

We therefore see that, for fixed T, both P and μ are constant for $v < v^*(T)$ so long as the present equations hold.

8·822. In the discussion of the Einstein-Bose condensation phenomenon, it was shown in § 6·431 that a wide variety of discontinuities in the various thermodynamic functions can occur, depending upon the nature of the singularities of the function $R_q(\lambda)$, which corresponds to the function $\xi\phi'(\xi)$. There will therefore be a similar variety of behaviour in the present theory, and unless we can calculate the dependence of b_l upon T and l for large values of l we can make no further progress. In the present state of the theory it is an impossible task to calculate the b_l's when l is large for any special model (calculations have been made of b_2 and b_3 only), but, if we assume that b_l is positive and proportional to an inverse power of l, we can apply the results of § 6·431. It is, however, by no means certain even that all the b_l's are positive for sufficiently large l, and, until more is known about the b_l's for special models, any conclusions drawn must be tentative ones. We shall therefore adopt the attitude that the present theory may give a satisfactory explanation of the condensation of an imperfect gas rather than that it definitely predicts it. We shall not discuss in detail the various cases that can arise from special choices of the b_l's, but we shall conclude this section with some general remarks concerning the isothermal curves.

In the first place, the isothermal curves may or may not have discontinuous gradients at $v = v^*(T)$. If $v > v^*(T)$, we have from (8·81·8) and (8·82·12)

$$\left(\frac{\partial P}{\partial v}\right)_T = -kT\frac{\xi}{v}\frac{(\Sigma l b_l \xi^l)^2}{\Sigma l^2 b_l \xi^l},$$

whereas $$(\partial P/\partial v)_T = 0 \quad \text{if} \quad v < v^*(T).$$

The isothermal curves therefore have continuous gradients if $\Sigma l^2 b_l \xi^{*l}$ is infinite and continuous gradients if $\Sigma l^2 b_l \xi^{*l}$ is finite. If $b_l \sim l^{-q}$ for large l, these conditions are equivalent to $q \leqslant 3$ and $q > 3$ respectively. Since b_l is a function of T it is possible for q to have different values for different temperature ranges.

Mayer has argued that the nature of the singularity of $F(z, b_l)$ must vary with the temperature in a continuous manner. In the limit of infinite T (or vanishingly small interactions between the particles) $F(z, b_l)$ is an integral function (actually $e^{vz} - 1$) with an essential singularity at infinity. For finite values of T the singularity occurs at a finite point on the real axis, and it might be expected that, when the temperature first reaches a sufficiently low value for the singularity to be determined by the condition (8·821·3), it is likely that $\Sigma l^2 b_l \xi^{*l}$ would be infinite and that it would only become finite at some lower temperature. If this is so, there would be two critical temperatures, the first giving the highest temperature for which $\partial P/\partial V$ can be zero, and the second the highest temperature for which $\partial P/\partial V$ can be discontinuous. Experiments showing such a behaviour have been reported from time to time, but it is generally believed that the results are vitiated by segregation effects in the fluids, and the facts are at present obscure.

Secondly, the whole of the preceding theory is based upon the assumption that the b_l's are independent of the volume. This is certainly true for sufficiently large volumes and untrue for small volumes, but it is impossible at the present time to say where the dividing line should be drawn. Clearly the assumption cannot be true for the liquid state or for highly compressed gases above their critical temperatures, but, unless the assumption remains true in a part of the heterogeneous region, the discussion of the condensation phenomenon in § 8·81 is invalid. Even if we assume that the b_l's are independent of V for volumes and temperatures lying in the heterogeneous region, it is not possible to lay down conditions which are both necessary and sufficient to ensure that the functions $\phi'(\xi)$ and $F(z, b_l)$ have branch points with the properties required to ensure the correctness of the arguments given in § 8·821. Attempts have been made by Born & Fuchs (1938) and by Mayer and his collaborators (see in particular Mayer & Mayer, 1940) to give rules governing the positions of the branch points and the values of the functions there, but considerable objections can be raised against their procedure, mainly on the grounds that the assumptions made are not well founded, and the subject is a highly controversial one. The reader is referred to the original papers for the details. Finally, although a great step forward has been taken by showing that the condensation phenomenon can be explained mathematically in terms of the equations derived here, it is impossible to be thoroughly satisfied until some physically acceptable model has been shown to possess the properties required by the theory.

REFERENCES

de Boer, J. (1949). Molecular distribution and equation of states of gases. *Rep. Progr. Phys.* **12**, 305.

de Boer, J. & Lunbeck, R. J. (1948). The properties of the condensed phase of the light helium isotope. *Physica*, **14**, 510.

Born, M. & Fuchs, K. (1938). The statistical mechanics of condensing systems. *Proc. Roy. Soc.* A, **166**, 391.

Buckingham, R. A. (1938). The classical equation of state of gaseous helium, neon and argon. *Proc. Roy. Soc.* A, **168**, 264.

Burnett, F. S. (1923). Experimental study of the Joule-Thomson effect in carbon dioxide. *Phys. Rev.* **22**, 590.

Callendar, H. L. (1920). *Properties of steam.* London.

Deming, W. E. & Shupe, L. E. (1931). The physical properties of compressed gases. *Phys. Rev.* **37**, 638.

Dodge, B. F. (1932). Physicochemical factors in high-pressure design. *Industr. Engng Chem.* **24**, 1353.

Friedman, A. S., White, D. & Johnston, H. L. (1951). Critical constants, boiling points, triple point constants and vapour pressures of the six isotopic hydrogen molecules, based on a simple mass relationship. *J. Chem. Phys.* **19**, 126.

Guggenheim, E. A. (1945). The principle of corresponding states. *J. Chem. Phys.* **13**, 253.

Hammel, E. F. (1950). Some calculated properties of tritium. *J. Chem. Phys.* **18**, 229.

Hirschfelder, J. O., Curtiss, C. F. & Bird, R. B. (1954). *Molecular theory of gases and liquids.* New York.

Kahn, B. & Uhlenbeck, G. E. (1938). The theory of condensation. *Physica*, **5**, 399.

Keenan, J. P. & Keyes, F. G. (1936). *Thermodynamic properties of steam.* New York.

Keesom, W. H. (1912). On the deduction from Boltzmann's entropy principle of the second virial coefficient for material particles which exert central forces upon each other and for rigid spheres of central symmetry containing an electric doublet at their centre. *Proc. Acad. Sci. Amst.* **15**, 256.

Lennard-Jones, J. E. (1924). On the determination of molecular fields. II. From the equation of state of a gas. *Proc. Roy. Soc.* A, **106**, 463.
Lennard-Jones, J. E. (1931). Cohesion. *Proc. Phys. Soc.* **43**, 461.
Lewis, G. N. (1901). The law of physico-chemical change. *Proc. Amer. Acad. Arts Sci.* **37**, 49.
Mayer, J. E. (1937). The statistical mechanics of condensing systems. I. *J. Chem. Phys.* **5**, 67.
Mayer, J. E. & Ackerman, P. G. (1937). The statistical mechanics of condensing systems. II. *J. Chem. Phys.* **5**, 74.
Mayer, J. E. & Harrison, S. F. (1938). The statistical mechanics of condensing systems. III. *J. Chem. Phys.* **6**, 87.
Mayer, J. E. & Mayer, M. G. (1940). *Statistical mechanics*. New York.
Newton, R. H. (1935). Activity coefficients of gases. *Industr. Engng Chem.* **27**, 302.
Pitzer, K. S. (1939). Corresponding states for perfect liquids. *J. Chem. Phys.* **7**, 583.
Ramsay, W. & Young, S. (1887). On the continuous transition from the liquid to the gaseous state of matter at all temperatures. *Phil. Mag.* (5), **23**, 435.
Roebuck, J. R. (1936). The Kelvin temperature of the icepoint. *Phys. Rev.* **50**, 370.
Roebuck, J. R. & Osterberg, H. (1935). The Joule-Thomson effect in nitrogen. *Phys. Rev.* **48**, 450.
Slater, J. C. (1931). The quantum theory of the equation of state. *Phys. Rev.* **38**, 237.
Sydoriak, S. G., Grilly, E. R. & Hammel, E. F. (1949). Condensation of pure He^3 and its vapour pressure between 1·2° K and its critical point. *Phys. Rev.* **75**, 303.
Titchmarsh, E. C. (1939). *The theory of functions*, 2nd ed. Oxford.
Ursell, H. D. (1927). The evaluation of Gibbs' phase integral for imperfect gases. *Proc. Camb. Phil. Soc.* **23**, 685.
Watson, K. M. & Smith, R. L. (1936). Generalised high pressure properties of gases. *Nat. Petrol. News*, **28**, No. 27, p. 29.

Chapter 9

THE HETEROGENEOUS EQUILIBRIUM OF A SINGLE SUBSTANCE

Clapeyron's Equation

9·1. The different states of aggregation in which a substance can exist are known as the phases of the substance. Every substance can exist in at least three phases, liquid, solid and gas, but many substances can exist in the solid state in several allotropic forms, well-known cases being ice, sulphur and tin. In the present chapter we discuss the phenomena associated with a change of phase, or with the heterogeneous state in which two or more phases coexist. The discussion is limited to the simplest case in which the phases are those of a single substance of invariable composition. More complicated cases are dealt with in Chapter 12.

9·11. *The vapour-pressure curve.* Consider a liquid in equilibrium with its vapour. At a given temperature, the two phases can only exist in contact at a definite pressure, the relative amounts of liquid and vapour being determined by the volume available to accommodate the substance. In the P, T diagram, therefore, the series of states representing coexistent liquid and vapour phases form a line, known as a phase-boundary curve. In the particular case when one of the phases is a vapour the curve is known as the vapour-pressure curve. The vapour-pressure curve has a positive gradient and it ends abruptly at the critical point, since above the critical point the vapour and liquid cannot exist in contact.

A differential equation defining the vapour-pressure curve can easily be found as follows. Consider the free energy per unit mass at constant pressure $G(T, P)$, expressed as a function of T and P. If we disregard metastable states, $G(T, P)$ is a single-valued function of T and P over the whole of the plane, though its functional form may be different for the different phases of the substance. Along a phase boundary $G(T, P)$, must be continuous, for, if it were discontinuous, its first derivatives would be infinite, and this would mean that the entropy and volume were infinite, which is impossible. Hence, if we use suffixes G and L to refer to the gaseous and liquid phases, the equation of the phase boundary curve is

$$G_G(T, P) = G_L(T, P). \qquad (9\cdot11\cdot1)$$

Since in each of the phases the fluid behaves as a simple substance, G is the same as the thermodynamic potential μ (equation (3·4·7)), so that equation (9·11·1) can be written as

$$\mu_G = \mu_L,$$

which is a particular case of the general condition of equilibrium discussed in § 3·61. Now $dg = -s\,dT + v\,dP$ for any reversible infinitesimal variation, so that if we restrict the variation to be along the phase-boundary curve, equation (9·11·1) gives $dg_G = dg_L$, which is

$$-s_G dT + v_G dP = -s_L dT + v_L dP. \tag{9·11·2}$$

The gradient of the phase-boundary curve is therefore given by

$$\frac{dP}{dT} = \frac{s_G - s_L}{v_G - v_L}, \tag{9·11·3}$$

which is a differential equation defining the phase-boundary curve. Now $T(s_G - s_L)$ is the heat which must be absorbed by the system to change unit mass of the liquid into vapour at the temperature T. That is, $T(s_G - s_L)$ is the latent heat of vaporization of the liquid, and is usually denoted by L_{LG}. We may therefore rewrite equation (9·11·3) as

$$\frac{dP}{dT} = \frac{L_{LG}}{T(v_G - v_L)}, \tag{9·11·4}$$

which is known as Clapeyron's equation. It may be used to calculate the vapour pressure when the latent heat is known or to calculate the latent heat from measurements of the vapour pressure. Since vaporization takes place at constant pressure, the latent heat of vaporization is given by $H_G - H_L$ (see § 3·1 (ii)). Hence equation (9·11·4) can be written in the alternative form

$$\frac{dP}{dT} = \frac{H_G - H_L}{T(v_G - v_L)}. \tag{9·11·5}$$

9·12. *The three phase-boundary curves.* There is nothing in the argument given in the preceding section which restricts us to liquids and gases. The equations (9·11·1)–(9·11·5) apply equally to the phase-boundary curves for equilibrium between vapour and solid, between solid and liquid or between two solid allotropic forms of a substance, when such exist. If we leave out of consideration for the moment the possibility of allotropic forms existing, there are three phase-boundary curves, the solid-liquid curve, the liquid-vapour curve and the solid-vapour curve, whose equations are respectively

$$g_S - g_L = 0, \quad g_L - g_G = 0, \quad g_S - g_G = 0,$$

where the suffix S refers to the solid. These three curves, which are often known as the melting curve, the vaporization curve and the sublimation curve respectively, intersect in the point given by

$$g_S(T, P) = g_L(T, P) = g_G(T, P), \tag{9·12·1}$$

which is known as the triple point. At the triple point the three phases can exist together in equilibrium in arbitrary proportions.

The general arrangement of the phase-boundary curves is shown in fig. 9·1, but various features are exaggerated in order to show some important points. (i) The vapour pressure of a solid decreases rapidly as the temperature is lowered, but it never actually vanishes except at the absolute zero. The sublimation curve therefore passes through the origin and touches the T axis there. (ii) The liquid-vapour curve ends abruptly at the critical point, so that there is no sharp division between the states that represent a liquid and those that represent a gas. We may, if we please, divide the portion of the P, T plane referring to liquid and gaseous states into three parts by the phase-boundary lines and by the line $T = T_c$, where

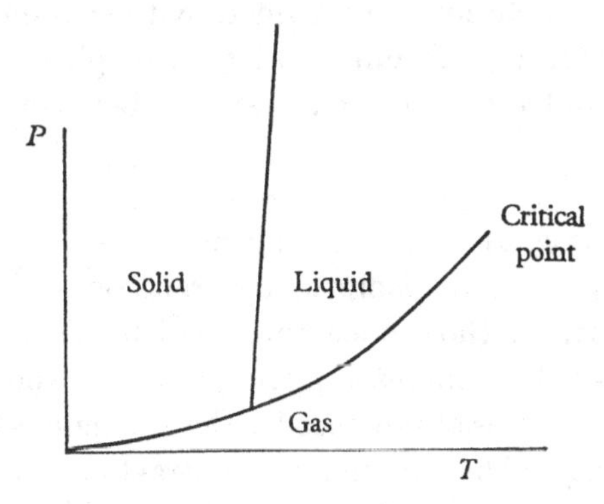

Fig. 9·1. The schematic arrangement of the phase-boundary curves.

T_c is the critical temperature. The states with $T > T_c$ can be called states of the gas, while those with $T < T_c$ belong either to the liquid or to the vapour, but this division is entirely arbitrary, and crossing the line $T = T_c$ does not correspond to any discontinuity in the properties of the substance. It is only when the liquid and the vapour exist in contact with one another that there is any absolute difference in their properties and that the liquid can be definitely distinguished from the vapour. Further, by continuity we can argue that states near the liquid-vapour boundary correspond to liquids when they lie above the boundary curve and to vapours when they lie below the boundary curve. But this distinction becomes completely vague and uninformative when we consider states very far removed from the phase-boundary curve, and there is no way in which we can distinguish between a highly compressed gas and a highly compressed liquid. It is always possible to transform a gas into a liquid without condensation taking place, that is, without a two-phase system being formed, by making the substance pass along a curve in the P, T plane which passes to the right of the critical point and which does not cross the liquid-vapour boundary curve. (iii) No critical point on the boundary curve separating the solid states from the fluid states has ever been found, and all the experimental evidence seems to show that this boundary curve continues indefinitely. This means that there is no possibility of transforming a fluid into a solid without two phases being formed.

It should be noted that if two phases can coexist down to $T=0$ the P, T curve must have a horizontal tangent there. For, by the third law, the entropies of the two phases must be the same at $T=0$, and hence $dP/dT=0$ at $T=0$ from (9·11·3). (An exceptional case could occur if the densities of the two phases were equal at $T=0$. The differential coefficient dP/dT would then have to be evaluated by l'Hôpital's rule. See § 9·41.)

Examples of this conclusion are provided by the vapour-pressure curves of all solids in stable equilibrium, and by the phase-boundary curves between graphite and diamond (fig. 9·3) and between solid and liquid helium (fig. 9·8).

9·13. *The phase diagram of water.* In general, solids expand on melting, but a few, notably ice and bismuth, contract on melting. Clapeyron's equation shows that, for those substances which expand on melting, the melting-point is raised by increasing the pressure, whereas the melting-point is lowered by increase of pressure for the anomalous substances which contract on melting. (The anomalous contraction can only occur for substances which have a very loose type of crystal structure in the solid state.)

The influence of pressure upon the melting-point of ice can be easily calculated from the following data: $L_{SL}=80$ cal./gram, $v_L=1$ cm.^{3}/gram, $v_S=1{\cdot}09$ cm.^{3}/gram at 0° C. These figures give $dT/dP=-0{\cdot}0074$° C./atmosphere; the measured value is $-0{\cdot}0075$° C./atmosphere. It also follows that the triple point of water is 0·0075° C., since it is the melting-point of ice at the natural vapour pressure, which is effectively zero.

To calculate the gradients of the sublimation curve and the vaporization curve at the triple point we require the following additional data: $L_{SG}=684$ cal./gram, $L_{LG}=604$ cal./gram, $v_G=205{,}000$ cm.^{3}/gram at the triple point, which is given by $P=4{\cdot}58$ mm. Hg and $T=0{\cdot}0075$° C. These give $dP_{SG}/dT=0{\cdot}384$ mm. Hg/degree and $dP_{LG}/dT=0{\cdot}339$ mm. Hg/degree.

The sublimation and vaporization curves for other substances are similar to those for water. The density of a vapour is always less than that of the solid or liquid from which it is derived, and dP/dT is always positive. Even for water the negative gradient of the melting curve does not continue indefinitely. (On *a priori* grounds it is extremely unlikely that the gradient of the melting curve should always be negative, since the melting curve would then cut the P axis, and at sufficiently large pressures the liquid would be the most stable form, however low the temperature.) At a pressure of 2115 kg./cm.^{2} a different kind of ice is formed, which has a melting curve of the normal type, and it has been found by Tammann (1900) and Bridgman (1937) that there are five and possibly six different kinds of ice, not all of which, however, can exist in contact with water. Ice, therefore, exhibits the phenomena of allotropy to a high degree.

9·14. *The phase diagram of sulphur.* As already mentioned in §7·61, sulphur has two crystalline forms, rhombic and monoclinic, the transition point for zero pressure occurring at 95·5° C. The full equilibrium diagram is shown schematically and much exaggerated in fig. 9·2, and it will be seen that there are three triple points which occur at $P = 10^{-6}$ atmospheres, $T = 95·5°$ C.; $P = 4 \times 10^{-6}$ atmospheres, $T = 119°$ C.; and $P = 1400$ atmospheres, $T = 153·7°$ C. It should be noted that the rhombic form cannot exist in contact with both the gas and the liquid in a stable state, but it is possible, as indicated by the dotted lines, to obtain the liquid and the monoclinic form in metastable supercooled states and for the rhombic form to exist in a superheated metastable state. There is, therefore, a metastable triple point where the rhombic form, the liquid and the gas are in equilibrium with one another.

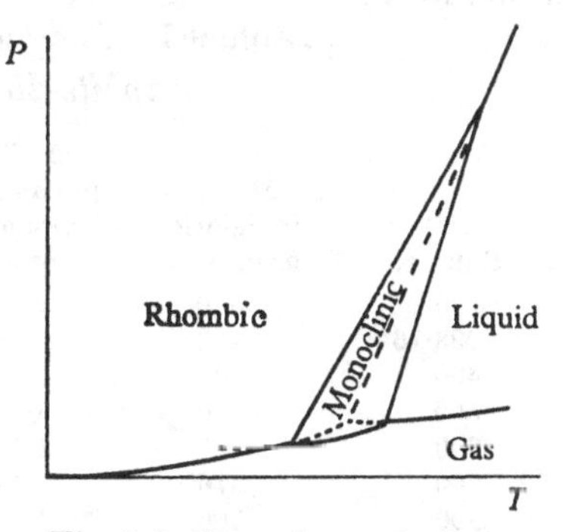

Fig. 9·2. The schematic phase diagram for sulphur.

9·15. *The phase diagram of carbon.* The possibility of transforming graphite, which is normally the stable form of carbon, into diamonds has for a very long time presented a problem of considerable scientific interest, on which there is an extensive literature. A survey of the previous results and a more accurate determination of the thermodynamic quantities involved have recently been published by Berman & Simon (1955). An outline of their conclusions is as follows.

The heat of transformation of graphite into diamond at 25° C. is 453 calories/gram atom at atmospheric pressure. This figure is obtained by measuring the heats of combustion at constant pressure of graphite and diamond to carbon dioxide. The difference in the two heats of combustion is the heat of transformation. Having obtained the difference

$$\Delta H(T, P) = H(T, P)_{\text{diamond}} - H(T, P)_{\text{graphite}}$$

for one particular temperature and for effectively zero pressure, we can obtain $\Delta H(T,0)$ for all temperatures from measurements of the specific heats of graphite and diamond over the temperature range concerned. At the same time, we can obtain the entropy functions $S(T,0)_{\text{diamond}}$ and $S(T,0)_{\text{graphite}}$ from the specific heats and from the fact that, by the third law of thermodynamics, we must have $S(0,P)_{\text{diamond}} = S(0,P)_{\text{graphite}}$. We can therefore also find $\Delta G(T,0) = \Delta H(T,0) - T\Delta S(T,0)$ as a function of T. The values of $\Delta H(T,0)$, $\Delta S(T,0)$ and $\Delta G(T,0)$ are given in table 9·1, and it will be seen that diamond becomes less stable as the temperature increases. On the other hand, since the atomic volume of diamond is considerably less than that of graphite, increasing pressure tends to make diamond the more stable form.

Table 9·1. *Thermodynamic data relating to the graphite-diamond transformation*

T in °K.	$\Delta H(T, 0)$ in calories/ gram atom	$-\Delta S(T, 0)$ in calories/ degree/gram atom	$\Delta G(T, 0)$ in calories/ gram atom	$V(T, 0)$ for graphite in cm.^{3}/gram atom	$V(T, 0)$ for diamond in cm.^{3}/gram atom
0	580	0	580	5·286	3·415
298·16	453	0·80	692	5·299	3·416
400	403	0·95	783	5·313	3·418
500	350	1·07	880	5·327	3·420
600	315	1·13	990	5·341	3·423
700	316	1·14	1110	5·356	3·427
800	300	1·15	1220	5·371	3·430
900	295	1·16	1340	5·387	3·434
1000	290	1·16	1450	5·404	3·438
1100	290	1·17	1570	5·421	3·443
1200	280	1·17	1685	5·438	3·447

In order to find $\Delta G(T,P)$ we have to compute the expression

$$\Delta G(T,P) = \Delta G(T,0) + \int_0^P \Delta V(T,P')\,dP',$$

where $\Delta V(T,0)$ is known from X-ray measurements and from determinations of the expansion coefficients (see the last two columns of table 9·1). The compressibilities cannot be ignored, and it is necessary to write

$$\Delta V(T,P) = \Delta V(T,0)\,(1 + AP + BP^2),$$

where A and B are known constants. A and B have been determined for pressures up to 100,000 atmospheres at room temperature, and it is assumed that they are independent of the temperature.

The phase-boundary curve is given by $\Delta G(T,P) = 0$, and the results obtained in this way are given in table 9·2. They are also given in fig. 9·3, extrapolated up to 3000° K. Above 1200° K. the equation of the curve is $P = 700 + 27T$, where P is measured in atmospheres.

Table 9·2. *The phase-boundary curve for the graphite-diamond transformation*

T in °K.	0	298·16	400	500	600	700
P in thousands of atmospheres	13·5	16·15	18·25	20·5	23	26
T in °K.	800	900	1000	1100	1200	
P in thousands of atmospheres	28·5	31·5	34	37	39·5	

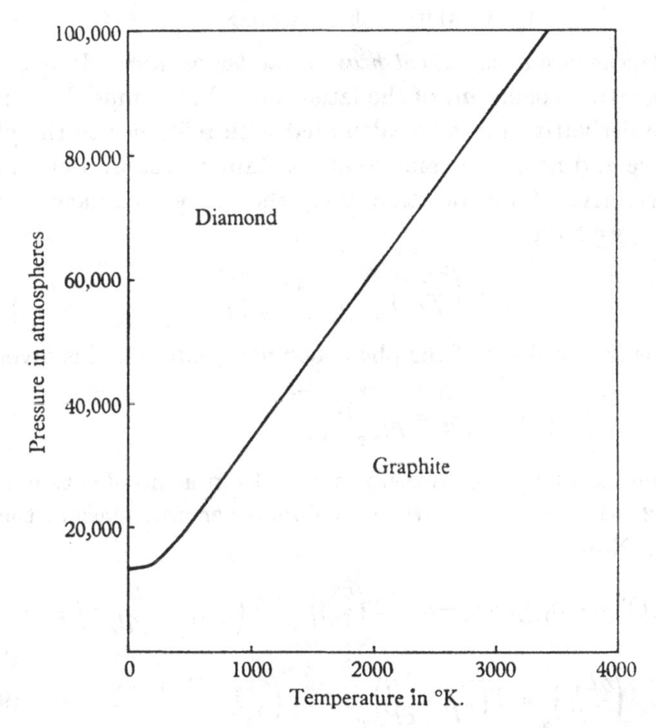

Fig. 9·3. The phase diagram of carbon.

Although diamond is thermodynamically stable at sufficiently high pressures, it by no means follows that at such pressures graphite will be transformed into diamond at a measurable rate, particularly at low temperatures. For example, Bridgman (1947) subjected graphite to pressures of the order of 400,000 atmospheres at room temperature, and to pressures of the order of 30,000 atmospheres at temperatures up to 3000° K. He was, however, only able to maintain the latter conditions for a few seconds, and in neither case did he succeed in making diamonds. The manufacture of diamonds can only be carried out if graphite can be subjected simultaneously to high pressures and to high temperatures for some considerable time. Improvements in the technique of contructing high-pressure vessels have recently made it possible to achieve a pressure of 100,000 atmospheres at temperatures of the order of 2300° K., and to subject graphite to these conditions for some hours (Bundy, Hall, Strong & Wentorf, 1955). In this way, small man-made diamonds have been obtained for the first time, a notable achievement more than a century after the first attempts to manufacture them were made.

Clausius's Equation

9·2. *The dependence of the latent heat on the temperature.* It is easy to find the temperature coefficient of the latent heat, but it must be borne in mind that the derivative must be calculated with reference to the phase-boundary curve and not, for example, at constant pressure. Denoting by D/DT the derivative of any function along the phase-boundary curve in the P, T plane, we have

$$\frac{DL_{12}}{DT}=\left(\frac{\partial L_{12}}{\partial T}\right)_P+\left(\frac{\partial L_{12}}{\partial P}\right)_T\frac{dP}{dT}, \tag{9·2·1}$$

where dP/dT is the gradient of the phase-boundary curve and is given by

$$\frac{dP}{dT}=\frac{L_{12}}{T(v_2-v_1)}. \tag{9·2·2}$$

Here L_{12} is the latent heat of transformation from a modification 1 to a modification 2, while v_1 and v_2 are the volumes per unit mass of the two modifications. Now

$$\left(\frac{\partial L_{12}}{\partial T}\right)_P=\frac{\partial}{\partial T}\{T(s_2-s_1)\}_P=s_2-s_1+T\left(\frac{\partial s_2}{\partial T}\right)_P-T\left(\frac{\partial s_1}{\partial T}\right)_P=\frac{L_{12}}{T}+C_{P,2}-C_{P,1}. \tag{9·2·3}$$

while

$$\left(\frac{\partial L_{12}}{\partial P}\right)_T=T\left(\frac{\partial s_2}{\partial P}-\frac{\partial s_1}{\partial P}\right)_T=-T\left(\frac{\partial v_2}{\partial T}-\frac{\partial v_1}{\partial T}\right)_P, \tag{9·2·4}$$

by the Maxwell equation (3·2·2). Hence

$$\frac{DL_{12}}{DT}-\frac{L_{12}}{T}=C_{P,2}-C_{P,1}-\frac{L_{12}}{v_2-v_1}\left\{\left(\frac{\partial v_2}{\partial T}\right)_P-\left(\frac{\partial v_1}{\partial T}\right)_P\right\}. \tag{9·2·5}$$

This equation is known as Clausius's equation. It may be used to calculate either DL_{12}/DT or one of the specific heats if all the other quantities are known.

To calculate C_P for water vapour at 100° C. and at the saturation pressure of 1 atmosphere, we require the following data: $L=539$ cal./gram, $DL/DT=-0{\cdot}64$ cal./gram/degree, $T=373°$ K., $v_G=1674$ cm.³/gram, $v_L=1$ cm.³/gram, $\partial v_G/\partial T=4{\cdot}813$ cm.³/gram/degree, $\partial v_L/\partial T=0{\cdot}001$ cm.³/

gram/degree, $C_{P,L}=1\cdot01$ cal./gram/degree. Equation (9·2·5) then gives $C_{P,G}=0\cdot47$ cal./gram/degree, which is the same as the value obtained by direct measurement.

9·21. *The specific heat of a saturated vapour.* In addition to the specific heats at constant volume and constant pressure, an important quantity for a vapour in contact with a liquid or a solid is the specific heat $C_{\text{sat.}}$ at constant saturation which is defined by

$$C_{\text{sat.}}=T\,Ds/DT. \tag{9·21·1}$$

If we distinguish the vapour by the suffix 2 and the other phase by the suffix 1, equation (9·21·1), in conjunction with (9·2·2), gives

$$C_{\text{sat.},2}=T\left(\frac{\partial s_2}{\partial T}\right)_P+T\left(\frac{\partial s_2}{\partial P}\right)_T\frac{dP}{dT}=C_{P,2}-\frac{L_{12}}{v_2-v_1}\left(\frac{\partial v_2}{\partial T}\right)_P. \tag{9·21·2}$$

There is a corresponding formula for $C_{\text{sat.},1}$. Equation (9·2·5) can therefore be put in the form

$$\frac{DL_{12}}{DT}-\frac{L_{12}}{T}=C_{\text{sat.},2}-C_{\text{sat.},1}, \tag{9·21·3}$$

as is obvious without calculation since

$$\frac{DL_{12}}{DT}=\frac{D}{DT}\{T(s_2-s_1)\}=s_2-s_1+T\left(\frac{Ds_2}{DT}-\frac{Ds_1}{DT}\right).$$

The two terms on the right of (9·21·3) are of the same order of magnitude for a vapour and are of opposite signs, and it is possible for $C_{\text{sat.},2}$ to be either positive or negative. The possibility of the occurrence of a negative specific heat is not surprising. If a saturated vapour is compressed adiabatically its temperature and pressure are both increased and its volume is diminished, but its state in the P, V diagram after the compression may lie either in the heterogeneous region or in the unsaturated region, depending upon whether the saturated-vapour curve in the P, V diagram is steeper or less steep than the adiabatic curve. In the first case it is necessary to add heat to the substance to keep the vapour saturated, and the specific heat is then positive, while in the latter case it is necessary to take away heat to keep the vapour saturated, and the specific heat is then negative. The data given in § 9·2 are sufficient to calculate $C_{\text{sat.},2}$ for steam at 100° C. from equation (9·21·2), or from equation (9·21·3), since for a liquid it can be assumed that $C_{\text{sat.}}=C_P$. The result is that $C_{\text{sat.},2}=-1\cdot07$ cal./gram/degree. Saturated water vapour therefore becomes supersaturated when it expands adiabatically.

These results are more easily understood by a consideration of the temperature-entropy diagram of the substance. It is clear from an inspection of fig. 8·17 that if the temperature of the saturated vapour increases its entropy must decrease, i.e. the specific heat of the saturated vapour is negative. If, on the other hand, the temperature-entropy curve is of the type shown in fig. 8·18, the specific heat of the saturated vapour will be positive for low temperatures and negative for high temperatures near the critical point.

The Vapour-Pressure Equation

9·3. For a vapour in contact with a solid (or with a liquid not too near the critical point) the Clapeyron equation (9·11·4) is usually integrated by making some simplifying assumptions. These are that the volume of the solid or liquid can be neglected in comparison with that of the vapour, and that the vapour behaves as a perfect gas. (It is essential to be clear as to what is meant by a vapour behaving as a perfect gas. It is obvious that a saturated vapour compressed isothermally cannot behave as a perfect gas since a portion of it would be condensed. But we are usually concerned with the behaviour of a vapour along the phase-boundary curve, and it is then not too bad an approximation to assume that the change in the properties of the vapour are given by the perfect gas laws. If necessary, corrections can be made by using one of the more exact equations of state such as Berthelot's, but we only apply them to the region of the P, T diagram corresponding to the vapour phase, or, in the limit, to the phase-boundary curve itself.)

It is usual to obtain the vapour-pressure equation by introducing an approximate expression for L based upon equation (9·2·5) into the Clapeyron equation (9·11·4) and integrating the latter. This is, however, a somewhat roundabout procedure (at any rate in the present exposition), since equation (9·11·4) is obtained by differentiating (9·11·1), so that by integrating again we can only arrive back at (9·11·1). It is therefore simpler to apply (9·11·1) direct. An explicit expression for G is given in equation (3·211·8), and this is evaluated for the gas phase (the perfect gas laws being assumed) in equations (3·52·4) or (3·52·5). In calculating G for the condensed phase, which, for definiteness, we consider to be a solid, it is sufficient to assume that C_P is independent of P and that v is independent of T and P. Also, since C_P is proportional to T^3 for solids for sufficiently small T (see § 6·24), we can take the lower limits of integration to be $T=0$ and $P=0$. We then obtain

$$G_S(T,P)=\int_0^T C_{P,S}(T')\left(1-\frac{T}{T'}\right)dT'+Pv_S+U_{0,S}-Ts_{0,S}, \qquad (9·3·1)$$

where $U_{0,S}$ and $s_{0,S}$ are the internal energy and entropy of unit mass of the solid when $T=0$, $P=0$. With $G_G(T,P)$ written in the form (3·52·4), the equilibrium condition $G_S(T,P)=G_G(T,P)$ gives the following equation for the vapour pressure P:

$$\log P = -\frac{L_0}{RT}+\frac{Pv_S}{RT}+\frac{C^0_{P,G}}{R}\log T+\frac{1}{RT}\int_0^T (C_{P,S}-C^1_{P,G})\left(1-\frac{T}{T'}\right)dT'+i, \tag{9·3·2}$$

where

$$L_0=U_{0,G}-U_{0,S}, \quad iR=s_{0,G}+R\log P^{\dagger}-C^0_{P,G}-C^0_{P,G}\log T^{\dagger}-s_{0,S}. \tag{9·3·3}$$

If we wish to take into account the departure of the vapour from the perfect gas laws we have merely to use the appropriate expression for G_G. For example, if the equation of state for the vapour is $Pv=RT+PB(T)$, then according to (8·5·3) and (8·8511·3) we have to replace $\log P$ in (9·3·2) by $\log P+PB/RT$.

Numerous empirical equations for the vapour-pressure curve have been proposed. Two of the simplest are

$$\log P = A - B/T \quad (B>0) \tag{9·3·4}$$

and

$$P = AT^r - B \quad (r>1). \tag{9·3·5}$$

Since these contain only two adjustable parameters, they can be expressed in reduced form. A variant upon (9·3·4), namely, the Antoine equation

$$\log P = A - B/(C+T), \tag{9·3·6}$$

is often used, but like the preceding equations it is only a good approximation over a limited temperature range, and for accurate work it is necessary to use an equation of the form

$$\log P = A - BT^{-1} + C\log T + DT. \tag{9·3·7}$$

The quantity i is called the vapour-pressure constant, and it depends upon the magnitudes of the constants $s_{0,G}$ and $s_{0,S}$. Now the vapour pressure of a solid is a perfectly definite physical quantity and so the expression (9·3·2), and consequently (9·3·3), cannot contain any arbitrary constants. This means that if we fix $s_{0,S}$, say, then $s_{0,G}$ cannot be assigned arbitrarily but must have a perfectly definite value such that i is independent of the particular choice of $s_{0,S}$. The value of i for any particular substance can be found by experiment or by a theoretical calculation based upon statistical mechanics.

9·31. *The statistical theory of the vapour-pressure equation.* Since we have already seen in Chapter 6 how to calculate the thermodynamic potential of a substance in the gaseous and in the solid phases, we can obtain an explicit expression for the vapour pressure which contains no arbitrary constants, provided that we can evaluate the partition functions. It is, however, essential that the thermodynamic potentials should depend in the correct manner upon the number of particles present, and though we automatically obtain the correct dependence if we start from equation (5·351·5) involving the partition function of the system, it is desirable to give an independent derivation of the formulae for the thermodynamic potentials.

Consider an assembly containing N molecules in a volume V, and let M_S of these molecules be in a solid phase and M_G in a gas phase, which we treat as perfect. Then, by (5·351·8) and (6·21·1), the free energy is of the form

$$F = -kT \log [\{z(T, V_S/M_S)\}^{M_s} Z_G(T, V-V_S)^{M_G}/M_G!] = -kT \log Z(M_S, M_G), \tag{9·31·1}$$

where Z_G is the partition function of a molecule in the gaseous phase and $z(T, V_S/M_S)^{M_s}$ is the partition function of the crystal. V_S is the volume of the crystal and $V-V_S$ is the volume available to the M_G molecules in the gaseous phase. The equilibrium value of F is determined, apart from terms which are insignificant for large M_S and M_G, by the maximum value of $Z(M_S, M_G)$ subject to the condition $M_S+M_G=N$. We therefore maximize $\log Z(M_S, M_G)+\lambda(M_S+M_G)$, which gives

$$\log z(T, V_S/M_S)+M_S\,\partial \log z/\partial M_S+\lambda=0, \quad \log Z_G-\log M_G+\lambda=0. \tag{9·31·2}$$

Hence, by (6·21·3) and (6·21·5),

$$kT\lambda = -kT\log z + PV_S/M_S, \quad kT\lambda = kT \log (M_G/Z_G), \tag{9·31·3}$$

and, by (5·34·7), (6·21·1) and (6·21·6), these equations are equivalent to

$$\mu_S = \mu_G. \tag{9·31·4}$$

This verifies that our definitions of μ_S and μ_G are correct.

The relation $\mu_S=\mu_G$, together with the condition $M_S+M_G=N$, determines all the properties of the combined assembly of the crystal and the vapour.

9·311. *The vapour-pressure equation for substances consisting of polyatomic molecules.* Although the preceding theory of the vapour-pressure equation applies to any type of molecule, the results are obtained in a form which is often inconvenient. Except for hydrogen, measurements are limited in practice to temperature regions where the rotational states of diatomic and of polyatomic molecules are fully excited, and there is therefore something to be said for regrouping the terms in equation (9·3·2) so as to give $C^0_{P,G}$ its normal high-temperature value. This involves making corresponding changes in $C^1_P(T)_G$ and i, so as to make the whole equation invariant. (We must, however, then be careful not to use the modified equation at temperatures lower than those for which it is valid.)

To make the changes required in equation (9·3·2) we split $C^1_P(T)_G$ into two parts, the rotational contribution $C(T)_{\text{rot.}}$ and the remainder. If T lies in the region where the rotations are classical and where $C(T)_{\text{rot.}}$ is a constant $C_{\text{rot.}}^{\text{class.}}$, we write

$$\int_0^T C(T')_{\text{rot.}}\left(1-\frac{T}{T'}\right)dT' = \int_0^{T_0} + \int_{T_0}^T C(T')_{\text{rot.}}\left(1-\frac{T}{T'}\right)dT', \quad (9\cdot311\cdot1)$$

where T_0 also lies in the region where the rotational specific heat is constant. The expression (9·311·1) is therefore of the form $A+BT-C_{\text{rot.}}^{\text{class.}}\,T\log T$, and if we substitute this back into (9·3·2) we obtain an equation of exactly the same form, but now $C_{P,G}^0$ must be taken as being the classical value of the translational and rotational specific heats, namely, $\frac{7}{2}k$ for diatomic and linear polyatomic molecules, and $4k$ for non-linear molecules, while $C_P^1(T)_G$ is the vibrational and electronic specific heat. The quantity i is also changed, but L_0 is not. It does not seem possible to deduce this last result, which is equivalent to putting $A=0$, for a rotational specific heat with an arbitrary temperature variation. This conclusion, however, follows from the known form of the rotational partition function (see below).

To make the corresponding changes in the theoretical vapour-pressure equation we use the fact that the internal partition function $Z_{\text{int.}}$ makes a contribution $-kT\log Z_{\text{int.}}$ to the thermodynamic potential μ per molecule. (If we measure μ per gram molecule we have to replace k by R.) If we write $Z_{\text{int.}}=Z_{\text{rot.}}Z_{\text{vib.}}Z_{\text{el.}}$ we can split the contribution to μ into two separate parts, and if the temperature is such that the rotations are classical we can use the formulae derived in § 6·17. In all cases $Z_{\text{rot.}}=\gamma T^r$, where γ is a constant, so that in the modified equation (9·3·2) we have to increase C_P^0 by rk and increase i by $\log\gamma$, while $C_P^1(T)$ now only refers to the vibrational specific heat (and the electronic specific heat if there is any). The various explicit formulae are as follows:

Monatomic gases:

$$C_P^0=\tfrac{5}{2}k, \quad i=\log\left(\frac{(2\pi m)^{\frac{3}{2}}k^{\frac{5}{2}}}{h^3}\frac{\varpi_G}{\varpi_S}\right). \quad (9\cdot311\cdot2)$$

Diatomic and linear polyatomic gases:

$$C_P^0=\tfrac{7}{2}k, \quad i=\log\left(\frac{(2\pi m)^{\frac{3}{2}}k^{\frac{5}{2}}}{h^3}\frac{8\pi^2 kI}{h^2}\frac{1}{\sigma}\frac{\varpi_G}{\varpi_S}\right). \quad (9\cdot311\cdot3)$$

Non-linear molecules:

$$C_P^0=4k, \quad i=\log\left(\frac{(2\pi m)^{\frac{3}{2}}k^{\frac{5}{2}}}{h^3}\frac{8\pi^2(2\pi k)^{\frac{3}{2}}}{h^3}\frac{(I_1I_2I_3)^{\frac{1}{2}}}{\sigma}\frac{\varpi_G}{\varpi_S}\right). \quad (9\cdot311\cdot4)$$

Here ϖ_G is the weight of the electronic ground state of the molecule while ϖ_S is defined by $s_{0,S}=k\log\varpi_S$.

9·32. *The vapour-pressure constant and the third law of thermodynamics.* The vapour-pressure equation involves two separate problems, the temperature and pressure dependence of the heat functions of the gaseous and condensed phases, and the entropy constants. The heat functions of the two phases can be measured by direct experiment, or they can be calculated to a high degree of accuracy by the methods given in Chapter 6. It therefore only remains to consider the vapour-pressure constant. The value of i for any substance can be found by an analysis of the experimental results, and a theoretical value of i is provided by the formulae of § 9·311, in which the only unknown quantity is ϖ_S. We can therefore calculate i if we know the entropy constant $s_{0,S}$ of the solid. Now the entropy constant of the gas has been chosen so as to make the entropy of the gas equal to zero at $T=0$ when the proper quantal formulae are applied (and these are the only formulae valid near $T=0$). Hence, by Nernst's principle, $s_{0,S}$ must be taken to zero for every solid in a state of absolute thermodynamic stability. The present problem is therefore a variation upon the comparison between the spectroscopic and the calorimetric entropies of a gas, which was discussed in detail in § 7·4. Nothing new can emerge from an examination of the vapour-pressure constant, but historically it preceded the comparison of the spectroscopic and the calorimetric entropies, and it was largely due to Nernst's preoccupation with the vapour-pressure equation and the related problem of the chemical equilibrium constants that he was led to formulate the third law of thermodynamics.

In general, we omit the nuclear spin weights since they are normally the same for the solid and for the gas. Then if we confine our attention to temperatures such that the molecular rotations in the gas are classical, the values of i are given by the expressions (9·311·2), (9·311·3) and (9·311·4). If $\exp i$ is measured in atmospheres, and if M^* is the molecular weight, (9·311·1) can be written as

$$i^* = i/\log_e 10 = -1{\cdot}587 + \tfrac{3}{2}\log_{10} M^* + \log_{10}(\varpi_G/\varpi_S), \qquad (9{\cdot}32{\cdot}1)$$

and i^* is called the vapour-pressure constant in practical units. This formula must be replaced by

$$i^* = 36{\cdot}815 + \tfrac{3}{2}\log_{10} M^* + \log_{10}(I/\sigma) + \log_{10}(\varpi_G/\varpi_S) \qquad (9{\cdot}32{\cdot}2)$$

for linear molecules, and by

$$i^* = 56{\cdot}265 + \tfrac{3}{2}\log_{10} M^* + \tfrac{1}{2}\log_{10}(I_1 I_2 I_3/\sigma^2) + \log_{10}(\varpi_G/\varpi_S) \quad (9{\cdot}32{\cdot}3)$$

for non-linear molecules.

Some measured values of i^* and the corresponding values of ϖ_S deduced from (9·32·1), (9·32·2) and (9·32·3) are given in table 9·3. The values of ϖ_S are the same as those discussed in § 7·5 and, with the exception of the value for Na, explanations have already been given. The sodium atom in the gaseous phase is in a $^2S_{\frac{1}{2}}$ state with weight 2. In the metallic state, the valency electrons are free and occupy conduction levels with paired spins (cf. § 6·32). We must therefore ascribe weight unity to the metal, in agreement with the value given in the table.

Table 9·3. *Values of the vapour-pressure constant* i^*

	i^*	M^*	ϖ_G	ϖ_S
He	−0·68	4	1	1
Ne	0·39	20	1	1
A	0·81	39	1	1
Na	0·76	23	2	1
N_2	−0·16	28	1	1
O_2	0·55	32	3	1
CO	−0·07	28	1	2

9·321. *The vapour-pressure constant of hydrogen.* Since the triple point of hydrogen occurs at $T = 13{\cdot}94°$ K., the vapour pressure of hydrogen is normally observed at temperatures where the molecules are in their lowest rotational states, with $j = 0$ for para- and $j = 1$ for ortho-molecules. (The vapour pressure of the solid can be measured at higher temperatures, but the pressures required are then very high and the gas cannot be treated as being nearly perfect.) In deriving the vapour-pressure equation for low temperatures it is therefore necessary to treat hydrogen as consisting of a mixture of two types of 'monatomic' molecules, and a special investigation is required.

According to the arguments given in § 7·52, the residual entropy of gaseous hydrogen is $k \log 4 + \frac{3}{4} k \log 3$ per molecule if the nuclear weights are included, and $\frac{3}{4} k \log 3$ if they are excluded. The weight to be assigned to the solid depends upon how its specific heat is defined near $T = 0$. Since the specific heat of a solid is small near $T = 0$, measurements of it are difficult and not very reliable. It is therefore usual to measure the specific heat down to say 15 or 12° K. and then to extrapolate it to $T = 0$ by using Debye's T^3 law. Now in the discussion of the entropy of solid hydrogen given in § 7·52, it appeared that at 12° K. the molecules are still rotating and that the ortho-molecules have weight 9, including the spin weight. If, therefore, we extrapolate the specific heat curve in fig. 7·5 from 12° K., the residual entropy of the solid is $k \log 4 + \frac{3}{4} k \log 3$ per molecule if the spin weight is included, and this is the same as that of the gas. It then follows that if we insert the specific heat so derived into equation (9·3·2) we must put $\varpi_G = \varpi_S$ in (9·311·2), and hence that i^* is given by $i^* = -1{\cdot}587 + \frac{3}{2} \log_{10} M^*$. A similar argument applies to deuterium (and also to HD, but here the argument is straightforward since there is only one species of molecule to consider, and ϖ_G and ϖ_S are both unity, excluding the nuclear spin weights). The calculated and observed values of i^* are in excellent agreement, being

$-1{\cdot}13$ and $-1{\cdot}09$ respectively for hydrogen, and $-0{\cdot}68$ and $-0{\cdot}67$ respectively for deuterium.

If the specific heat of solid hydrogen is measured below 12° K. and not merely extrapolated to zero by means of Debye's formula, the hump in the specific heat due to the splitting of the ortho-rotational levels by the crystalline field will appear (cf. § 7·521 and fig. 7·6). If we use this specific heat in the vapour-pressure equation, we must then ascribe a lower residual entropy to the solid than to the gas, and we should have to take $\varpi_S = 4$ for both hydrogen and deuterium, whereas the weights in the gas phase are $4 \times 3^{\frac{3}{4}}$ and $4 \times 3^{\frac{1}{3}}$ respectively. In this case, therefore, we have $i^* = -0{\cdot}77$ for hydrogen and $i^* = -0{\cdot}52$ for deuterium.

At high temperatures, gaseous hydrogen behaves as if it were composed of diatomic molecules with weight unity, excluding the nuclear spin weight 4, and with a symmetry number $\sigma = 2$. Its vapour-pressure constant at temperatures above 300° K. is therefore quite different from its low-temperature value. If i^* is obtained by extrapolating the specific heat of the solid to zero from 12° K., the high-temperature values of i^* (for a perfect gas) would be $-3{\cdot}73$ for hydrogen, $-2{\cdot}77$ for deuterium and $-2{\cdot}68$ for HD. At such high temperatures, however, the physically significant quality is the chemical constant and not the vapour pressure constant (see § 11·61).

The Properties of Helium

9·4. Of all the gases, helium is the one that most nearly approaches being a perfect gas, but at very low temperatures it possesses some very peculiar properties. In particular, a new type of change of state occurs, called a transition of the second order.

Helium is the most difficult of all gases to liquefy, its critical temperature being 5·20° K. and its critical pressure being 2·26 atmospheres. At atmospheric pressure it boils at 4·216° K. Liquid helium was first prepared by Kamerlingh Onnes in 1908, but all his attempts to solidify it failed completely, and it was not until 1926 that solid helium was obtained by Keesom, who made the remarkable discovery that solid helium can only exist under a pressure of approximately 22 atmospheres at least, and that liquid and not solid helium is the stable form at the absolute zero.

The usual method of solidifying a liquid at low temperatures is to make it boil under reduced pressure in a thermally insulated vessel. Since the latent heat of vaporization has to be supplied by the liquid itself, its temperature falls and the liquid that remains unevaporated finally solidifies. The methods employed to solidify helium are quite different. One method is to force the liquid slowly through a fine capillary under a steadily increasing pressure and to observe the pressures at both ends of the capillary. When the pressure at the far end of the capillary remains constant when the pressure at the inlet end is increased, the capillary must be blocked by solid helium. By such experiments, Simon, Ruhemann & Edwards (1929) were able to observe the melting curve of solid helium up to a pressure of 5400 atmospheres, and Holland, Hugill & Jones (1951) up to 7500 atmospheres, its melting-point being then about 50° K.

Since the solid-fluid boundary curve of helium in the P, T plane always lies above the liquid-vapour curve, there is no triple point at which the solid, liquid and vapour can coexist. There is, however, another modification of helium to be considered. When liquid helium is cooled by boiling under reduced pressure, a peculiar change takes place at about 2·2° K. Above this temperature boiling takes place normally, but at lower temperatures no bubbles are formed although the liquid continues to evaporate rapidly. It was suggested by Keesom & Wolfke in 1927 that a phase change of a peculiar kind takes place at this temperature, ordinary liquid helium (liquid helium I) changing into another modification of liquid helium (liquid helium II). Since that date the properties of liquid helium II have been investigated by many workers, and the results of their investigations may be summed up in the statement that liquid helium II is a superfluid.

The transition between liquid helium I and liquid helium II in the presence of helium vapour, takes place at a pressure of 38·3 mm. Hg and a temperature of 2·186° K. This point in the P, T diagram is called the λ-point. The transition can, however, take place without the vapour being present, and the transition points form a curve in the P, T diagram, known as the λ-curve, extending from the liquid-vapour curve to the solid-liquid curve and intersecting the latter at $P = 29{\cdot}96$ atmospheres, $T = 1{\cdot}778°$ K. The complete phase diagram is shown, schematically and much exaggerated, in fig. 9·4. It will be seen that, though no triple point exists in the ordinary sense, there are two points at which three phases are in equilibrium. At the first, the three phases are helium vapour, liquid helium I and liquid helium II, while at the second the three phases are solid helium, liquid helium I and liquid helium II.

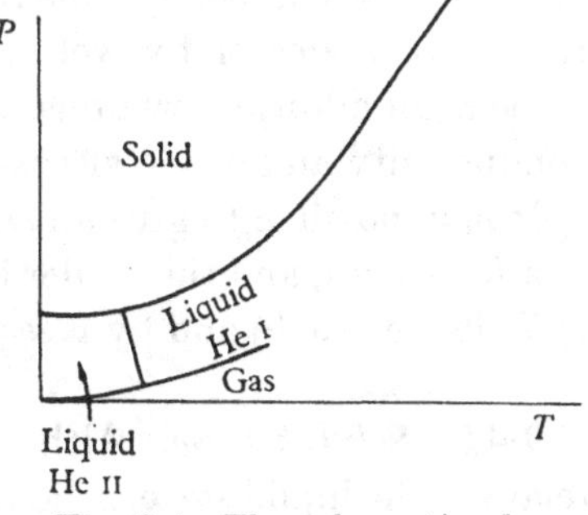

Fig. 9·4. The schematic phase diagram of helium.

The main peculiarity in the transition from liquid helium I to liquid helium II at any point of the λ-curve is that it is not associated with a latent heat or with a change in volume. Normally, when a substance is being cooled and is changing from modification 1 to modification 2 the transition does not take place instantaneously. When the temperature drops to the transition temperature a small amount of modification 2 appears and a heterogeneous mixture is formed. The transition then takes place at constant temperature, the composition of the mixture gradually changing as the latent heat of transition is given out until the substance consists entirely of modification 2, when the temperature begins to fall once more. With the transition from liquid helium I to liquid helium II the phenomenon is quite different. Since there is no latent heat to be given out, the cooling takes place continuously and no heterogeneous mixture is formed. All that happens is that the rate of cooling changes abruptly as the λ-curve is crossed, but there is no period of constant temperature, and at any moment the fluid consists either wholly of liquid helium I or wholly of liquid helium II. The two triple points are therefore not triple points in the ordinary sense, since, although theoretically three phases can coexist in equilibrium, in practice there will only be two phases, and liquid helium I and liquid helium II are never found in contact with one another.

Since the transition from liquid helium I to liquid helium II is not accompanied by a latent heat or by a change in volume it cannot be considered to be a normal phase transition. But although the entropy S and the volume V are continuous, it is found that their derivatives, the specific heat at constant pressure and the expansion coefficient, are discontinuous. As a consequence, other related properties are discontinuous when the λ-curve is crossed, a list of these being as follows: the compressibility, the velocity of sound, and the temperature coefficients of the latent heats of fusion and vaporization, of the dielectric constant and of the refractive index. The discontinuities in all these quantities can be related to the discontinuities in the specific heat and the expansion coefficient, either by thermodynamics or by well-established theories of the phenomena concerned. In addition, there is a very marked discontinuity in the thermal conductivity and in the viscosity.

It is impossible to give an adequate account of the properties of helium in a few pages, and the reader is referred to the treatises by Burton, Smith & Wilhelm (1940) and by Keesom (1942) for fuller details.

9·41. *Solid helium.* Although the more unusual features of helium relate to the liquid state, solid helium has a considerable interest of its own. In many respects it is the simplest solid known, but, on the other hand, quantum effects, and in particular the zero-point energy of the lattice vibrations, play a dominant part in determining the properties of the solid and of the liquid. The various thermodynamic properties of solid helium have recently been measured with considerable accuracy by Dugdale & Simon (1953), whose results supplement those given in Keesom's book. Some of the more important properties are given below.

(i) The melting temperature as a function of the pressure is given in table 9·4. It will be seen that, in accordance with the third law of thermodynamics, $dP/dT \to 0$ as $T \to 0$.

Table 9·4. *The melting temperature of helium as a function of the pressure*

P in atmospheres	25·27	25·32	25·50	25·81	26·32	27·13	28·39	29·3
T in °K.	1·15	1·20	1·30	1·40	1·50	1·60	1·70	1·75
P in atmospheres	48·5	108·8	141	285	423	590	790	1160
T in °K.	2·29	3·59	4·23	6·5	8·20	10·17	12·2	15·5
P in atmospheres	1800	2380	2850	3540	4700	5400	6170	7270
T in °K.	20·4	24·1	27·9	32·4	38·4	42·0	45	50

(ii) The entropies of solid and liquid helium along the melting curve are shown in fig. 9·5. It will be seen that there is an entropy difference of about 1·5 calories/degree/gram atom between solid and liquid helium for temperatures above 2° K., but this rapidly drops to zero as soon as liquid helium II becomes the relevant liquid phase.

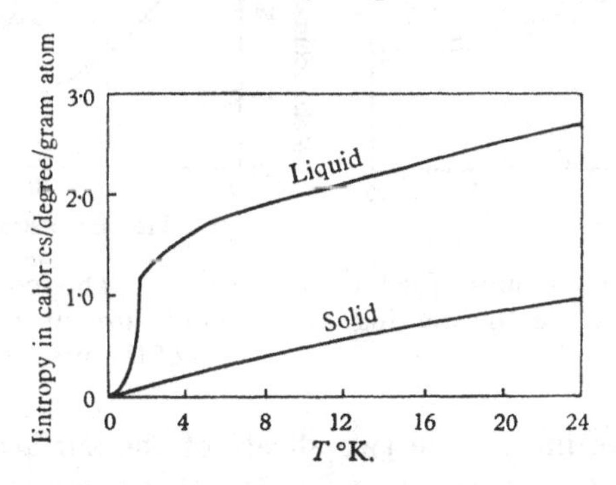

Fig. 9·5. The entropy of solid and liquid helium along the melting curve.

(iii) The molar volumes of the solid and of the liquid along the melting curve are shown in fig. 9·6.

(iv) There is a phase transition in the solid, the triple point for the two solid phases and the liquid phase occurring at 14·9° K. and at about 1100 atmospheres. It is probable, but not certain, that the high-temperature form has a face-centred cubic structure and that the low-temperature modification has a hexagonal close-packed structure.

(v) Solid helium is highly compressible, the molar volume at 0° K. as a function of the pressure being as shown in fig. 9·7.

9·42. *The thermodynamics of the transition from liquid helium I to liquid helium II.* As discussed in § 9·11 the normal curve defining a phase boundary between the regions in the P, T plane belonging to two modifications I and II is given by $G_{\mathrm{I}}(T,P)=G_{\mathrm{II}}(T,P)$, the differential form of which is Clapeyron's equation (9·11·4) or (9·11·5). For the transition between the two modifications of liquid helium, however, Clapeyron's equation becomes nugatory, since it gives dP/dT as the undetermined form 0/0. We now assume that the λ-curve is defined by the conditions that

$$s_{\mathrm{I}}(T,P)=s_{\mathrm{II}}(T,P), \quad v_{\mathrm{I}}(T,P)=v_{\mathrm{II}}(T,P) \tag{9·42·1}$$

all along the λ-curve (Ehrenfest, 1933; Keesom, 1933). That is, G and its first derivatives are continuous across the λ-curve but the second derivatives are discontinuous. If we draw the surface $G=G(T,P)$ with G as a

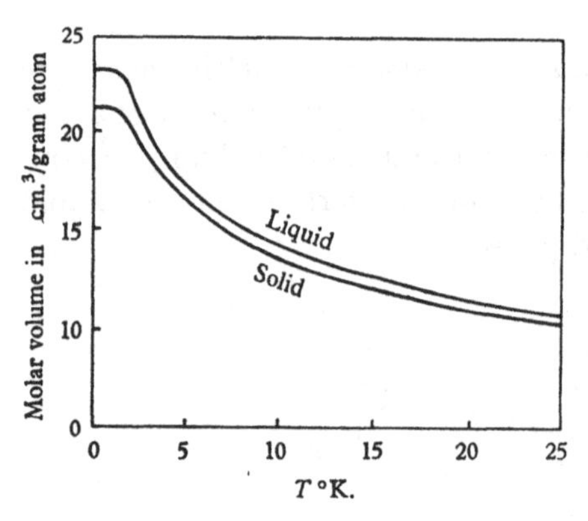

Fig. 9·6. The molar volumes of solid and liquid helium along the melting curve.

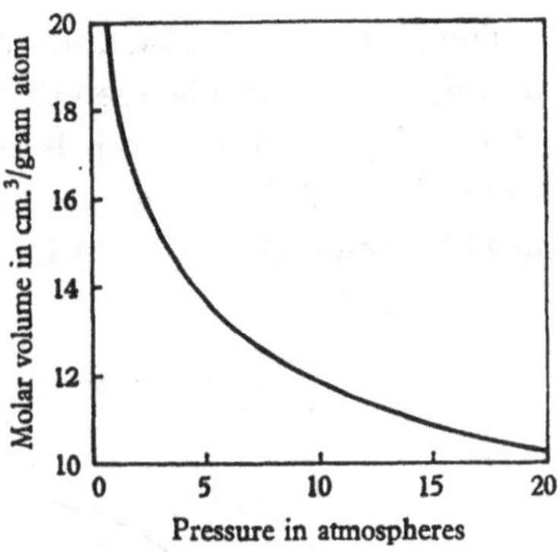

Fig. 9·7. The molar volume of solid helium at 0° K. as a function of the pressure.

third rectangular coordinate, the two sheets of the surface have contact of the second order all along the λ-curve. For this reason the transition is said to be of the second order.

If we consider an infinitesimal displacement dT, dP along the λ-curve, the first of the equations (9·42·1) give

$$ds_{\mathrm{I}}=\left(\frac{\partial s_{\mathrm{I}}}{\partial T}\right)_P dT+\left(\frac{\partial s_{\mathrm{II}}}{\partial P}\right)_T dP=ds_{\mathrm{II}}=\left(\frac{\partial s_{\mathrm{II}}}{\partial T}\right)_P dT+\left(\frac{\partial s_{\mathrm{II}}}{\partial P}\right)_T dP. \tag{9·42·2}$$

Now $$(\partial s/\partial T)_P=C_P/T, \quad (\partial s/\partial P)_T=-(\partial v/\partial T)_P,$$

by equation (3·2·2). Hence the differential equation giving the λ-curve is

$$\frac{dP}{dT}=\frac{\{\partial(s_{\mathrm{II}}-s_{\mathrm{I}})/\partial T\}_P}{\{\partial(v_{\mathrm{II}}-v_{\mathrm{I}})/\partial T\}_P}=\frac{C_{P,\mathrm{II}}-C_{P,\mathrm{I}}}{vT(\alpha_{\mathrm{II}}-\alpha_{\mathrm{I}})}, \tag{9·42·3}$$

where
$$\alpha = \frac{1}{v}\left(\frac{\partial v}{\partial T}\right)_P \tag{9·42·4}$$

is the expansion coefficient. The equation (9·42·3) takes the place of Clapeyron's equation (9·11·3). It can be derived directly from the latter by treating the right-hand side of (9·11·3) as an undetermined form and applying l'Hôpital's rule.‡

The fact that v is continuous across the λ-curve gives us another equation for dP/dT. For an infinitesimal displacement along the λ-curve,

$$dv_{\mathrm{I}} = \left(\frac{\partial v_{\mathrm{I}}}{\partial T}\right)_P dT + \left(\frac{\partial v_{\mathrm{I}}}{\partial P}\right)_T dP = dv_{\mathrm{II}} = \left(\frac{\partial v_{\mathrm{II}}}{\partial T}\right)_P dT + \left(\frac{\partial v_{\mathrm{II}}}{\partial P}\right)_T dP. \tag{9·42·5}$$

Hence
$$\frac{dP}{dT} = \frac{\alpha_{\mathrm{II}} - \alpha_{\mathrm{I}}}{\kappa_{\mathrm{II}} - \kappa_{\mathrm{I}}}, \tag{9·42·6}$$

where
$$\kappa = -\frac{1}{v}\left(\frac{\partial v}{\partial P}\right)_T \tag{9·42·7}$$

is the isothermal compressibility.

Combining the two equations (9·42·3) and (9·42·6) we obtain the relation

$$C_{P,\mathrm{II}} - C_{P,\mathrm{I}} = vT(\alpha_{\mathrm{II}} - \alpha_{\mathrm{I}})^2/(\kappa_{\mathrm{II}} - \kappa_{\mathrm{I}}) \tag{9·42·8}$$

connecting the discontinuities in the specific heat, the expansion coefficient and the compressibility across the λ-curve.

9·421. The experimental arrangements are such that at the λ-point the measurements are carried out at the normal vapour pressure of helium and not at constant pressure. To derive the formulae in this case, let γ denote any line in the P, T plane. Then the derivative $(D\Phi/DT)_\gamma$ of any function Φ in the direction γ is given by

$$\left(\frac{D\Phi}{DT}\right)_\gamma = \left(\frac{\partial \Phi}{\partial T}\right)_P + \left(\frac{\partial \Phi}{\partial P}\right)_T \left(\frac{dP}{dT}\right)_\gamma, \tag{9·421·1}$$

where $(dP/dT)_\gamma$ is the gradient of the line γ. Hence if λ denotes the direction of the λ-curve, then by eliminating $\partial\Phi/\partial P$ from equation (9·421·1) and the corresponding equation for the direction λ, we have

$$\left(\frac{D\Phi}{DT}\right)_\gamma = \frac{(dP/dT)_\gamma}{(dP/dT)_\lambda}\left(\frac{D\Phi}{DT}\right)_\lambda + \left(1 - \frac{(dP/dT)_\gamma}{(dP/dT)_\lambda}\right)\left(\frac{\partial \Phi}{\partial T}\right)_P. \tag{9·421·2}$$

Now since s and v are continuous across the λ-curve, we have

$$(D\Delta s/DT)_\lambda = 0, \quad (D\Delta v/DT)_\lambda = 0,$$

‡ If $f(0) = g(0) = 0$, and if one of $f'(0)$ and $g'(0)$ is not zero, then

$$\lim_{x\to 0} \frac{f(x)}{g(x)} = \frac{f'(0)}{g'(0)}.$$

where $\Delta s = s_{II} - s_I$, $\Delta v = v_{II} - v_I$, and so

$$\left(\frac{D\Delta s}{DT}\right)_\gamma = \left(1 - \frac{(dP/dT)_\gamma}{(dP/dT)_\lambda}\right)\left(\frac{\partial \Delta s}{\partial T}\right)_P, \quad \left(\frac{D\Delta v}{DT}\right)_\gamma = \left(1 - \frac{(dP/dT)_\gamma}{(dP/dT)_\lambda}\right)\left(\frac{\partial \Delta v}{\partial T}\right)_P. \tag{9·421·3}$$

We may therefore write (9·42·3) in the form

$$\left(\frac{dP}{dT}\right)_\lambda = \frac{(D\Delta s/DT)_\gamma}{(D\Delta v/DT)_\gamma}. \tag{9·421·4}$$

This equation applies to any direction γ and to any point on the λ-curve. If we apply it at the λ-point, then, since the gradient of the vapour-pressure curve is continuous there‡ we may take γ to be the direction of the vapour-pressure curve and in this case (9·421·4) becomes

$$\left(\frac{dP}{dT}\right)_\lambda = \frac{\Delta C_{\text{sat.}}}{vT\Delta\alpha_{\text{sat.}}}. \tag{9·421·5}$$

This equation can be used instead of (9·42·3), but the equations (9·42·6) and (9·42·7) cannot be put into a similar form.

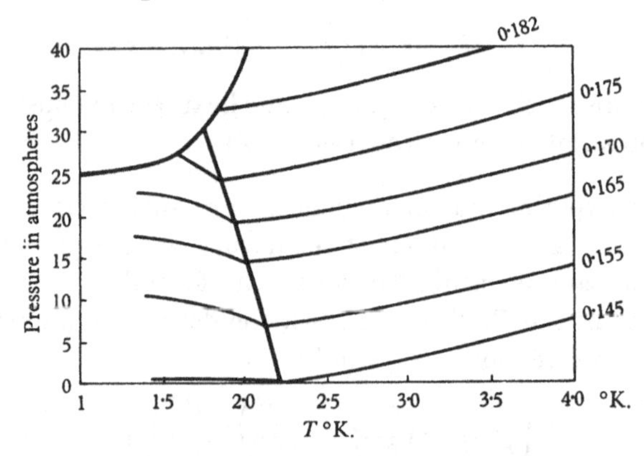

Fig. 9·8. The isochores of liquid helium. Densities in grams/cm.³.

9·43. Some of the more unusual properties of liquid helium are shown in figs. 9·8 and 9·9. The curves in fig. 9·8 show that $(\partial P/\partial T)_V < 0$ for liquid helium II, while those in fig. 9·9 show that the thermal expansion coefficient is also negative, as it must be. The fact that $(\partial P/\partial T)_V$ and $(\partial v/\partial T)_P$ are negative to the left of the λ-line and are positive to the right makes it impossible for us to represent the equation of state in a single P, V diagram, and, as explained at the end of § 4·2, we must consider the P, V diagram as consisting of two sheets which are joined along the λ-line. This multi-

‡ This follows from Clapeyron's equation since the latent heat is continuous; but d^2P/dT^2 is discontinuous.

valuedness can be avoided by using an entropy-volume diagram, or, better, by constructing the $U(S, V)$ surface.

Fig. 9·10 gives some values of the specific heat, and the curves clearly show the presence of the discontinuity, though its magnitude is by no means

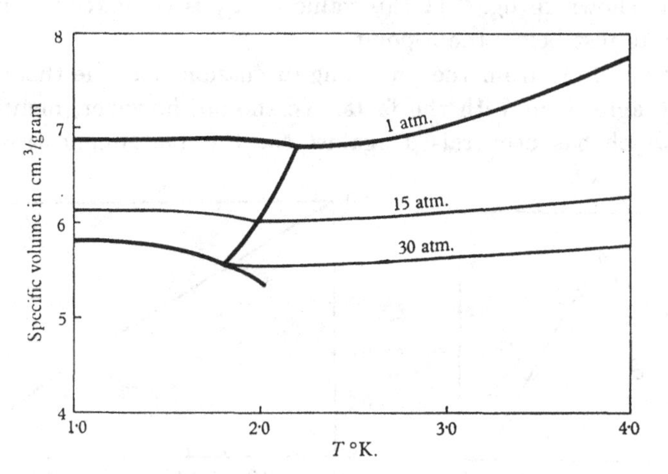

Fig. 9·9. The isobars of liquid helium.

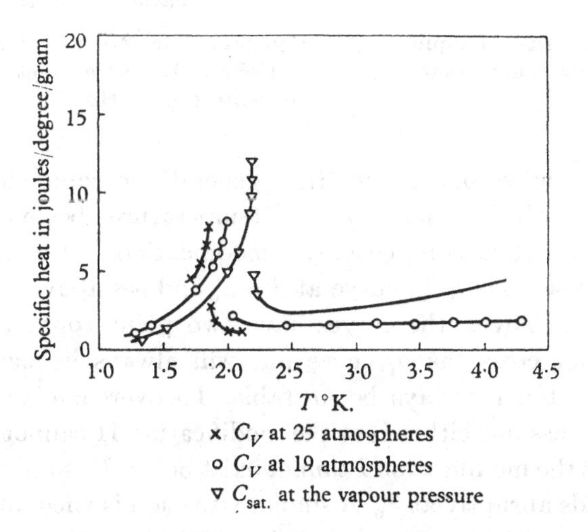

Fig. 9·10. The specific heat of liquid helium.

easy to determine accurately. The value $\Delta C_{sat.} = 7{\cdot}95$ joules/gram/degree is the generally accepted value at the λ-point, though in some experiments a much higher value was obtained.

Fig. 9·11 gives the density of liquid helium at the saturation vapour pressure. The density of liquid helium II near the λ-point is not easy to measure, but equation (9·421·5) can be used to calculate $\Delta\alpha_{sat.}$ and hence α_{II}. The measured value of dP/dT is $-8{\cdot}19\times10^7$ dynes cm.^{2} degree^{-1}, and $1/v=0{\cdot}1462$ g. cm.$^{-3}$. Hence $\Delta\alpha_{sat.}=-0{\cdot}0648$ degree^{-1}. The measured value of α_I is 0·0222 degree^{-1}, so that the calculated value of α_{II} is $-0{\cdot}0426$ degree^{-1}. As shown in fig. 9·11 this value of α_{II} is compatible with the measurements just below the λ-point.

We may conclude from the preceding discussion that the theory is in satisfactory agreement with the facts. We should, however, mention an objection which has been raised against the thermodynamic treatment

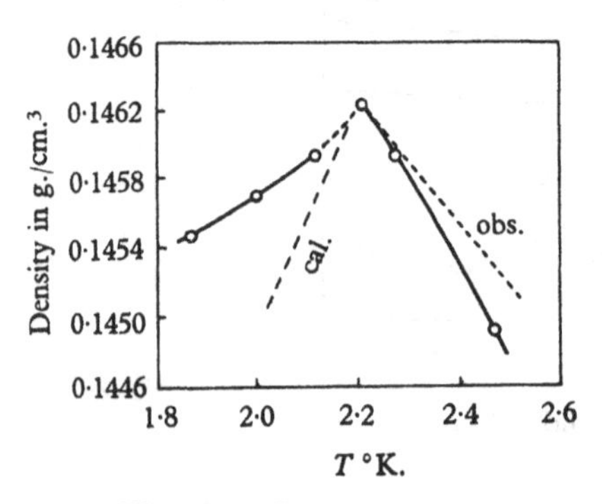

Fig. 9·11. The density of liquid helium at the saturation pressure.

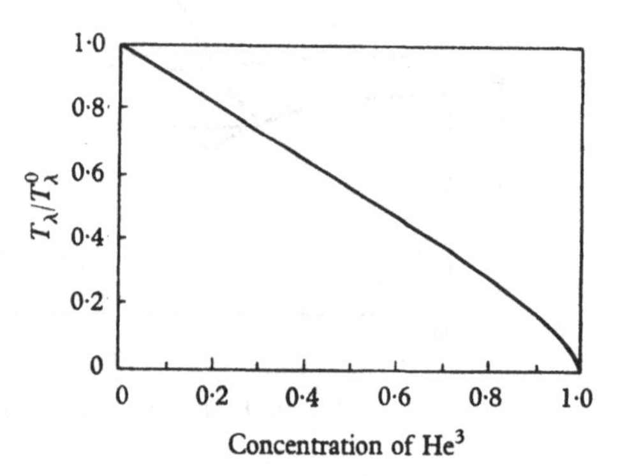

Fig. 9·12. The λ-point of mixtures of He^3 and He^4 as a function of the concentration of He^3.

of transitions of the second order. In a general allotropic change when a modification II which is stable at low temperatures becomes unstable at a temperature T_0 and is replacedby a modification I, the $G_{II}(T)$ curve (for fixed P) crosses the $G_I(T)$ curve at $T=T_0$ and lies above it for greater values of T. If, however, the curves have two point contact at T_0, the G_{II} curve will not cross the G_I curve but will always lie below it, and the modification I will always be unstable. To overcome this difficulty it is necessary to assume either that the modification II cannot exist at all above T_0 or that the modification I cannot exist below T_0, so that one of the two G curves ends abruptly at T_0. A similar situation is encountered in the order-disorder transitions in metals (see § 14·212).

We have at the moment no soundly based theory of liquid helium II, though many suggestions have been put forward, mainly based upon the condensation phenomenon which occurs in the theory of perfect gases obeying Einstein-Bose statistics (§ 6·42) and similar systems. In the Einstein-Bose theory the transition is normally of the third order, not the second, but it was shown in §§ 6·433 and 6·434 that, for systems with a certain distribution of energy levels, the specific heat is discontinuous. Whether a liquid could have such a distribution of energy levels is by no means clear, but the temperature at which the discontinuity would theoretically occur and its amount are of the same order of magnitude as those which occur in liquid helium. It is therefore generally felt that there is some connexion, at present obscure, between the λ-transition in liquid helium and the Einstein-Bose condensation. This conjecture is strengthened by the fact that helium consisting of the isotope He^3 of mass 3, which obeys Fermi-Dirac statistics, behaves like a normal fluid and does not show any λ-transition. Mixtures of He^4 and He^3 have a λ-point, the temperature of which decreases as the concentration of He^3 increases (fig. 9·12). The reader is referred to the review articles by Atkins (1952), Daunt (1952) and Dingle (1952) and to London's book (1954) for further details of the properties of liquid helium and of the theories that have been put forward to account for them.

REFERENCES

Atkins, K. R. (1952). Wave propagation and flow in liquid helium II. *Advanc. Phys.* **1**, 169.

Berman, R. & Simon, F. E. (1955). On the graphite-diamond equilibrium. *Z. Electrochem.* **59**, 333.

Bridgman, P. W. (1937). The phase diagram of water to 45,000 kg./cm.^{2}. *J. Chem. Phys.* **5**, 964.

Bridgman, P. W. (1947). An experimental contribution to the problem of diamond synthesis. *J. Chem. Phys.* **15**, 92.

Bundy, F. P., Hall, H. T., Strong, H. M. & Wentorf, R. H. (1955). Man-made diamonds. *Nature, Lond.*, **176**, 51.

Burton, E. F., Smith, H. G. & Wilhelm, J. O. (1940). *Phenomena at the temperature of liquid helium.* New York.

Daunt, J. G. (1952). Properties of helium three at low temperatures. *Advanc. Phys.* **1**, 209.

Dingle, R. B. (1952). Theories of helium II. *Advanc. Phys.* **1**, 111.

Dugdale, J. S. & Simon, F. E. (1953). Thermodynamic properties and melting of solid helium. *Proc. Roy. Soc.* A, **218**, 291.

Ehrenfest, P. (1933). Phase transitions in the normal and generalised sense classified according to the singularities of the thermodynamic functions. *Proc. Acad. Sci. Amst.* **36**, 153.

Holland, F. A., Hugill, J. A. W. & Jones, G. O. (1951). The solid-fluid equilibrium of helium above 5000 atmospheres pressure. *Proc. Roy. Soc.* A, **207**, 268.

Keesom, W. H. & Wolfke, M. (1927). Two different states of liquid helium. *C.R. Acad. Sci., Paris*, **185**, 1465.

Keesom, W. H. (1933). On the jump in the expansion coefficient of liquid helium in passing the lambda-point. *Proc. Acad. Sci. Amst.* **36**, 147.

Keesom, W. H. (1942). *Helium.* Amsterdam.

London, F. (1954). *Superfluids: macroscopic theory of superfluid helium.* New York.

Simon, F. E., Ruhemann, M. & Edwards, W. A. M. (1929). The melting curve of helium. *Z. phys. Chem.* B, **6**, 62.

Tammann, G. (1900). The limits of the solid state. *Ann. Phys., Lpz.*, (4), **2**, 1, 424.

Chapter 10

ELECTRIC AND MAGNETIC PHENOMENA

The Force and Energy Relations in an Electrostatic Field

10·1. The various force and energy relations which are derived in works on electrostatics and magnetostatics make no reference to thermal effects. To obtain a more general theory we may start from the relation

$$dQ = dU + \Sigma A_i da_i \tag{10·1·1}$$

connecting the heat absorbed by a system in an infinitesimal quasi-static change with the increase in the internal energy and the mechanical work done by the system on external bodies, A_i being the generalized force exerted by the system on its surroundings corresponding to the generalized coordinate a_i. Equation (10·1·1) is effectively the definition of dU as the expression $-\Sigma A_i da_i$ when the system is thermally insulated from its surroundings; and, if the only forces on the system are electric and magnetic forces, we have

$$(dU)_S = -(\Sigma A_i da_i)_e - (\Sigma A_i da_i)_m = dU_e + dU_m, \tag{10·1·2}$$

where U_e and U_m are the electric and magnetic energies of the material system. There are so many differing accounts of magnetic energy in the literature that the theory is apt to be confusing, particularly when statements are not considered carefully with reference to the context in which they occur. For example, the energy of a magnetic field may be said to be potential energy when the sources of the field are permanent magnets, and to be kinetic energy when the field is produced by electric currents. But such considerations are irrelevant in the present exposition where the magnetic energy associated with a particular system is defined by the relation $dU_m = -(\Sigma A_i da_i)_m$, the A_i's being actual ponderomotive forces. It is clear from this definition that the value of dU_m is independent of any hypothesis concerning the nature and molecular constitution of the body which produces the field, and that dU_m refers only to the particular system under consideration, while its relation to the change in the magnetic energy of the external bodies remains undefined by the definition (10·1·1).

The most general case of anisotropic dielectrics or magnets is complicated and can scarcely be discussed profitably without going into much greater detail than is possible here. We shall therefore confine our attention to the simpler aspects of the theory. From a formal point of view the macroscopic theory of dielectrics is the same as that of magnets, and we shall set up the theory for dielectrics only, but we shall apply it to either type of body with only minor modifications.

10·11. Consider an electric dipole of moment $\mathscr{P}$ in a given electrostatic field $\mathscr{E}$, which is a function of position. Suppose the dipole to consist of a charge $-\epsilon$ at the point $\mathbf{r}$ and a charge ϵ at the point $\mathbf{r}+\delta\mathbf{s}$, where $\mathscr{P}=\lim \epsilon\delta\mathbf{s}$. Then the force $\mathscr{F}$ on the dipole is

$$\mathscr{F}=\lim \epsilon\{\mathscr{E}(\mathbf{r}+\delta\mathbf{s})-\mathscr{E}(\mathbf{r})\}=(\mathscr{P}.\mathrm{grad})\,\mathscr{E}, \tag{10·11·1}$$

and in addition there is a couple whose moment is $\mathscr{P}\times\mathscr{E}$. If therefore, we have a dielectric whose polarization is $\mathscr{P}$ in an electric field, and if $\mathbf{r}$ is the position vector of an element of volume $d\tau$ relative to a point O, fixed in the body, whose position vector relative to a fixed origin O' is $\mathbf{R}$, the forces on the dielectric due to the electric field may be reduced to a force $\mathscr{F}(\mathbf{R})$ at O given by

$$\mathscr{F}(\mathbf{R})=\int\{\mathscr{P}(\mathbf{R}+\mathbf{r}).\mathrm{grad}_{\mathbf{r}}\}\,\mathscr{E}(\mathbf{R}+\mathbf{r})_{\mathrm{ext.}}\,d\tau, \tag{10·11·2}$$

together with a couple which is irrelevant to the calculation. $\mathscr{E}(\mathbf{R}+\mathbf{r})_{\mathrm{ext.}}$ is the electric force at the point $\mathbf{R}+\mathbf{r}$ of the dielectric due to the other bodies in the system, and the integration is taken over the region of space occupied by the dielectric when the reference point O is at $\mathbf{R}$. It should be noted that $\mathscr{E}_{\mathrm{ext.}}$ is not the total electric force at the point $\mathbf{R}+\mathbf{r}$ since each element of the dielectric exerts a force on every other element, but these internal forces give zero contribution to the total force and couple on the dielectric. Now, if the dielectric is in equilibrium under the action of the electric field and of mechanical constraints, the force exerted by the dielectric on the constraints is given by (10·11·2). Hence, in an infinitesimal translation of the dielectric in which O is displaced $\delta\mathbf{R}$ and in which the body is not rotated, the mechanical work done by the dielectric against the constraints is $\mathscr{F}(\mathbf{R})\cdot\delta\mathbf{R}$, and we therefore have

$$\delta U_e=-\mathscr{F}(\mathbf{R}).\delta\mathbf{R}. \tag{10·11·3}$$

Now since $\mathscr{E}_{\mathrm{ext.}}$ is an electrostatic field it is derivable from a potential, so that $\partial\mathscr{E}_{i,\mathrm{ext.}}/\partial x_j=\partial\mathscr{E}_{j,\mathrm{ext.}}/\partial x_i$ (in other words, $\mathrm{curl}\,\mathscr{E}_{\mathrm{ext.}}=0$). We may therefore write

$$(\mathscr{P}.\mathrm{grad}_{\mathbf{r}})\,\mathscr{E}_{j,\mathrm{ext.}}=\sum_i \mathscr{P}_i\frac{\partial\mathscr{E}_{j,\mathrm{ext.}}}{\partial x_i}=\sum_i \mathscr{P}_i\frac{\partial\mathscr{E}_{i,\mathrm{ext.}}}{\partial x_j} \tag{10·11·4}$$

and

$$\delta U_e=-\delta\int d\tau\int_0^{\mathscr{E}_{\mathrm{ext.}}}\mathscr{P}(\mathbf{R}+\mathbf{r}).d\mathscr{E}(\mathbf{R}+\mathbf{r})_{\mathrm{ext.}}, \tag{10·11·5}$$

where $\mathscr{P}(\mathbf{R}+\mathbf{r})$ is ordinarily a known function (determined by the constitution of the dielectric) of the field $\mathscr{E}(\mathbf{R}+\mathbf{r})_{\mathrm{ext.}}$ at the point $\mathbf{R}+\mathbf{r}$, so that

$$\mathscr{P}(\mathbf{R}+\mathbf{r}).d\mathscr{E}(\mathbf{R}+\mathbf{r})_{\mathrm{ext.}}$$

is the perfect differential of a function of $\mathscr{E}(\mathbf{R}+\mathbf{r})_{\mathrm{ext.}}$.

To prove that this expression for δU_e is the same as that given by (10·11·2) and (10·11·3), we have to calculate the difference of $\int_0^{\mathscr{E}_{\text{ext.}}} \mathscr{P}.d\mathscr{E}_{\text{ext.}}$ integrated over the dielectric after and before the displacement. Now in the displacement the element of volume $d\tau$ situated at the point $\mathbf{R}+\mathbf{r}$ is transformed into the element of volume $d\tau$ situated at the point $\mathbf{R}+\mathbf{r}+\delta\mathbf{R}$. Hence

$$\delta\mathscr{E}_{\text{ext.}} = \delta\mathbf{R}.\text{grad}_{\mathbf{r}}\,\mathscr{E}(\mathbf{R}+\mathbf{r})_{\text{ext.}}$$

and

$$d\tau\,\delta\int_0^{\mathscr{E}_{\text{ext.}}} \mathscr{P}.d\mathscr{E}_{\text{ext.}} = d\tau \sum_i \mathscr{P}_i \sum_j \frac{\partial \mathscr{E}_{i,\text{ext.}}}{\partial x_j}\,\delta R_j$$

for any element of volume $d\tau$, which proves the correctness of the expression (10·11·5) for δU_e. (There is no loss in generality in restricting the displacement of O to be a translation. For if we wish to calculate U_e for any configuration we may suppose that the dielectric is initially at infinity, i.e. in zero electric field, in such an orientation that it can be brought into the desired configuration by a pure translation, and, by the conservation of energy, ΔU_e must be independent of the exact manner in which the dielectric is moved from its initial to its final position.)

In general, U_e as defined above is not the electrostatic energy of the whole assembly including the sources of the field. If the relation between $\mathscr{P}$ and $\mathscr{E}$ is linear, the total electrostatic energy of an assembly is given by the familiar expression

$$U_{e,\,\text{total}} = \frac{1}{8\pi}\int \mathscr{E}.\mathscr{D}\,d\tau,$$

where $\mathscr{D}$ is the electric displacement and where the integral is taken over the whole of space. To illustrate the difference between U_e and $U_{e,\text{total}}$, we consider the simple case of a parallel plate condenser for which the volume between the plates is V. We neglect end-effects and suppose that the lines of force are straight lines perpendicular to the plates. Then, if the surface densities on the plates are $\pm\sigma$, and if the condenser is in a vacuum, we have $\mathscr{E}=\mathscr{D}=4\pi\sigma$ and $U^{(1)}_{e,\text{total}}=2\pi\sigma^2 V$. If we now bring up from infinity a uniform dielectric of dielectric constant K which just fills the condenser, and keep the charges unaltered, we have $\mathscr{E}=4\pi\sigma/K$, $\mathscr{D}=4\pi\sigma$ and $U^{(2)}_{e,\text{total}}=2\pi\sigma^2 V/K$. Now, since the initial and final positions of the charges on the condenser plates are the same, the only mechanical work done is that due to the displacement of the dielectric. (Note that external work is only done when the dielectric is in the inhomogeneous field at the entrance to the condenser.) Hence the decrease in $U_{e,\,\text{total}}$ must be equal to the mechanical work done by the dielectric, which is $-U_e$. Therefore

$$U_e = -2\pi\sigma^2 V(1-1/K).$$

But, according to (10·11·5), $U_e = -\frac{1}{2}V\mathscr{P}\mathscr{E}_{\text{ext.}}$ when $\mathscr{P}$ is proportional to $\mathscr{E}_{\text{ext.}}$, and since $4\pi\mathscr{P} = \mathscr{D} - \mathscr{E} = 4\pi\sigma(1-1/K)$ and $\mathscr{E}_{\text{ext.}} = 4\pi\sigma$ (the value when the dielectric is absent) we have $U_e = -2\pi\sigma^2(1-1/K)$, which agrees with the value deduced from the energy balance.

We may, if we please, always consider the energy of the whole assembly, but in many cases this introduces unnecessary complications, and it is usually simpler to discuss the energy relations of a particular dielectric body and hence to base the theory on (10·11·5) in which the integration only needs to be taken over the body concerned and not over the whole of space. In applying (10·11·5), however, we meet difficulties in the general case, since for a given material $\mathscr{P}$ is a known function of the total field $\mathscr{E}$ and not of the external field $\mathscr{E}_{\text{ext.}}$. Therefore, unless the relation between $\mathscr{E}$ and $\mathscr{E}_{\text{ext.}}$ can be found for the particular system under consideration, the integral cannot be evaluated explicitly. In practice it is usually necessary to deal only with those cases in which the relation between $\mathscr{E}$ and $\mathscr{E}_{\text{ext.}}$ is especially simple. One such case is that considered above of a parallel plate condenser. A still simpler case is that of a dielectric in the form of a long rod parallel to the field, since $\mathscr{E}$ and $\mathscr{E}_{\text{ext.}}$ are then identical. In many problems of practical interest, however, the shape of the body is important. It is then necessary to take the demagnetizing field explicitly into account (see e.g. § 10·722), and the theory is often difficult.

We shall require to use (10·11·5) when, instead of the dielectric being displaced, it is held fixed and the external field is varied. The expression (10·11·5) is still the correct one for this case since, if $-\Delta U_e$ is the work done when the external field varies from $\mathscr{E}^{(1)}_{\text{ext.}}$ to $\mathscr{E}^{(2)}_{\text{ext.}}$, we may calculate $-\Delta U_e$ as the difference $-(U_e^{(2)} - U_e^{(1)})$ between the works done in bringing up the dielectric from infinity in the two cases when the field is $\mathscr{E}^{(2)}_{\text{ext.}}$ and $\mathscr{E}^{(1)}_{\text{ext.}}$ respectively.

For a more detailed discussion of the different ways in which it is possible to derive suitable expressions for the electric (and magnetic) energy the reader is referred to the papers by Guggenheim (1936*a*, *b*) and by Stoner (1937).

10·12. If the system is not restricted to be adiabatic and if the shape of a dielectric is such that both $\mathscr{E}$ and $\mathscr{E}_{\text{ext.}}$ are constant over the volume occupied by the dielectric in a given configuration (the fields cannot be constant everywhere since they must vanish at infinity, and the whole of the work done in bringing up the dielectric from infinity relates to regions of space where the field is inhomogeneous), then at constant volume V and at constant composition we have from (10·1·1) and (10·11·5), provided that hysteresis effects are negligible,

$$TdS = dU + V\mathscr{P}.d\mathscr{E}_{\text{ext.}}. \tag{10·12·1}$$

As already mentioned, this is really the definition of U, but, if we please, we may define U in an infinite number of different ways by adding any perfect differential to the above expression for dU. In particular we may put

$$\left.\begin{aligned} U_1 &= U + V\mathscr{P}.\mathscr{E}_{\text{ext.}}, & dU_1 &= TdS + \mathscr{E}_{\text{ext.}}.d(V\mathscr{P}),\\ U_2 &= U - V\int \mathscr{P}.d\mathscr{E}_{\text{int.}}, & dU_2 &= TdS - V\mathscr{P}.d\mathscr{E},\\ U_3 &= U_2 + V\mathscr{P}.\mathscr{E}, & dU_3 &= TdS + \mathscr{E}.d(V\mathscr{P}), \end{aligned}\right\} \tag{10·12·2}$$

and we may base the theory on any of these relations. If we use U_2 we have

$$TdS = \frac{\partial U_2}{\partial T}dT + \left(\frac{\partial U_2}{\partial \mathscr{E}} + V\mathscr{P}\right).d\mathscr{E}. \tag{10·12·3}$$

Since dS must be a perfect differential this gives

$$\frac{\partial U_2}{\partial \mathscr{E}} = VT^2 \frac{\partial}{\partial T}\frac{\mathscr{P}}{T} \tag{10·12·4}$$

and

$$U_2 = U_2^0 + VT^2 \frac{\partial}{\partial T}\left(\frac{1}{T}\int_0^{\mathscr{E}} \mathscr{P}.d\mathscr{E}\right), \tag{10·12·5}$$

where U_2^0 is the value of U_2 when $\mathscr{E} = 0$. There are corresponding expressions for U, U_1 and U_3. We may also introduce the free energy $F = U - ST$, and corresponding expressions for F_1, F_2, F_3. Then

$$dF = -SdT - V\mathscr{P}.d\mathscr{E}_{\text{ext.}} \tag{10·12·6}$$

and

$$F = F_0(T) - V\int_0^{\mathscr{E}_{\text{ext.}}} \mathscr{P}.d\mathscr{E}_{\text{ext.}}, \tag{10·12·7}$$

while

$$dF_2 = -SdT - V\mathscr{P}.d\mathscr{E}. \tag{10·12·8}$$

We therefore see that, unless $\mathscr{P}$ is independent of T, the internal energy and the electrostatic energy are not directly related and that the usual expressions for the electrostatic energy are expressions for the free energy and not for the internal energy.

10·13. *Electrostriction.* If we consider more general variations than those dealt with so far we can write

$$dF = -SdT - PdV - V\mathscr{P}.d\mathscr{E}_{\text{ext.}} + \Sigma\mu_i dn_i \tag{10·13·1}$$

for a substance of variable composition subject to an isotropic pressure P. We may define a large number of thermodynamic functions by taking different sets of independent variables to specify the state of the substance, and we can obtain corresponding sets of Maxwell relations between the various quantities which occur in the differential forms for the thermodynamic functions. These enable us to calculate such quantities as the difference of the specific heats at constant polarization and at constant field, or the change in pressure due to a change in the electric field. For example, (10·13·1) gives

$$\left(\frac{\partial S}{\partial \mathscr{E}_{\text{ext.}}}\right)_{T,V,n_i} = V\left(\frac{\partial \mathscr{P}}{\partial T}\right)_{\mathscr{E}_{\text{ext.}},V,n_i}, \quad \left(\frac{\partial P}{\partial \mathscr{E}_{\text{ext.}}}\right)_{T,V,n_i} = \left(\frac{\partial V\mathscr{P}}{\partial V}\right)_{T,\mathscr{E}_{\text{ext.}},n_i}, \tag{10·13·2}$$

while the corresponding equation for dG gives

$$\left(\frac{\partial V}{\partial \mathscr{E}_{\text{ext.}}}\right)_{T,P,n_i} = -\left(\frac{\partial V\mathscr{P}}{\partial P}\right)_{T,\mathscr{E}_{\text{ext.}},n_i} \tag{10·13·3}$$

The change in volume due to the presence of an electric field is the phenomenon known as electrostriction.

THE STATISTICAL MECHANICS OF POLAR SUBSTANCES

10·2. *A polar gas.* Consider a gas consisting of molecules each of which carries a permanent electric moment $\boldsymbol{\mu}$.‡ If there is an electric field $\boldsymbol{\mathscr{E}}$ and if the molecular polarizability is α, the potential energy of a molecule is $-\boldsymbol{\mathscr{E}}.\boldsymbol{\mu}-\frac{1}{2}\alpha\mathscr{E}^2$. For not too high temperatures we can assume that the internal kinetic energy is entirely rotational, and we can write the expression for the internal energy of a molecule in the form

$$\mathscr{H} = \tfrac{1}{2}(I_1\omega_1^2 + I_2\omega_2^2 + I_3\omega_3^2) - \boldsymbol{\mathscr{E}}.\boldsymbol{\mu} - \tfrac{1}{2}\alpha\mathscr{E}^2, \tag{10·2·1}$$

where I_1, I_2, I_3 are the principal moments of inertia and where ω_1, ω_2, ω_3 are the components of the angular velocity referred to the principal axes. We can evaluate the internal partition function Z if classical statistics are applicable, by using the change of variables given in § 6·13, and we find that

$$Z = \frac{\Pi\rho}{\sigma}\frac{2\pi}{h^3}(I_1I_2I_3)^{\frac{1}{2}}(2\pi kT)^{\frac{3}{2}}\iint \exp(-\boldsymbol{\mathscr{E}}.\boldsymbol{\mu} - \tfrac{1}{2}\alpha\mathscr{E}^2)\sin\theta\, d\theta\, d\phi, \tag{10·2·2}$$

where $\Pi\rho$ denotes the nuclear spin weights, σ is the symmetry number of the molecule, and where θ and ϕ are the polar angles of one of the principal axes of the molecule. If we write $Z = Z_{\text{rot.}} Z_o$, where $Z_{\text{rot.}}$ is the rotational partition function in the absence of a field and where Z_o is an 'orientational partition function', we can evaluate Z_o by changing to polar coordinates θ', ϕ' such that $\theta' = 0$ is along the direction of $\boldsymbol{\mu}$. We then have

$$\begin{aligned} Z_o &= \int_0^{2\pi} d\phi' \int_0^{\pi} \exp(-\mathscr{E}\mu\cos\theta' - \tfrac{1}{2}\alpha\mathscr{E}^2)\sin\theta'\, d\theta' \\ &= x^{-1}\sinh x \exp(-\tfrac{1}{2}\alpha\mathscr{E}^2/kT), \end{aligned} \tag{10·2·3}$$

‡ It is customary to denote an electric or magnetic moment by the symbol μ. No confusion can arise as to whether μ denotes an electric moment of a thermodynamic potential. Except in equation (10·13·1), μ is not used as a thermodynamic potential in the present chapter.

where $$x = \mathscr{E}\mu/kT. \tag{10·2·4}$$

Since the 'orientational' free energy F_o is $-NkT \log Z_o$, where N is the number of molecules in volume V, the total electric moment $V\mathscr{P}$ is given by

$$V\mathscr{P} = -\frac{\partial F_o}{\partial \mathscr{E}} = N\mu L\left(\frac{\mathscr{E}\mu}{kT}\right) + N\alpha\mathscr{E}, \tag{10·2·5}$$

where $$L(x) = \coth x - 1/x. \tag{10·2·6}$$

L is known as the Langevin function (Langevin, 1905; Debye, 1912).

For small values of x, we have $L(x) = \frac{1}{3}x$ while $L(x) \to 1$ as $x \to \infty$. $\mathscr{P}$ is therefore proportional to $\mathscr{E}$ for small values of $\mathscr{E}\mu/kT$, and we can define the susceptibility χ by the relation $\chi = \mathscr{P}/\mathscr{E}$ with

$$\chi = \frac{K-1}{4\pi} = \frac{N}{V}\left(\alpha + \frac{1}{3}\frac{\mu^2}{kT}\right). \tag{10·2·7}$$

For large values of $\mathscr{E}\mu/kT$, on the other hand, $V\mathscr{P}$ tends to $N(\mu + \alpha\mathscr{E})$, and, if the polarizability α is zero, $V\mathscr{P}$ has the saturation value $N\mu$ (see fig. 10·1).

10·21. The formula (10·2·7) is true whether we assume that the Hamiltonian is given by the classical expression (10·2·1) or derive it by a detailed quantal calculation (see, for example, van Vleck, 1932). It is therefore possible to determine the dipole moments and polarizabilities of molecules from measurements of the dielectric constants of gases. It is only necessary to plot χ as a function of $1/T$ to determine both α and μ, and the value of μ gives valuable quantitative information concerning the structure of the molecule concerned. For example, the heteropolar CO_2 molecule has zero dipole moment, which is only explicable if the molecule is linear. Ammonia, on the other hand, has a dipole moment of $1{\cdot}47 \times 10^{-18}$ e.s.u., and this rules out a planar structure for the NH_3 molecule but is consistent with a pyramidal structure. The susceptibilities of methane and its chlorine derivatives, shown in fig. 10·2, show that CH_4 and CCl_4 are non-polar and therefore highly symmetrical molecules, whereas CH_3Cl, CH_2Cl_2 and $CHCl_3$ are polar molecules, all of which is readily understandable if the molecules are tetrahedral with the carbon atom at the centroid of the tetrahedron.

Much of the evidence provided by the investigation of dipole moments can be obtained by other methods. In general, however, the dipole moment of a molecule gives more direct information concerning the structure than many of the classical methods of chemistry, and it has the further advantage that, if the forces between the atoms are known, the dipole moment of any assumed structure can be calculated and compared with the experimental value. A list of dipole moments and polarizabilities of a number of molecules of various types is given in table 10·1.

Table 10·1. *Dipole moments and polarizabilities of simple molecules*

	H_2	HCl	HBr	HI	H_2O	H_2S	CO
$\mu \times 10^{18}$ e.s.u.	0	1·03	0·79	0·38	1·84	0·93	0·12
$\alpha \times 10^{24}$ cm.^{3}/molecule	0·79	2·63	3·58	5·4	1·68	3·78	1·95
	CO_2	N_2O	NH_3	SO_2	CH_4	CH_3Cl	
$\mu \times 10^{18}$ e.s.u.	0	0	1·47	1·61	0	1·86	
$\alpha \times 10^{24}$ cm.^{3}/molecule	2·65	3·0	2·26	3·72	2·61	4·56	
	CH_2Cl_2	$CHCl_3$	CCl_4	CH_3OH	$(CH_3)_2O$	$(CH_3)_2CO$	
$\mu \times 10^{18}$ e.s.u.	1·59	0·95	0	1·68	1·34	2·94	
$\alpha \times 10^{24}$ cm.^{3}/molecule	(7)	8·23	10·5	3·23	5·16	6·33	

10·22. *The local field.* The preceding formulae can only strictly apply to a gas whose density is sufficiently low for the dielectric constant not to differ too greatly from unity. In dense gases, liquids or solids, the local field acting upon a molecule due to the dipole moments of the other molecules

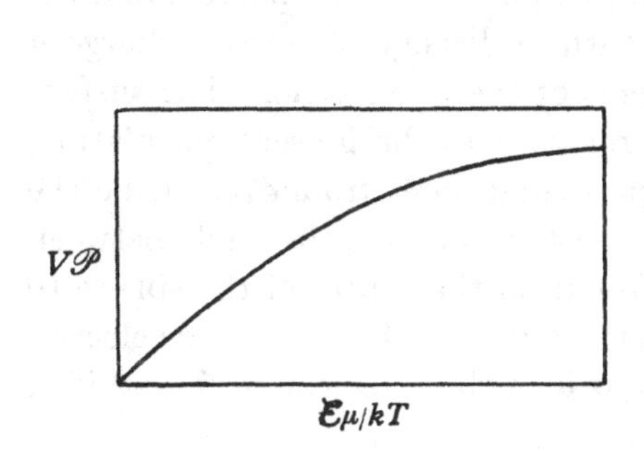

Fig. 10·1. The polarization of a gas as a function of the field.

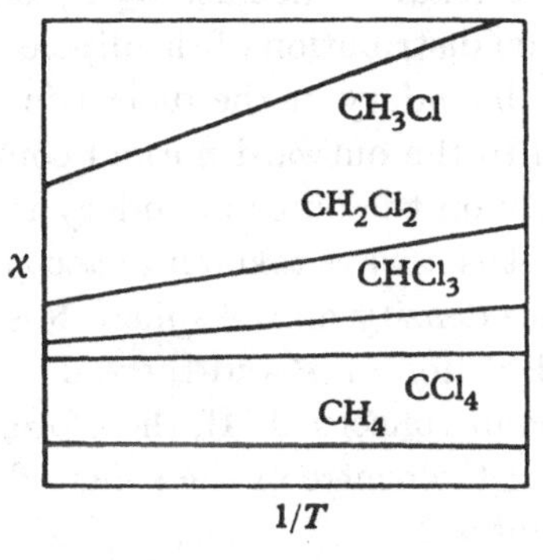

Fig. 10·2. The susceptibility of the chloromethanes.

may be appreciable, and the formulae must be generalized to include it. The field at a given point due to an infinite three-dimensional array of dipoles can be expressed as a three-dimensional sum, and it is well known that this sum is not absolutely convergent. It is therefore necessary to consider carefully how to carry out the summation properly in order to obtain the correct physical result. We denote the total field acting upon a dipole by $\mathscr{E} = \mathscr{E}_1 + \mathscr{E}_{\text{local}}$. Here $\mathscr{E}_1$ is the electric field inside the dielectric calculated by the ordinary rules of electrostatics, i.e. it is $-\operatorname{grad} \mathscr{V}$, where $\mathscr{V}$ is the electrostatic potential. $\mathscr{E}_1$ is not necessarily equal to the external field, since in calculating it we make allowance for the influence of the shape of the specimen concerned. The susceptibility is always defined in terms of $\mathscr{E}_1$ and not $\mathscr{E}_{\text{ext.}}$, since otherwise it would be a function of the shape of the material. We therefore have to determine χ as the ratio $\mathscr{P}/\mathscr{E}_1$.

An elementary method of avoiding the convergence difficulties was given by Lorentz (1909, p. 138). To calculate the field acting on a given dipole in a dielectric, consider a sphere S with its centre at the dipole. The sphere must be physically small but of such a radius that it contains many dipoles. Also the dipoles outside the sphere must be sufficiently far away from the dipole under consideration for them to be considered as forming a continuous distribution, so that they can be replaced by the equivalent Poisson distribution. (In practice there is no difficulty in fulfilling the different criteria. Stated in mathematical terms, the complete summation is split up into two parts, referring to dipoles which are near to or distant from the dipole under consideration. These two summations are calculated separately, the second being replaced by an integration. The value of the complete sum is of course independent of the method of division.) If we denote the electric fields due to the dipoles inside and outside the sphere by $\mathscr{E}_2$ and $\mathscr{E}_3$, we can readily calculate $\mathscr{E}_3$ by elementary electrostatics. The equivalent Poisson distribution of the dipoles contained between the sphere S and the outer boundary of the dielectric is a surface density of electric charge σ equal to the outward normal component of the polarization. The surface density on the outer boundary is not relevant to the present calculation, since it is always taken into account in calculating $\mathscr{E}_1$ from $\mathscr{E}_{\text{ext.}}$, while the surface density on the sphere S is $-\mathscr{P}\cos\theta$, where θ is the angle between the direction of $\mathscr{P}$ and the radius vector from the centre of the sphere to the point concerned. If, therefore, a is the radius of the sphere, the electric field at the centre of the sphere due to σ is in the direction of $\mathscr{P}$ and is of amount

$$\mathscr{E}_3 = \int \frac{\sigma\cos\theta}{a^2}\,dS = \frac{\mathscr{P}}{a^2}\int \cos\theta\,dS = \tfrac{4}{3}\pi\mathscr{P}.$$

The calculation of $\mathscr{E}_2$ is in general a much more difficult matter, but it was pointed out by Lorentz that $\mathscr{E}_2$ vanishes identically for an array of parallel dipoles which has cubic symmetry. To show this we consider a dipole with moment μ_i at the point $\mathbf{r}_i$, directed along the z direction. Then the electric field produced at the origin by the array is

$$\mathscr{E}_x = \sum_i 3\mu\frac{x_i y_i}{r_i^5}, \quad \mathscr{E}_y = \sum_i 3\mu\frac{y_i z_i}{r_i^5}, \quad \mathscr{E}_z = \sum_i \mu\frac{3z_i^2 - r_i^2}{r_i^5},$$

where the summation is taken over all the dipoles except the one at the origin. These expressions are zero by symmetry for an array with cubic symmetry. They are also zero for a random distribution of atoms, provided that all the dipoles are parallel to one another.

When the above conditions are satisfied, the polarizing force acting on a molecule is‡

$$\mathscr{E}=\mathscr{E}_1+\tfrac{4}{3}\pi\mathscr{P}. \tag{10·22·1}$$

We also have

$$\mathscr{P}=N\gamma\mathscr{E}/V, \tag{10·22·2}$$

where γ is a constant, and since

$$\mathscr{P}=\chi\mathscr{E}_1=(K-1)\,\mathscr{E}_1/4\pi, \tag{10·22·3}$$

we find that

$$\frac{K-1}{K+2}=\tfrac{4}{3}\pi\frac{N}{V}. \tag{10·22·4}$$

This formula is usually known as the Lorenz-Lorentz formula. It reduces to (10·2·7) if $K-1$ is small and if we put $\gamma=\alpha+\frac{1}{3}\mu^2/kT$; but we may not conclude without further investigation that we can in fact substitute this expression for γ into (10·22·4). We may certainly substitute $\gamma=\alpha$ into (10·22·4), but the term $\frac{1}{3}\mu^2/kT$ does not arise from a system of parallel dipoles and hence is not covered by the conditions under which (10·22·4) has been derived

The derivation given above of the value of the local field is elementary, but from time to time it has caused considerable controversy, and the Lorenz-Lorentz formula has often been used outside the range of its validity. In the next section we give a much more sophisticated proof of (10·22·1) which can be generalized to give the corresponding results for non-cubic arrays of parallel dipoles.

10·221. A simple crystal lattice is characterized by the vectors $\mathbf{a}_1$, $\mathbf{a}_2$, $\mathbf{a}_3$ giving the edges of the unit cell. If we take the origin at one particular lattice point, all the lattice points are given by the vectors $\mathbf{g}_a$, where

$$\mathbf{g}_a=g_1\mathbf{a}_1+g_2\mathbf{a}_2+g_3\mathbf{a}_3, \tag{10·221·1}$$

g_1, g_2, g_3 being positive or negative integers. We suppose that at each lattice point there is a positive charge ϵ, and in addition that there is a uniform volume charge of density $-\epsilon/\Delta$, where $\Delta\equiv\mathbf{a}_1.(\mathbf{a}_2\times\mathbf{a}_3)$ is the volume of the unit cell. The whole system is electrically neutral, and the electrostatic potential $\mathscr{V}$ must satisfy the following conditions.

First $\mathscr{V}(\mathbf{r})$ has the average value zero; secondly, it must satisfy Poisson's equation

$$\nabla^2\mathscr{V}(\mathbf{r})=4\pi\epsilon/\Delta; \tag{10·221·2}$$

thirdly, it is periodic with the unit cell as its fundamental period; and fourthly, it becomes infinite like ϵ/r at the origin.

If $\mathscr{V}(\mathbf{r})$ is known for any lattice consisting of point charges, the potential due to an array of equal parallel dipoles on the same lattice can be obtained

‡ In general it is sufficient to consider a specimen in the form of a long rod parallel to the field, and it is then unnecessary to distinguish between $\mathscr{E}_1$ and $\mathscr{E}_{\text{ext.}}$.

at once by differentiation. If we consider two lattices of positive and negative charges respectively, whose relative displacement is $\delta\mathbf{s}$, the potential at the point $\mathbf{r}$ is $\mathscr{V}(\mathbf{r}-\delta\mathbf{s})-\mathscr{V}(\mathbf{r})=-\delta\mathbf{s}.\mathrm{grad}\,\mathscr{V}$, and so the potential of a lattice of equal parallel dipoles is $\mathscr{V}_\mu(\mathbf{r})=-\boldsymbol{\mu}.\mathrm{grad}\,\mathscr{V}(\mathbf{r})/\epsilon$, where $\boldsymbol{\mu}$ is the vector defining the magnitude and direction of the dipoles.

The field $\mathscr{E}$ due to the array of dipoles is given by

$$\mathscr{E}=-\mathrm{grad}\,\mathscr{V}_\mu(\mathbf{r})=\mathrm{grad}\,\{\boldsymbol{\mu}.\mathrm{grad}\,\mathscr{V}(\mathbf{r})/\epsilon\},$$

and it can be found at any point if $\mathscr{V}(\mathbf{r})$ is known. As a consequence of the first condition imposed upon $\mathscr{V}(\mathbf{r})$, the average value of the electric field is zero. If we wish to calculate $\mathscr{E}$ at the origin, we have to use $\mathscr{V}'(\mathbf{r})=\mathscr{V}(\mathbf{r})-\epsilon/r$ instead of $\mathscr{V}(\mathbf{r})$, and if the dipoles are directed along the z axis we have

$$\mathscr{E}_x=\frac{\mu}{\epsilon}\frac{\partial^2\mathscr{V}'}{\partial x\,\partial z},\quad \mathscr{E}_y=\frac{\mu}{\epsilon}\frac{\partial^2\mathscr{V}'}{\partial y\,\partial z},\quad \mathscr{E}_z=\frac{\mu}{\epsilon}\frac{\partial^2\mathscr{V}'}{\partial z^2}. \tag{10·221·3}$$

For a cubic array we can calculate $\mathscr{E}$ at the origin without using the explicit expression for $\mathscr{V}'$. By symmetry

$$\mathscr{E}_x=\mathscr{E}_y=0,\quad \mathscr{E}_z=\tfrac{1}{3}(\mu/\epsilon)\,\nabla^2\mathscr{V}'. \tag{10·221·4}$$

But, by (10·221·2), $\nabla^2\mathscr{V}'=\nabla^2(\mathscr{V}-\epsilon/r)=4\pi\epsilon/\Delta$, and so

$$\mathscr{E}_z=\tfrac{4}{3}\pi\mu/\Delta=\tfrac{4}{3}\pi\mathscr{P}, \tag{10·221·5}$$

since μ/Δ is the polarization per unit volume. This completes the alternative proof of (10·22·1) for a cubic array of parallel dipoles.

In the general case we need the explicit expression for $\mathscr{V}(\mathbf{r})$ in order to calculate the local field. We therefore expand $\mathscr{V}(\mathbf{r})$ as a triple Fourier series. To do this we have to introduce the reciprocal vectors $\mathbf{b}_1$, $\mathbf{b}_2$, $\mathbf{b}_3$, which are defined by the relations

$$\mathbf{a}_i.\mathbf{b}_j=\delta_{ij}\quad (i,j=1,2,3), \tag{10·221·6}$$

where $\delta_{ij}=0$ if $i\neq j$, $\delta_{ij}=1$ if $i=j$. The reciprocal lattice is the set of points defined by the vectors $\mathbf{g}_b\equiv g_1\mathbf{b}_1+g_2\mathbf{b}_2+g_3\mathbf{b}_3$, where the g's are integers. The unit cell of the reciprocal lattice has the volume

$$\mathbf{b}_1.(\mathbf{b}_2\times\mathbf{b}_3)=1/\{\mathbf{a}_1.(\mathbf{a}_2\times\mathbf{a}_3)\}=1/\Delta.$$

With this notation, the expansion of $\mathscr{V}(\mathbf{r})$ is given by

$$\mathscr{V}(\mathbf{r})=\sum_{\mathbf{g}}\mathscr{V}_{\mathbf{g}}\exp(2\pi i\mathbf{g}_b.\mathbf{r}), \tag{10·221·7}$$

since this ensures that $\mathscr{V}(\mathbf{r}+\mathbf{a}_i)=\mathscr{V}(\mathbf{r})$. Now if $\mathscr{U}$ and $\mathscr{V}$ are two twice-differentiable functions and if the only singularity of either function is a pole ϵ/r of $\mathscr{V}$, Green's theorem takes the form

$$\int(\mathscr{U}\nabla^2\mathscr{V}-\mathscr{V}\nabla^2\mathscr{U})\,d\tau=\int\left(\mathscr{U}\frac{\partial\mathscr{V}}{\partial n}-\mathscr{V}\frac{\partial\mathscr{U}}{\partial n}\right)dS+4\pi\epsilon\mathscr{U}_0, \tag{10·221·8}$$

where the volume integration is over any region of space, $\partial/\partial n$ denotes differentiation along the outward normal to the bounding surface and where $\mathscr{U}_0$ is the value of $\mathscr{U}$ at the origin. If we take $\mathscr{V}$ as the potential of the three-dimensional lattice defined above, put

$$\mathscr{U} = \exp(-2\pi i \mathbf{g}_b . \mathbf{r}) \tag{10·221·9}$$

and integrate over a unit cell, we obtain

$$\pi |\mathbf{g}_b|^2 \int \mathscr{V} \exp(-2\pi i \mathbf{g}_b . \mathbf{r})\, d\tau = \epsilon. \tag{10·221·10}$$

Now since the average value of $\mathscr{V}$ is zero, we have $\mathscr{V}_0 = 0$, and multiplying (10·221·7) by $\exp(-2\pi i \mathbf{g}_b . \mathbf{r})$ and integrating over the unit cell, we have

$$\mathscr{V}_{\mathbf{g}} \Delta = \int \mathscr{V} \exp(-2\pi i \mathbf{g}_b . \mathbf{r})\, d\tau. \tag{10·221·11}$$

Hence
$$\mathscr{V}(\mathbf{r}) = \frac{\epsilon}{\pi\Delta} \sum_{\mathbf{g} \neq 0} \frac{\exp(-2\pi i \mathbf{g}_b . \mathbf{r})}{|\mathbf{g}_b|^2}. \tag{10·221·12}$$

In general this expression for $\mathscr{V}(\mathbf{r})$ converges slowly. It diverges on the cell edges and it cannot be differentiated term by term.

A method of obtaining (10·221·12) in a rapidly convergent form has been given by Ewald (1921). By use of the identity

$$\frac{1}{\alpha} = \frac{1}{\alpha} \exp(-\pi^2\alpha/\eta^2) + \int_0^{\pi^2/\eta^2} e^{-\alpha\xi}\, d\xi \tag{10·221·13}$$

we can transform (10·221·12) into

$$\mathscr{V}(\mathbf{r}) = \frac{\epsilon}{\pi\Delta} \sum_{\mathbf{g} \neq 0} \frac{1}{|\mathbf{g}_b|^2} \exp\{2\pi i \mathbf{g}_b . \mathbf{r} - \pi^2 |\mathbf{g}_b|^2/\eta^2\}$$
$$+ \frac{\epsilon}{\pi\Delta} \int_0^{\pi^2/\eta^2} \sum_{\mathbf{g} \neq 0} \exp(2\pi i \mathbf{g}_b . \mathbf{r} - \xi |\mathbf{g}_b|^2)\, d\xi. \tag{10·221·14}$$

By using the transformation theory of triple theta functions, the integral, which we denote by I, can be expressed in terms of the error function

$$G(x) = \frac{2}{\sqrt{\pi}} \int_x^\infty e^{-\beta^2} d\beta = \sqrt{\frac{\lambda}{\pi}} \int_0^{\lambda/x^2} \frac{e^{-\lambda/t}}{t^{\frac{3}{2}}}\, dt, \tag{10·221·15}$$

and it is shown below in § 10·222 that

$$\mathscr{V}(\mathbf{r}) = \frac{\epsilon}{\pi\Delta} \sum_{\mathbf{g} \neq 0} \frac{1}{|\mathbf{g}_b|^2} \exp[2\pi i \mathbf{g}_b . \mathbf{r} - \pi^2 |\mathbf{g}_b|^2/\eta^2] + \epsilon \sum_{\mathbf{g}} \frac{G(\eta |\mathbf{g}_a - \mathbf{r}|)}{|\mathbf{g}_a - \mathbf{r}|} - \frac{\pi\epsilon}{\Delta\eta^2}. \tag{10·221·16}$$

By choosing η suitably, both series can readily be calculated. If we wish to find the potential at a lattice point, say the origin, it is necessary to subtract the term ϵ/r in $\mathscr{V}(\mathbf{r})$ arising from the point charge there. We can do this by using the identity

$$\frac{\epsilon}{r}=\frac{2\epsilon}{\sqrt{\pi}\,r}\int_0^{\eta r} e^{-\beta^2}d\beta+\frac{2\epsilon}{\sqrt{\pi}\,r}\int_{\eta r}^{\infty} e^{-\beta^2}d\beta. \tag{10·221·17}$$

The second integral then cancels the term in (10·221·16) arising from $\mathbf{g}_a=0$. We now have all the formulae required to calculate the local field of any array of parallel dipoles arranged on a crystal lattice either at a general point or at a lattice point (Kornfeld, 1924). Some results for non-cubic lattices have been given by McKeehan (1933*b*) and Mueller (1938).

10·222. *Proof of the theta function transformation.* Instead of dealing directly with the integral I in (10·221·14), we consider the expression

$$\sum_{\mathbf{g}} \exp\{-\xi\,|\,\mathbf{g}_b+\mathbf{s}\,|^2+2\pi i(\mathbf{g}_b+\mathbf{s}).\mathbf{r}\}, \tag{10·222·1}$$

where $\mathbf{s}$ is an arbitrary vector in the reciprocal space. This expression is a triply periodic function of $\mathbf{s}$ with the unit cell of the reciprocal lattice as its period. We can therefore expand it as a triple Fourier series of the form

$$\sum_{\mathbf{h}} A_{\mathbf{h}} \exp(2\pi i\mathbf{h}_a.\mathbf{s}), \tag{10·222·2}$$

where

$$A_{\mathbf{h}}=\Delta\int \sum_{\mathbf{g}} \exp\{-\xi\,|\,\mathbf{g}_b+\mathbf{s}'\,|^2+2\pi i(\mathbf{g}_b+\mathbf{s}').\mathbf{r}-2\pi i\mathbf{h}_a.\mathbf{s}'\}\,d\tau', \tag{10·222·3}$$

the integration being over a unit cell (which has volume $1/\Delta$) in the reciprocal space. Now in the last term in the exponential we can replace $\mathbf{s}'$ by $\mathbf{g}_b+\mathbf{s}'$ since $\mathbf{h}_a.(\mathbf{g}_b+\mathbf{s}')=\mathbf{h}.\mathbf{g}+\mathbf{h}_a.\mathbf{s}'$, and since $\mathbf{h}.\mathbf{g}$ is an integer. Also, instead of integrating over a unit cell and summing with respect to $\mathbf{g}$ in (10·222·3), we can omit the sum and integrate over all (reciprocal) space since the summation with respect to $\mathbf{g}$ is equivalent to summing over all the cells. Hence $A_{\mathbf{h}}$ can be written as

$$\begin{aligned}
A_{\mathbf{h}}&=\Delta\int_{\text{all space}} \exp[-\xi\,|\,\mathbf{g}_b+\mathbf{s}'\,|^2+2\pi i(\mathbf{g}_b+\mathbf{s}').(\mathbf{r}-\mathbf{h}_a)]\,d\tau'\\
&=\Delta\int_{\text{all space}} \exp[-\xi\,|\,\mathbf{s}'\,|^2+2\pi i\mathbf{s}'.(\mathbf{r}-\mathbf{h}_a)]\,d\tau'\\
&=\Delta\int_{\text{all space}} \exp[-\xi\,|\,\mathbf{s}'-\pi i(\mathbf{r}-\mathbf{h}_a)/\xi\,|^2-\pi^2\,|\,\mathbf{r}-\mathbf{h}_a\,|^2/\xi]\,d\tau'\\
&=\Delta(\pi/\xi)^{\frac{3}{2}}\exp(-\pi^2\,|\,\mathbf{r}-\mathbf{h}_a\,|^2/\xi).
\end{aligned} \tag{10·222·4}$$

But, by putting $\mathbf{s}=0$ in (10·222·1) and (10·222·2), we obtain the relation

$$\sum_{\mathbf{g}} \exp[-\xi\,|\,\mathbf{g}_b\,|^2+2\pi i\mathbf{g}_b.\mathbf{r}]=\sum_{\mathbf{h}} A_{\mathbf{h}},$$

and so the integral I can be written as

$$I=\pi^{\frac{1}{2}}\epsilon\int_0^{\pi^2/\eta^2}\sum_{\mathbf{h}}\exp(-\pi^2|\mathbf{r}-\mathbf{h}_a|^2/\xi)\frac{d\xi}{\xi^{\frac{3}{2}}}-\frac{\epsilon}{\pi\Delta}\int_0^{\pi^2/\eta^2}d\xi$$

$$=\epsilon\sum_{\mathbf{h}}\frac{G(\eta\,|\,\mathbf{h}_a-\mathbf{r}\,|)}{|\,\mathbf{h}_a-\mathbf{r}\,|}-\frac{\pi\epsilon}{\eta^2\Delta}, \qquad (10\cdot222\cdot5)$$

by (10·221·15), which proves the result quoted in the preceding section.

10·223. *The Lorenz-Lorentz formula.* We should expect the formula (10·22·4) to apply to non-polar isotropic substances, since the electric moments of the atoms are then entirely induced and are necessarily all parallel to the inducing field. It is in fact obeyed to a high order of accuracy by non-polar gases for a wide range of densities, as shown by the data given in table 10·2 for air at room temperature and at pressures from 1 to 200 atmospheres. If, however, it is applied to polar substances whose atoms have permanent dipoles, it can lead to fallacious results. If we write (10·22·4) as

$$K=\frac{1+\frac{8}{3}\pi N\gamma/V}{1-\frac{4}{3}\pi N\gamma/V}, \qquad (10\cdot223\cdot1)$$

we see that K becomes infinite if the denominator vanishes. Now for a polar substance where $\gamma=\alpha+\frac{1}{3}\mu^2/kT$, the denominator would in fact become infinite for a sufficiently small value of T no matter how small μ might be. In practice, the critical value of T would be very high, and for water, for example, where $\mu=1\cdot84\times10^{-18}$ e.s.u., the denominator would be negative for $T<1200°$ K. We must therefore conclude that the Lorentz formula (10·22·1) grossly exaggerates the local field when it is due to rotating permanent dipoles. The calculation of the local field when the dipoles are not assumed to be parallel is a matter of extreme complexity, and no rigorous solution of the problem has yet been given.

Table 10·2. *The dielectric constant of air as a function of the density*

Relative density	1	14·84	42·13	69·24	96·16	123·04	176·27
$\frac{1}{\rho}\frac{K-1}{K+2}\times10^7$	1953	1947	1959	1961	1961	1956	1953

10·224. *Onsager's formula.* A tentative expression for the local field, which may be valid in extreme conditions where Lorentz's formula does not apply, has been derived by Onsager (1936) in the following way. Consider a spherical cavity in a dielectric just large enough to contain one molecule. Then if there is a uniform electric field $\mathscr{E}_1$ in the dielectric in the absence of the cavity, and if we may calculate the field in the presence of the cavity by applying the laws of macroscopic electrostatics, the field inside the cavity is

$$\mathscr{E} = 3K\mathscr{E}_1/(2K+1), \tag{10·224·1}$$

which is the force polarizing the molecule there. If the polarization is given by (10·22·2), then

$$K - 1 = 4\pi \frac{3K}{2K+1} \frac{N\gamma}{V}. \tag{10·224·2}$$

This is a quadratic equation for K whose solution is

$$K = \tfrac{1}{4}(3x+1) + \tfrac{1}{8}\sqrt{(9x^2+6x+9)}, \tag{10·224·3}$$

where

$$x = 4\pi N\gamma/V. \tag{10·224·4}$$

Onsager's theory is by no means soundly based, since it extrapolates macroscopic results to cavities of molecular dimensions. It must therefore be considered as a speculative hypothesis. It is probable that for polar substances Onsager's formula gives a better description of the facts than Lorentz's, and his formula does not predict an infinite dielectric constant. There have been several attempts to generalize Onsager's formula or to give it a better theoretical basis, the most important being by van Vleck (1937*b*) and by Kirkwood (1939). The reader is referred to the original papers or to Fröhlich's monograph (1949) for details of these theories, and their application to the dielectric constants of liquids.

Ferroelectricity

10·3. A number of substances possess a permanent electric moment in the absence of an electric field, a phenomenon known as ferroelectricity, as being analogous to ferromagnetism. (The title is a misnomer, since the substances concerned contain no iron.) The simplest of these substances is barium titanate $BaTiO_3$, which has a spontaneous polarization of 48×10^3 e.s.u. at room temperature. At high temperatures the crystal structure of barium titanate is cubic, but at 120° C. it undergoes a phase transition and becomes slightly tetragonal, the sides of the unit cell being a, a, c ($c > a$), and at the same time it becomes spontaneously polarized. At about $-5°$ C. the structure becomes monoclinic with axes c', b', c' ($c' > b'$) and with the angle β between the first and the third axis very slightly less than 90° (by about 15′). Such a structure is usually described by an orthorhombic unit cell whose edges are either parallel to the diagonals of the rhombic faces of the monoclinic cell or perpendicular to these faces. At about $-80°$ C. the structure becomes slightly rhombohedral, the edges of the unit cell being a'', a'', a''. The parameters of the various unit cells as functions of the temperature are shown in fig. 10·3 (Kay & Vousden, 1949), the central line being the cube root of the volume of the unit cell. The variation of the spontaneous polarization is given in fig. 10·4 (Merz, 1949), and it will be seen that both the lattice parameters and the polarization show hysteresis effects.

It is supposed that the polarization is primarily due to a displacement of the Ti^{4+} ion through a distance of about $0{\cdot}1 \times 10^{-8}$ cm., and the occurrence

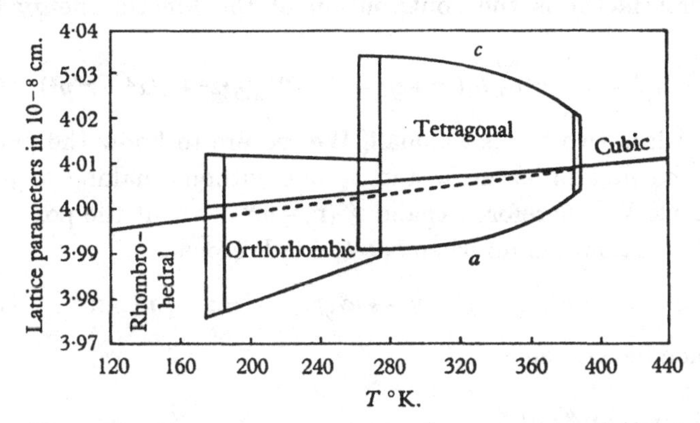

Fig. 10·3. The parameters of the cubic or pseudo-cubic unit cells of barium titanate, showing hysteresis effects.

of a spontaneous polarization in a crystal such as barium titanate can be explained on the basis of the Lorenz-Lorentz formula if the electric moment per atom, for a given external electric field, increases slowly as the temperature decreases. Such a variation of the electric moment will occur if

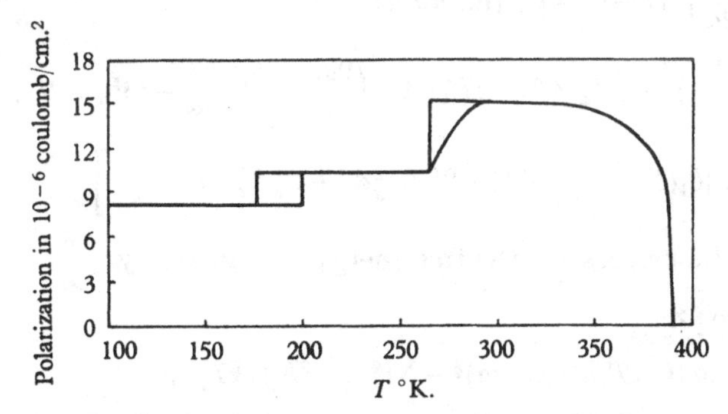

Fig. 10·4. The spontaneous polarization of barium titanate as a function of the temperature, showing hysteresis effects.

the restoring force on an ion is anharmonic (Devonshire, 1949, 1951, 1954; Slater, 1950). It is sufficient for our present purpose to neglect the coupling of the lattice vibrations of the crystal, and we therefore need only consider a single ion subject to an electric field $\mathscr{E}$, the potential energy $\mathscr{V}(\mathbf{r})$ of the restoring forces having the symmetry of the crystal lattice, which we take to be cubic. If the charge and mass of the ion are q and m respectively, the partition function of the ion is

$$Z=\frac{(2\pi mkT)^{\frac{3}{2}}}{h^3}\int\exp\{[-\mathscr{V}(\mathbf{r})+q\mathscr{E}.\mathbf{r}]/kT\}\,d\tau, \qquad (10\cdot3\cdot1)$$

where the first factor is the contribution of the kinetic energy terms. We write

$$\mathscr{V}(\mathbf{r})=a(x^2+y^2+z^2)+b_1(x^4+y^4+z^4)+2b_2(y^2z^2+z^2x^2+x^2y^2) \quad (10\cdot3\cdot2)$$

and treat the last two terms as small. We require to know the partition function to the first order in b_1 and b_2 but without making any other approximations. We therefore expand $\mathscr{V}(\mathbf{r})-q\mathscr{E}.\mathbf{r}$ about the point where this expression is a minimum when $b_1=b_2=0$. We put

$$x'=x-\tfrac{1}{2}q\mathscr{E}_x/a,\quad y'=y-\tfrac{1}{2}q\mathscr{E}_y/a,\quad z'=z-\tfrac{1}{2}q\mathscr{E}_z/a, \qquad (10\cdot3\cdot3)$$

and Z can then be written as

$$\begin{aligned}Z=&\frac{(2\pi mkT)^{\frac{3}{2}}}{h^3}\exp\left(\tfrac{1}{4}q^2\mathscr{E}^2/akT\right)\\
&\times\int_{-\infty}^{\infty}\int_{-\infty}^{\infty}\int_{-\infty}^{\infty}\exp\left(-ar'^2/kT\right)\left[1-\frac{b_1}{kT}\left\{\left(x'+\frac{q\mathscr{E}_x}{2a}\right)^4+\left(y'+\frac{q\mathscr{E}_y}{2a}\right)^4\right.\right.\\
&+\left.\left(z'+\frac{q\mathscr{E}_z}{2a}\right)^4\right\}-\frac{2b_2}{kT}\left\{\left(y'+\frac{q\mathscr{E}_y}{2a}\right)^2\left(z'+\frac{q\mathscr{E}_z}{2a}\right)^2+\left(z'+\frac{q\mathscr{E}_z}{2a}\right)^2\left(x'+\frac{q\mathscr{E}_x}{2a}\right)^2\right.\\
&+\left.\left.\left(x'+\frac{q\mathscr{E}_x}{2a}\right)^2\left(y'+\frac{q\mathscr{E}_y}{2a}\right)^2\right\}\right]dx'\,dy'\,dz'\\
=&\left(\frac{\pi kT}{h}\right)^3\left(\frac{2m}{a}\right)^{\frac{3}{2}}\exp\left(\tfrac{1}{4}q^2\mathscr{E}^2/akT\right)\left[1-\left(\frac{9b_1}{4}+\frac{3b_2}{2}\right)\frac{kT}{a^2}-(3b_1+2b_2)\frac{q^2\mathscr{E}^2}{4a^3}\right.\\
&\left.-\frac{q^4}{16a^4kT}\{b_1(\mathscr{E}_x^4+\mathscr{E}_y^4+\mathscr{E}_z^4)+2b_2(\mathscr{E}_y^2\mathscr{E}_z^2+\mathscr{E}_z^2\mathscr{E}_x^2+\mathscr{E}_x^2\mathscr{E}_y^2)\}\right]. \qquad (10\cdot3\cdot4)\end{aligned}$$

If there are N ions present, the free energy F_2 is given by

$$\begin{aligned}F_2=&-NkT\log Z\\
=&-NkT\log\{(\pi kT/h)^3(2m/a)^{\frac{3}{2}}-N(\tfrac{9}{4}b_1+\tfrac{3}{2}b_2)(kT/a)^2\\
&-V\{\tfrac{1}{2}(\chi_0-\chi_1T)\mathscr{E}^2-\tfrac{1}{4}\xi(\mathscr{E}_x^4+\mathscr{E}_y^4+\mathscr{E}_z^4)-\tfrac{1}{2}\eta(\mathscr{E}_y^2\mathscr{E}_z^2+\mathscr{E}_z^2\mathscr{E}_x^2+\mathscr{E}_x^2\mathscr{E}_y^2)\}, \qquad (10\cdot3\cdot5)\end{aligned}$$

where $\mathscr{E}$ includes the local field, so that

$$\mathscr{E}=\mathscr{E}_{\text{ext.}}+\lambda\mathscr{P}, \qquad (10\cdot3\cdot6)$$

and where

$$\chi_0=\frac{1}{2}\frac{N}{V}\frac{q^2}{a},\quad \chi_1=(\tfrac{3}{2}b_1+b_2)\frac{Nk}{V}\frac{q^2}{a^3},\quad \xi=\frac{1}{4}\frac{N}{V}\frac{q^4b_1}{a^4},\quad \eta=\frac{1}{4}\frac{N}{V}\frac{q^4b_2}{a^4}. \quad (10\cdot3\cdot7)$$

The components of the polarization are $\mathscr{P}_x$, $\mathscr{P}_y$, $\mathscr{P}_z$, where

$$\left.\begin{aligned}\mathscr{P}_x &= -\partial F_2/\partial(\mathscr{E}_x V) = (\chi_0-\chi_1 T)\,\mathscr{E}_x - \xi\mathscr{E}_x^3 - \eta\mathscr{E}_x(\mathscr{E}_y^2+\mathscr{E}_z^2),\\ \mathscr{P}_y &= -\partial F_2/\partial(\mathscr{E}_y V) = (\chi_0-\chi_1 T)\,\mathscr{E}_y - \xi\mathscr{E}_y^3 - \eta\mathscr{E}_y(\mathscr{E}_z^2+\mathscr{E}_x^2),\\ \mathscr{P}_z &= -\partial F_2/\partial(\mathscr{E}_z V) = (\chi_0-\chi_1 T)\,\mathscr{E}_z - \xi\mathscr{E}_z^3 - \eta\mathscr{E}_z(\mathscr{E}_x^2+\mathscr{E}_y^2).\end{aligned}\right\} \qquad (10\cdot3\cdot8)$$

If there is no external field, the equations (10·3·6) and (10·3·8) always have the solution $\mathscr{P}=0$, but if λ is sufficiently large and if T is sufficiently small, there may be a solution with $\mathscr{P}_x \neq 0$, $\mathscr{P}_y = \mathscr{P}_z = 0$. The value of $\mathscr{P}_x$ will then be given by

$$\mathscr{P}_x^2 = (\lambda\chi_0 - 1 - \lambda\chi_1 T)/(\xi\lambda^3) = (\chi_1/\xi\lambda^2)\,(T_0 - T) \quad (T \leqslant T_0) \qquad (10\cdot3\cdot9)$$

for small values of $T_0 - T$, where

$$\frac{\chi_1}{\xi\lambda^2} = \frac{2(3+2b_2/b_1)\,ka}{\lambda^2 q^2}, \quad kT_0 = k\frac{\lambda\chi_0 - 1}{\lambda\chi_1} = \frac{a^2}{3b_1+2b_2} - \frac{2a^3}{\lambda(3b_1+2b_2)}\frac{V}{Nq^2}. \qquad (10\cdot3\cdot10)$$

If T_0 is positive there is a real transition point, below which the crystal is spontaneously polarized, the polarization being proportional to $(T_0 - T)^{\frac{1}{2}}$ for small values of $T_0 - T$. A necessary condition for the occurrence of a spontaneous polarization is therefore

$$\lambda q^2 N/V > 2a. \qquad (10\cdot3\cdot11)$$

Above the transition point, the polarization is small if the field is small, and we may neglect the cubic terms in (10·3·8). The susceptibility is then given by

$$\chi = \frac{\mathscr{P}}{\mathscr{E}_{\text{ext.}}} = \frac{\chi_0}{\lambda\chi_1}\frac{1}{T-T_0} = \frac{a^2}{\lambda(3b_1+2b_2)}\frac{1}{k(T-T_0)}. \qquad (10\cdot3\cdot12)$$

The predictions of the formulae derived above in are qualitative agreement with the observations. We still have to verify that the local field is sufficiently large to give rise to a real transition point, but before doing so it is convenient to consider the effect of the spontaneous polarization on the crystal structure.

10·31. *The crystal structure of a ferroelectric crystal.* If the crystal becomes spontaneously polarized in the x direction, the x axis is no longer equivalent to the y and z axes, and the crystal only has tetragonal symmetry. The crystal will in addition be slightly distorted and the sides of the unit cell no will longer be all equal. (The effect on the lattice constants could be calculated by including the contribution of the deformations to the elastic free energy, but for our present purpose it is unnecessary to introduce such complications.) The polarization could equally well occur along the y or z axes, and the crystal actually becomes polarized in domains which are orientated in such a manner that in the absence of an external field the macroscopic polarization is zero. The deformation of the unit cell can, however, be determined from the X-ray diffraction patterns. The change in crystal structure from cubic to tetragonal, then to orthorhombic and finally to rhombohedral can be explained as follows (Devonshire, 1949).

We introduce a further term $\frac{1}{6}V\zeta(\mathscr{E}_x^6+\mathscr{E}_y^6+\mathscr{E}_z^6)$ into the expression (10·3·5) for the free energy. Then, if $\mathscr{E}_{\text{ext.}}=0$, the equations for the polarization have the four following types of solution:

$$\mathscr{P}_x=\mathscr{P}_y=\mathscr{P}_z=0, \tag{10·31·1}$$

$$\mathscr{P}_y=\mathscr{P}_z=0, \quad \lambda^5\zeta\mathscr{P}_x^4+\lambda^3\xi\mathscr{P}_x^2-\lambda(\chi_0-\chi_1 T)+1=0, \tag{10·31·2}$$

$$\mathscr{P}_z=0, \quad \mathscr{P}_y=\mathscr{P}_x, \quad \lambda^5\zeta\mathscr{P}_x^4+\lambda^3(\xi+\eta)\mathscr{P}_x^2-\lambda(\chi_0-\chi_1 T)+1=0, \tag{10·31·3}$$

$$\mathscr{P}_x=\mathscr{P}_y=\mathscr{P}_z, \quad \lambda^5\zeta\mathscr{P}_x^4+\lambda^3(\xi+2\eta)\mathscr{P}_x^2-\lambda(\chi_0-\chi_1 T)+1=0. \tag{10·31·4}$$

At any particular temperature the stable state will be the one with the smallest value of the free energy, and, by choosing the constants properly, it is possible to arrange for each type of solution to be the relevant one in turn as the temperature is lowered. The crystal will then be successively unpolarized, polarized along an edge of the unit cell, polarized along the diagonal of a face of the unit cell and polarized along a cell diagonal, and it will therefore exhibit the observed symmetries.

10·32. *The magnitude of the local field.* The barium titanate lattice has a somewhat open structure due to the presence of the large barium ions. As a result the restoring force on the titanium ion is comparatively small (i.e. the constant a is small), and a large ionic polarization is possible. In spite of this, it is found that the inequality (10·3·11) cannot be satisfied with a reasonable value of a if we take λ to have the normal value $\frac{4}{3}\pi$. A much larger value of λ can, however, be introduced into the theory by taking account of the fact that, although the whole structure has cubic symmetry, the fields acting on the individual ions have a lower symmetry (van Santen & Opechowski, 1948; Slater, 1950).

In barium titanate the barium ions occupy the corners of the unit cells, the titanium ions the cube centres and the oxygen ions the face centres. We may therefore consider the structure to be composed of five simple cubic lattices with the same lattice constant. Relative to the barium lattice, the coordinates of the other ions are as follows (expressed in submultiples of the lattice constant):

$$Ti^{4+}, (\tfrac{1}{2},\tfrac{1}{2},\tfrac{1}{2}); \quad O^{2-}, (\tfrac{1}{2},0,\tfrac{1}{2}), (0,\tfrac{1}{2},\tfrac{1}{2}), (\tfrac{1}{2},\tfrac{1}{2},0).$$

The coordinates of the other ions relative to the titanium lattice are

$$Ba^{2+}, (\tfrac{1}{2}, \tfrac{1}{2}, \tfrac{1}{2}); \quad O^{2-}, (0, \tfrac{1}{2}, 0), (\tfrac{1}{2}, 0, 0), (0, 0, \tfrac{1}{2}),$$

while relative to the three oxygen lattices they are

$$Ba^{2+}, (0, \tfrac{1}{2}, \tfrac{1}{2}); \quad Ti^{4+}, (\tfrac{1}{2}, 0, 0); \quad O^{2-}, (\tfrac{1}{2}, \tfrac{1}{2}, 0), (\tfrac{1}{2}, 0, \tfrac{1}{2}),$$

and two other sets of coordinates obtained by permuting the coordinates cyclically. The local fields at all these points in the various lattices can be computed from the formulae established in § 10·221 (McKeehan, 1933*a*; Luttinger & Tisza, 1946). If the polarization is along the z axis and if the local field is $\lambda \mathscr{P}_z$, the values of λ are as follows:

$$\left.\begin{aligned} \lambda(0,0,0) = \lambda(\tfrac{1}{2}, \tfrac{1}{2}, \tfrac{1}{2}) = \tfrac{4}{3}\pi, \\ \lambda(\tfrac{1}{2}, \tfrac{1}{2}, 0) = \tfrac{4}{3}\pi - 8{\cdot}668, \quad \lambda(0, \tfrac{1}{2}, \tfrac{1}{2}) = \lambda(\tfrac{1}{2}, 0, \tfrac{1}{2}) = \tfrac{4}{3}\pi + 4{\cdot}334, \\ \lambda(\tfrac{1}{2}, 0, 0) = \lambda(0, \tfrac{1}{2}, 0) = \tfrac{4}{3}\pi - 15{\cdot}040, \quad \lambda(0, 0, \tfrac{1}{2}) = \tfrac{4}{3}\pi + 30{\cdot}080. \end{aligned}\right\} \quad (10{\cdot}32{\cdot}1)$$

This means that the local field exerted by the barium lattice on a barium ion or a titanium ion has the Lorentz value (and similarly for the local field exerted by the titanium lattice on a titanium ion or a barium ion), while the local fields exerted by or on the oxygen ions differ considerably from the Lorentz values.

If we denote the three oxygen lattices by the suffixes 1, 2, 3, where 3 refers to the lattice containing those oxygen ions which have the same x and y coordinates as the titanium ions, we may write the local fields acting on the various ions in terms of the polarizations of the five lattices as follows (all the fields and polarizations are parallel to the z axis, and we may therefore drop the suffix z without ambiguity):

$$\begin{aligned} \mathscr{E}(\mathrm{Ti})_{\mathrm{local}} = {} & \lambda(0,0,0)\,\mathscr{P}(\mathrm{Ti}) + \lambda(\tfrac{1}{2}, \tfrac{1}{2}, \tfrac{1}{2})\,\mathscr{P}(\mathrm{Ba}) \\ & + \lambda(\tfrac{1}{2}, 0, 0)\,\{\mathscr{P}(\mathrm{O})_1 + \mathscr{P}(\mathrm{O})_2\} + \lambda(0, 0, \tfrac{1}{2})\,\mathscr{P}(\mathrm{O})_3, \end{aligned} \quad (10{\cdot}32{\cdot}2)$$

$$\begin{aligned} \mathscr{E}(\mathrm{Ba})_{\mathrm{local}} = {} & \lambda(\tfrac{1}{2}, \tfrac{1}{2}, \tfrac{1}{2})\,\mathscr{P}(\mathrm{Ti}) + \lambda(0,0,0)\,\mathscr{P}(\mathrm{Ba}) \\ & + \lambda(\tfrac{1}{2}, 0, \tfrac{1}{2})\,\{\mathscr{P}(\mathrm{O})_1 + \mathscr{P}(\mathrm{O})_2\} + \lambda(\tfrac{1}{2}, \tfrac{1}{2}, 0)\,\mathscr{P}(\mathrm{O})_3, \end{aligned} \quad (10{\cdot}32{\cdot}3)$$

$$\begin{aligned} \mathscr{E}(\mathrm{O})_{1,\mathrm{local}} = {} & \lambda(\tfrac{1}{2}, 0, 0)\,\mathscr{P}(\mathrm{Ti}) + \lambda(\tfrac{1}{2}, 0, \tfrac{1}{2})\,\mathscr{P}(\mathrm{Ba}) \\ & + \lambda(0,0,0)\,\mathscr{P}(\mathrm{O})_1 + \lambda(\tfrac{1}{2}, \tfrac{1}{2}, 0)\,\mathscr{P}(\mathrm{O})_2 + \lambda(0, 0, \tfrac{1}{2})\,\mathscr{P}(\mathrm{O})_3, \end{aligned} \quad (10{\cdot}32{\cdot}4)$$

$$\begin{aligned} \mathscr{E}(\mathrm{O})_{2,\mathrm{local}} = {} & \lambda(\tfrac{1}{2}, 0, 0)\,\mathscr{P}(\mathrm{Ti}) + \lambda(\tfrac{1}{2}, 0, \tfrac{1}{2})\,\mathscr{P}(\mathrm{Ba}) \\ & + \lambda(\tfrac{1}{2}, \tfrac{1}{2}, 0)\,\mathscr{P}(\mathrm{O})_1 + \lambda(0,0,0)\,\mathscr{P}(\mathrm{O})_2 + \lambda(0, 0, \tfrac{1}{2})\,\mathscr{P}(\mathrm{O})_3, \end{aligned} \quad (10{\cdot}32{\cdot}5)$$

$$\begin{aligned} \mathscr{E}(\mathrm{O})_{3,\mathrm{local}} = {} & \lambda(0, 0, \tfrac{1}{2})\,\mathscr{P}(\mathrm{Ti}) + \lambda(\tfrac{1}{2}, \tfrac{1}{2}, 0)\,\mathscr{P}(\mathrm{Ba}) \\ & + \lambda(0, 0, \tfrac{1}{2})\,\{\mathscr{P}(\mathrm{O})_1 + \mathscr{P}(\mathrm{O})_2\} + \lambda(0,0,0)\,\mathscr{P}(\mathrm{O})_3. \end{aligned} \quad (10{\cdot}32{\cdot}6)$$

Now for each ion we have an equation of the form (10·3·8) connecting $\mathscr{P}_z$ with $\mathscr{E}_z$, where $\mathscr{E}_z = \mathscr{E}_{\mathrm{ext.}} + \mathscr{E}_{\mathrm{local}}$, and by using the above expressions for $\mathscr{E}_{\mathrm{local}}$ we can determine the polarizations as functions of $\mathscr{E}_{\mathrm{ext.}}$. In particular, if we wish to determine the condition for a transition point we can restrict ourselves to terms linear in the field, and put

$$\mathscr{E}_{\text{ext.}} = 0 \quad \text{and} \quad \mathscr{E}_{\text{local}} = \mathscr{P}/(\chi_0 - \chi_1 T)$$

(with appropriate constants for each type of ion). The equations (10·32·2) to (10·32·6) are then four linear equations for the four independent polarizations (owing to symmetry we necessarily have $\mathscr{P}(\text{O})_1 = \mathscr{P}(\text{O})_2$), and the critical temperature is the temperature at which the determinant of the coefficients vanishes.

A detailed analysis of the predictions of the equations derived above has been carried out by Slater, on the assumption that the polarization of the barium and oxygen ions is entirely due to the displacement of their electrons by the electric field, the ions themselves remaining in their equilibrium positions, whereas the polarization of the titanium ions is due partly to the electronic polarizability but mainly to the displacement of the ions. He found that, to reproduce the correct transition temperature, the ionic polarizability of the titanium ions need only be about one-sixth as great as that required according to the simple Lorentz theory. In addition, he found that the barium ions and the oxygen ions of types 1 and 2 scarcely contribute to the polarization, while the titanium ions contribute about 37 % and the oxygen ions of type 3 about 60 % of the polarization. Thus, although the polarization arises primarily from the displacement of the loosely bound titanium ions, it is greatly enhanced by the moments induced in those oxygen ions which can form straight chains of dipoles with the titanium ions.

The theory given here does not apply to the more complex ferroelectric crystals, such as Rochelle salt, which have both an upper and a lower transition temperature. The theory of these substances is not at present well understood, but the reader will find an account of some of the hypotheses which have been put forward in the review articles by von Hippel (1950) and by Devonshire (1954).

The Statistical Mechanics of Paramagnetic Bodies

10·4. The magnetic susceptibility of a molecule is due partly to the orbital motions of the electrons and partly to their intrinsic magnetic moments. Owing to coupling between the electrons and to the 'spin-orbit' interactions, the general problem of the effect of a magnetic field upon the energy of a molecule is a complicated one, but it is considerably simplified if the orbital angular momentum vanishes. The permanent magnetic moment is then entirely due to the electron spins, while the effect of the magnetic field on the orbital motion of the electrons is to produce a diamagnetic moment. For an atom, the diamagnetic susceptibility is given by

$$\chi = -\frac{N}{V}\frac{\epsilon^2}{6mc^2}\Sigma\overline{r^2}, \tag{10·4·1}$$

where $-\epsilon$ is the charge and m the mass of an electron, c is the velocity of light, $\overline{r^2}$ is the mean value of the square of an electron from the nucleus and the summation is over all the electrons in the atom. There is no such simple formula for a molecule, but as a first approximation the diamagnetic susceptibility of a molecule is the sum of the susceptibilities of the atoms of which it is composed.

The intrinsic magnetic moments of the electrons give rise to the paramagnetic susceptibility which is of a similar nature to the electric susceptibility of a polar gas and which can readily be calculated. If the total spin angular momentum of a molecule is $\mathrm{S}h/2\pi$, where h is Planck's constant and where $2\mathrm{S}+1$ is an integer, the component of the spin angular momentum in any prescribed direction can only take the values

$$(\mathrm{S}, \mathrm{S}-1, \ldots, -\mathrm{S}+1, -\mathrm{S})\,h/2\pi,$$

and, due to the presence of an external magnetic field H, the energy levels of a molecule are $-(\mathrm{S}, \mathrm{S}-1, \ldots, -\mathrm{S}+1, -\mathrm{S})\,2\mu_0 H$, where $\mu_0 = \epsilon h/(4\pi mc)$ is the magnetic moment of an electron‡ (μ_0 is the Bohr magneton $9{\cdot}27 \times 10^{-21}$ erg gauss^{-1}). The partition function of a paramagnetic gas in a magnetic field therefore contains the factor

$$Z = \sum_{r=-\mathrm{S}}^{\mathrm{S}} \exp(-2r\mu_0 H/kT) = \frac{\sinh\{(2\mathrm{S}+1)\,\mu_0 H/kT\}}{\sinh(\mu_0 H/kT)}, \tag{10·4·2}$$

and the total magnetic moment VI of the gas is given by

$$VI = -NkT\,\partial \log Z/\partial H = 2N\mathrm{S}\mu_0 B_{\mathrm{S}}(\mu_0 H/kT), \tag{10·4·3}$$

where

$$2\mathrm{S}B_{\mathrm{S}}(x) = (2\mathrm{S}+1)\coth\{(2\mathrm{S}+1)\,x\} - \coth x. \tag{10·4·4}$$

The difference between (10·4·3) and the Langevin-Debye formula (10·2·5) is due to the fact that, according to quantum theory, the possible orientations of a magnetic dipole relative to a magnetic field form a discrete set, whereas the orientations of an electric dipole are not subject to such a restriction. If $\mathrm{S}\to\infty$ and $\mu_0 \to 0$ in such a way as to keep $2\mu_0\mathrm{S}$ finite and equal to M_0, $B_{\mathrm{S}}(\mu_0 H/kT)$ tends to $L(M_0/kT)$, and we regain the Langevin-Debye theory (as we must do, since the orientations of the magnetic dipole

‡ In order to conform to the standard notation we use μ_0 and S in this chapter to denote the Bohr magneton and the spin quantum number respectively. In some formulae both the entropy and the spin quantum number occur, but confusion between the symbols is unlikely, especially since different founts are used.

then form a continuous set). If, on the other hand, there is only one electron per molecule we have $S=\frac{1}{2}$ and

$$VI=N\mu_0 \tanh(\mu_0 H/kT). \tag{10·4·5}$$

10·41. For small values of $\mu_0 H/kT$, I is proportional to H and we can therefore define a paramagnetic susceptibility, which is given by

$$\chi=\frac{4N}{3V}\frac{S(S+1)\mu_0^2}{kT}. \tag{10·41·1}$$

The approximate proportionality of χ to $1/T$ was established as an experimental fact by Curie in 1895, and an ideal paramagnetic substance is one which obeys Curie's law exactly.

If the orbital angular momentum is not zero there are a number of possible complications to be considered. The simplest and most important case is that of an atom in which the spin-orbit interaction is of the Russell-Saunders type. If the orbital and spin angular momenta of the ground state of the atom are characterized by the quantum numbers L and S, and if J is the total quantum number of the ground state, the energy levels in the presence of a magnetic field are $-(J, J-1, \ldots, -J+1, -J)\,g\mu_0 H$, where g is the Landé g-factor

$$g=1+\frac{J(J+1)+S(S+1)-L(L+1)}{2J(J+1)}. \tag{10·41·2}$$

In this case (10·4·2) and (10·41·1) are replaced by

$$Z=\frac{\sinh\{(2J+1)\,g\mu_0 H/kT\}}{\sinh(g\mu_0 H/kT)}, \quad \chi=\tfrac{1}{3}g^2\frac{N}{V}\frac{J(J+1)\mu_0^2}{kT}. \tag{10·41·3}$$

If, however, the multiplet separation is of the order of kT, it is necessary to include the effect of states other than the ground state, and a more complicated formula is required. We then have to determine the energy levels of all the states concerned as functions of the magnetic field. It is necessary to calculate the energy to order H^2, and it can be shown (van Vleck, 1932, p. 235) that

$$\chi=\frac{N}{V}\frac{\Sigma\{\alpha_J+\frac{1}{3}g_J^2 J(J+1)\mu_0^2/kT\}(2J+1)\exp(-E_J^0/kT)}{\Sigma(2J+1)\exp(-E_J^0/kT)}, \tag{10·41·4}$$

where the summation is taken over the possible values of J, namely, L+S, L+S−1, ..., |L−S|. The Landé g-factor is given by (10·41·2), E_J^0 is the energy of the state with quantum number J in the absence of a magnetic field, and α_J is a quantity determined by the second-order Zeeman effect of the multiplet. The susceptibility will have an anomalous temperature variation for temperature ranges such that kT is of the same order as the multiplet separations.

For a more detailed discussion of the general formulae for the susceptibility of atoms and of molecules the reader is referred to the comprehensive treatises by van Vleck (1932) and by Stoner (1934). The data which follow are taken from these sources.

10·411. The only paramagnetic gas in which the susceptibility is due to the electron spins alone is oxygen, and the experimental results are consistent with the value $S=1$. The molecule of the other well-known paramagnetic gas NO has an odd electron and therefore S must be $\frac{1}{2}$, but in this case the orbital angular momentum is not zero, and a complete theory of the paramagnetism of nitric oxide requires a detailed analysis of the molecular states (van Vleck, 1932, p. 269).

10·412. *The susceptibilities of salts of the rare earths.* Most of the measurements of paramagnetic susceptibilities have been carried out on solutions of salts of the rare earths and of the transition elements or on

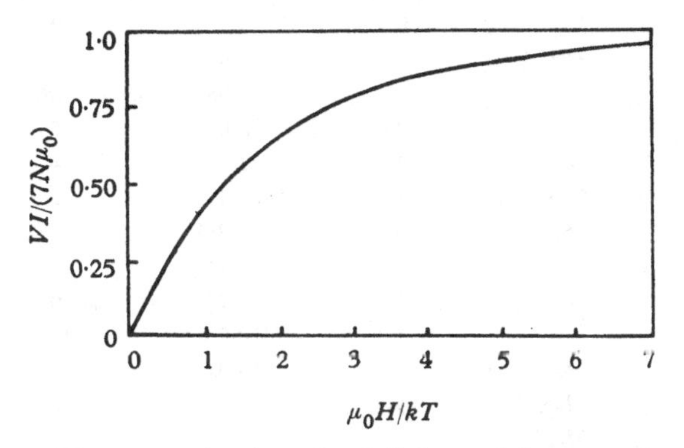

Fig. 10·5. The magnetization of gadolinium sulphate as a function of $\mu_0 H/kT$.

their crystalline salts in a highly hydrated form. The above theory should apply reasonably well to these salts, since the ions should behave roughly as a paramagnetic gas. A typical example is provided by gadolinium sulphate $Gd_2(SO_4)_3.8H_2O$, where the gadolinium ion Gd^{3+} is in an $^8S_{\frac{7}{2}}$ state, so that the orbital angular momentum is zero and the spin quantum number S is $\frac{7}{2}$. The measurements for this substance have been carried out at such high fields and at such low temperatures (Woltjer & Kamerlingh Onnes, 1923) that the saturation effects can be observed, and it will be seen from the data shown in fig. 10·5 that the experimental data are in complete agreement with equation (10·4·3), with $S=\frac{7}{2}$.

The other rare-earth ions, which are normally triply ionized, have non-zero orbital angular momenta, and χ is given by (10·41·3) instead of by

(10·41·1). The quantity $\sqrt{(3kT\chi V/N\mu_0^2)}$, which is often known as the effective magneton number, should therefore be equal to $g\{J(J+1)\}^{\frac{1}{2}}$ if (10·41·3) is correct, and it is found that this is true for all the rare-earth ions with the exception of samarium and europium. These two ions, particularly the latter, have low-lying excited states. Their susceptibilities are therefore greater than the values given by (10·41·3) and they must be calculated from van Vleck's generalized equation. All the quantities required in the calculation are, however, not known with great accuracy, and only upper and lower limits can be obtained for the susceptibilities. The theoretical values of $\sqrt{(3kT\chi V/N\mu_0^2)}$ obtained in this way (from van Vleck's formula for Sm^{3+} and Eu^{3+}, and from $g\{J(J+1)\}^{\frac{1}{2}}$ for the other ions) are given in table 10·3, and it will be seen that there is good agreement with the experimental values.

Table 10·3. *Calculated and observed effective magneton numbers of the rare earth ions at room temperature*

Ion	Normal state	g	Calculated	Observed
La^{3+}	$^1S_0$	0	0	0
Ce^{3+}	$^2F_{\frac{5}{2}}$	$\frac{6}{7}$	2·54	2·4
Pr^{3+}	$^3H_4$	$\frac{4}{5}$	3·58	3·5
Nd^{3+}	$^4I_{\frac{9}{2}}$	$\frac{8}{11}$	3·62	3·5
Il^{3+}	$^5I_4$	$\frac{3}{5}$	2·68	—
Sm^{3+}	$^6H_{\frac{5}{2}}$	$\frac{2}{7}$	1·55–1·65	1·5
Eu^{3+}	$^7F_0$	0	3·4 –3·5	3·6
Gd^{3+}	$^8S_{\frac{7}{2}}$	2	7·94	8·0
Tb^{3+}	$^7F_6$	$\frac{3}{2}$	9·72	9·5
Ds^{3+}	$^6H_{\frac{15}{2}}$	$\frac{4}{3}$	10·65	10·7
Ho^{3+}	$^5I_8$	$\frac{5}{4}$	10·61	10·3
Er^{3+}	$^4I_{\frac{15}{2}}$	$\frac{6}{5}$	9·58	9·5
Tu^{3+}	$^3H_6$	$\frac{7}{6}$	7·56	7·3
Yb^{3+}	$^2F_{\frac{7}{2}}$	$\frac{8}{7}$	4·54	4·5
Lu^{3+}	$^1S_0$	0	0	0

10·413. *The susceptibilities of salts of the transition elements.* It is found that the susceptibilities of salts of the transition metals are not given by equation (10·41·3) but are in better agreement with the values given by equation (10·41·1). In the salts of the rare earths, the electrons responsible for the magnetization are the $4f$ electrons, and these, being buried in the atomic cores, are sufficiently shielded from the effect of the surrounding ions for them to be considered as having more or less the same properties as if they were in the gaseous state. In the salts of the transition metals of the iron group, on the other hand, the electrons concerned are the $3d$ electrons which form the outer shells of the ions, and we cannot treat these electrons as if they were under no forces except those of the ion to which they belong. It is then necessary to take into account the effect of the electrostatic fields produced by the other ions present, and it is no longer necessarily possible to use the orbital quantum number L and the components of the orbital angular momentum to classify the various energy states in the absence of spin-orbit forces. For example, if the electrostatic field has rhombic symmetry, a possible electrostatic potential is

$$\mathscr{V} = Ax^2 + By^2 - (A+B)z^2 \quad (A \neq B),$$

since this satisfies $\nabla^2\mathscr{V} = 0$. In such a field the three standard p wave functions, which are of the form $xf(r)$, $yf(r)$, $zf(r)$, will have different energies, and the energy levels with $l = 1$ are no longer degenerate. They cannot, therefore, be further split by the presence of a magnetic field, and in such a case there can be no magnetic moment arising from the orbital angular momentum. This state of affairs is usually described by saying that the orbital angular momentum is quenched by the anisotropic electric fields present. The actual arrangement of the energy levels, and the absence or presence of an orbital magnetic moment, will depend upon the symmetry and magnitude of the electrostatic fields, and a considerable number of calculations of special cases have been carried out. We shall, however, ignore these refinements and shall adopt the simple assumption that the orbital angular momentum is completely quenched and that only the ground state need be considered. In this case the susceptibility is due entirely to the resultant spin of the ion, and (10·41·1) should apply. The theoretical and experimental results are compared in table 10·4, and it will be seen that the agreement is fair.

Table 10·4. *Calculated and observed susceptibilities of ions of the transition elements as they occur in salts at room temperature*

Ion	Normal state if free	$2\{S(S+1)\}^{\frac{1}{2}}$	$\sqrt{(3kT\chi V/N\mu_0^2)}$ observed
Sc^{2+}, Ti^{3+}, V^{4+}	${}^2D_{\frac{3}{2}}$	1·73	1·75
Ti^{2+}, V^{3+}	${}^3F_2$	2·83	2·75–2·85
V^{2+}, Cr^{3+}, Mn^{4+}	${}^4F_{\frac{3}{2}}$	3·87	3·7 –4·0
Cr^{2+}, Mn^{3+}	${}^5D_0$	4·9	4·8
Mn^{2+}, Fe^{3+}	${}^6S_{\frac{5}{2}}$	5·92	5·2 –5·95
Fe^{2+}	${}^5D_4$	4·9	5·35
Co^{2+}	${}^4F_{\frac{9}{2}}$	3·87	4·6 –5·0
Ni^{2+}	${}^3F_4$	2·83	3·25
Cu^{2+}	${}^2D_{\frac{5}{2}}$	1·74	1·8 –2·0

10·414. *The effect of the crystalline field.* The foregoing theory gives much too simplified an account of the effect of the crystalline field upon the properties of magnetic ions in a solid. A much more elaborate theory has been developed which is sufficiently general to enable us to calculate the energy levels of an ion taking into account the spin-orbit forces and the crystalline field of the surrounding atoms and ions. In recent years the technique of paramagnetic resonance has also been developed by means of which the separations between the various energy levels can be determined experimentally as functions of an external magnetic field by observing the absorption of centimetre electromagnetic waves. (See, for example, the review articles by Bleaney & Stevens, 1953 and by Bowers & Owen, 1955.) It has, therefore, proved possible to analyse in considerable detail the energy-level systems of salts of the rare earths and of the transition metals. If the crystalline field produces a large splitting in the degenerate levels of the metallic ions, and if the spin-orbit and spin-spin interactions can be considered to be small perturbations, the theory can be formally generalized by introducing a tensor g-factor g_{ij}. The partition function (10·41·3) is then replaced by

$$Z = \frac{\sinh\left[(2S+1)\mu_0\left(\sum_{i,j,k} g_{ij}g_{ik}H_jH_k\right)^{\frac{1}{2}}/kT\right]}{\sinh\left[\mu_0\left(\sum_{i,j,k} g_{ij}g_{ik}H_jH_k\right)^{\frac{1}{2}}/kT\right]}, \qquad (10·414·1)$$

so that the susceptibility is anisotropic and depends upon the direction of the magnetic field relative to the crystal axes. A discussion of these extensions of the theory would, however, take us too far afield. They are particularly important at very low temperatures.

10·42. *The adiabatic demagnetization of paramagnetic salts.* It was pointed out by Debye (1926) and by Giauque (1927) that, if the electron spins in a paramagnetic salt could be alined by a sufficiently strong magnetic field, very low temperatures could be reached by demagnetizing the salts adiabatically. This method of obtaining low temperatures is now extensively used. With fields of the order of 10,000 gauss and initial temperatures of the order of 1° K., final temperatures of the order of 0·001° K. can be attained. A purely thermodynamic theory of the phenomena can be given, based upon the formulae which are derived later in § 10·53. We adopt here a slightly different method of approach and consider from the outset the explicit formula for the entropy of an ideal paramagnetic salt.

According to (10·4·2) the entropy of an ideal paramagnetic salt is given by

$$S = S_0(T) + Nk\log\frac{\sinh\{(2S+1)\mu_0H/kT\}}{\sinh(\mu_0H/kT)}$$
$$-(2S+1)\frac{N\mu_0H}{T}\coth\frac{(2S+1)\mu_0H}{kT} + \frac{N\mu_0H}{T}\coth\frac{\mu_0H}{kT}, \qquad (10·42·1)$$

where $S_0(T)$ is a function of T only. For large fields this reduces to

$$S = S_0(T) + Nk(1 + 2\mu_0 H/kT)\exp(-2\mu_0 H/kT) + \dots \quad (\mu_0 H \gg kT), \quad (10{\cdot}42{\cdot}2)$$

and for low fields to

$$S = S_0(T) + Nk\log(2\mathrm{S}+1) - \tfrac{2}{3}\mathrm{S}(\mathrm{S}+1)Nk(\mu_0 H/kT)^2 + \dots \quad (kT \gg \mu_0 H). \quad (10{\cdot}42{\cdot}3)$$

Now at temperatures of the order of 1° K. the contribution of the lattice vibrations to the entropy is negligible and we can therefore reduce the total entropy to as small a value as we please by applying a large enough field. It would then appear from (10·42·1) that we could reach arbitrarily near to the absolute zero, since in an adiabatic change H/T is constant if $S_0(T) = 0$. On the other hand, (10·42·3) shows that in the absence of a field the entropy S is $Nk\log(2\mathrm{S}+1)$ at $T = 0$, and not zero. This paradox arises because of the double limit $H \to 0$, $T \to 0$. If $H = 0$, the entropy S decreases monotonically to $Nk\log(2\mathrm{S}+1)$ as $T \to 0$, but, at $T = 0$, the presence of an external magnetic field, however small, will reduce the entropy to zero. In actual practice, of course, no substance behaves as an ideal paramagnetic at extremely low temperatures and (10·42·1) breaks down. At temperatures of the order of 1° K. and in the absence of a magnetic field, the entropy is correctly given by (10·42·3) as $Nk\log(2\mathrm{S}+1)$. This means that at about 1° K. there are $(2\mathrm{S}+1)^N$ possible orientations of the spins, i.e. that the system has a degeneracy of order $(2\mathrm{S}+1)^N$. This degeneracy must, however, be only apparent and the energy levels must be arranged in groups, the energy differences between the groups being of the order of $k\Theta_m$, where Θ_m is of the order of 0·001–0·01. At very low temperatures, however, the entropy must be zero, and the substance must become either ferromagnetic with all the electron spins parallel or antiferromagnetic with half the spins in one direction and half in the other (see §§ 10·5 and 10·6). In both cases the configuration is such that only one arrangement of spins is possible.

There are three main factors responsible for the splitting of the $(2\mathrm{S}+1)^N$ orientational energy levels, namely, the anisotropic electric fields which exist in a crystal, the magnetic interactions between the spins, and the electrostatic exchange forces. In a highly hydrated salt (and those which are not highly hydrated cannot be considered to be even approximately ideal paramagnetics) the magnetic interactions and the exchange forces are small, and the main splitting of the levels is due to the crystalline field. In potassium chrome alum $KCr(SO_4)_2 . 12H_2O$, which has been extensively studied, the lowest state of the Cr^{3+} ion is effectively a ^4S state. (The ground state of the free ion is ^4F, but in the crystal the orbital angular momentum is quenched by the crystalline field.) This ^4S state actually consists of two doublets with an energy separation of the order of 0·3 cm.$^{-1}$. Therefore, as the temperature is lowered, the entropy decreases from $Nk \log 4$ to $Nk \log 2$ in a temperature range centred round 0·5° K. At a considerably lower temperature the magnetic interaction of the spins and the exchange forces come into play, and the substance appears to become antiferromagnetic. The general trend of the entropy as a function of the temperature is shown in fig. 10·6, and it will be seen that temperatures of the order of 0·01° K. can be reached by the adiabatic demagnetization of this alum. To attain very low temperatures it is essential to use substances for which the entropy variation and the specific heat anomalies occur at as low a temperature as possible.

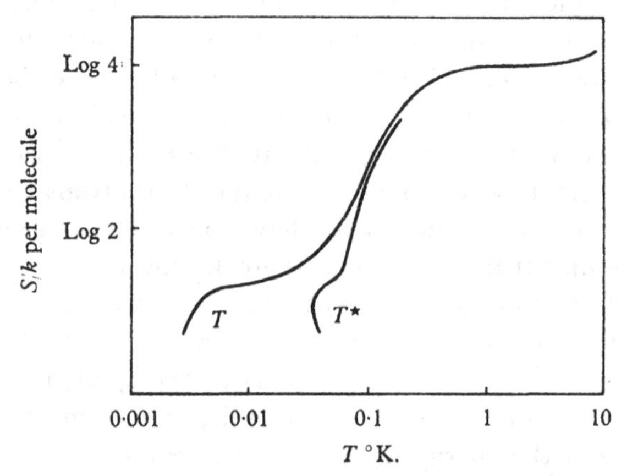

Fig. 10·6. The entropy of potassium chrome alum as a function of T^* and of T.

10·421. If we know the system of energy levels either by paramagnetic resonance methods or by theoretical calculations of the effect of the crystalline fields, the determination of the entropy and of the specific heat is straightforward. The partition function can be written down at once, and, if there are only two levels with weights ϖ_0 and ϖ_1 to be considered, we obtain the formulae derived in § 6·16 when considering the electronic excitation functions in a gaseous molecule. The specific heat is therefore of the general form shown in fig. 6·6, p. 147. For large values of T the specific heat is proportional to T^{-2}, while it tends to zero like $e^{-\Delta E/kT}$ below the temperature corresponding to the maximum. The variation of the entropy from $k \log (\varpi_0 + \varpi_1)$ for large T to $k \log \varpi_0$ for small T is given by equation (6·16·2). The variation of the susceptibility with temperature is given by the analogue of van Vleck's formula (10·41·4).

The influence of the magnetic forces and of the exchange interactions in splitting the energy levels still further is much more difficult to calculate, and no satisfactory theory exists. An attempt has been made by van Vleck

(1937*a*) to calculate these effects, and his results have been tentatively applied to an analysis of the data by Hebb & Purcell (1937). But the theory can at best only apply to temperatures which are high relative to the energies involved, and many features of the phenomena are therefore obscured. If we assume that the effect of the magnetic and exchange forces can be described as being equivalent to the existence of an inner field, we can apply the formal theories of ferromagnetism and antiferromagnetism given in §§ 10·5 and 10·6, though at the moment it is difficult to calculate what the internal field should be.

10·422. *The absolute temperature scale below* 1° K. The usual method of determining the absolute temperature below 1° K. is to introduce an empirical temperature T^* which is proportional to the reciprocal of the magnetic susceptibility. If Curie's law is obeyed, T^* is the same as T, but if there are departures from Curie's law it is necessary to use the second law of thermodynamics to calculate T from T^*. We have first of all to obtain the entropy S as a function of T^*. In principle this is straightforward, for, if T^* is reached by adiabatic demagnetization from temperature T_i and field H, $S(T^*)$ is the same as $S(H, T_i)$, and since T_i is in the temperature region for which (10·42·1) holds, we can calculate $S(H, T_i)$ (or measure it by the heat evolved during the isothermal magnetization of the substance at T_i). We must then determine the internal energy $U(T^*)$ as a function of T^*, which can be done by measuring the rate of heating on the empirical T^* scale, of the substance by γ-rays or by other methods. Then from the relation $dU(T^*) = T\,dS(T^*)$, we have

$$T = \frac{\partial U(T^*)}{\partial T^*} \bigg/ \frac{\partial S(T^*)}{\partial T^*}.$$

In practice the measurements are not easy to carry out, particularly since the thermal conductivity is so low that it is difficult to ensure uniformity of temperature. The entropy of potassium chrome alum is shown in fig. 10·6 as a function of T^* as well as of T, and, since the susceptibility of this substance passes through a maximum at $T^*_{\text{min.}}$, there are two branches of the S, T^* curve which must be considered separately (de Klerk, Steenland & Gorter, 1949). For absolute temperatures above that corresponding to $T^*_{\text{min.}}$, T is an increasing function of T^*, while for lower absolute temperatures T is a decreasing function of T^*. In view of the existence of a maximum in the susceptibility, it seems likely that below $T^*_{\text{min.}}$ the substance is in an antiferromagnetic state (see § 10·6).

For a more detailed description of the properties of the typical salts of interest in low-temperature work, the reader is referred to the monographs by Casimir (1940) and by Garrett (1954).

10·423. *Nuclear paramagnetism and negative absolute temperatures.* Many atomic nuclei have non-zero spins, and their magnetic moments can therefore be alined by a magnetic field. But since the magnetic moment of a nucleus is only of the order of 10^{-4} of a Bohr magneton, nuclear magnetic effects in solids normally only become apparent at temperatures of the order of 10^{-3} °K. or lower. At such temperatures the thermal vibrations of a solid are negligible, and the nuclear spins can be considered as forming an isolated thermodynamic system. The spins are quasi-independent of one another, but they are weakly coupled together by the spin-spin interactions, which ensure that any departure from thermodynamic equilibrium dies out in about 10^{-5} sec. (see, for example, Bloembergen, 1949). The interactions between the nuclear spins and the lattice vibrations, on the other hand, are so small that a non-equilibrium distribution of energy between the spins and the vibrations can exist for times of the order of some minutes even at room temperature.

If we consider N nuclei with spin $\frac{1}{2}$ and magnetic moment μ_n in a magnetic field H, each nucleus has two spin states with energies $\pm\mu_n H$. The partition function per nucleus is

$$Z = \exp(\mu_n H\vartheta) + \exp(-\mu_n H\vartheta) = 2\cosh(\mu_n H\vartheta), \qquad (10\cdot423\cdot1)$$

where we have used the empirical temperature ϑ instead of the more normal $1/(kT)$, while the energy U of the N nuclei is

$$U = N\frac{-\mu_n H\exp(\mu_n H\vartheta) + \mu_n H\exp(\mu_n H\vartheta)}{\exp(\mu_n H\vartheta) + \exp(-\mu_n H\vartheta)} = -N\mu_n H\tanh(\mu_n H\vartheta). \qquad (10\cdot423\cdot2)$$

Now the minimum value that U can have is $-N\mu_n H$, which occurs when all the nuclei are in the spin state with the energy $-\mu_n H$, i.e. when $\vartheta = \infty$. The normal maximum value of U is zero, which occurs when half the nuclei have the energy $-\mu_n H$ and half have the energy $\mu_n H$. In this case we have $\vartheta = 0$. There is, however, no reason to exclude the possibility that more than half of the nuclei have energies $\mu_n H$, but such an arrangement is normally considered to be unstable. If, however, we disregard this objection for the moment, such a state of affairs can only be described by the thermodynamic equations if we allow ϑ to take on negative values (Ramsey, 1956).

In the discussion of the second law of thermodynamics in Chapters 2 and 4, it was pointed out that there are three separate conclusions which can be drawn from it. These are first the existence of the absolute temperature and of the entropy, second, the fact that the entropy varies in one direction only in an irreversible change, and third, that this direction is such that the entropy change is an increase. Now the first conclusion is a consequence of the properties of reversible changes only, as given, for example, in generalization C on p. 78, and it is independent of the other two conclusions. The third conclusion in particular could be modified by a change in the definitions of the absolute temperature and of the entropy. It is therefore by no means impossible for systems to exist for which some of the usual laws of thermodynamics require modification. The analysis in § 4·8, of how far it is possible to develop the laws of thermodynamics using only the properties of reversible processes, revealed that exceptional cases could arise when the empirical temperature and the empirical entropy curves (or surfaces) do not give a single-valued mapping of the parameter space of the system concerned. In such exceptional cases, the thermodynamic states of the system are divided into two and separated by a barrier which is such that two states, one on either side of the barrier, cannot be connected by a reversible process. Whether such exceptional cases can occur in nature is not predicted by the theory, and, if they do, all the requirements of the theory would be met if generalizations C and C′ were valid for both sets of states but with a different generalization C″ for the two sets.

In the present problem we can include the states with $U>0$ by modifying generalization C″ or the definitions of some of the thermodynamic parameters in various ways. If we wish to retain the principle that the entropy always increases in an irreversible change, which is equivalent to retaining the concept of entropy as measuring the degree of randomness of a system, the changes required can be found as follows.

To determine the entropy and the absolute temperature we have as usual to find an integrating factor for the differential form

$$dU + VI\,dH, \tag{10·423·3}$$

where VI is the total magnetization of the body concerned. Now

$$VI = N\frac{\mu_n \exp(\mu_n H\vartheta) - \mu_n \exp(-\mu_n H\vartheta)}{\exp(\mu_n H\vartheta) + \exp(-\mu_n H\vartheta)} = N\mu_n \tanh(\mu_n H\vartheta), \tag{10·423·4}$$

and so (10·423·3) is equal to $dU + N\mu_n \tanh(\mu_n H\vartheta)\,dH$, which has the familiar integrating factor $\kappa\vartheta$, where κ is a constant. We may therefore write

$$\begin{aligned} dS &= \kappa d(\vartheta U) - \kappa U\,d\vartheta + \kappa N\mu_n \tanh(\mu_n H\vartheta)\,\vartheta\,dH \\ &= \kappa d[\vartheta U + N\log\{2\cosh(\mu_n H\vartheta)\}], \end{aligned}$$

and so

$$dS = \kappa\vartheta(dU + VI\,dH), \quad S = \kappa[\vartheta U + N\log\{2\cosh(\mu_n H\vartheta)\}]. \tag{10·423·5}$$

(The additive constant $N\kappa\log 2$ cannot be found by this simple argument. To determine it rigorously we require to consider a variation in which N

varies, but here it is sufficient to note that $N\kappa \log 2$ is the value required to make $S=0$ for $\vartheta=\infty$.)

The expression (10·423·5) for S is an even function of ϑ. It is zero for $\vartheta=\infty$, and, if κ is positive, it reaches its maximum value of $N\kappa \log 2$ at $\vartheta=0$. For negative values of ϑ, it decreases steadily to zero as ϑ decreases

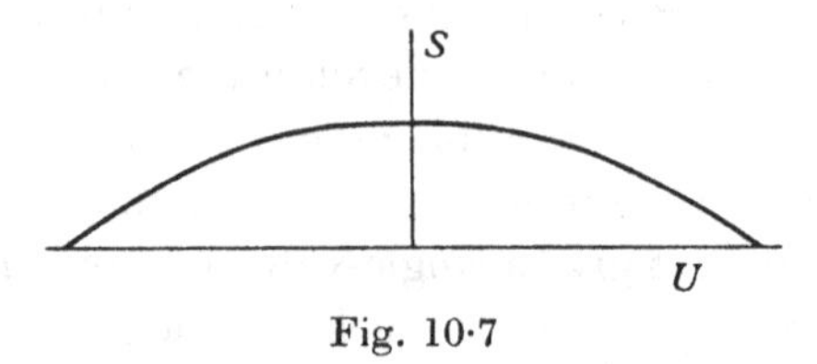

Fig. 10·7

from zero to minus infinity. In order to obtain the usual thermodynamic formulae applicable when the expression (10·423·2) for U is negative, we have to take κ equal to k, Boltzmann's constant, and define the absolute temperature T as $1/(k\vartheta)$ $(0<\vartheta<\infty)$. The absolute temperature is then

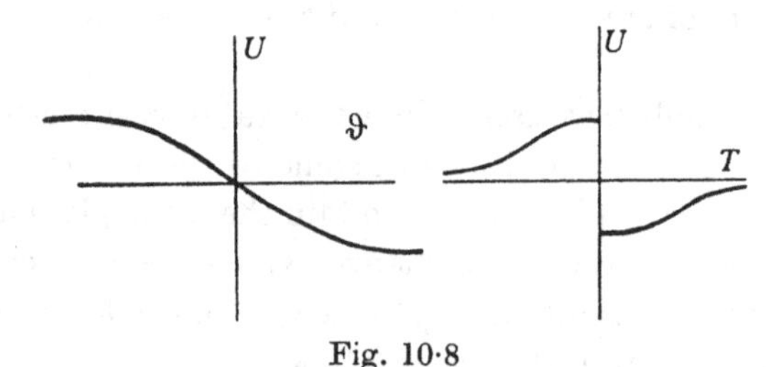

Fig. 10·8

positive. To obtain the states in which U is positive, we have to take ϑ to be negative, but there is no need for κ to be the same for the two sets of states $U<0$ and $U>0$, and we can choose κ to be positive or negative as we please. If, however, we adopt the convention that κ is always positive and equal to k, the entropy is a continuous function of ϑ and U, but the

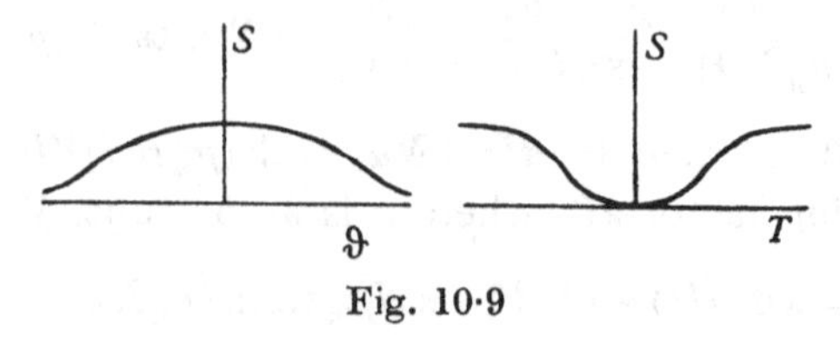

Fig. 10·9

absolute temperature T, defined by $T=1/(k\vartheta)$, is discontinuous and must be allowed to take negative values.

The general behaviour of S as a function of U is shown in fig. 10·7, and of U and S as functions of ϑ and of T in figs. 10·8 and 10·9. It will be seen that S correctly represents the randomness of the system, in that S is zero when the spins of the nuclei are either all parallel or all antiparallel to the magnetic field, while S is a maximum when half the spins are parallel and half are antiparallel to the field. We have therefore to make no changes in the properties usually associated with the entropy function, and, if we take it as an axiom that any spontaneous processes increases the randomness of a system, the increasing property of entropy applies to systems with negative temperatures. Some of the other properties of thermodynamic systems do, however, require reinterpreting.

If we define a system in state A to be hotter than the same system in state B if $U_A > U_B$, we see from fig. 10·8 that all negative absolute temperatures are hotter than all positive absolute temperatures. It also follows that, if T_A and T_B are both positive or both negative, A is hotter than B if $T_A > T_B$. Further, since the hotter of two bodies with negative temperatures has the smaller entropy, it is necessary to restate Kelvin's principle as follows: *No cyclic process exists which produces no other effect than the conversion of work into heat which is absorbed by a body with a negative temperature.* Clausius's principle (see p. 18), on the other hand, remains unchanged.

10·424. The arguments put forward in the preceding section concerning the possible existence of systems with negative absolute temperatures are of theoretical interest only unless some mechanism can be found by which such systems can be physically realized. The first experiments in which negative temperatures were obtained seem to be those carried out on lithium fluoride by Purcell & Pound (1951). A pure crystal of lithium fluoride at room temperature was placed in a strong magnetic field, and the spins alined themselves, with more nuclei in the lower than in the higher energy levels. The magnetic field was then suddenly reversed in so short a time that the populations of the various levels were unable to change. The magnetic energy of each level was, however, reversed in sign, and therefore in the reversed field the higher energy levels were the more densely populated, a state of affairs which, as we have seen, can be described by attributing a negative temperature to the spin system. In this way Purcell & Pound were able to achieve a negative temperature of the order of $-350°$ K.

Up to the present, no other systems than spin systems have been discovered which have negative temperatures. One criterion which is necessary is that the energy levels should have an upper bound, for only if this condition is satisfied can the internal energy increase while the entropy decreases, if no external work is done. A further criterion is that the system must be capable of being thermodynamically isolated from all bodies with positive temperatures, and it is unlikely that there are many systems which satisfy both criteria.

10·43. *Weak paramagnetism in metals.* Apart from the transition metals and the ferromagnetic metals which owe their special properties to the existence of appreciable internal fields, most metals are either diamagnetic or weakly paramagnetic. As pointed out in § 6·32, the conduction electrons in a metal obey Fermi-Dirac statistics and not classical statistics, and it can be readily shown that this results in a small temperature-independent spin paramagnetism. If $f_0(E)$ is the Fermi function, and if we use the notation of § 6·3, the total magnetic moment of the electrons is

$$VI = \mu_0 \sum_i \{f_0(E_i - \mu_0 H) - f_0(E_i + \mu_0 H)\} = 2\mu_0^2 H \sum_i \partial f_0(E_i)/\partial\zeta + O(H^2). \tag{10·43·1}$$

If the volume V of the assembly is sufficiently large we can replace the summation by an integration as in §§ 5·4 and 6·3 and, using the notation of the latter section, we have

$$VI = 4\pi(2mkT/h^2)^{\frac{3}{2}} H\mu_0^2 V \partial F_{\frac{1}{2}}(\xi)/\partial\zeta. \tag{10·43·2}$$

Hence, if we use the expression (6·3·6) for N/V, the number of free electrons per unit volume, the susceptibility per gram atom can be written as

$$\chi_A = \frac{\mathscr{L} z\mu_0^2}{kT} \frac{1}{F_{\frac{1}{2}}(\xi)} \frac{dF_{\frac{1}{2}}(\xi)}{d\xi}, \tag{10·43·3}$$

where $\mathscr{L}$ is Avogadro's number and z is the valency of the metal.

If T is very much larger than T_0 the degeneracy temperature of the electron gas, we have $F_{\frac{1}{2}}(\xi) = \Gamma(\frac{3}{2})\, e^{\xi}$ and

$$\chi_A = \mathscr{L} z\mu_0^2/kT, \tag{10·43·4}$$

which is the same as (10·41·1) with $S = \frac{1}{2}$. If, however, $T_0 \gg T$, we have $F_{\frac{1}{2}}(\xi) = \frac{2}{3}\xi^{\frac{3}{2}} + \ldots$, by (6·31·6), and

$$\chi_A = \tfrac{3}{2}\mathscr{L} z\mu_0^2/\zeta_0, \tag{10·43·5}$$

which, on substituting from (6·311·2) for ζ_0, gives

$$\chi_A = 4\pi\mathscr{L} m\frac{\mu_0^2}{h^2}\left(\frac{3z}{\pi n_a^2}\right)^{\frac{1}{3}}, \tag{10·43·6}$$

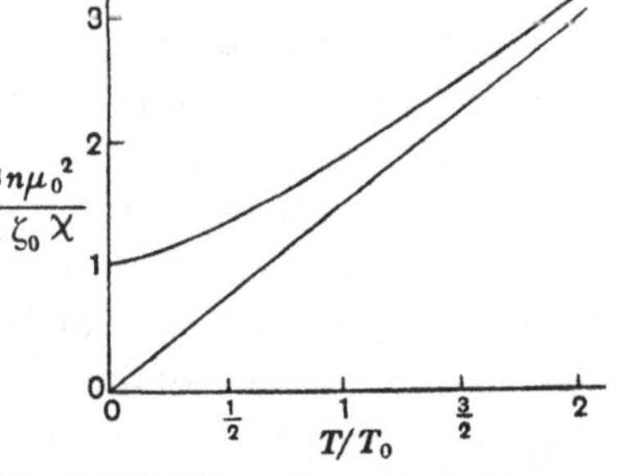

Fig. 10·10. The reciprocal of the paramagnetic susceptibility as a function of the temperature for a gas of free electrons. The straight line gives the classical value of $1/\chi$.

where n_a is the number of atoms per cm.³. For intermediate temperatures, χ is most conveniently obtained numerically. The general behaviour of χ is shown in fig. 10·10.

In order to compare the theoretical predictions with the observed values it is first necessary to correct for the diamagnetism of the metallic ions, which can be done by assuming that the ionic cores are the same for the metallic state as for the free atoms. Secondly, the full theory of the effect of a magnetic field upon a free electron gas shows that there is a diamagnetic effect which is exactly one-third of the paramagnetic effect due to the electron spin (see, for example, Wilson, 1953, pp. 160 ff.). If both these corrections are made, the results for the monovalent metals are as given in table 10·5. A more elaborate theory is required for multivalent metals.

Table 10·5. *Susceptibilities per gram atom of the monovalent metals in units of* 10^{-6}

	$n_a \times 10^{-21}$	Total χ_A observed	χ_A due to ions	χ_A due to valency electrons	χ_A calculated
Li	48·3	(3·5)	− 0·7	(4·2)	6·6
Na	26·2	12 15	− 5·5	17·5 20·5	10·0
K	13·8	15·5 21	− 13·5	29 34·5	15·5
Rb	11·3	6 18	− 23	29 41	17·5
Cs	9·0	− 13 29	− 35	22 64	20
Cu	85	− 5·5	− 14	8·5	4·5
Ag	59	− 21	− (25 to 30)	(4 to 9)	6
Au	59	− 29	− 43	14	6

FERROMAGNETISM

10·5. When an apparently unmagnetized piece of iron is placed in a magnetic field of a few gauss a magnetic moment is induced which is proportional to the field. If the magnetic field is increased the magnetic moment increases rapidly and reaches an approximately constant value for fields of the order of a few hundred gauss, after which there is a slow rise in the magnetization with increasing field. These magnetization phenomena are exceedingly complex and depend upon the previous history and the state of purity of the specimen, and also, if the specimen is a single crystal, upon the relation of the inducing field to the crystal axes. They are, however, explicable on the assumption that a ferromagnet consists of a number of domains of microscopic dimensions each exhibiting the same degree of spontaneous magnetism, which depends only on the temperature. In an apparently unmagnetized ferromagnet the moments of the various domains cancel one another on a macroscopic scale, and the phenomena which occur on applying a magnetic field are due to variations in the sizes of the domains and in the orientations of their directions of magnetization. The most convincing argument for the existence of the microscopic spontaneous magnetism is that a ferromagnetic body possesses an excess specific heat compared with normal metals; that is, a ferromagnet has an excess internal energy which, to a first approximation at least, does not depend upon the apparent magnetization but which is a function of the temperature only.

The form of the magnetization curve is much simplified if the specimen is a single crystal of ellipsoidal form, so that if it is placed in a uniform magnetic field the demagnetizing coefficient can be calculated and the magnetizing field in the specimen is uniform and known. (It is assumed in what follows either that the specimen is in the form of a long rod parallel to the external field or that the necessary correction is introduced to take care of the demagnetizing effects.) If an external field is applied along a cubic axis of a single crystal of iron the magnetization increases very rapidly

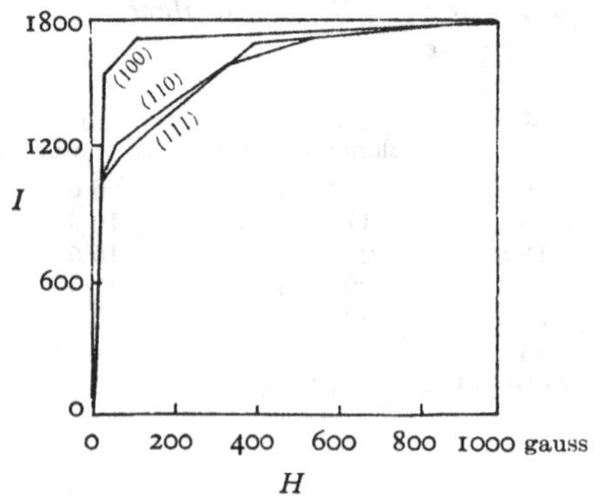

Fig. 10·11. The magnetization curves for single crystals of iron.

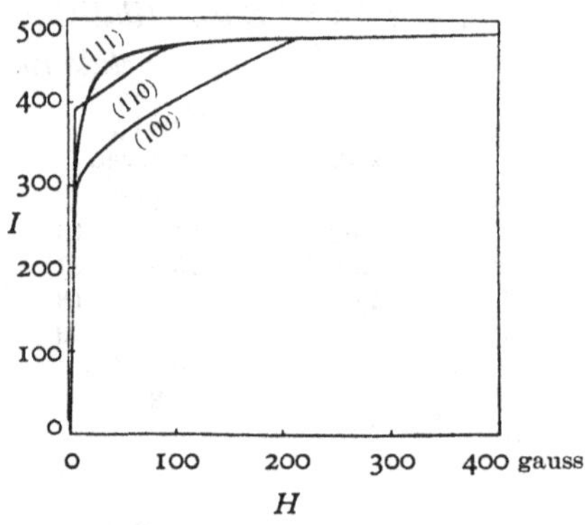

Fig. 10·12. The magnetization curves for single crystals of nickel.

as the field is increased, and reaches a constant value (the saturation value appropriate to the temperature of the specimen) at a few tens of gauss (see fig. 10·11). For other directions the magnetization curve is more complex, though the saturation value in sufficiently high fields is the same as for a cubic axis. We may therefore say that the cubic axes are 'directions of easy magnetization'. In nickel the cube diagonals are the directions of easy magnetization, and the magnetization curves for the other principal directions are shown in fig. 10·12. In a macroscopically unmagnetized body the various domains are magnetized to the saturation value along one or other of the directions of easy magnetization. For more information on all these subjects the reader is referred to the review articles by Stoner (1948, 1950).

It was first suggested by Weiss in 1907 that a ferromagnet is a paramagnetic body in which there is an internal magnetic field, proportional to the magnetization, which is sufficiently strong to aline the spins of the electrons in spite of the disordering effect of the thermal motion. If we postulate the existence of such a field we can explain qualitatively many of the most important features of ferromagnetism, but the origin of the internal field remained obscure until 1928, when Heisenberg showed that in certain circumstances the exchange forces which are responsible for the valency forces between atoms can introduce a coupling between the electron spins in a metal tending to aline them. He also showed that the exchange forces are equivalent to an internal magnetic field of the order of 10^7 gauss, which is of the correct order of magnitude.

The experimental data suggest that the magnetization of ferromagnets is entirely due to the spins of the electrons and that the average number of electrons per atom which contribute to the magnetization is not integral. If there are N such electrons in a ferromagnet of volume V, the total magnetic moment at $T=0$ will be $VI=N\mu_0$, where μ_0 is the Bohr magneton, but to determine I at any other temperature we require to know the structure (or at least the energy levels) of the ferromagnet. Many theories of the structure of ferromagnets, some of them simple and some very complex, have been put forward. The application of thermodynamics and statistical mechanics to the problem can, however, be illustrated by dealing with any of the theories, and we therefore choose the simplest, which is essentially that proposed by Weiss.

We shall assume that the relation between I and the total magnetic field H is that given by (10·4·4), and that

$$H=H_{\text{ext.}}+a\rho I, \tag{10·5·1}$$

where $H_{\text{ext.}}$ is the applied magnetic field, ρ is the density and a is a constant. The spontaneous magnetization is determined by putting $H_{\text{ext.}}=0$, and, if the equations (10·4·4) and (10·5·1) are to have a non zero solution for I, the gradient of the I, H curve (10·4·4) for $H=0$ must be greater than $1/a\rho$. The condition for the existence of spontaneous magnetization is therefore

$$T<\theta, \quad \theta=N\mu_0^2 a\rho/kV. \tag{10·5·2}$$

The temperature θ at which ferromagnetism disappears is known as the Curie temperature. Its value is 1043° K. for iron, 1393° K. for cobalt and 631° K. for nickel. If $T<\theta$, the spontaneous magnetization is given by the implicit equation

$$\frac{VI}{N\mu_0}=\tanh\left(\frac{VI}{N\mu_0}\frac{\theta}{T}\right). \tag{10·5·3}$$

As $x\to\infty$, $\tanh x\sim 1-2e^{-2x}$, and since $VI\to N\mu_0$ as $T\to 0$, we have

$$VI/N\mu_0=1-2e^{-2\theta/T} \quad (T\to 0). \tag{10·5·4}$$

Also $\tanh x=x-\frac{1}{3}x^3+\dots$ for small x, and so, near the Curie point,

$$VI/N\mu_0=\sqrt{3}\,(1-T/\theta)^{\frac{1}{2}} \quad (T\to\theta-0). \tag{10·5·5}$$

The general form of the I, T curve is as shown in fig. 10·13, where the relative magnetization $\sigma=VI/N\mu_0$ is given as a function of T/θ. When

$H_{\text{ext.}} \neq 0$, I is a steadily decreasing function of T. It is small above the Curie point for any physically realizable field, but I never becomes zero for any finite value of T.

10·51. When $T > \theta$, the only solution of (10·5·3) is $I = 0$, and the body is merely paramagnetic with I proportional to $H_{\text{ext.}}$. Equations by (10·4·4) and (10·5·1) then give

$$\frac{VI}{N\mu_0} = \frac{\mu_0 H}{kT} + \ldots = \frac{\mu_0 H_{\text{ext.}}}{kT} + \frac{\theta}{T}\frac{VI}{N\mu_0} + \ldots,$$

so that

$$\chi = \frac{I}{H_{\text{ext.}}} = \frac{N}{V}\frac{\mu_0^2}{k(T-\theta)}. \tag{10·51·1}$$

This is known as the Curie-Weiss law.

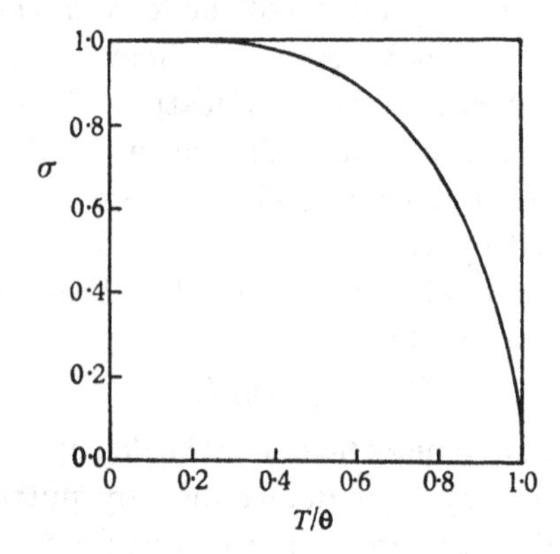

Fig. 10·13. The relative magnetization as a function of T/θ, where θ is the Curie temperature, according to Weiss's theory.

10·52. To calculate the thermal effects in a ferromagnet it is simplest to start from one of the expressions for the free energy. If we take the partition function to be given by (10·4·2) with $S = \frac{1}{2}$, the independent variables are T and H, so that the free energy concerned is F_2 (cf. equation (10·12·8)). Hence

$$F_2 = F_0(T) - NkT \log\{2\cosh(\mu_0 H/kT)\}, \quad H = H_{\text{ext.}} + a\rho I, \tag{10·52·1}$$

where $F_0(T)$ is a function of T only, and where

$$dF_2 = -SdT - VIdH. \tag{10·52·2}$$

Differentiation of F_2 with respect to H gives

$$VI = N\mu_0 \tanh(\mu_0 H/kT), \tag{10·52·3}$$

which is the same as (10·4·5) except that H here includes the internal field, while differentiation with respect to T gives

$$S = S_0(T) + Nk\log\{2\cosh(\mu_0 H/kT)\} - VHI/T. \qquad (10\cdot52\cdot4)$$

To determine the specific heat we have to keep $H_{\text{ext.}}$ constant and not H, so that the specific heat C per unit mass is given by

$$\rho VC = T\left(\frac{\partial S}{\partial T}\right)_{H_{\text{ext.}}} = T\frac{dS_0}{dT} - V(H_{\text{ext.}} + a\rho I)\left(\frac{\partial I}{\partial T}\right)_{H_{\text{ext.}}}, \qquad (10\cdot52\cdot5)$$

all the other terms vanishing on account of (10·52·3). Since $\partial I/\partial T < 0$, the magnetic specific heat is positive and it is a continuous function of T for fixed $H_{\text{ext.}}$ unless $H_{\text{ext.}} = 0$. If, however, there is no external magnetic field, $\partial I/\partial T$ is discontinuous at the Curie point and there is a discontinuity in the specific heat. Thus the transition from ferromagnetism to paramagnetism in the absence of an external field is a second-order transition.

If we write (10·52·3) in the form

$$\frac{T}{\theta} = \frac{\sigma}{\tanh^{-1}\sigma}, \quad \sigma = \frac{VI}{N\mu_0},$$

we have

$$\frac{d(T/\theta)}{d\sigma} = \frac{(T/\theta)(1-\sigma^2 - T/\theta)}{\sigma(1-\sigma^2)}, \qquad (10\cdot52\cdot6)$$

and so the 'magnetic' part of the specific heat is

$$C_M = -aI\left(\frac{\partial I}{\partial T}\right)_{H_{\text{ext.}}} = -\frac{N}{V\rho}\sigma\frac{d\sigma}{dT}k\theta = \frac{\sigma^2(1-\sigma^2)}{(T/\theta)\{(T/\theta)-1+\sigma^2\}}\frac{Nk}{\rho}. \qquad (10\cdot52\cdot7)$$

The magnetic specific heat increases steadily from zero at $T=0$ to a maximum of $\frac{3}{2}Nk/V\rho$ at $T=\theta$, being given by

$$C_M = \tfrac{3}{2}(3 - 2\theta/T)\,Nk/V\rho \quad (T \to \theta - 0) \qquad (10\cdot52\cdot8)$$

just below the Curie point. Above the Curie point C_M is, of course, zero.

10·521. We may derive expressions for the other thermodynamic functions from (10·52·1). For example, we have

$$U_2 = -T^2\left(\frac{\partial}{\partial T}\frac{F_2}{T}\right)_H = U_0(T) - N\mu_0 H\tanh(\mu_0 H/kT) = U_0(T) - VIH. \qquad (10\cdot521\cdot1)$$

Also, by the analogue of (10·12·2),

$$U = U_2 + V\int I\,d(H - H_{\text{ext.}}) = U_2 + a\rho V\int I\,dI = U_2 + \tfrac{1}{2}a\rho VI^2, \qquad (10\cdot521\cdot2)$$

so that

$$U = U_0(T) - VIH_{\text{ext.}} - \tfrac{1}{2}a\rho VI^2. \qquad (10\cdot521\cdot3)$$

This equation for U is often made the starting point of the theory.

10·53. *The magneto-caloric effect.* We have already discussed the magneto-caloric effect in paramagnetic salts in § 10·42, the treatment there being based upon the explicit expression for the entropy as a function of the magnetic field and of the temperature. A similar treatment is possible for ferromagnets since we have the explicit expression (10·52·4), with $H = H_{\text{ext.}} + a\rho I$, for the entropy. This, however, only applies to a particular model, and an alternative method of discussing the interrelation of changes in the temperature and changes in the magnetic field is to derive a general formula for $(\partial T/\partial H_{\text{ext.}})_S$. This is particularly convenient for ferromagnetic bodies where the temperature change is small. The formula can be obtained as follows. The equation

$$dU = T\,dS - VI\,dH_{\text{ext.}} \tag{10·53·1}$$

shows that

$$\left(\frac{\partial T}{\partial H_{\text{ext.}}}\right)_S = -\left(\frac{\partial VI}{\partial S}\right)_{H_{\text{ext.}}}. \tag{10·53·2}$$

If we neglect volume changes, then since

$$\left(\frac{\partial I}{\partial S}\right)_{H_{\text{ext.}}} = \left(\frac{\partial I}{\partial T}\right)_{H_{\text{ext.}}}\left(\frac{\partial T}{\partial S}\right)_{H_{\text{ext.}}} = \left(\frac{\partial I}{\partial T}\right)_{H_{\text{ext.}}}\frac{T}{V\rho C_H}, \tag{10·53·3}$$

where C_H is the specific heat for constant external field, we can write (10·43·2) as

$$\left(\frac{\partial T}{\partial H_{\text{ext.}}}\right)_S = -\frac{T}{\rho C_H}\left(\frac{\partial I}{\partial T}\right)_{H_{\text{ext.}}}. \tag{10·53·4}$$

Since $\partial I/\partial T$ is negative, an increase in the external magnetic field produces an increase in the temperature of an isolated ferromagnet, a phenomenon known as the magneto-caloric effect. The effect is not large, but may amount to about 1° C. near the Curie point. Some experimental values for nickel are shown in fig. 10·14.

10·54. *The magnetization curves of ideal single crystals.* As already mentioned on p. 317, an apparently unmagnetized piece of iron consists of a large number of domains each of which is magnetized along one or other of the directions of easy magnetization to the saturation value corresponding to the temperature of the specimen. In the absence of a magnetic field all the six directions of easy magnetization are equivalent, but in the presence of a field this is no longer true. If the specimen is in the form of an ideal single crystal, and if a magnetic field is applied along the (1, 0, 0) direction, the magnetic moments of all the domains will swing over and aline themselves along the (1, 0, 0) direction, and the magnetic moment of the specimen will reach its saturation value $I_0(T)$ at a very low value of the external field (ideally in a vanishingly small field). If, however, the field is applied along the (1, 1, 0) direction, half the domains will be magnetized along the $(1, 0, 0)$ direction and half along the $(0, 1, 0)$ direction, so that the value of the apparent magnetization will be $I_0(T)/\sqrt{2}$. But, if the field is sufficiently strong, the magnetization of a domain need not be along a direction of easy magnetization and the apparent magnetization can exceed $I_0(T)/\sqrt{2}$.

To give a quantitative explanation of this phenomenon it is necessary to assume that the energy of a domain depends upon the orientation of the magnetization $\mathbf{I}(T)$ relative to the crystal axes. The simplest hypothesis that gives any directional effect in a cubic crystal is to assume that the

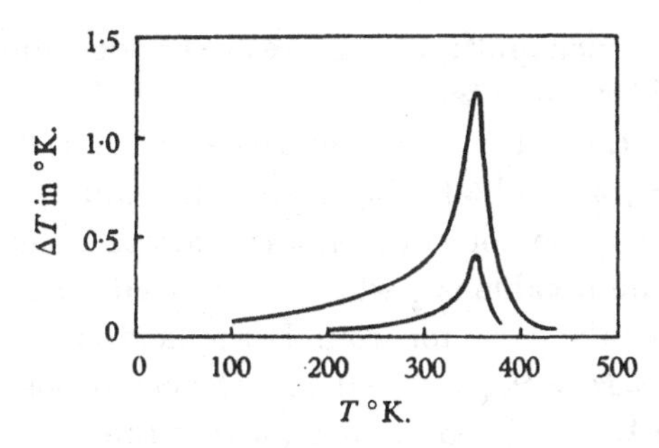

Fig. 10·14. The magneto-caloric effect for nickel. Upper curve, $\Delta H = 17{,}775$ gauss; lower curve, $\Delta H = 4160$ gauss.

magnetization of any domain is $\mathbf{I}_0(T)$, and that the energy per unit volume of a domain depends upon the orientation of $\mathbf{I}_0$ relative to the crystal axes and that its variable part is given by‡

$$-\mathbf{H}.\mathbf{I}_0 + CI_0^4(l^4 + m^4 + n^4), \qquad (10\cdot54\cdot1)$$

where

$$\mathbf{I}_0 = (l, m, n)\, I_0.$$

An energy function of this type cannot be obtained merely by considering the exchange forces between the atoms, and it is necessary to include the effect of the crystalline field, the spin-orbit forces and the spin-spin forces, which can give rise to an anisotropic energy function. The calculation of the effect of these anisotropic forces on the energy levels of special models is extremely difficult, and although many investigations have been carried out, it cannot be said that any of them are entirely satisfactory (see, for example, van Vleck, 1937*a*). For our present purpose it is sufficient to postulate the existence of the energy function (10·54·1) and to consider C as an adjustable parameter (Gans & Czerlinski, 1933). In simple atoms the

‡ The anisotropic part of the energy can be written in several equivalent forms. In a cubic crystal the fourth order term is often written as $K\,(m^2n^2 + n^2l^2 + l^2m^2)$. Since $(l^2 + m^2 + n^2)^2$ is isotropic, this is equivalent to $CI_0^4\,(l^4 + m^4 + n^4)$ with $K = -2CI_0^4$.

exchange energy is of the order of 10^3 times the spin-spin and spin-orbit energies, and although it is not to be expected that their ratio would be the same in a complex ferromagnetic metal,‡ we may hope to obtain an approximate theory by assuming that the anisotropic energy function is small compared with the exchange energy.

If we make (10·54·1) a minimum subject to $l^2+m^2+n^2=1$, we have to satisfy the equations

$$\left.\begin{aligned} -H_x+4CI_0^3l^3+\lambda I_0 l=0, \quad -H_y+4CI_0^3m^3+\lambda I_0 m=0,\\ -H_z+4CI_0^3n^3+\lambda I_0 n=0, \end{aligned}\right\} \quad (10\cdot54\cdot2)$$

where λ is an undetermined multiplier. There are various important special cases, which we consider separately.

(i) For iron the directions of easy magnetization are $(\pm 1, 0, 0)$, $(0, \pm 1, 0)$, $(0, 0, \pm 1)$. Now if $l^2+m^2+n^2=1$, the maximum value of $l^4+m^4+n^4$ is unity, and it is attained when one of l, m, n is unity and the other two are zero, while the mimimum value is $\frac{1}{3}$ which is attained for $l=m=n=1/\sqrt{3}$. We therefore have to take $C<0$ for iron. If the field is along the $(1,1,0)$ direction we have $H_x=H_y=H/\sqrt{2}$, $H_z=0$, and the equations (10·54·2) have one solution in which $\mathbf{I}_0$ is inclined at an angle θ to the x axis, and another solution in which $\mathbf{I}_0$ is inclined at the same angle θ to the y axis. In both cases $\mathbf{I}_0$ is in the xy-plane and each solution applies to half the domains. Therefore

$$I=I_0(\cos\theta+\sin\theta)/\sqrt{2}=I_0(l+m)/\sqrt{2}, \quad (10\cdot54\cdot3)$$

and if we eliminate λ from the first two of the equations (10·54·2) we have

$$H/\sqrt{2}=-4CI_0^3lm(l+m). \quad (10\cdot54\cdot4)$$

But since $n=0$, we have $(l+m)^2=1+2lm$, and hence $2lm=2I^2/I_0^2-1$, which gives

$$\frac{I}{I_0}\left(\frac{2I^2}{I_0^2}-1\right)=-\frac{H}{4CI_0^3} \quad (0<H<-4CI_0^3,\ I>I_0/\sqrt{2}). \quad (10\cdot54\cdot5)$$

For $H>-4CI_0^3$, we have $I=I_0$.

If the field is along the $(1,1,1)$ direction we have $H_x=H_y=H_z=H/\sqrt{3}$, and there are three solutions of (10·54·2) in each of which $\mathbf{I}_0$ is inclined at an angle ϵ to the direction of $\mathbf{H}$, namely, the cube diagonal, and each solution applies to one-third of the domains. Therefore $I=I_0\cos\epsilon$, where $\cos\epsilon=(l+m+n)/\sqrt{3}$. Consider the solution which corresponds to $\mathbf{I}=(I_0, 0, 0)$

‡ In a ferromagnet the orbital angular momentum is quenched by the crystalline field. This does not mean that the orbital augular momentum vanishes identically, but merely that its average value is zero, and the spin-orbit energy may be considerable. The dependence of the spin-orbit energy on the direction of I_0 is determined by the symmetry of the crystal lattice.

when $H=0$. Then $m=n$ by symmetry, and if we write $l=\cos\theta, m=n=\cos\phi$, we have $\cos^2\theta=1-2\cos^2\phi$, i.e. $\sin\theta=\surd 2\cos\phi$. Also, if we eliminate λ from the equations (10·54·2), we have

$$H+4\surd 3\, CI_0^3(\cos\theta+\cos\phi)\cos\theta\cos\phi=0.$$

This equation can be put into various forms, of which the following is one. If we write $\cos\alpha=1/\surd 3$, we have $\epsilon=\alpha-\theta$, $\cos\phi=\cot\alpha\sin\theta$, and after some simplification we obtain

$$\cos\epsilon-\cos(4\alpha-3\epsilon)+\tfrac{2}{3}H/CI_0^3=0 \quad (0<H<-\tfrac{8}{3}CI_0^3,\ I/I_0>1/\surd 3). \tag{10·54·6}$$

The magnetization curves given by (10·54·5) and (10·54·6) are in good qualitative agreement with the observed curves, which are shown in fig. 10·11. It must, however, be emphasized that the values of $-CI_0^4$ obtained from different experiments differ considerably, and that better agreement between the calculated and observed magnetization curves for the (1, 1, 1) direction can be obtained by including a sixth-order term in the anisotropic energy. The value of $-CI_0^4$ deduced from a comparison of the observed and calculated curves is about 2×10^5 ergs/cm.^{3} at room temperature. The energy due to the exchange interactions, on the other hand, is given by $\frac{1}{2}a\rho I_0^2=\frac{1}{2}Nk\theta/V$, which is of the order of 10^9 ergs/cm.^{3}, so that we are justified in considering the anisotropic forces to be small in comparison with the exchange forces.

(ii) We have to take $C>0$ for nickel, since the easy directions of magnetization are the cube diagonals. If the magnetic field is along the (1, 0, 0) direction, we have $\mathbf{I}_0=I_0(l, \pm m, \pm n)$, and the last two equations (10·54·2) give $\lambda=-2CI_0^2(1-l^2)$. Since $I=lI_0$, the first equation reduces to

$$2\frac{I}{I_0}\left(3\frac{I^2}{I_0^2}-1\right)=\frac{H}{CI_0^3} \quad (0<H<4CI_0^3,\ I>I_0/\surd 3). \tag{10·54·7}$$

If the field is in the (1, 1, 0) direction, we have $H_x=H_y=H/\surd 2$ and $I=I_0(l+m)/\surd 2=\surd 2\, lI_0$, since $\mathbf{I}_0=I_0(l, l, \pm n)$. The equations (10·54·2) then reduce to

$$-H/\surd 2+4CI_0^3l^3+\lambda I_0 l=0, \quad \lambda=-4CI_0^2n^2=-4CI_0^2(1-2l^2),$$

which give

$$2\frac{I}{I_0}\left(3\frac{I^2}{I_0^2}-2\right)=\frac{H}{CI_0^3} \quad (0<H<2CI_0^3,\ I>\surd\tfrac{2}{3}I_0). \tag{10·54·8}$$

The value of CI_0^4 deduced by a comparison between the observed and calculated magnetization curves is $2·5\times 10^4$ ergs/cm.^{3} at room temperature.

(iii) Cobalt has a hexagonal structure and the direction of easy magnetization is along the hexagonal axis, which we take to be the z axis. In this case it is possible to include a quadratic as well as a quartic term in the expression for the orientational energy, and instead of (10·54·1) we have

$$-\mathbf{H}.\mathbf{I}_0+CI_0^2n^2+C'I_0^2n^4 \quad (C<0,\ C+2C'<0). \tag{10·54·9}$$

If $\mathbf{I}_0$ makes an angle θ with the z axis and if $\mathbf{H}$ is perpendicular to it, (10·54·9) can be written $-HI_0\sin\theta+I_0^2\cos^2\theta(C+C'\cos^2\theta)$, which is a minimum when $HI_0=-2I_0^2\sin\theta(C+2C'\cos^2\theta)$. But $I=I_0\sin\theta$, and so

$$\frac{H}{I_0}=-2(C+2C')\frac{I}{I_0}+4C'\frac{I^3}{I_0^3}. \tag{10·54·10}$$

For cobalt, I/I_0 is a steadily increasing function of H as shown in fig. 10·15, and we must therefore have $4C'-C>0$. Equation (10·54·10) is then applicable if $0<H<-2CI_0$. The values of CI_0^2 and $C'I_0^2$ deduced from a comparison between the calculated and observed magnetization curves are -6×10^6 ergs/cm.³ and 10^6 ergs/cm.³ respectively.

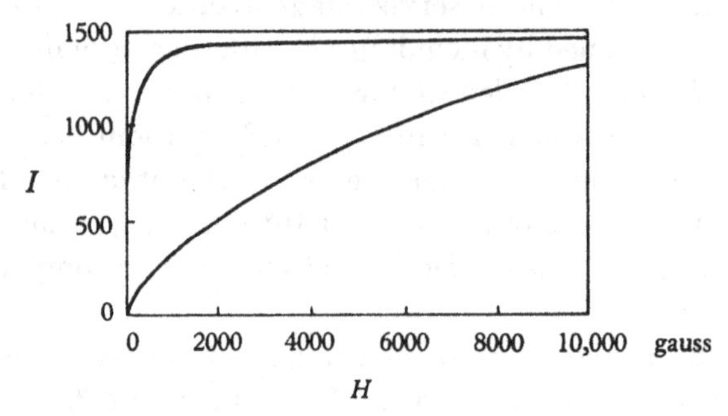

Fig. 10·15. The magnetization curves for cobalt. Upper curve, along the hexagonal axis. Lower curve, perpendicular to the hexagonal axis.

The preceding theory only applies to those simple cases in which the direction of magnetization and the shape of the body are such that the demagnetizing field need not be taken explicitly into account. An extended form of the theory which applies to more complex cases has been given by Néel (1944) and by Lawton & Stewart (1948).

The anisotropy constants vary considerably with temperature, the general trend being as shown in fig. 10·16. As mentioned above, there is no really satisfactory quantitative theory of the anisotropic constants, and more particularly of their temperature variation. In view of the complexity of the problem, great improvements in the theory are unlikely, but even the present theory is satisfactory in its qualitative aspects.

10·55. *The quantum theory of ferromagnetism.* The first attempt to set up a theory of ferromagnetism based upon the quantum theory of the interaction of an assembly of atoms was made by Heisenberg in 1928. If we try to determine the energy levels of such an assembly by the method of successive approximations, and we take as the zero approximation the state in which all the atoms are fixed on their lattice points but do not interact with one another, the first approximation to the energy contains the terms

$$\mathscr{H}_{\text{exch.}} = -2\Sigma J_{ij}\,\mathbf{S}_i\,.\,\mathbf{S}_j, \tag{10·55·1}$$

where the summation is taken over all pairs of atoms. The quantity $\mathbf{S}_i$ is the spin angular momentum operator of the atoms i and J_{ij} is the exchange integral relating to the pair of atoms i and j (see, for example, Dirac, 1935, p. 228). These contributions to the energy are known as the exchange

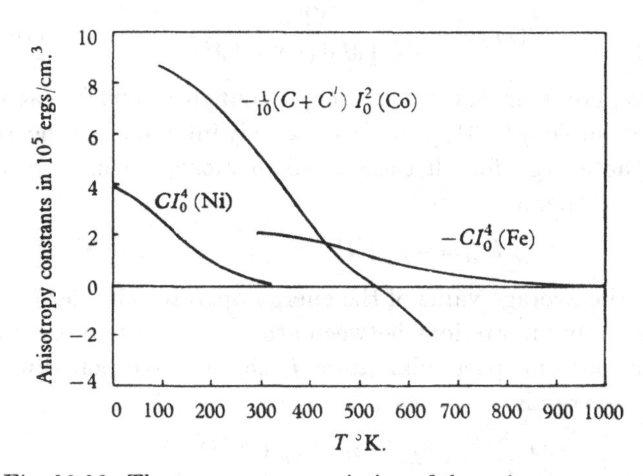

Fig. 10·16. The temperature variation of the anisotropy constants.

energies. They are of electrostatic origin and are of the order of 0·01–0·1 eV., but they result in an apparent coupling of the spins of the various atoms, which is some 10^3–10^4 times larger than that due directly to the magnetic moments of the atoms. The magnitude of the exchange energy falls off rapidly as the distance apart of the atoms increases, and for most purposes it is sufficient to restrict the summation in (10·55·1) to pairs of nearest neighbours. The reduction of the energy operator to the form (10·55·1) involves certain assumptions which are certainly not rigorously justifiable for a system containing a large number of atoms, but even if we assume that the energy levels can be correctly derived from (10·55·1) they can only be obtained explicitly by using approximate methods of calculation, the limits of validity of which are very difficult to estimate. In the present chapter we shall describe some of the simpler theories that have been put forward, while a discussion of the more difficult points is postponed until §§ 14·3 and 14·4.

10·551. *The high-temperature magnetization.* If the temperature is sufficiently high, the partition function will be determined by the mean energy of the system and will be independent of the distribution of the energy levels round the mean. In this case the partition function can be calculated by an elementary method, which is essentially a simplified version of Heisenberg's original theory.

If there are N atoms, each with spin $\frac{1}{2}\hbar$ and zero orbital angular momentum, and if $\frac{1}{2}N+\frac{1}{2}M$ atoms have their spins positively while $\frac{1}{2}N-\frac{1}{2}M$ atoms have their spins negatively directed along an external magnetic field H, the total magnetic moment is $\mu_0 M$. For a given value of the magnetic moment, the spins can be arranged in

$$W(M)=\frac{N!}{(\frac{1}{2}N+\frac{1}{2}M)!\,(\frac{1}{2}N-\frac{1}{2}M)!} \tag{10·551·1}$$

ways. There are, however, many microscopic configurations for each value of M (see, for example, § 14·31), but if we are only interested in the average value E_M of the energy for all these configurations, we may group the various states together and write

$$E_M=-\mu_0 MH+\bar{E}_{\text{exch.}}, \tag{10·551·2}$$

where $\bar{E}_{\text{exch.}}$ is the average value of the energy operator (10·55·1).

If we include only interactions between atoms which are nearest neighbours and consider one particular atom i, the average contribution of (10·55·1) to the energy is

$$-2J\sum_j (S_{i,x}\bar{S}_{j,x}+S_{i,y}\bar{S}_{j,y}+S_{i,z}\bar{S}_{j,z}), \tag{10·551·3}$$

where the summation is taken over the neighbours of the atom i. If the resultant magnetization is along the z axis, we have

$$\bar{S}_{j,x}=\bar{S}_{j,y}=0, \quad \bar{S}_{j,z}=\tfrac{1}{2}M/N. \tag{10·551·4}$$

Hence the value of (10·551·3) is $-zMJS_{i,z}/N$, where z is the number of nearest neighbours. The energy E_M is therefore given by

$$E_M=-\mu_0 MH-\tfrac{1}{4}zJM^2/N, \tag{10·551·5}$$

it being necessary to introduce a factor $\frac{1}{2}$ into the second term to avoid double counting of the nearest neighbours. With the assumptions made, the logarithm of the partition function is given by

$$\log Z=\log W(M)-E_M/kT, \tag{10·551·6}$$

and, treated as a function of M, this is a maximum when

$$\log\frac{N+M}{N-M}=\frac{zJM}{NkT}+\frac{2\mu_0 H}{kT}. \tag{10·551·7}$$

The average magnetic moment is therefore given by

$$M\mu_0 = N\mu_0 \tanh\left(\frac{zJM}{2NkT} + \frac{\mu_0 H}{kT}\right), \tag{10·551·8}$$

which is the same as the basic formula of the Weiss theory, with the Curie temperature θ given by

$$\theta = \tfrac{1}{2}zJ/k. \tag{10·551·9}$$

10·552. According to the preceding theory, ferromagnetism will only occur if the exchange integral J is positive. For simple molecules the spins of the electrons are normally coupled in such a way as to make the ground state the state of minimum multiplicity, and J is negative. The conditions in which J is positive are somewhat unusual, and, in spite of a considerable number of calculations, no very definite criteria have emerged. In the present state of the theory, therefore, we have to introduce the condition $J > 0$ as an additional assumption, which is plausible theoretically for elements of the iron group but not rigorously established. If we derive the values of J from (10·551·9) and from the measured values of θ (with $z = 8$ for iron, and $z = 12$ for cobalt and nickel), we find that J is $2{\cdot}2 \times 10^{-2}$ eV. for iron, $2{\cdot}7 \times 10^{-2}$ eV. for cobalt and 9×10^{-3} eV. for nickel. These are of the same order of magnitude as exchange forces between simple atoms.

In deriving equation (10·551·7), a number of averaging processes have been used whose validity is by no means certain (see, for example, § 14·32). A number of attempts have been made to introduce elaborations into the theory, but these have not been particularly successful, and the main significance of Heisenberg's theory is that it provides a rational explanation of the Weiss internal field.

10·553. *The low-temperature magnetization.* At low temperatures the partition function is determined by the lowest energy levels of the system and not by the mean energy. In this case it is necessary to obtain explicit expressions for the energy levels, and, provided that the exchange integral is positive, the necessary calculations can be carried out approximately by a method due to Slater (1929) and first applied to the problem of ferromagnetism by Bloch (1930). It consists of a generalization of Heitler and London's treatment of the valency forces in molecules. In the zero approximation the ferromagnet is considered to consist of N separate atoms, each situated at a lattice point and each containing one valency electron in its ground state. The correct zero-order antisymmetrical wave functions for the assembly are constructed as sums of determinants, and the energies of the various states can then be calculated to the first order. The lowest state is the one in which all the spins are positively directed, and the states with one, two, etc., reversed spins are successively determined. The problem can be solved exactly for a linear chain if certain simplifying assumptions are made (Bethe, 1931), but, with the same assumptions, only an approximate solution can be obtained for a three-dimensional lattice. For a simple cubic lattice, Bloch found that the magnetization is given by

$$VI = N\mu_0\{1-\gamma(kT/J)^{\frac{3}{2}}\} \quad (kT \ll J), \tag{10·553·1}$$

where

$$\gamma = \frac{1}{2\pi^2}\int_0^\infty \frac{u^{\frac{1}{2}}du}{e^u-1} = 0·117. \tag{10·553·2}$$

The temperature variation of I given by (10·553·1) is more in accordance with the experimental facts than the experimental variation predicted by (10·5·4). The approximations made in deriving (10·552·1) are, however, so difficult to justify that it is impossible to say what the range of validity of the formula is, and whether it applies in the temperature region where the temperature variation of I is readily measurable. Moreover, it is known that Bloch's method gives incorrect results when applied to a two-dimensional square lattice. According to Bloch's theory such a lattice should be only paramagnetic, but the exact theory shows that a ferromagnetic transition point actually exists (see § 14·35).

Several attempts have been made to improve Bloch's result, but they have not been conspicuously successful, and the problem is a most difficult one.

10·554. *The collective electron theory of ferromagnetism.* The preceding theories of ferromagnetism are in strong disagreement with the experimental facts, in that the magneton numbers (the saturation moments per atom at $T=0$) of the ferromagnetic metals are not integral. Also, by forming alloys of the various ferromagnetic metals, the magneton number can be made to vary continuously with composition (see fig. 10·17). To explain such phenomena by means of the preceding theories it is necessary to postulate that a ferromagnetic metal consists of a mixture of atoms with different spins, and the theory becomes exceedingly complex. If, in addition, we wish to explain the conduction properties of ferromagnetic materials, further complications have to be introduced by considering atoms in polar states.

An alternative theory is based upon the approximation of treating the valency electrons as quasi-free, in that they are considered to be capable of moving over the whole of the metal, subject to the binding forces due to the metal ions. The theory based upon this approximation is often known as the collective electron theory. A general description of the wave functions and of the energy levels of such a model has been briefly described in § 6·321, where it was pointed out that the energy levels resemble those of free electrons but are grouped into distinct bands. We may therefore take as our starting point the theory of the weak paramagnetism of metals which is given in § 10·43, and generalize it so as to include the effect of the exchange forces and of the detailed structure of the energy bands, which need not each contain an integral number of electrons per atom.

An investigation of the properties of magnetic bodies containing quasi-free electrons would take us too far afield, and the reader is referred to books dealing with the theory of metals for details of the theory (see, for example, Wilson, 1953, Chapter 7). We shall therefore merely describe in a general way the properties of nickel.

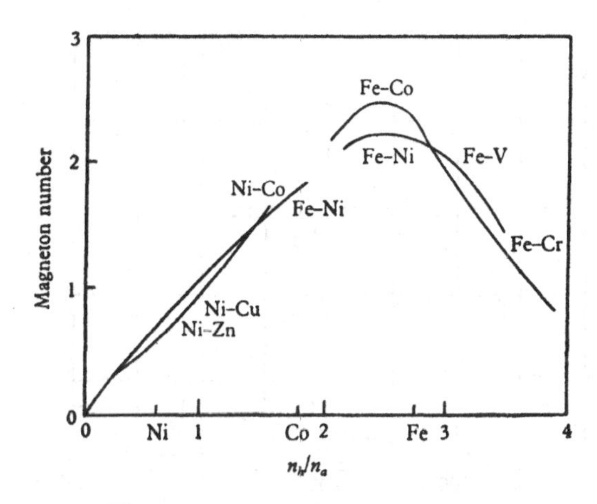

Fig. 10·17. The magneton number of ferromagnetic alloys as a function of the number of holes per atom in the 3*d* band.

The behaviour of nickel suggests that in the metallic state the outer electrons are distributed between two bands, which we may describe as the 3*d* and 4*s* bands, and the average number of electrons per atoms in the 4*s* band is 0·6, while there is the same number of vacancies in the 3*d* band. The electrons which are mainly responsible for the ferromagnetic properties of nickel are those in the 3*d* band. We may formally take into account the binding of the electrons by introducing an effective mass into the formula for the energy levels for free electrons, as in § 6·321, and if we generalize the calculation in § 10·43 by introducing an effective mass and an internal field we obtain formulae which are very similar to those given by Weiss's theory. The main difference is that the electrons contribute to the specific heat even above the Curie point. The observed variation of the relative magnetization with temperature is qualitatively in agreement with the predictions of equation (10·5·3) and of fig. 10·13, and of their generalizations, but the experimental values of I decrease more rapidly with T than the theoretical ones both near $T=0$ and near the Curie point.

The specific heat of nickel as a function of the temperature is shown in fig. 10·18, and the discontinuity at the Curie point is evident. It is, however, by no means easy to disentangle the magnetic specific heat from the other contributions, particularly since the electronic specific heat (see § 6·32) is appreciable. The Weiss theory can only give a qualitative account of the temperature variation of the specific heat and a detailed analysis of the experimental data requires a knowledge of the full theory. The most probable values for the lattice specific heat C_L, the electronic specific heat

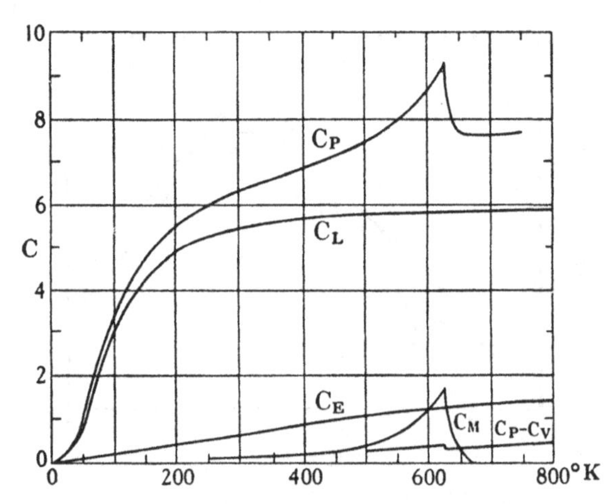

Fig. 10·18. The specific heat of nickel in cal./degree/gram atom. C_P is the observed specific heat; C_L is the lattice specific heat; C_E is the electronic and C_M is the magnetic specific heat. The last three are at constant volume.

C_E and the magnetic specific heat C_M are shown in the diagram, but it should be borne in mind that this dissection depends upon the correctness of the theory on which it is based.

The susceptibility per gram atom of nickel above the Curie point is shown in fig. 10·19 and it will be seen that $1/\chi$ diverges considerably from the straight line given by Weiss's theory. Such a variation in $1/\chi$ is predicted by the collective electron theory.

None of the present theories of ferromagnetism is entirely satisfactory, since each of them neglects some important characteristic of the many-body problem, and the various approximations cannot be rigorously justified. In the present state of the theory which approach one considers to be most satisfactory is largely a matter of intuition rather than of logic, but the collective electron theory has the advantage that it can be used to discuss the transport properties as well as the equilibrium behaviour of ferromagnets.

ANTIFERROMAGNETISM

10·6. Although ferromagnets have received more study than any other highly magnetic bodies, they are not the only solids in which large internal fields exist. In recent years increasing attention has been paid to two other types of magnetic bodies, known as antiferromagnets and ferrimagnets. In the present section we give a short account of the simpler properties of these substances.

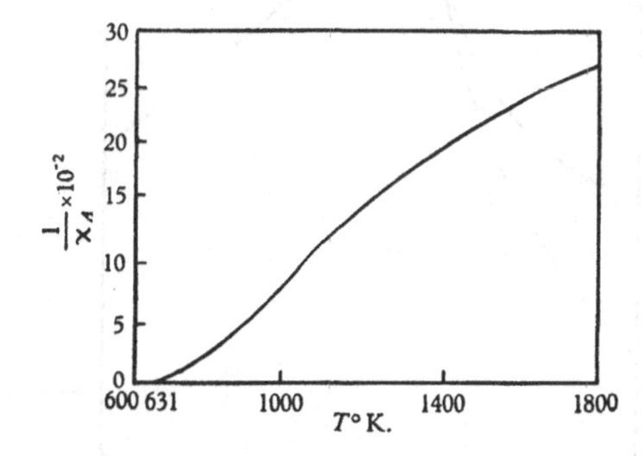

Fig. 10·19. Paramagnetic susceptibility of nickel above the Curie point.

If the exchange energy between the atoms in a solid is negative, there will be a tendency for the spins of neighbouring atoms to be anti-parallel to one another; and if the thermal agitation is sufficiently small, there will be an ordered arrangement of the spins into antiparallel sets. The existence of substances with such an arrangement of spins was first suggested by Néel in 1932. They are called antiferromagnets, and many examples are now known. Those which have been most studied are compounds of the various transition metals, particularly with oxygen, sulphur and the halogens. They differ from ferromagnetic bodies in that they cannot possess a permanent magnetic moment, and their most obvious magnetic characteristic is a maximum in the susceptibility at the transition temperature between the ordered and disordered states. The susceptibility of manganese monoxide, which is a typical antiferromagnet, is shown in fig. 10·20 as a function of the temperature (Bizette, Squire & Tsaï, 1938). The specific heat also shows an anomaly at the transition temperature (Millar, 1928; see fig. 10·21).

A detailed theory of antiferromagnetism, starting from the interactions between the individual atoms, encounters even more difficulties than the microscopic theory of ferromagnetism, and methods which give reasonable approximations for the latter problem are by no means so readily applicable to the former. We shall therefore base the account which follows upon obvious generalizations of the formal Weiss theory (Néel, 1932, 1936; Bitter, 1937; van Vleck, 1941).

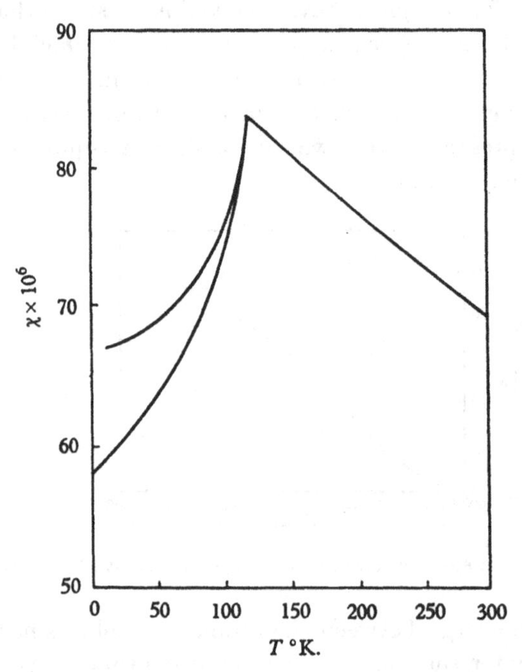

Fig. 10·20. The susceptibility of MnO. Lower curve, $H = 5000$ gauss; upper curve, $H = 24{,}000$ gauss.

10·61. The elementary theory of ferromagnetism given in § 10·5 can be generalized to deal with antiferromagnetism in the following way. We suppose that the atoms can be divided into two sets on two equivalent interpenetrating lattices which are such that, at $T = 0$, all the spins on one lattice are positively, and all the spins on the other lattice are negatively, directed. We assume that the magnetization of each lattice is given by (10·4·3) with $\frac{1}{2}N$ substituted for N and with the appropriate suffix $+$ or $-$, so that

$$V\mathbf{I}_+ = N\mathrm{S}\mu_0 B_\mathrm{S}(\mu_0 \mathbf{H}_+/kT), \quad V\mathbf{I}_- = N\mathrm{S}\mu_0 B_\mathrm{S}(\mu_0 \mathbf{H}_-/kT). \quad (10\cdot61\cdot1)$$

We also assume that the effect of the exchange forces can be taken into account by writing

$$\mathbf{H}_+ = \mathbf{H}_{\mathrm{ext.}} - \alpha\rho\mathbf{I}_+ - \beta\rho\mathbf{I}_-, \quad \mathbf{H}_- = \mathbf{H}_{\mathrm{ext.}} - \beta\rho\mathbf{I}_+ - \alpha\rho\mathbf{I}_-, \quad (10\cdot61\cdot2)$$

where α and β are positive constants. We must assume that the forces between the two lattices are larger than the forces within either lattice, so that $\beta > \alpha$. (Since there are two magnetization vectors to be considered, we cannot assume that $\mathbf{I}_+$ and $\mathbf{I}_-$ are parallel to one another or to $\mathbf{H}_{\text{ext.}}$.) The free energy F_2 is given by

$$F_2 = F_0(T) - \tfrac{1}{2}NkT \log \frac{\sinh\{(2S+1)\mu_0 H_+/kT\}}{\sinh(\mu_0 H_+/kT)} - \tfrac{1}{2}NkT \log \frac{\sinh\{(2S+1)\mu_0 H_-/kT\}}{\sinh(\mu_0 H_-/kT)}.$$

If $\mathbf{H}_{\text{ext.}} = 0$, we must have $\mathbf{I}_+ = \mathbf{I}_-$, where $|\mathbf{I}_+| = |\mathbf{I}_-| = I_0(T)$ and

$$VI_0 = NS\mu_0 B_S\{\mu_0 \rho(\beta - \alpha) I_0/kT\} \quad (\beta > \alpha). \tag{10·61·3}$$

This has a non-zero solution if

$$T < T_0 = \tfrac{2}{3}N\mu_0^2 S(S+1)\rho(\beta - \alpha)/kV, \tag{10·61·4}$$

and T_0 is the transition point at which the magnetization of the two sublattices becomes zero. Since each sublattice is separately magnetized to the

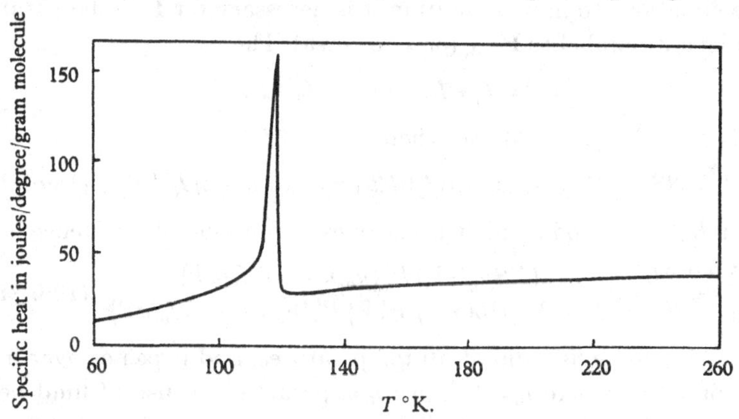

Fig. 10·21. The specific heat of MnO.

intensity of magnetization $I_0(T)$, the specific heat of an antiferromagnet will have the same temperature variation as a ferromagnet with the same value of S. As in the case of ferromagnets, there will be certain directions of easy magnetization which give the direction of the total magnetization when the external field is sufficiently small, and the magnetic behaviour will be different according as $\mathbf{H}_{\text{ext.}}$ is parallel or perpendicular to the direction of easy magnetization.

Case (i). $\mathbf{H}_{\text{ext.}}$ *is parallel to a direction of easy magnetization.* Above the transition point, $\mathbf{I}_+$ and $\mathbf{I}_-$ are small, and we may therefore write

$$SB_S(x) = \tfrac{1}{3}S(S+1)x.$$

We then have

$$\left.\begin{aligned} V\mathbf{I}_+ &= \tfrac{2}{3}NS(S+1)\mu_0^2(\mathbf{H}_{\text{ext.}} - \alpha\rho\mathbf{I}_+ - \beta\rho\mathbf{I}_-)/kT, \\ V\mathbf{I}_- &= \tfrac{2}{3}NS(S+1)\mu_0^2(\mathbf{H}_{\text{ext.}} - \beta\rho\mathbf{I}_+ - \alpha\rho\mathbf{I}_-)/kT, \end{aligned}\right\} \tag{10·61·5}$$

of which the solution is

$$V\mathbf{I}_+ = V\mathbf{I}_- = \tfrac{2}{3}N\mathrm{S}(\mathrm{S}+1)\,\mu_0^2\mathbf{H}_{\mathrm{ext.}}/\{k(T+\theta)\}, \qquad (10\cdot61\cdot6)$$

where

$$\theta = \tfrac{2}{3}N\mu_0^2\mathrm{S}(\mathrm{S}+1)\,\rho(\alpha+\beta)/kV. \qquad (10\cdot61\cdot7)$$

Hence the susceptibility $\chi_{\parallel}$ is given by

$$\chi_{\parallel} = \frac{\mathbf{I}_+ + \mathbf{I}_-}{\mathbf{H}_{\mathrm{ext.}}} = \frac{4}{3}\frac{N}{V}\frac{\mu_0^2\mathrm{S}(\mathrm{S}+1)}{k(T+\theta)} \quad (T \geqslant T_0). \qquad (10\cdot61\cdot8)$$

Below the transition temperature, $\mathbf{H}_{\mathrm{ext.}}$ can be treated as small but $\mathbf{I}_+$ and $\mathbf{I}_-$ cannot. We therefore expand (10·61·1) in powers of $\mathbf{H}_{\mathrm{ext.}}$ obtaining

$$\left.\begin{aligned} V\mathbf{I}_+ &= -N\mathrm{S}\mu_0 B_{\mathrm{S}}(x) - N\mathrm{S}\mu_0^2(\mathbf{H}_{\mathrm{ext.}}/kT)\,B'_{\mathrm{S}}(x),\\ V\mathbf{I}_- &= -N\mathrm{S}\mu_0 B_{\mathrm{S}}(y) - N\mathrm{S}\mu_0^2(\mathbf{H}_{\mathrm{ext.}}/kT)\,B'_{\mathrm{S}}(y),\\ x = \mu_0\rho(\alpha\mathbf{I}_+ &+ \beta\mathbf{I}_-)/kT, \quad y = \mu_0\rho(\beta\mathbf{I}_+ + \alpha\mathbf{I}_-)/kT. \end{aligned}\right\} \qquad (10\cdot61\cdot9)$$

For these equations to have a solution it is necessary for $\mathbf{I}_+$ to be parallel and $\mathbf{I}_-$ to be antiparallel to $\mathbf{H}_{\mathrm{ext.}}$ (or vice versa). Then

$$|\mathbf{I}_+| = I_0 + I_1, \quad |I_-| = I_0 - I_1,$$

where I_0 is given by (10·61·3) and where

$$VI_1 = N\mathrm{S}\mu_0^2[\{H_{\mathrm{ext.}} - \rho(\alpha+\beta)\,I_1\}/kT]\,B'_{\mathrm{S}}\{\mu_0\rho(\beta-\alpha)\,I_0/kT\}. \quad (10\cdot61\cdot10)$$

(Note that $B_{\mathrm{S}}(x)$ is an odd and $B'_{\mathrm{S}}(x)$ is an even function of x.) Hence

$$\chi_{\parallel} = \frac{\mathbf{I}_+ + \mathbf{I}_-}{\mathbf{H}_{\mathrm{ext.}}} = \frac{2I_1}{H_{\mathrm{ext.}}} = \frac{(N\mathrm{S}\mu_0^2/kV)\,B'_{\mathrm{S}}\{\mu_0\rho(\beta-\alpha)\,I_0/kT\}}{T + \{N\mathrm{S}\mu_0^2\rho(\alpha+\beta)/kV\}\,B'_{\mathrm{S}}\{\mu_0\rho(\beta-\alpha)\,I_0/kT\}}. \quad (10\cdot61\cdot11)$$

This expression for $\chi_{\parallel}$ is valid at all temperatures, and it passes over into (10·61·8) for $T \geqslant T_0$ where $I_0 = 0$. Since I_0 approaches a constant limit near $T = 0$, whereas B'_{S} tends exponentially to zero, we see that $\chi_{\parallel}$ is zero at $T = 0$. It increases as T increases and reaches its maximum value $\chi_{\parallel,\mathrm{max.}}$ at $T = T_0$, where

$$\chi_{\parallel,\mathrm{max.}} = \frac{4}{3}\frac{N}{V}\frac{\mu_0^2\mathrm{S}(\mathrm{S}+1)}{k(T_0+\theta)} = \frac{1}{\beta\rho}. \qquad (10\cdot61\cdot12)$$

Case (ii). $\mathbf{H}_{\mathrm{ext.}}$ *is perpendicular to a direction of easy magnetization.* If the two sublattices are spontaneously magnetized (which can only occur if $T < T_0$), and, in the case of cubic crystals, if a vanishingly small magnetic field is imposed along the particular direction of easy magnetization concerned, the individual magnetizations are equal and opposite. If a magnetic field $\mathbf{H}_{\mathrm{ext.}}$ is imposed at right angles to the direction of easy magnetization, both $\mathbf{I}_+$ and $\mathbf{I}_-$ will orient themselves so that the couples on them vanish. If then, $\mathbf{I}_+$ and $\mathbf{I}_-$ make angles ϕ and $\pi - \phi$ with the direction of easy magnetization, the total magnetization $\mathbf{I} = \mathbf{I}_+ + \mathbf{I}_-$ is in the direction of $\mathbf{H}_{\mathrm{ext.}}$ and is of magnitude $2\,|\,\mathbf{I}_+\,|\sin\phi = 2I_0\sin\phi$. (We assume here that the anisotropic part of the magnetic energy is negligible. See § 10·62.) The couples will vanish if

$$\mathbf{I}_+ \times \mathbf{H}_+ = \mathbf{I}_- \times \mathbf{H}_- = H_{\text{ext.}} I_0 \cos\phi - \beta\rho I_0^2 \sin 2\phi = 0. \qquad (10\cdot61\cdot13)$$

The susceptibility $\chi_\perp$ is therefore given by

$$\chi_\perp = \frac{2I_0 \sin\phi}{H_{\text{ext.}}} = \frac{1}{\beta\rho} \quad (T \leqslant T_0) \qquad (10\cdot61\cdot14)$$

which is equal to $\chi_{\parallel,\,\text{max.}}$.

Case (iii). *The susceptibility of an amorphous specimen.* If the specimen consists of a mixture of crystals having all orientations, the average susceptibility will be given by

$$\chi = \tfrac{1}{3}\chi_\parallel + \tfrac{2}{3}\chi_\perp \quad (T \leqslant T_0), \qquad \chi = \chi_\parallel \quad (T \geqslant T_0). \qquad (10\cdot61\cdot15)$$

The limiting value $\chi(0)$ of the susceptibility as $T \to 0$ is therefore $2/(3\beta\rho)$, while the maximum value $\chi_{\text{max.}}$ is $1/\beta\rho$. Also

$$\chi(0)/\chi(T_0) = \tfrac{2}{3}. \qquad (10\cdot61\cdot16)$$

10·62. *Anisotropic effects.* Just as in § 10·54 we can generalize the theory to include effects which depend upon the orientation of the magnetization vectors relative to the crystal axes. Here we shall confine our attention to the simplest case of a crystal with only one axis of symmetry in order to discuss a phenomenon which does not occur in ferromagnets. We suppose that the axis of symmetry is the z axis and that the 'anisotropic energy' is Cn^2, where l, m, n are the direction cosines of the total magnetization **I**. The inclusion of the anisotropic energy changes the formula (10·61·14) for $\chi_\perp$. If $U(\phi)$ is the energy as a function of ϕ, the couple acting on either $\mathbf{I}_+$ or $\mathbf{I}_-$ is $-\tfrac{1}{2}\partial U/\partial\phi$, and a term $C \sin\phi \cos\phi$ must therefore be added to (10·61·13). This gives

$$\chi_\perp = \frac{1}{\beta\rho - \tfrac{1}{2}C/I_0^2}. \qquad (10\cdot62\cdot1)$$

A more striking effect is the following. If the external field is along the z axis, then by symmetry **I** must be either along or perpendicular to this axis, and the corresponding energies due to the external fields are $-\tfrac{1}{2}\chi_\parallel H^2_{\text{ext.}}$ and $-\tfrac{1}{2}\chi_\perp H^2_{\text{ext.}}$. Hence the total energies in the two cases are

$$U_\parallel = C - \tfrac{1}{2}\chi_\parallel H^2_{\text{ext.}}, \quad U_\perp = -\tfrac{1}{2}\chi_\perp H^2_{\text{ext.}}. \qquad (10\cdot62\cdot2)$$

If $C > 0$, the easy directions of magnetization are perpendicular to the crystal axis. If, on the other hand, $C < 0$, the direction of easy magnetization is the crystal axis and in general the spins point along it. But, if

$$H_{\text{ext.}} > H_c = \{2|C|/(\chi_\perp - \chi_\parallel)\}^{\frac{1}{2}}, \qquad (10\cdot62\cdot3)$$

$U_\perp$ is less than $U_\parallel$, since $\chi_\perp > \chi_\parallel$, and there is therefore a critical field at which the spins swing over from being parallel to being perpendicular to the magnetic field (Gorter & Haanties, 1952). This phenomenon has been observed in the crystal $CuCl_2.2H_2O$, which has an orthorhombic structure and a transition point at 4·3° K. (Gorter, 1953).

10·63. *The transition point in strong fields.* If the external magnetic field is not zero, the transition between the disordered and partially ordered states takes place at a temperature lower than T_0. In general there are several different cases to consider, but for simplicity we shall confine the discussion to the simplest case in which $S=\frac{1}{2}$, $\alpha=0$, and when the external field is parallel to the magnetizations of the two sublattices (Garrett, 1951 *b*). We then have

$$\left.\begin{aligned} VI_+ &= \tfrac{1}{2}N\mu_0 \tanh\{(\mu_0 H_{\text{ext.}} - \mu_0\beta\rho I_-)/kT\}, \\ VI_- &= \tfrac{1}{2}N\mu_0 \tanh\{(\mu_0 H_{\text{ext.}} - \mu_0\beta\rho I_+)/kT\}. \end{aligned}\right\} \quad (10{\cdot}63{\cdot}1)$$

Just below the transition point we have $I_- = I_+ - \delta I$, where $\delta I/I_+$ is small, and on subtracting the two equations (10·63·1) we have, to the first order in $\delta I/I_+$,

$$V = \frac{N\mu_0^2\beta\rho}{2kT}\left(1 - \tanh^2\frac{\mu_0 H_{\text{ext.}} - \mu_0\beta\rho I_+}{kT}\right). \quad (10{\cdot}63{\cdot}2)$$

Also, at the transition point itself where $I_+ = I_-$, we have

$$VI_+ = \tfrac{1}{2}N\mu_0 \tanh\{(\mu_0 H_{\text{ext.}} - \mu_0\beta\rho I_+)/kT\}, \quad (10{\cdot}63{\cdot}3)$$

and the equations (10·63·2) and (10·63·3) determine the transition point T as a function of $H_{\text{ext.}}$. Now (10·63·3) can be written as

$$2VI_+/N\mu_0 = (1 - T/T_0)^{\frac{1}{2}}, \quad (10{\cdot}63{\cdot}4)$$

where

$$T_0 = N\mu_0^2\beta\rho/(2kV), \quad (10{\cdot}63{\cdot}5)$$

and the explicit equation for the transition curve is

$$\mu_0 H_{\text{ext.}} = kT_0(1 - T/T_0)^{\frac{1}{2}} + kT \tanh^{-1}(1 - T/T_0)^{\frac{1}{2}}. \quad (10{\cdot}63{\cdot}6)$$

The transition curve is shown in fig. 10·22. For $T=0$ we have $\mu_0 H_{\text{ext.}} = kT_0$, while near $T = T_0$ we have $\mu_0 H_{\text{ext.}} = 2kT_0(1 - T/T_0)^{\frac{1}{2}}$. At $T=0$ half the spins are parallel and half are antiparallel to the field if $\mu_0 H_{\text{ext.}} < kT_0$, while if $\mu_0 H_{\text{ext.}} > kT_0$ all the spins point in the same direction.

10·631. The states described by the equations derived in the preceding section have some distinctly peculiar properties, and the theory is not entirely satisfactory. The difficulties can be most simply illustrated by considering the more general problem in which $\alpha \neq 0$, but restricting the discussion to the particular case of $T=0$. Then if $\mu_0 H_{\text{ext.}} < \frac{1}{2}N\mu_0^2(\beta - \alpha)\rho/V$

half the spins are parallel and half are antiparallel to the field, while if $\mu_0 H_{\text{ext.}} > \frac{1}{2}N\mu_0^2(\beta+\alpha)\rho/V$ all the spins point in the same direction. For fields satisfying $\frac{1}{2}N\mu_0^2(\beta-\alpha)\rho/V < \mu_0 H_{\text{ext.}} < \frac{1}{2}N\mu_0^2(\beta+\alpha)\rho/V$, however, the obvious solution of the equations is $VI_+ = \frac{1}{2}N\mu_0$ and $\mu_0 H_- = 0$. But for this solution the entropy is not zero, and, by the third law of thermodynamics, it cannot correspond to the most stable state. There must therefore be other solutions to the problem which are not necessarily describable by the equations so far set up. It might, for example, be possible for the crystal to split up into a number of microscopic domains in each of which either half or all the spins are parallel to one another, the numbers of the two

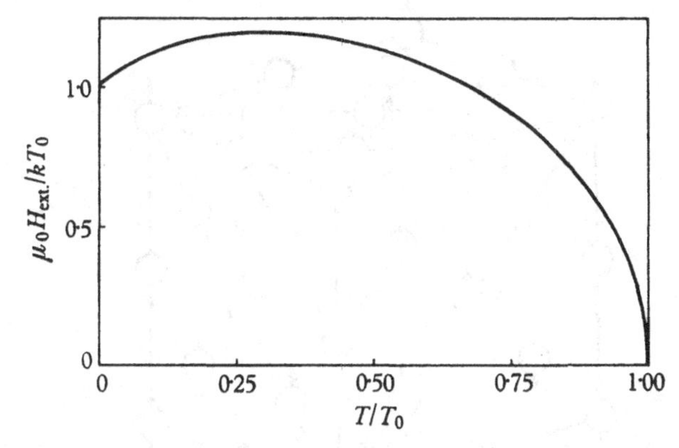

Fig. 10·22. The transition curve for an antiferromagnet.

different types of domain varying with the external magnetic field. Whether the most stable state is of this type or has some entirely different configuration is not known.

When $\alpha = 0$, the anomalous range of values of $H_{\text{ext.}}$ disappears, but in spite of this the thermodynamic functions show other anomalous features. This raises still further doubts about the validity of the whole of the preceding theory, and at the present time these are unresolved.

10·64. The foregoing elementary theory gives a surprisingly good description of the main experimental facts which were described in § 10·6, and more elaborate theories give very much the same results. Although initially the existence of antiferromagnetism was inferred from the presence of a maximum in the susceptibility-temperature curve, in recent years the spin structure of antiferromagnets has been investigated by neutron diffraction methods (Shull, Strauser & Wollam, 1951), which can be used to supplement the data on the atomic arrangement furnished by X-ray measurements.

The structure of MnO, which has a rock-salt crystal lattice, is shown in fig. 10·23. The manganese and oxygen atoms each form a face-centred cubic lattice, while the oxygen atoms are at the centres of the unit cells of the manganese lattice, and vice versa. The spin arrangement in the ordered state, however, is such that the unit 'magnetic' cell has twice the linear dimensions of the ordinary crystallographic cell, and it contains 32 manganese and 32 oxygen atoms.

Since the magnetic ions are separated from one another by oxygen, sulphur or halogen ions it seems unlikely that forces tending to aline the spins are the ordinary exchange forces such as occur in ferromagnetic

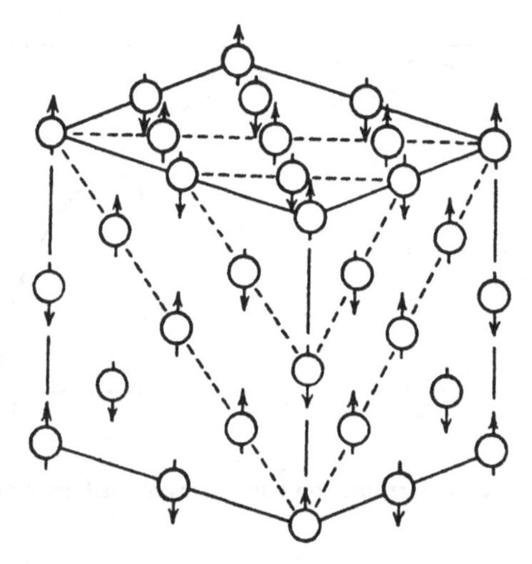

Fig. 10·23. The crystal structure of MnO, showing the spins on the manganese ions in one unit magnetic cell.

materials. Various mechanisms which would produce a coupling between the spins of next-nearest-neighbour ions have been investigated, the most important being those classified as super-exchange and double-exchange effects. These are similar to the configuration interactions which are common in complex atomic spectra. A general account of these difficult calculations can be found in the review articles by Lidiard (1954) and by Nagamiya, Yosida & Kubo (1955).

Antiferromagnetic bodies can be roughly classified into two main types according as the transition temperature occurs at fairly high or at very low temperatures. Some typical data for antiferromagnetics with high transition temperatures are given in table 10·6. (It should be borne in mind that the experimental data are subject to large uncertainties.) The values of $\chi(0)/\chi(T_0)$ given refer to randomly orientated crystal mixtures, and should, according to (10·61·16), be equal to 0·67. In substances of this type it is probable that the super-exchange forces are responsible for the ordering.

Table 10·6. *Parameters of some antiferromagnets*

	V_2O_3	V_2O_4	Cr_2O_3	MnO	MnO_2	MnS
T_0	173	340	310	116	84	165
θ	?	720	1070	610	?	528
$\chi(0)/\chi(T_0)$	?	?	0·76	0·69	0·93	0·82

	MnF_2	FeO	FeS	FeF_2	CoO	NiO
T_0	72	198	613	79	293	(520)
θ	113	570	857	117	280	?
$\chi(0)/\chi(T_0)$	0·72	0·78	?	0·72	?	?

The behaviour of substances in which the spins only become regularly arranged at very low temperatures is less well understood than that of substances with high transition temperatures. In most cases there is still more uncertainty whether the ordered arrangement is antiferromagnetic or ferromagnetic, though the balance of the evidence favours the former. The substances in question are necessarily magnetically dilute, and as already mentioned in § 10·42 the spin-spin forces and the exchange forces are then of the same order of magnitude, and the problem of determining their spin structure is exceedingly complicated. The transition temperatures of some substances which are probably antiferromagnetic near the absolute zero are given in Table 10·7.

Table 10·7. *The transition temperatures of some low temperature antiferromagnets*

	T_0
$CrK(SO_4)_2 . 12H_2O$	0·004
$Mn(NH_4)_2(SO_4)_2 . 6H_2O$	0·15
$Co(NH_4)_2(SO_4)_2 . 6H_2O$	0·084
$Fe(NH_4)_2(SO_4)_2 . 12H_2O$	0·043
$CuK_2(SO_4)_2 . 6H_2O$	0·05

In general, transition temperatures have only been measured in weak magnetic fields, but more extensive measurements exist for cobalt ammonium sulphate (Garrett, 1951*a*). This substance is monoclinic, and the susceptibility is different for the three different axes. Garrett determined the locus of the transition points in the H, T plane, and his results for one particular crystallographic axis are shown in fig. 10·24. This equilibrium diagram is in general agreement with the theoretical predictions of equation (10·63·6).

10·65. *Ferrimagnetism.* A generalization of the theory of § 10·61 has been given by Néel (1948) in which it is assumed that the atoms with unpaired spins are arranged on two non-equivalent lattices. In this case, the equations (10·61·1) and (10·61·2) have to be replaced by

$$V\mathbf{I}_+ = MS\mu_0 B_S(\mu_0 \mathbf{H}_+/kT), \quad V\mathbf{I}_- = NS\mu_0 B_S(\mu_0 \mathbf{H}_-/kT), \quad (10·65·1)$$

where $\frac{1}{2}M$ and $\frac{1}{2}N$ are the numbers of atoms on the two lattices and where

$$\mathbf{H}_+ = \mathbf{H}_{\text{ext.}} - \alpha\rho\mathbf{I}_+ - \beta\rho\mathbf{I}_-, \quad \mathbf{H}_- = \mathbf{H}_{\text{ext.}} - \beta\rho\mathbf{I}_+ - \gamma\rho\mathbf{I}_-. \quad (10·65·2)$$

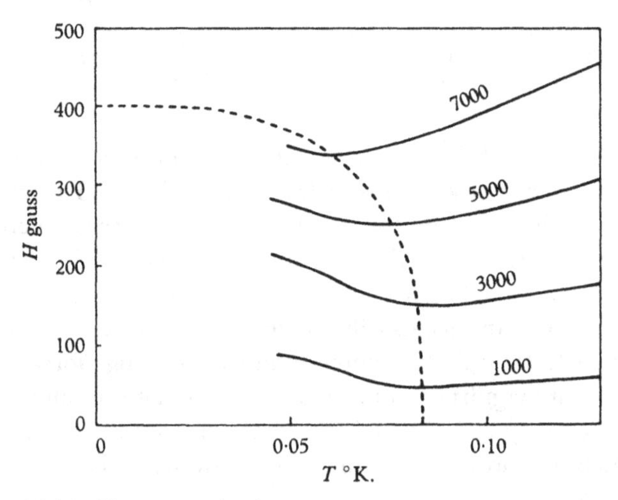

Fig. 10·24. The magnetic phase diagram of cobalt ammonium sulphate. Lines of constant magnetization are also shown.

The constants α, β and γ are assumed to be positive and to satisfy the conditions $\alpha \neq \gamma$, $\beta > \alpha$, $\beta > \gamma$. The free energy F_2 is given by

$$F_2 = F_0(T) - \tfrac{1}{2}MkT \log \frac{\sinh\{(2S+1)\mu_0 H_+/kT\}}{\sinh(\mu_0 H_+/kT)}$$
$$- \tfrac{1}{2}NkT \log \frac{\sinh\{(2S+1)\mu_0 H_-/kT\}}{\sinh(\mu_0 H_-/kT)}. \quad (10·65·3)$$

If $T=0$ and $H_{\text{ext.}}=0$ there is the usual solution $I_+ = I_- = 0$, $I = I_+ + I_- = 0$, but if β is sufficiently large it is possible to have

$$VI_+ = \tfrac{1}{2}MS\mu_0, \quad VI_- = -\tfrac{1}{2}NS\mu_0, \quad VI = \tfrac{1}{2}(M-N)S\mu_0. \quad (10·65·4)$$

For, by the analogue of (10·521·1), the energy U_2 is

$$U_2 = -VI_+H_+ - VI_-H_- = \rho V(\alpha I_+^2 + 2\beta I_+ I_- + \gamma I_-^2), \quad (10·65·5)$$

and for the solution (10·65·4) we have

$$U_2 = \tfrac{1}{4}\rho S^2\mu_0^2(\alpha M^2 - 2\beta MN + \gamma N^2)/V. \quad (10·65·6)$$

If this is negative (for which a necessary condition is $\beta^2 > \alpha\gamma$) there will be a spontaneous magnetization in spite of the ordering forces being antiferromagnetic in sign. The magnetization of each sublattice will decrease steadily to zero as the temperature increases from zero to the Curie temperature T_0.

In addition to the above solution of the equations (10·65·1) and (10·65·2), Néel considered other solutions which exist for various ranges of the parameters. If we confine our attention to $T=0$, these solutions are such that either $VI_+ = \frac{1}{2}M\mathrm{S}\mu_0$, $H_- = 0$ or $H_+ = 0$, $VI_- = -\frac{1}{2}N\mathrm{S}\mu_0$, and they correspond to configurations with non-zero entropies, since the spins on one of the sublattices are not fully alined. They therefore cannot be considered as defining stable configurations, and, as in the corresponding problem discussed in § 10·631, there must be more stable configurations. No theory of these has yet been given.

The word ferrimagnetism was introduced by Néel to describe the occurrence of a spontaneous magnetization when the spins of the atoms are not all parallel at $T=0$ and when the ordering forces tend to aline neighbouring spins antiparallel to one another. This can only occur when the atoms concerned can be split up into two (or more) sets which are such that the ordering forces between two atoms, one in either set, are greater than the ordering forces between two atoms in the same set. If the two sets are equivalent, the resultant magnetic moment is zero and the substance is antiferromagnetic, but in the general case the total moment will not vanish.

10·651. If the temperature is greater than the Curie temperature, we can treat both the external field and the magnetizations as small and write $B_\mathrm{S}(x) = \frac{2}{3}(\mathrm{S}+1)x$. The equations (10·65·1) then become

$$VI_+ = \tfrac{2}{3}M\mathrm{S}(\mathrm{S}+1)\mu_0^2 H_+/kT, \quad VI_- = \tfrac{2}{3}N\mathrm{S}(\mathrm{S}+1)\mu_0^2 H_-/kT, \qquad (10\cdot651\cdot1)$$

which give

$$\rho\chi = C\frac{(M+N)T + MN(\alpha+\gamma-2\beta)C}{(T+MC\alpha)(T+NC\gamma) - MNC^2\beta^2}, \qquad (10\cdot651\cdot2)$$

where

$$C = \frac{2\mathrm{S}(\mathrm{S}+1)\mu_0^2\rho}{3Vk}. \qquad (10\cdot651\cdot3)$$

This may be written as

$$\frac{1}{\chi} = \frac{3Vk(T+\theta_1)}{2(M+N)\mathrm{S}(\mathrm{S}+1)\mu_0^2} - \frac{MN\rho^2}{(M+N)^3}\{M(\beta-\alpha) - N(\beta-\gamma)\}^2 \frac{2\mathrm{S}(\mathrm{S}+1)\mu_0^2}{3Vk(T-\theta_2)}, \qquad (10\cdot651\cdot4)$$

where

$$\left.\begin{aligned} \theta_1 &= \rho\frac{M^2\alpha + 2MN\beta + N^2\gamma}{M+N}\frac{2\mathrm{S}(\mathrm{S}+1)\mu_0^2}{3Vk}, \\ \theta_2 &= \rho\frac{MN(2\beta-\alpha-\gamma)}{M+N}\frac{2\mathrm{S}(\mathrm{S}+1)\mu_0^2}{3Vk}. \end{aligned}\right\} \qquad (10\cdot651\cdot5)$$

The susceptibility as a function of the temperature is shown in fig. 10·25. It becomes infinite at the Curie temperature T_0, which, according to (10·651·2), is given by

$$T_0 = \frac{S(S+1)\mu_0^2\rho}{3Vk}[\sqrt{\{(M\alpha - N\gamma)^2 + 4MN\beta^2\}} - (M\alpha + N\gamma)]. \quad (10\cdot651\cdot6)$$

If the substance is ever to become spontaneously magnetized, the parameters must be such as to make T_0 positive.

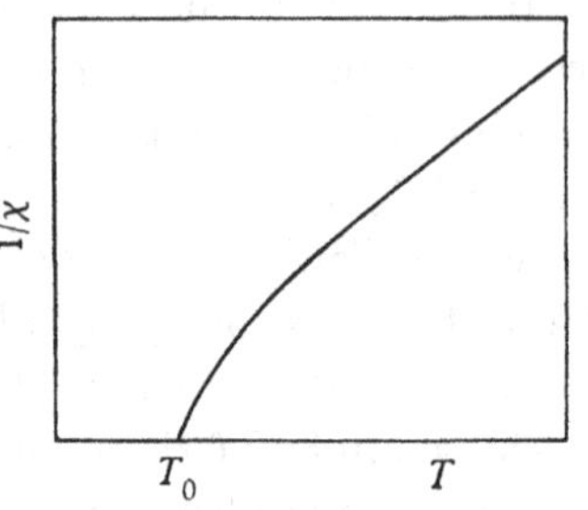

Fig. 10·25. The susceptibility of a ferrite above the Curie point.

10·652. *The magnetism of the ferrites.* The earliest known ferromagnet, namely, natural magnetite, Fe_3O_4, is an example of a large class of magnetic substances known as ferrites, which have the general chemical formula XY_2O_4. In all these substances the metallic ions contribute a much smaller magnetic moment than they do in paramagnetic salts, the magneton number of the Fe_3O_4 molecule, for example, being only 4·2 as compared with the maximum possible value of 14 if all the spins of the Fe^{2+} ion and the two Fe^{3+} ions were parallel. It was suggested by Néel that the ferrites are ferrimagnetic and that the spins are arranged in two or more antiparallel but inequivalent sets. It should be noted that this explanation of the magneton number being less than the maximum possible value is quite different in character from that put forward in § 10·554 in connexion with the magneton numbers of the ferromagnetic metals themselves. In § 10·554 the ordering forces considered are such as to tend to aline neighbouring spins parallel to one another, but, since the electrons concerned are quasi-free, complete alinement of the spins may be inhibited by the increase in the kinetic energy of the electrons which would be necessitated by the alinement. The ferrites, on the other hand, are ionic compounds and contain no free electrons at $T = 0$, so that the theory of § 10·554 cannot apply to them. Their properties can, however, be reasonably well described by the theory of the preceding section or by obvious generalizations of it. In particular, the susceptibility above the Curie point varies with temperature roughly in the manner shown in fig. 10·25.

Those ferrites in which X is doubly and Y is triply ionized have crystal structures of the spinel type. The unit cell contains eight XY_2O_4 molecules, the thirty-two oxygen atoms being arranged in such a way that there are twenty-four interstices which will accommodate the twenty-four metal ions. Eight of the interstices are each surrounded by four oxygen ions arranged tetrahedrally, while sixteen are each surrounded by six oxygen ions arranged octahedrally. If we denote the interstices of the tetrahedral and of the octahedral types by the symbols A and B, the normal spinel structure is such that all the X ions are on A sites and all the Y ions are on B sites. Another possible structure is that in which the X atoms are on B sites and the Y atoms are equally divided between A and B sites. This structure is known as the inverse spinel structure, and the chemical formula for the ferrite can then be written as $Y(XY)O_4$.

From measurements of the lattice spacing it seems probable, but by no means fully established, that ferrites with the normal spinel structure are paramagnetic, and that only ferrites with the inverse structure can be spontaneously magnetized. Thus $ZnFe_2O_4$ and $CdFe_2O_4$ have normal structures and are paramagnetic, while $Fe(MgFe)O_4$ has an inverse structure and is ferromagnetic, or rather ferrimagnetic, since its magneton number is 1·2 instead of 10.

The simple theory of § 10·65 does not apply in all its details to most ferrites since they contain ions of three and not of two types, namely, triply ionized Y atoms on A sites, doubly ionized X atoms on B sites and triply ionized Y atoms on B sites. We should, however, expect the qualitative aspects of the theory to remain unchanged when applied to more complex systems than those envisaged in § 10·65. The magneton numbers of some ferrites are given in table 10·8, and it will be seen that they agree reasonably well with the hypothesis that the spins of the two Fe^{3+} lattices are antiparallel, the whole of the resultant magnetization coming from the spins of the doubly ionized atoms. A general account of the properties of ferrites will be found in the review article by Fairweather, Roberts & Welch (1952).

Table 10·8. *Magneton numbers of certain ferrites*

$Fe(MnFe)O_4$	$Fe(FeFe)O_4$	$Fe(CoFe)O_4$	$Fe(NiFe)O_4$
5·0	4·2	3·3	2·3

Superconductivity‡

10·7. The electrical resistivity ρ of a metal consists of two parts, one of which decreases rapidly as the temperature is lowered while the other is independent of the temperature. At very low temperatures practically the whole resistivity consists of this second portion, which is therefore called the residual resistivity. The residual resistivity differs from sample to

‡ For a more detailed account see Burton, Smith & Wilhelm (1940) and more particularly Shoenberg (1952). A full list of references will be found in Shoenberg's book.

sample, and is supposed to be due to impurities, cracks, strains, etc. In an ideally pure substance the residual resistivity would be zero. If any sample of a metal is taken and if its residual resistivity ρ_0 is determined by measurements taken at very low temperatures (less than 10° K.), the temperature-dependent part ρ_i can be found by subtraction as $\rho_i = \rho - \rho_0$. It is found that ρ_i is the same for all samples and can be thought of as the resistivity of the pure metal; ρ_i is therefore called the ideal resistivity. The general behaviour of ρ_i as a function of temperature is shown in fig. 10·26. For a sample of metal of reasonable purity the residual resistivity is of the order of one-thousandth of the total resistivity at 0° C.

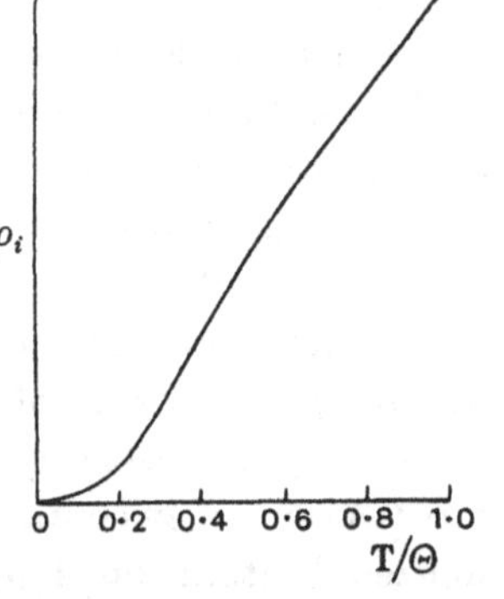

Fig. 10·26. The ideal resistance of a metal as a function of the temperature.

In 1911 Kamerlingh Onnes discovered the remarkable phenomenon of superconductivity when investigating the resistance of mercury at temperatures obtainable by the use of liquid helium. At the temperatures in question the resistance is entirely the residual resistance, due to impurities, etc., but, on lowering the temperature below 4·3° K., Kamerlingh Onnes found that the resistance rapidly decreased and had become completely negligible at 4·2° K. Later experiments have shown that with 'chemically pure' metals the transition to the resistanceless state takes place in a very small temperature interval, but with alloys the transition may be spread over a temperature range of the order of 1°. It is therefore customary to define the transition temperature as the temperature at which the resistance has dropped to one-half of the value of the residual resistance.

At temperatures above the transition temperature the metal is said to be in the normal state, while at temperatures lower than the transition temperature it is said to be in the superconducting state. All the evidence points to the resistance being identically zero in the superconducting state, since persistent currents have been produced in superconducting rings of metal which have remained unchanged for many hours. (Persistent currents are most easily produced by cooling a ring of metal in a not too large magnetic field to a temperature below the transition temperature, and then reducing the field to zero. A current is induced in the ring, and persists so long as the temperature remains below the transition temperature.)

Superconductivity is a comparatively rare phenomenon amongst metals, only twenty-one pure metals having been found to be superconductors up to date. These are shown in table 10·9 together with their transition temperatures. On the other hand, a large number of alloys and semi-metallic compounds are superconductors. In the majority of the alloys at least one of the components is a superconductor, but the superconducting alloy Au_2Bi, with a transition temperature of 1·95° K., is formed from two elements neither of which separately has yet been found to be a superconductor at the lowest temperatures now available. Many semi-metallic compounds, borides, carbides, nitrides and sulphides are superconductors. Copper sulphide and molybdenum carbide are examples of compounds formed out of non-superconducting elements, while vanadium silicide, V_3Si, is notable in having a transition temperature of 17·1° K., the highest yet found.

Table 10·9. *The transition temperatures T_0 of superconducting elements*

	Tc	Nb	Pb	V	Ta	La	Hg	Sn
T_0 in °K.	11·2	8	7·22	5·1	4·4	4·37	4·152	3·73
	In	Tl	Th	Al	Ga	Re	Zn	
T_0 in °K.	3·37	2·38	1·39	1·20	1·10	1·0	0·91	
	U	O	Zr	Cd	Ti	Ru	Hf	
T_0 in °K.	(0·8)	0·71	0·70	0·56	0·53	0·47	0·35	

10·71. *The magnetic properties of superconductors.* It was found by Kamerlingh Onnes in 1914 that superconductivity can be destroyed by an external magnetic field. The phenomenon is in general a complicated one and is only relatively simple for a chemically pure metal in the form of a wire with the magnetic field parallel to the wire. The magnetic field required to destroy the superconductivity is a function of the temperature, being zero at the normal transition temperature and increasing as the temperature is lowered. The resistance does not usually increase discontinuously up to the value for the normal state when the magnetic field reaches a certain value, but for chemically pure metals the increase is rapid and the transition to the normal state takes place at a fairly definite value of the magnetic field. For impure metals and alloys, or for transverse fields, on the other hand, the transition takes place more or less gradually, the resistance reappearing when the magnetic field reaches a certain value but not attaining the resistance of the normal state until the field has reached a somewhat higher value. The threshold field H_c for the destruction of superconductivity at temperature T is defined to be the value of the magnetic field required to restore the resistance to the value it has when the metal is in its normal state.

If a metal is in the superconducting state it is impossible to change the magnetic induction at any point of it. For if the induction were changed, an electromotive force would be induced which would set up an infinite electric current, and this is clearly impossible. If, therefore, a superconducting piece of metal is introduced into a uniform magnetic field, the lines of magnetic force will be distorted and none will enter the metal. On the other hand, if a metal in the normal state is placed in a uniform magnetic field and cooled to a temperature below the transition temperature, the magnetic field being less than the threshold field at the temperature concerned, we should expect the lines of magnetic induction to be 'frozen' in the metal; and if the field were removed we should expect that the magnetic induction at any point in the metal would remain the same as it was before the metal became superconducting. However, in 1933 Meissner & Ochsenfeld, investigating the magnetic properties of a superconducting cylinder of tin placed in a transverse magnetic field, found a result which was quite different from that described above. They found that when the cylinder, in its normal state, was cooled in the magnetic field below the transition temperature, the lines of magnetic induction, instead of being frozen in, were pushed out of the cylinder, and they concluded that in an ideal superconductor the magnetic induction is always zero.

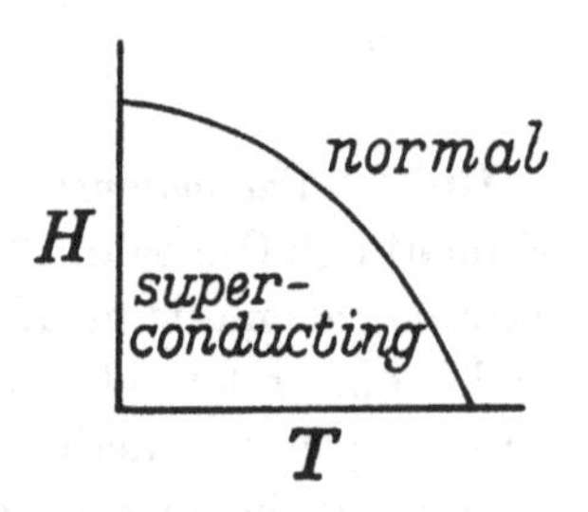

Fig. 10·27. The phase diagram of a superconductor.

This property of a superconductor strongly suggests that the transition of a body from the normal to the superconducting state can be treated as a reversible phase change, with a phase diagram of the general type shown in fig. 10·27. The H, T curve is assumed to have zero gradient at $T=0$, and it is found that in all cases the gradient is finite when $H=0$. If we assume that a superconductor can be described as a body for which $\mathbf{B}=0$, we can apply the thermodynamic theory of § 10·1 to derive relationships between the various physical quantities (Gorter & Casimir, 1934), and the theory takes its simplest form if we assume that a superconductor has an intensity of magnetization $\mathbf{I}$ which is related to the magnetic field $\mathbf{H}$ by the equation $\mathbf{I}=-\mathbf{H}/4\pi$. In fact, our knowledge of the structure of metals would lead us to suppose that the magnetic moment of a superconductor is due to surface currents and not to magnetic dipoles distributed throughout the body of the superconductor. But from measurements of the field outside the superconductor we can discover nothing about the sources of the field inside or on the surface of the superconductor, and from the formal point of view it is immaterial whether we put $\mathbf{B}=0$, $\mathbf{I}=-\mathbf{H}/4\pi$ inside a superconductor, or put $\mathbf{B}=\mathbf{H}=0$ and assume that there is a current density $\mathbf{j}$ which in electrostatic units is given by $\mathbf{j}/c=-\operatorname{curl}\mathbf{I}$, where c is the velocity of light.‡ These different hypotheses give the same field at all external points, and since only the external field is observable, either description of the internal field will do equally well if we follow a consistent treatment.

10·72. *The thermodynamics of the superconducting state.* If we have a superconductor in the form of a long rod, and if there is a uniform magnetic field H parallel to the rod, the free energy F is determined by the equation analogous to (10·12·6), namely,

$$dF = -S\,dT - VI\,dH. \tag{10·72·1}$$

Changes of volume are not taken into account so that it is unnecessary to include the term $-P\,dV$, and, since the superconductor is in the form of a long rod, the magnetic field is unaltered by the presence of the superconductor, and H and $H_{\text{ext.}}$ are the same. Now $I = -H/4\pi$, and the specific heat $C = T(\partial S/\partial T)_H$ is independent of H since $(\partial S/\partial H)_T = V(\partial I/\partial T)_H = 0$. We may therefore write

$$S = \int^T \frac{C(T')}{T'}\,dT'$$

and

$$\left.\begin{aligned} F &= -\int^T dT' \int^{T'} \frac{C(T'')}{T''}\,dT'' + \frac{V}{4\pi}\int_0^H H'\,dH' \\ &= \int^T C(T')\,dT' - T\int^T \frac{C(T')}{T'}\,dT' + \frac{VH^2}{8\pi}. \end{aligned}\right\} \tag{10·72·2}$$

If we denote the superconducting and normal phases by the suffixes s and n we must have $F_s = F_n$ along the transition curve separating the two phases in the H, T diagram. For we have

$$(\partial F/\partial T)_H = -S, \quad (\partial F/\partial H) = -VI, \tag{10·72·3}$$

and if F_s and F_n were not equal along the transition curve there would be an infinite discontinuity in S and I across the curve. Now F_n is given by (10·72·2) with the last term omitted and with $C = C_n$. Also when $H = 0$ we have $F_s = F_n$ at $T = T_0$, the transition temperature in zero field. Hence the equation of the transition curve is

$$\frac{VH_c^2}{8\pi} = \int_T^{T_0} \{C_s(T') - C_n(T')\}\,dT' - T\int_T^{T_0} \frac{C_s(T') - C_n(T')}{T'}\,dT' = F_n(T) - F_s(T)_{H=0}. \tag{10·72·4}$$

‡ The Maxwell equations for a steady field are div $\mathbf{B} = 0$, curl $\mathbf{H} = 4\pi\mathbf{j}/c$, curl $\mathbf{B} = 0$. If $\mathbf{B} = \mathbf{H} + 4\pi\mathbf{I}$, we can rewrite the equations as div $\mathbf{B} = 0$, curl $\mathbf{B} = 4\pi$ curl $\mathbf{I} + 4\pi\mathbf{j}/c$, curl $\mathbf{H} = -4\pi$ curl $\mathbf{I}$, which shows that $\mathbf{I}$ is equivalent to a current density $-c$ curl $\mathbf{I}$. If $\mathbf{I}$ is constant inside the superconductor, the 'equivalent current' is a surface current $c\mathbf{H} \times \mathbf{n}$, where $\mathbf{H}$ is the field (necessarily tangential) just outside the superconductor and $\mathbf{n}$ is the unit vector along the outward normal.

Also since $(\partial S/\partial H)_T = 0$, the difference in the entropies of the two phases is given by

$$S_n - S_s = -\frac{\partial F_n}{\partial T} + \left(\frac{\partial F_s}{\partial T}\right)_{H=0} = -\frac{VH_c}{4\pi}\frac{dH_c}{dT}, \qquad (10\cdot72\cdot5)$$

while Q, the heat absorbed in the transition from the superconducting to the normal state, is given by

$$Q = T(S_n - S_s) = -\frac{VTH_c}{4\pi}\frac{dH_c}{dT}. \qquad (10\cdot72\cdot6)$$

It should be noted that by the third law of thermodynamics we must have $dH_c/dT = 0$ at $T = 0$.

Equation (10·72·4) can be obtained in differential form by differentiating it twice with respect to T. This gives

$$\left(\frac{dH_c}{dT}\right)^2 + H_c\frac{d^2H_c}{dT^2} = 4\pi\frac{C_s - C_n}{vT}, \qquad (10\cdot72\cdot7)$$

and in particular

$$\left(\frac{dH_c}{dT}\right)^2_{T=T_0} = 4\pi\frac{C_s - C_n}{vT_0}, \qquad (10\cdot72\cdot8)$$

which is known as Rutgers's formula (Rutgers, 1934). Equation (10·72·6) is the analogue of Clapeyron's equation (9·11·4). It should be noted that the transition from the superconducting to the normal state is in general a normal phase change, but that it degenerates into a transition of the second order at $T = T_0$, since $H_c = 0$ there and since dH_c/dT is finite.

Table 10·10. *The discontinuity in the specific heat of superconductors in zero fields*

	T_0 in °K.	v cm.³/gram atom	dH_c/dT gauss/°K.	$\Delta C \times 10^3$ calculated joules/°K.	$\Delta C \times 10^3$ observed joules/°K.
Pb	7·22	17·8	200	41·8	52·5
Ta	4·40	10·9	320	39·3	34–38
Sn	3·73	16·1	151	10·9	10–12
In	3·37	15·2	146	8·7	9·5
Tl	2·38	16·8	139	6·1	6·5
Al	1·20	9·9	177	3·0	(2)

The temperature variation of the specific heat of tin in the absence of a magnetic field is shown in fig. 10·28, and the discontinuity is much more clearly defined than the corresponding discontinuity in the specific heat of liquid helium (fig. 9·1). The numerical values for a number of metals are given in table 10·10, and it will be seen that the observed values of $\Delta C = C_s - C_n$ are in good agreement with those calculated from equation (10·72·8).

Equation (10·72·6) for the latent heat when the transition takes place in the presence of a magnetic field has been verified for tin down to 1° K.

by Keesom & van Laer (1938), but the experimental data for other metals are not very extensive. However, the entropy differences for a number of metals have been calculated by Daunt & Mendelssohn (1937) from the H_c, T curve, and their results are shown in fig. 10·29. It will be seen that, in accordance with the third law of thermodynamics, the entropy differences tend to zero as $T \to 0$.

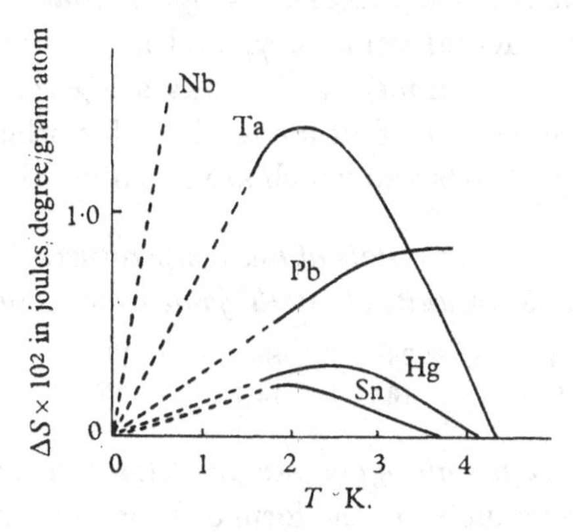

Fig. 10·28. The entropy difference between the normal and superconducting states.

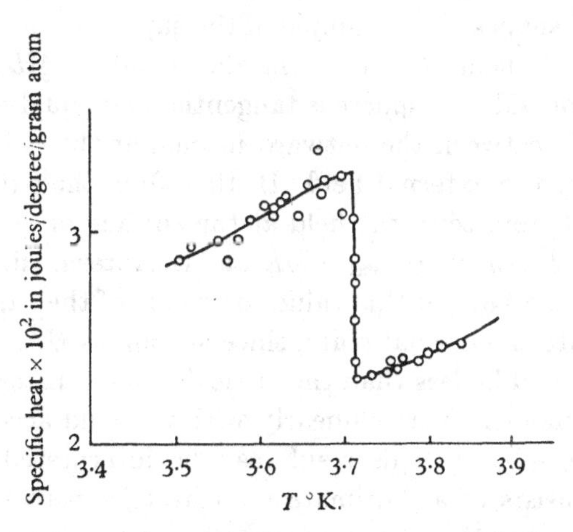

Fig. 10·29. The specific heat of tin in zero magnetic field.

10·721. For many metals the transition curve is approximately parabolic. We may therefore write

$$H_c = H_m(1 - T^2/T_0^2), \tag{10·721·1}$$

and equation (10·72·7) then gives

$$\frac{1}{V}(C_s - C_n) = -\frac{H_m^2 T}{2\pi T_0^2} + \frac{3H_m^2 T^3}{2\pi T_0^4} \tag{10·721·2}$$

Now the specific heat of a metal is of the form $C/V = \gamma T + \alpha T^3$ at low temperatures, where γT is the contribution due to the conduction electrons and αT^3 is the contribution due to the lattice vibrations. Hence

$$\gamma_n = \gamma_s + H_m^2/(2\pi T_0^2) \geqslant H_m^2/(2\pi T_0), \qquad (10\cdot721\cdot3)$$

and we can therefore obtain a lower limit to γ_n from measurements of H_m and T_0 (Kok, 1934). Such evidence as exists suggests that $\gamma_s = 0$, and in this case (10·721·2) gives the actual value of γ_n and not merely a lower limit to it. Some values of γ_n determined in this way are given in table 10·11. They are of the same order of magnitude as those for other metals determined by calorimetric methods and which are given in table 6·8, p. 172.

Table 10·11. *Specific heats of metals at low temperatures determined from equation* (10·721·3) *in units of* $10^{-4}T$ *joule/degree/gram atom*

	Hg	Al	Tl	Sn	Pb	Nb	Ta
$C_{V,A}$	15·6	14·7	16	14·7	30	250	82

10·722. *The intermediate state of a superconductor.* The foregoing theory only applies to a superconductor in the form of a long rod in a longitudinal magnetic field, and the phenomena are much more complex for superconductors of other shapes. For example, if the superconductor is a sphere the internal magnetic field H is uniform and equal to $\frac{3}{2}H_{\text{ext.}}$, while the magnetic field just outside the sphere is tangential and equal to $\frac{3}{2}H_{\text{ext.}} \sin\theta$, where θ is the angle between the outward normal at the point concerned and the direction of the external field. If, therefore, the external field is gradually increased from zero, the field at the surface of the sphere first becomes critical at $\theta = \frac{1}{2}\pi$ when $H_{\text{ext.}} = \frac{2}{3}H_c$, and the internal field is then H_c. If the field is increased beyond this value the whole of the superconductor cannot pass over into the normal state, since as long as $H_{\text{ext.}} < H_c$ the field inside the sphere would be less than the critical value. Instead it is found that the magnetic moment varies linearly with the field according to the relation $4\pi I = -(H_c - H_{\text{ext.}})$. This result can be interpreted as meaning that the sphere consists of an intimate mixture of superconducting and normal regions which are individually so small that the state of the sphere is macroscopically homogeneous. In this state, which is known as the intermediate state, the induction B is equal to $3H_{\text{ext.}} - 2H_c$ and it varies from 0 to H_c as $H_{\text{ext.}}$ increases from $\frac{2}{3}H_c$ to H_c, while the internal magnetic field remains constant and equal to H_c. The relation between H and B, the magnetic field and magnetic induction inside the sphere, is therefore as shown in fig. 10·30. If we assume that this is a general relation we can determine the thermodynamic properties of any body for which we can calculate the internal field. In particular, if the body is an ellipsoid in a uniform external field, the internal field and the intensity of magnetization are always uniform and

$$H = H_{\text{ext.}} - 4\pi r I, \tag{10·722·1}$$

where r is the demagnetizing coefficient which depends upon the shape of the ellipsoid (e.g. $r = \frac{1}{3}$ for a sphere). The magnetic state of the ellipsoid can therefore be defined by

$$B = 0, \quad H = -4\pi I = H_{\text{ext.}}/(1-r) \tag{10·722·2}$$

for the superconducting state, where $H_{\text{ext.}} < (1-r)\,H_c$, and by

$$H = H_c, \quad B = H_{\text{ext.}} + H_c(1 - 1/r), \quad 4\pi r I = H_{\text{ext.}} - H_c \tag{10·722·3}$$

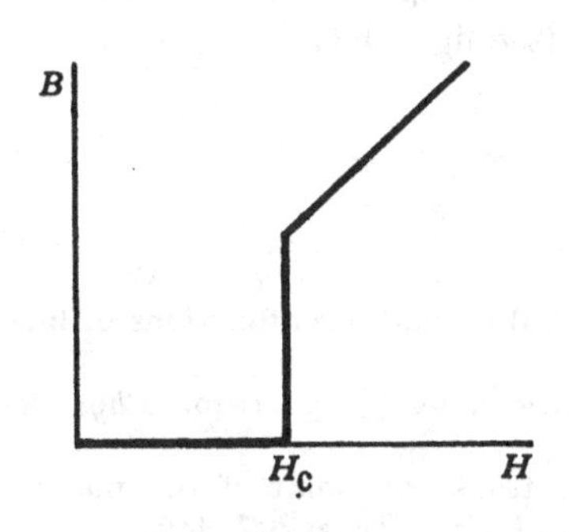

Fig. 10·30. The magnetic induction as a function of the magnetic field.

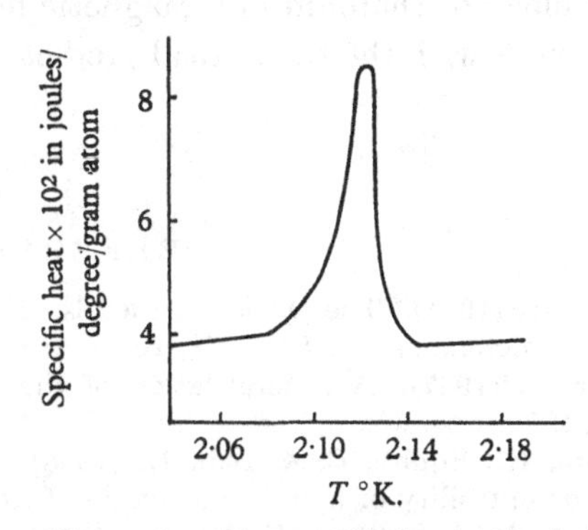

Fig. 10·31. The specific heat of a thallium block in a magnetic field of 33·6 gauss.

for the intermediate state, where $(1-r)\,H_c < H_{\text{ext.}} < H_c$. Hence for the intermediate state

$$F = F_0(T) - V\int_0^{H_{\text{ext.}}} I\,dH_{\text{ext.}} = F_0(T) + \frac{V}{8\pi}\left\{H_c^2 - \frac{1}{r}(H_c - H_{\text{ext.}})^2\right\}, \tag{10·722·4}$$

where $F_0(T)$ is a function of T only. But $F = F_n$ when $H_{\text{ext.}} = H_c$, and this condition determines $F_0(T)$. Therefore

$$F = F_n(T) - (V/8\pi r)\,(H_c - H_{\text{ext.}})^2, \tag{10·722·5}$$

from which we can determine all the thermodynamic properties of the intermediate state (Peierls, 1936). The entropy S is a continuous function of $H_{\text{ext.}}$ and T, and is given by

$$S = -\left(\frac{\partial F}{\partial T}\right)_{H_{\text{ext.}}} = S_n(T) + \frac{V}{4\pi r}(H_c - H_{\text{ext.}})\frac{dH_c}{dT}, \tag{10·722·6}$$

and, on using (10·72·7) to eliminate d^2H_c/dT^2, the specific heat is given by

$$C = T\frac{\partial s}{\partial T} = C_n + \frac{1}{r}\left(1 - \frac{H_{\text{ext.}}}{H_c}\right)(C_s - C_n) + \frac{vT}{4\pi r}\frac{H_{\text{ext.}}}{H_c}\left(\frac{dH_c}{dT}\right)^2. \tag{10·722·7}$$

At constant temperature the specific heat increases discontinuously from C_s to $C_s + \frac{vT}{4\pi r}(1-r)\left(\frac{dH_c}{dT}\right)^2$ when $H_{\text{ext.}} = (1-r)H_c$. It increases linearly with $H_{\text{ext.}}$ in the intermediate state and decreases discontinuously from $C_n + \frac{vT}{4\pi r}\left(\frac{dH_c}{dT}\right)^2$ to C_n when $H_{\text{ext.}} = H_c$. Similarly, if the external field is constant, the specific heat has two discontinuities at the temperatures from which $(1-r)H_c = H_{\text{ext.}}$ and $H_c = H_{\text{ext.}}$.

No measurements have been carried out on the specific heat of ellipsoids in the intermediate state, but the results obtained by Keesom & Kok (1934) for a block of thallium in a magnetic field of 33·6 gauss are in qualitative agreement with the theoretical predictions (see fig. 10·31).

REFERENCES

Bethe, H. (1931). The theory of metals. Eigenvalues and eigenfunctions of linear atomic chains. *Z. Phys.* **71**, 205.

Bitter, F. (1937). A generalisation of the theory of ferromagnetism. *Phys. Rev.* **54**, 79.

Bizette, H., Squire, C. & Tsaï, B. (1938). The transition point of the magnetic susceptibility of manganous oxide. *C.R. Acad. Sci., Paris*, **207**, 449.

Bleaney, B. & Stevens, K. W. H. (1953). Paramagnetic resonance. *Rep. Progr. Phys.* **16**.

Bloch, F. (1930). On the theory of ferromagnetism. *Z. Phys.* **61**, 206.

Bloembergen, N. (1949). The interaction of nuclear spins in a crystalline lattice. *Physica*, **15**, 386.

Bowers, K. D. & Owen, J. (1955). Paramagnetic resonance. II. *Rep. Progr. Phys.* **18**.

Burton, E. F., Smith, H. G. & Wilhelm, J. O. (1940). *Phenomena at the temperature of liquid helium.* New York.

Casimir, H. B. G. (1940). *Magnetism and very low temperatures.* Cambridge.

Daunt, J. G. & Mendelssohn, K. (1937). Equilibrium curve and entropy difference between the supraconductive and the normal state in Pb, Hg, Sn, Ta and Nb. *Proc. Roy. Soc.* A, **160**, 127.

Debye, P. (1912). Some results of a kinetic theory of insulators. *Phys. Z.* **13**, 97.

Debye, P. (1926). Some remarks on magnetisation at low temperatures. *Ann. Phys., Lpz.*, (4), **81**, 1154.

Devonshire, A. F. (1949). Theory of barium titanate. I. *Phil. Mag.* (7), **40**, 1040.

Devonshire, A. F. (1951). Theory of barium titanate. II. *Phil. Mag.* (7), **42**, 1065.

Devonshire, A. F. (1954). Theory of ferroelectrics. *Advanc. Phys.* **3**, 85.

Dirac, P. A. M. (1935). *The principles of quantum mechanics*, 2nd ed. Oxford.

Ewald, P. P. (1921). The calculation of optical and electrostatic lattice potentials. *Ann. Phys., Lpz.*, (4), **64**, 253.

Fairweather, A., Roberts, F. F. & Welch, A. J. E. (1952). Ferrites. *Rep. Progr. Phys.* **15**.

Fröhlich, H. (1949). *Theory of dielectrics, dielectric constants and dielectric loss.* Oxford.

Garrett, C. G. B. (1951*a*). Experiments with an anisotropic magnetic crystal at temperatures below 1° K. *Proc. Roy. Soc.* A, **206**, 242.

Garrett, C. G. B. (1951*b*). The critical field curve in an antiferromagnetic crystal. *J. Chem. Phys.* **19**, 1154.

Garrett, C. G. B. (1954). *Magnetic cooling.* New York.

Gans, R. & Czerlinski, E. C. (1933). Additions to the theory of the magnetisation curves of ferromagnetic single crystals. *Ann. Phys., Lpz.*, (5), **16**, 625.

Giauque, W. R. (1927). A thermodynamic treatment of certain magnetic effects. A proposed method of producing temperatures considerably below 1° absolute. *J. Amer. Chem. Soc.* **49**, 1864.

Gorter, C. J. (1953). Observations on antiferromagnetic $CuCl_2.2H_2O$ crystals. *Rev. Mod. Phys.* **25**, 332.

Gorter, C. J. & Casimir, H. B. G. (1934). Superconductivity. *Physica*, **1**, 305.

Gorter, C. J. & Haanties, J. (1952). Antiferromagnetism at the absolute zero of temperature in the case of rhombic symmetry. *Physica*, **18**, 285.

Guggenheim, E. A. (1936*a*). On magnetic and electrostatic energy. *Proc. Roy. Soc.* A, **155**, 49.

Guggenheim, E. A. (1936*b*). The thermodynamics of magnetisation. *Proc. Roy. Soc.* A, **155**, 70.

Hebb, M. H. & Purcell, E. M. (1937). A theoretical study of magnetic cooling experiments. *J. Chem. Phys.* **5**, 342.

Heisenberg, W. (1928). The theory of ferromagnetism. *Z. Phys.* **49**, 619.

von Hippel, A. (1950). Ferroelectricity, domain structure and phase transitions of barium titanate. *Rev. Mod. Phys.* **22**, 221.

Kay, H. F. & Vousden, P. (1949). Symmetry changes in barium titanate and their relation to its ferroelectric properties. *Phil. Mag.* (7), **40**, 1019.

Keesom, W. H. & Kok, J. A. (1934). Measurements of the latent heat of thallium in relation to the transition in a constant external magnetic field from the superconducting to the non-superconducting state. *Physica*, **1**, 175.

Keesom, W. H. & van Laer, P. H. (1938). Measurements of the atomic heats of tin in the superconductive and in the non-superconductive states. *Physica*, **5**, 193.

Kirkwood, J. G. (1939). The dielectric polarisation of polar liquids. *J. Chem. Phys.* **7**, 911.

de Klerk, D., Steenland, M. J. & Gorter, C. J. (1949). Determination of very low thermodynamic temperatures in chromium potassium alum. *Physica*, **15**, 649.

Kok, J. A. (1934). Superconductivity and Fermi-Dirac statistics. *Physica*, **1**, 1103.

Kornfeld, H. (1924). The calculation of electrostatic potentials and the energy of dipole and quadrupole lattices. *Z. Phys.* **22**, 27.

Langevin, P. (1905). The theory of magnetism. *J. Phys. Radium*, **4**, 678.

Lawton, H. & Stewart, K. H. (1948). Magnetisation curves for ferromagnetic single crystals. *Proc. Roy. Soc.* A, **193**, 72.

Lidiard, A. B. (1954). Antiferromagnetism. *Rep. Progr. Phys.* **17**.

Lorentz, H. A. (1909). *The theory of electrons.* Leipzig.

Luttinger, J. M. & Tisza, L. (1946). Theory of dipole interaction in crystals. *Phys. Rev.* **70**, 954.

McKeehan, L. W. (1933*a*). Magnetic dipole fields in unstrained cubic crystals. *Phys. Rev.* **43**, 913.

McKeehan, L. W. (1933*b*). Magnetic dipole energy in hexagonal crystals. *Phys. Rev.* **43**, 1025.

Meissner, W. & Ochsenfeld, R. (1933). A new effect on the occurrence of superconductivity. *Naturwissenschaften*, **21**, 787.

Merz, W. (1949). The electrical and optical behaviour of single-domain crystals. *Phys. Rev.* **76**, 1221.

Millar, R. W. (1928). The specific heats at low temperatures of manganous oxide, manganous-manganic oxide and manganese dioxide. *J. Amer. Chem. Soc.* **50**, 1875.

Mueller, H. (1938). Theory of the photoelastic effect of cubic crystals. *Phys. Rev.* **47**, 947.

Nagamiya, T., Yosida, K. & Kubo, R. (1955). Antiferromagnetism. *Advanc. Phys.* **4**, 1.

Néel, L. (1932). Influence of the fluctuations of the molecular field on the magnetic properties of bodies. *Ann. Phys., Paris,* (12), **18**, 1.

Néel, L. (1936). Magnetic properties of the metallic state and energy of interaction between magnetic atoms. *Ann. Phys., Paris,* (11), **5**, 232.

Néel, L. (1944). The laws of the magnetisation and of the subdivision into elementary domains of a single crystal of iron. *J. Phys. Radium,* **5**, 241.

Néel, L. (1948). Magnetic properties of the ferrites. *Ann. Phys., Paris,* (12), **3**, 137.

Onsager, L. (1936). Electric moments of molecules in liquids. *J. Amer. Chem. Soc.* **58**, 1486.

Peierls, R. E. (1936). Magnetic transition curves of supraconductors. *Proc. Roy. Soc.* A, **155**, 613.

Purcell, E. M. & Pound, R. V. (1951). A nuclear spin system at negative temperature. *Phys. Rev.* **81**, 279.

Ramsey, N. F. (1956). Thermodynamics and statistical mechanics at negative absolute temperatures. *Phys. Rev.* **103**, 20.

Rutgers, A. J. (1934). Superconductivity. *Physica,* **1**, 1055.

van Santen, J. H. & Opechowski, W. (1948). A generalisation of the Lorentz-Lorenz formula. *Physica,* **14**, 545.

Shoenberg, D. (1952). *Superconductivity.* Cambridge.

Shull, C. G., Strauser, W. A. & Wollan, E. Q. (1951). Neutron diffraction by paramagnetic and antiferromagnetic substances. *Phys. Rev.* **83**, 333.

Slater, J. C. (1929). The theory of complex spectra. *Phys. Rev.* **34**, 1293.

Slater, J. C. (1950). The Lorentz correction in barium titanate. *Phys. Rev.* **78**, 748.

Stoner, E. C. (1934). *Magnetism and matter.* London.

Stoner, E. C. (1937). Magnetic energy and the thermodynamics of magnetisation. *Phil. Mag.* (7), **23**, 833.

Stoner, E. C. (1948). Ferromagnetism. *Rep. Progr. Phys.* **11**.

Stoner, E. C. (1950). Ferromagnetism; magnetisation curves. *Rep. Progr. Phys.* **13**.

van Vleck, J. H. (1932). *The theory of electric and magnetic susceptibilities.* Oxford.

van Vleck, J. H. (1937*a*). The influence of dipole-dipole coupling on the specific heat and susceptibility of a paramagnetic salt. *J. Chem. Phys.* **5**, 320.

van Vleck, J. H. (1937*b*). On the role of dipole-dipole coupling in dielectric media. *J. Chem. Phys.* **5**, 556.

van Vleck, J. H. (1941). On the theory of antiferromagnetism. *J. Chem. Phys.* **9**, 85.

Weiss, P. (1907). The molecular field hypothesis and ferromagnetism. *J. Phys. Radium,* **6**, 661.

Wilson, A. H. (1953). *The theory of metals,* 2nd ed. Cambridge.

Woltjer, H. R. & Onnes, H. Kamerlingh (1923). Magnetisation of gadolinium sulphate at temperatures obtainable with liquid helium. *Proc. Acad. Sci. Amst.* **26**, 626.

Chapter 11

GAS MIXTURES AND CHEMICAL REACTIONS

Perfect Gas Mixtures in which no Chemical Reactions Occur

11·1. Gases containing more than one constituent are usually called mixtures, while the term 'solution' is normally applied only to liquid or solid phases, though logically gas mixtures could also be termed solutions. It is, however, convenient to give separate accounts of gas mixtures and of liquid or solid solutions, since the former are simpler and since much of the interest in solutions centres around the heterogeneous equilibrium of two or more phases. We shall therefore confine our attention in the present chapter to gases, and devote Chapter 12 to the theory of solutions.

11·11. *Dalton's law.* It is possible to obtain the properties of mixtures of perfect gases in various ways, depending upon the particular experimental facts that we take as fundamental. The course adopted here is to introduce the assumptions in two stages, the first of which enables us to determine the internal energy and the second the entropy of a mixture of perfect gases.

If two gases at the same temperature and pressure in different vessels are brought into communication, the gases will diffuse into one another. If no chemical reaction takes place it is found that the volume and energy changes are very small. We therefore assume that ideal gases would have the following property:

If two or more perfect gases mix slowly by diffusion at a given temperature and pressure, and if no chemical reaction takes place, the total volume of the gases remain constant and no heat is taken in or given out.

This generalization from experience enables us to introduce the very useful concept of the partial pressure of a gas in a mixture. Consider r perfect gases at the temperature T and the pressure P, and let their volumes be V_i $(i = 1, 2, \ldots, r)$. If each gas were expanded isothermally to volume V, where $V = \Sigma V_i$, the pressure of the ith gas would be p_i, where $p_i V = PV_i$. Hence $\Sigma p_i = P$. Now if the gases are allowed to mix at the pressure P, the mixture has pressure P and volume V, so that if we define p_i as the partial pressure of the ith gas, the total pressure is the sum of the partial pressures of the various gases. This is known as Dalton's law, and can be stated in words as follows:

The pressure of a mixture of perfect gases is the sum of the partial pressures which the various gases would separately produce if at the same temperature each occupied the whole volume occupied by the mixture.

11·12. *The equation of state.* We next consider the equation of state of a mixture of perfect gases. If n_i gram molecules of the ith gas when present alone at pressure P and temperature T occupy a volume V_i, then $PV_i = n_i RT$. If the gases are mixed, the pressure and temperature are still P and T, the volume is $V = \Sigma V_i$, and so

$$PV = \sum_{i=1}^{r} PV_i = nRT, \tag{11·12·1}$$

where n is the total number of gram molecules present and is given by

$$n = \Sigma n_i. \tag{11·12·2}$$

If M_i^* is the molecular weight of the ith gas, the total mass is $\Sigma n_i M_i^*$ grams, and the mean molecular weight M^* is

$$M^* = \Sigma n_i M_i^* / \Sigma n_i. \tag{11·12·3}$$

The equation of state of a mixture of perfect gases is therefore the same as for a single perfect gas provided that the mass is measured in units of the mean gram molecular weight.

11·13. *The specific heat.* Since the internal energy of the gases is unaltered by mixing, we have

$$U = \Sigma U_i. \tag{11·13·1}$$

Also $U_1 = n_i(C_{V,i} T + v_{0,i})$, and, if we put $U = n(C_V T + v_0)$, we have

$$v_0 = \Sigma x_i v_{0,i} \tag{11·13·2}$$

and

$$C_V = \Sigma x_i C_{V,i}, \tag{11·13·3}$$

where x_i, the mole fraction of the ith gas, is defined by the equation

$$x_i = n_i / \Sigma n_i \quad (\Sigma x_i = 1). \tag{11·13·4}$$

Further, since the equation of state is the same as for a single gas we have $C_P = C_V + R$. Hence

$$C_P = \Sigma x_i C_{P,i} + R = \Sigma x_i (C_{P,i} + R) = \Sigma x_i C_{P,i}. \tag{11·13·5}$$

Therefore, the whole heat capacity of a mixture of perfect gases is equal to the sum of the separate heat capacities of the individual gases before mixing, either at constant volume or at constant pressure.

11·131. In addition to the mole fractions defined above, there are several other quantities which can be used to define the composition. In the first place we can express the masses of the constituents in grams instead of in terms of the number of gram molecules. Secondly, we may use the masses (or the numbers of gram molecules) per unit volume, and thirdly, in the case of solutions, we may denote one of the constituents as the solvent and the others as the solutes, and give the ratios of the masses (or of the numbers of gram molecules) of the solutes to that of the solvent. Since we shall have little occasion to use these alternative definitions we shall not write the equations in terms of them.

11·14. *The entropy of a mixture.* Just as for a single gas, the entropy of a perfect gas mixture is given by equation (3·51·4). Hence, if as usual $T^\dagger$ and $P^\dagger$ are the temperature and pressure of a standard state,

$$S = n\left(C_P \log\frac{T}{T^\dagger} - R\log\frac{P}{P^\dagger} + s_0\right) \quad (11\cdot14\cdot1)$$

$$= \Sigma n_i\left(C_{P,i} \log\frac{T}{T^\dagger} - R\log\frac{P}{P^\dagger} + s_{0,i}\right) + n(s_0 - \Sigma x_i s_{0,i}), \quad (11\cdot14\cdot2)$$

by (11·13·5). This can be written

$$S(T,P) = \Sigma n_i s_i(T,P) + n(s_0 - \Sigma x_i s_{0,i}). \quad (11\cdot14\cdot3)$$

Now $n_i s_i(T,P)$ is the entropy of the ith gas before mixing, and we therefore see that the entropy of the mixture can only differ from the sum of the entropies before mixing by a constant. Since diffusion is a spontaneous process, this constant must be positive or zero, but to determine its exact value we require further experimental evidence.

11·141. *Semi-permeable membranes.* In order to determine the entropy of a mixture of gases it is necessary to find some method of separating them reversibly. This can be done, at least in theory, by means of a semi-permeable membrane which is such that it will allow molecules of one gas to pass freely through it while it is impermeable to molecules of all other gases. Semi-permeable membranes which are permeable to water, but not to large organic molecules such as sugar, are fairly common, and approximations to membranes which are permeable to certain gases but not to others also exist. The best known example is palladium, which, when heated to redness, is very permeable to hydrogen but quite impervious to all other gases.

As a result of the somewhat meagre experimental evidence on the behaviour of real gases, we assume that, if a mixture of perfect gases is in equilibrium with another mixture of perfect gases through a membrane permeable only to some of the gases, the partial pressures of the gases to which the membrane is permeable are the same on both sides of the membrane. The temperature is necessarily the same throughout.

Consider a mixture of two perfect gases contained in a cylinder with four pistons, of which A and B are fixed while A' and B' are movable and at a fixed distance apart equal to AB (fig. 11·1). The pistons A and B' are permeable to the gases 1 and 2 respectively, while the pistons A' and B are impermeable to both gases. Originally let A' coincide with A and B' with B. Then raise A' and B' slowly so that the gas 1 passes into the space AA' and the gas 2 into the space BB', until finally B' is in contact with A and the space $A'A$ contains the whole of gas 1 while the space BB' contains the whole of gas 2.

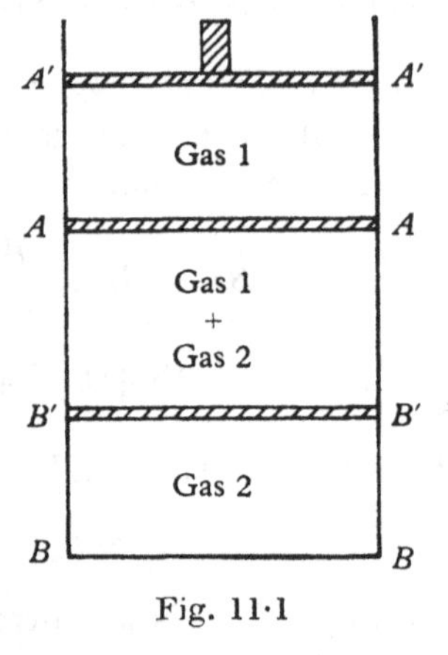

Fig. 11·1

The total upthrust on the pistons A' and B' is zero since the partial pressures of gas 1 on either side of A' are the same and the partial pressures of gas 2 on either side of B' are the same. Hence no external work is done in the process. Also, since the process is isothermal, the internal energy is unaltered, and since $Q = \Delta U + W$ no heat enters or leaves the system. Further, the process is reversible and so $T\Delta S = Q = 0$. The entropy is therefore unaltered. This important result can be generalized to any number of gases and can be stated as follows:

The entropy of a mixture of perfect gases is the sum of the entropies which each of the constituents would have, if at the same temperature, it occupied the whole volume of the mixture.

11·142. The result given in the preceding section is the natural complement of Dalton's law. It enables us to calculate the entropy of a mixture in terms of the entropies of its constituents. Applying the result, we find at once for the entropy of a mixture the expression

$$S = \Sigma n_i \left(C_{P,i} \log \frac{T}{T^\dagger} - R \log \frac{p_i}{P^\dagger} + s_{0,i} \right), \tag{11·142·1}$$

where

$$p_i = x_i P \tag{11·142·2}$$

by Dalton's law. Comparing this with (11·14·2) we see that

$$s_0 - \Sigma x_i s_{0,i} = -R \Sigma x_i \log x_i. \tag{11·142·3}$$

This is the increase in entropy per gram molecule of the mixture when the gases diffuse into one another. Since the x_i's are positive fractions less than unity, (11·142·3) is positive, as it must be. The increase in entropy is a constant independent of T and P, as was found in § 11·14, but we now see that it is independent of the nature of the gases forming the mixture and is determined entirely by the numbers of the different molecules present.

11·143. The result of the preceding section shows that, when 1 gram molecule of a gas diffuses into 1 gram molecule of another gas, the increase in entropy is $2R \log 2$ and is quite independent of the nature of the gases. We might therefore try to argue that when 2 gram molecules of the same gas mix by diffusion the entropy increases by $2R \log 2$, whereas in fact the entropy is clearly unaltered. This is known as Gibbs's paradox. The fallacy lies in assuming that we can pass from the case of dissimilar gases to the case of similar gases by a continuous variation of the properties of the gases. This assumption is not true, since, however little the gases differ, two dissimilar gases can always be separated from one another after they have been mixed by using a method of separation which depends upon the difference in the properties of the gases. For example, if the gases consist of two different isotopes of an element, their chemical properties are identical but some of their physical properties differ slightly, and the isotopes can be separated, though with difficulty, by distillation, by thermal diffusion, by diffusion through a membrane or by any process which is different for two molecules which differ only in mass. On the other hand, if we allow two portions of the same gas to diffuse into one another, there is no way in which we can separate them again so as to obtain in one vessel all the molecules that composed one of the original portions of the gas. It is therefore meaningless to talk about a change in entropy in this case, since there is no possible way in which it could be measured. To sum up, molecules are either dissimilar or similar. They are dissimilar if they differ in any property by any amount, however small, and, in calculating the entropy of a mixture, the dissimilarity must be taken into account. The entropy of a mixture is a discontinuous function of the difference in properties of the constituents. It has one value when the constituents are dissimilar and changes discontinuously to another value when the difference between the constituents vanishes.

The Fundamental Thermodynamic Equations of a Mixture of Perfect Gases

11·2. The simplest way of obtaining the fundamental thermodynamic equation of a mixture is to construct the G function from its definition $G = U - ST + PV$ by means of equations (11·12·1), (11·13·1) and (11·142·1). We find

$$G(T, P, n_i) = \Sigma n_i \left\{ C_{P,i}\left(T - T \log \frac{T}{T^\dagger}\right) + RT \log \frac{n_i P}{n P^\dagger} + v_{0,i} - T s_{0,i} \right\}. \quad (11\cdot 2\cdot 1)$$

For a perfect gas in the wide sense, the expression corresponding to that given in equation (3·52·4) is

$$G(T,P,n_i)=\Sigma n_i\left\{\int_0^T C_{P,i}(T')\,dT' - C^0_{P,i}\,T\log\frac{T}{T^\dagger} - T\int_0^T \frac{C^1_{P,i}(T')}{T'}\,dT' + RT\log\frac{n_i P}{nP^\dagger} + v_{0,i} - Ts_{0,i}\right\}, \quad (11\cdot 2\cdot 2)$$

where C^0_P is the value of $C_P(T)$ for $T=0$ and $C^1_P(T)=C_P(T)-C^0_P$. Differentiating with respect to n_i we obtain for the thermodynamic potential the expression

$$\mu_i=\left(\frac{\partial G}{\partial n_i}\right)_{T,P,n_j}=C_{P,i}\left(T-T\log\frac{T}{T^\dagger}\right)+RT\log\frac{n_i P}{nP^\dagger}+v_{0,i}-Ts_{0,i} \quad (11\cdot 2\cdot 3)$$

from (11·2·1), and

$$\mu_i=\int_0^T C_{P,i}(T')\,dT' - C^0_{P,i}\,T\log\frac{T}{T^\dagger} - T\int_0^T \frac{C^1_{P,i}(T')}{T'}\,dT' + RT\log\frac{n_i P}{nP^\dagger} + v_{0,i} - Ts_{0,i} \quad (11\cdot 2\cdot 4)$$

from (11·2·2). (In differentiating the expression $\Sigma n_i \log(n_i P/n)$ it must be borne in mind that n is a function of n_i. The derivative of this expression with respect to n_i is $\log(n_i P/n)+1-\Sigma n_i/n=\log(n_i P/n)$.) Equation (11·2·4) may be written as

$$\mu_i(T,P,n_j)=\mu^0_i(T,p_i), \quad (11\cdot 2\cdot 5)$$

where $p_i=n_i P/n$ is the partial pressure of the ith gas and $\mu^0_i(T,P)$ is its thermodynamic potential in the pure state at temperature T and pressure P.

By solving equation (11·2·3) for $n_i P/n$ and summing over i we can find the $P(T,\mu_i)$ equation for the mixture. It is

$$P=P^\dagger\Sigma\exp\left[(s_{0,i}-C_{P,i})/R\right](T/T^\dagger)^{C_{P,i}/R}\exp\left[(\mu_i-v_{0,i})/RT\right]. \quad (11\cdot 2\cdot 6)$$

The content of this equation, taken in conjunction with equation (3·51·18), which refers to a single perfect gas, can be expressed in words as follows:

The pressure in a mixture of perfect gases is equal to the sum of the partial pressures which each gas would separately produce at the same temperature and with the same value of its potential.

We may, if we please, take this statement as defining the properties of a mixture of perfect gases, as was done by Gibbs. It is equivalent to the two statements made in §§ 11·1 and 11·141, and, since it defines the $P(T,\mu_i)$ fundamental equation, all the properties of a perfect gas mixture can be deduced from it.

It is a straightforward matter to construct the other fundamental equations for a perfect gas mixture by the methods given in §§ 3·51 and 3·52, but care is necessary to ensure that the correct entropy constants are chosen. To construct the $F(T, V, n_i)$ equation, which, apart from $G(T, P, n_i)$, is the only one that we shall have occasion to use, it is simplest to write down the entropy as a function of V and T, starting from the statement given at the end of § 11·141. By (2·72·9), the entropy of the mixture is given by

$$S = \Sigma n_i \left(C^0_{V,i} \log \frac{T}{T^\dagger} + \int_0^T \frac{C^1_{V,i}(T')}{T'} dT' + R \log \frac{V}{n_i v^\dagger} + s'_{0,i} \right), \quad (11\cdot2\cdot7)$$

where $T^\dagger$, $v^\dagger$ and $s'_{0,i}$ are independent of n_i. The free energy at constant volume is therefore given by

$$F(T, V, n_i) = \Sigma n_i \left(\int_0^T C_{V,i}(T')\, dT' - C^0_{V,i}\, T \log \frac{T}{T^\dagger} - T \int_0^T \frac{C^1_{V,i}(T')}{T'} dT' - RT \log \frac{V}{n_i v^\dagger} + v_{0,i} - T s'_{0,i} \right) \quad (11\cdot2\cdot8)$$

$$= \Sigma F_i(T, V, n_i), \quad (11\cdot2\cdot9)$$

where $F_i(T, V, n_i)$ is the free energy at constant volume of n_i gram molecules of the ith gas, occupying the volume V at the temperature T in the absence of the other gases. This result may be stated as follows:

The free energy at constant volume of a mixture of perfect gases is the sum of the free energies at constant volume which each constituent would have if, at the same temperature, it occupied the whole volume of the mixture.

11·21. *The statistical mechanics of perfect gas mixtures.* The results of the preceding sections can be readily deduced from the fundamental formula (5·351·5) for the free energy of a system in terms of its partition function, namely,

$$F = -kT \log Z_{\text{system}}. \quad (11\cdot21\cdot1)$$

According to the atomic model of a perfect gas, the molecules of a perfect gas move entirely independently of one another, so that the partition function of a system of such molecules is the product of the partition functions of the separate molecules, due allowance being made for the indistinguishability of the molecules (see § 5·35). Hence if we have N_1 and N_2 molecules of two different kinds in an enclosure of volume V and of the temperature T, we have

$$Z_{\text{system}} = \frac{Z_1(V, T)^{N_1}}{N_1!} \frac{Z_2(V, T)^{N_2}}{N_2!}, \quad (11\cdot21\cdot2)$$

where $Z_1(V, T)$ and $Z_2(V, T)$ are the partition functions for an individual molecule of type 1 and type 2 respectively. The factors $1/N_1!$ and $1/N_2!$ have to be introduced to ensure that the interchange of two indistinguishable molecules does not give rise to a different molecular complexion. The combination of (11·21·1) and (11·21·2) now shows that

$$F(T, V, n_1, n_2) = -kT\log\frac{Z_1(V,T)^{N_1}}{N_1!} - kT\log\frac{Z_2(V,T)^{N_2}}{N_2!}$$

$$= F_1(T, V, n_1) + F_2(T, V, n_2), \qquad (11\cdot21\cdot3)$$

and when this result is generalized to r sets of gases we regain equation (11·2·9), from which all the properties of the gas mixture follow at once.

Mixtures of Imperfect Gases

11·3. Having established the fundamental equations for a mixture of perfect gases, we can extend them at once to a mixture of imperfect gases by the method used in § 8·5 for a simple imperfect gas. We write the integrated form of the relation $(\partial G/\partial P)_{T,n_i} = V$ as

$$G(T,P,n_i) = G_0(T,n_i) + nRT\log\frac{P}{P^\dagger} + \int_0^P \left(V(T,P',n_i) - \frac{nRT}{P'}\right)dP', \qquad (11\cdot3\cdot1)$$

where $n = \Sigma n_i$ and where the function $G_0(T, n_i)$ is determined by the condition that (11·3·1) should pass over into (11·2·2) as $P \to 0$. This gives

$$G(T,P,n_i) = \Sigma n_i\left[\int_0^T C_{P,i}(T')\,dT' - C^0_{P,i}\,T\log\frac{T}{T^\dagger} - T\int_0^T \frac{C^1_{P,i}(T')}{T'}dT'\right]_{\text{ideal}}$$
$$+ \Sigma n_i RT\log\frac{n_i P}{nP^\dagger} + \int_0^P\left(V - \frac{nRT}{P'}\right)dP' + \Sigma n_i(\upsilon_{0,i} - Ts_{0,i}) \qquad (11\cdot3\cdot2)$$

$$= G(T,P,n_i)_{\text{ideal}} + \int_0^P\left(V - \frac{nRT}{P'}\right)dP'. \qquad (11\cdot3\cdot3)$$

We can now obtain all the other thermodynamic functions by differentiation. In particular we have

$$H(T,P,n_i) = -T^2\frac{\partial}{\partial T}\frac{G}{T} = \Sigma n_i \mathrm{H}_i(T)_{\text{ideal}} - \int_0^P T^2\frac{\partial}{\partial T}\frac{V}{T}dP', \qquad (11\cdot3\cdot4)$$

$$S(T,P,n_i) = -\frac{\partial G}{\partial T} = S(T,P,n_i)_{\text{ideal}} + \int_0^P\left(\frac{nR}{P'} - \frac{\partial V}{\partial T}\right)dP', \qquad (11\cdot3\cdot5)$$

$$C_P(T,P,n_i) = \frac{1}{n}\frac{\partial H}{\partial T} = \frac{T}{n}\frac{\partial S}{\partial T} = \Sigma x_i C_{P,i}(T)_{\text{ideal}} - \int_0^P T\frac{\partial^2 v}{\partial T^2}dP', \qquad (11\cdot3\cdot6)$$

where $\mathrm{H}_i(T)_{\text{ideal}}$ and $C_{P,i}(T)_{\text{ideal}}$ are the heat function per gram molecule and the specific heat of the pure ith component in the ideal state. Also

$$\mu_i(T,P,n_j) = \frac{\partial G}{\partial n_i} = \mu_i(T,p_i)_{\text{ideal}} + \int_0^P\left(\frac{\partial V}{\partial n_i} - \frac{RT}{P'}\right)dP', \qquad (11\cdot3\cdot7)$$

where $p_i = n_i P/n$ is the partial pressure of the ith gas and where $\mu_i(T, P)_{\text{ideal}}$ is its thermodynamic function in the pure state at temperature T and pressure P when it is considered to be an ideal gas (i.e. in the limit of very small P).

There are corresponding formulae in terms of the variables T, V, n_i. They are

$$F(T, V, n_i) = F(T, V, n_i)_{\text{ideal}} + \int_V^\infty \left(P - \frac{nRT}{V'}\right) dV', \qquad (11\cdot3\cdot8)$$

$$U(T, V, n_i) = \Sigma n_i U_i(T)_{\text{ideal}} - \int_V^\infty T^2 \frac{\partial}{\partial T}\frac{P}{T} dV', \qquad (11\cdot3\cdot9)$$

$$S(T, V, n_i) = S(T, V, n_i)_{\text{ideal}} + \int_V^\infty \left(\frac{nR}{V'} - \frac{\partial P}{\partial T}\right) dV', \qquad (11\cdot3\cdot10)$$

$$C_V(T, V, n_i) = \Sigma x_i C_{V,i}(T)_{\text{ideal}} - \int_V^\infty T \frac{\partial^2 P}{\partial T^2} dV'. \qquad (11\cdot3\cdot11)$$

11·31. *The fugacities.* We define the fugacity $f_i(T, P, n_j)$ of the ith component by the relations

$$\mu_i(T, P, n_j) = \mu_i^*(T) + RT \log f_i(T, P, n_j), \quad f_i \to n_i P/n \text{ as } P \to 0. \quad (11\cdot31\cdot1)$$

Then, since $\qquad \mu_i(T, p_i)_{\text{ideal}} = \mu_i^*(T) + RT \log p_i,$

where $p_i = n_i P/n$, we obtain from (11·3·7) the explicit relation

$$RT \log f_i = RT \log \frac{n_i P}{n} + \int_0^P \left(\frac{\partial V}{\partial n_i} - \frac{RT}{P'}\right) dP', \qquad (11\cdot31\cdot2)$$

which can be evaluated if the equation of state is known.

11·32. *A slightly imperfect gas mixture.* Much less is known about imperfect gas mixtures than single imperfect gases, and we therefore confine our attention to cases where the pressure is small enough for the third and higher virial coefficients to be neglected.

The equation of state can be written in the form

$$PV = \Sigma n_i RT + RT \sum_{i,j} n_i n_j B_{ij}(T) V^{-1} + \ldots, \qquad (11\cdot32\cdot1)$$

since, in the expansion of PV in descending powers of V, the coefficient of V^{-r+1} must be a homogeneous function of degree r in the n_i's, the coefficients in this homogeneous function being functions of T only. Inversion of (11·32·1) gives the alternative form

$$V = nRTP^{-1} + \sum_{i,j} \frac{n_i n_j}{n} B_{ij}(T) + \ldots. \qquad (11\cdot32\cdot2)$$

We now summarize the more important new features which depend upon the gas mixture being imperfect.

11·321. *The changes in volume and pressure due to mixing.* The corrections to Dalton's law can be expressed either as the change in volume when the gases diffuse into one another at constant pressure, or as the difference between the pressure of the mixture and the sum of the pressures which the individual constituents would have if they occupied the whole volume of the mixture. The deviations from the law of the additivity of volumes can be calculated from equation (11·32·2). The volumes of the individual gases are V_i $(i=1, \ldots, r)$, where $V_i = n_i RTP^{-1} + n_i B_{ii}(T)$. Therefore

$$V - \Sigma V_i = \sum_{i,j} \frac{n_i n_j}{n} B_{ij}(T) - \sum_j n_i B_{ii}(T). \tag{11·321·1}$$

This becomes more symmetrical if the last term is replaced by

$$-\tfrac{1}{2}\Sigma n_i n_j (B_{ii} + B_{jj})/n,$$

and we then have

$$\frac{\Delta V}{V} = \frac{V - \Sigma V_i}{V} = -\frac{P}{2RT} \sum_{i,j} x_i x_j (B_{ii} - 2B_{ij} + B_{jj}). \tag{11·321·2}$$

Similarly, the pressure P_i which the ith gas would have if its volume were V is $P_i = n_i RTV^{-1} + n_i^2 RTB_{ii} V^{-2}$. Therefore

$$\frac{\Delta P}{P} = \frac{P - \Sigma P_i}{P} = \frac{P}{RT} \sum_{i \neq j} x_i x_j B_{ij}. \tag{11·321·3}$$

In general neither $\Delta V/V$ nor $\Delta P/P$ will be so small as to be negligible at high pressures. Empirically, however, it seems that $\Delta P/P$ is much smaller than $\Delta V/V$ for gases well above their critical points, but neither quantity is negligible when the temperature is not very much greater than the critical temperature of one of the components, as is shown by the curves given in fig. 11·2 for the argon-ethylene system at 25° C. (Masson & Dolley, 1923; see also Gibby, Tanner & Masson, 1929; Tanner & Masson, 1930). (The critical constants are as follows. For argon $T_c = 151°$ K., $P_c = 48$ atmospheres; for ethylene $T_c = 282°$ K., $P_c = 50·7$ atmospheres.) It will be noted that in general there is an increase in volume on mixing and that the deviations reach a maximum at about 75 atmospheres. For the same range of pressures, $\Delta P/P$, on the other hand, never exceeds 8 %.

11·322. *The heat of mixing.* The heat function H is given by

$$H(T, P, n_i) = \Sigma n_i H_i(T)_{\text{ideal}} - PT^2 \sum_{i,j} \frac{n_i n_j}{n} \frac{d}{dT} \frac{B_{ij}}{T}, \tag{11·322·1}$$

while the heat function of the r separate gases is

$$\Sigma n_i H_i(T, P) = \Sigma \left(n_i H_i(T)_{\text{ideal}} - PT^2 \frac{d}{dT} \frac{B_{ii}}{T} \right). \tag{11·322·2}$$

The heat of mixing Q_P, defined as the heat which must be absorbed, per gram molecule of the mixture, during the process of mixing in order to keep the temperature and the pressure constant, is equal to the increase in H. It is therefore given by

$$Q_P = -PT^2\left(\sum_{i,j} x_i x_j \frac{d}{dT}\frac{B_{ij}}{T} - \sum_i x_i \frac{d}{dT}\frac{B_{ii}}{T}\right). \qquad (11\cdot322\cdot3)$$

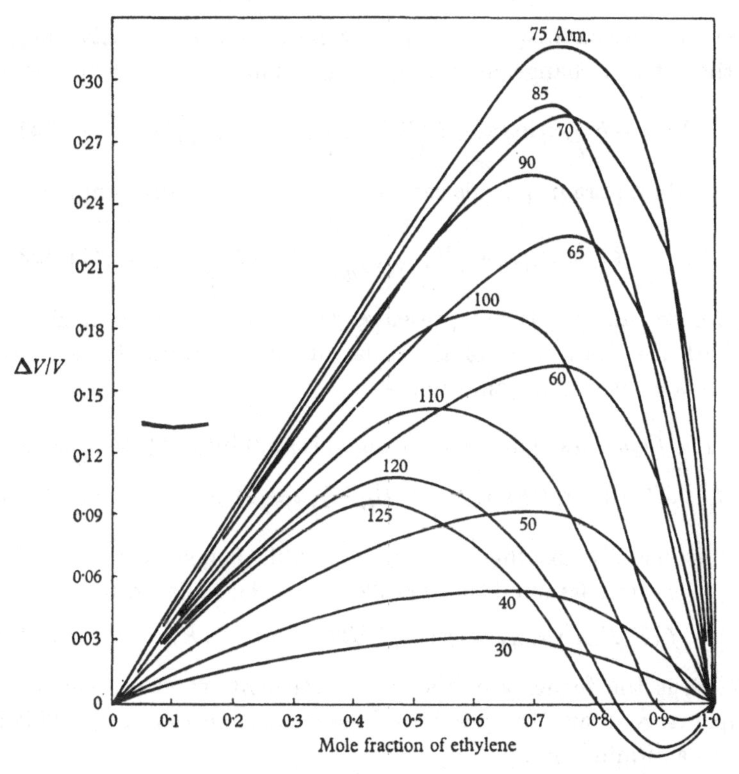

Fig. 11·2. Deviations from Dalton's law for the argon-ethylene system.

This can be written more symmetrically as follows. The expression inside the bracket is

$$\frac{1}{2}\left\{2\sum_{i,j} x_i x_j \frac{d}{dT}\frac{B_{ij}}{T} - \sum_{i,j} x_i x_j\left(\frac{d}{dT}\frac{B_{ii}}{T} + \frac{d}{dT}\frac{B_{jj}}{T}\right)\right\} = -\tfrac{1}{2}\sum_{i,j} x_i x_j \frac{d}{dT}\frac{B_{ii} - 2B_{ij} + B_{jj}}{T}.$$

Hence

$$Q_P = \tfrac{1}{2}PT^2 \sum_{i,j} x_i x_j \frac{d}{dT}\frac{B_{ii} - 2B_{ij} + B_{jj}}{T}. \qquad (11\cdot322\cdot4)$$

For gases which do not react chemically and which are far from their condensation points, the heat of mixing is small and of the order of P to $10P$ calories per gram molecule if P is measured in atmospheres.

11·323. *The entropy of mixing.* The entropy change on mixing is mainly determined by the number of molecules present, but there is a small correction term proportional to P. We have

$$S(T,P,n_i) = S(T,P,n_i)_{\text{ideal}} - P\sum_{i,j}\frac{n_i n_j}{n}\frac{dB_{ij}}{dT}. \qquad (11\cdot323\cdot1)$$

The entropy of the r separate gases is

$$\Sigma n_i s_i = \Sigma n_i\{s_i(T,P)_{\text{ideal}} - P\,dB_{ii}/dT\}. \qquad (11\cdot323\cdot2)$$

But, from § 11·142, we have $S(T,P,n_i)_{\text{ideal}} - \Sigma n_i s_i(T,P)_{\text{ideal}} = -R\Sigma n_i \log x_i$, and hence the entropy change on mixing, per gram molecule of the mixture, is

$$\Delta s = -R\sum_i x_i \log x_i - P\left(\sum_{i,j} x_i x_j \frac{dB_{ij}}{dT} - \sum_i x_i \frac{dB_{ii}}{dT}\right). \qquad (11\cdot323\cdot3)$$

As in the preceding paragraph this can be written in the more symmetrical form

$$\Delta s = -R\sum_i x_i \log x_i + \tfrac{1}{2}P\sum_{i,j} x_i x_j \frac{d}{dT}(B_{ii} - 2B_{ij} + B_{jj}). \qquad (11\cdot323\cdot4)$$

For normal gases and moderate pressures the second term is negligible compared with the first, being of the order of $10^{-2}P$ calorie/degree/gram molecule if P is measured in atmospheres.

11·324. *The fugacities.* On substituting (11·32·2) into (11·31·2) we find that

$$f_i(T,P,n_j) = x_i P \exp\left[\left(2\sum_j x_j B_{ij} - \sum_{j,k} x_j x_k B_{jk}\right)P/RT\right]. \qquad (11\cdot324\cdot1)$$

Since there is usually insufficient data available concerning the virial coefficients, it is often assumed that (11·324·1) can be replaced by

$$f_i(T,P,n_j) = x_i P \exp(B_{ii}P/RT) = x_i f_i^0(T,P), \qquad (11\cdot324\cdot2)$$

where $f_i^0(T,P)$ is the fugacity of the ith constituent when present alone at the temperature T and the pressure P (see equation (8·511·3)). This is equivalent to assuming that

$$B_{ii} - 2B_{ij} + B_{jj} = 0 \quad (\text{all } i, j), \qquad (11\cdot324\cdot3)$$

a relation which is not in general true.

11·33. *Corresponding states for mixtures of imperfect gases.* Since the days of van der Waals there have been considerable speculations concerning the possibility of deducing the equation of state for a mixture from the equations of state of the pure constituents. If A denotes any physical constant (such as a virial coefficient) describing the properties of a gas mixture, it will in general be a quadratic function of the mole fractions of the constituents. Thus for a binary mixture we may write

$$A = x_1^2 A_{11} + 2x_1 x_2 A_{12} + x_2^2 A_{22}, \qquad (11\cdot33\cdot1)$$

and the constants A_{11} and A_{22} can be found from the properties of the pure constituents. But it is by no means obvious that there is any possibility of determining A_{12} from the values of A_{11} and A_{22}, and in fact there is in general no connexion between the three constants. Various hypotheses have, however, been put forward which might be expected to be roughly true for mixtures of gases having similar molecular configurations. The simplest of these are the following:

$$A_{12} = \tfrac{1}{2}(A_{11} + A_{22}), \tag{11·33·2}$$

$$A_{12} = (A_{11}A_{22})^{\frac{1}{2}}, \tag{11·33·3}$$

and

$$A_{12}^{\frac{1}{3}} = \tfrac{1}{2}(A_{11}^{\frac{1}{3}} + A_{22}^{\frac{1}{3}}). \tag{11·33·4}$$

A discussion of the empirical data will be found, for example, in the review articles by Beattie & Stockmayer (1940) and by Beattie (1949), and it would appear that the various constants should be treated differently. Thus the constant a in van der Waals's formula should be calculated from (11·33·3), while b should be calculated from (11·33·2). An alternative method of approach is to try to set up a reduced equation for gas mixtures. This can only be expected to be successful if the laws of force between the various molecules concerned have the same general form. A recent discussion along these lines of some simple binary mixtures has been given by Guggenheim & McGlashan (1951).

11·331. It was shown in §8·423 that the equation of state of many simple gases may be written in a dimensionless form by expressing the pressure, volume and temperature as ratios with respect to a standard pressure, volume and temperature, characteristic of each gas, which may either be values derived from calculations of the interaction energies between the molecules or may be the experimental values at, say, the critical point. The existence of this reduced equation of state for a single gas is a consequence of the fact that for simple molecules the mutual potential energy of two molecules can be written as $E(r_{ik}) = E_0 f(r_{ik}/\sigma)$, where f is a universal function. We can then define v^* and T^*, where v^* is proportional to σ^3 and T^* is proportional to E_0, so that the second virial coefficient B is $v^*\psi(T/T^*)$, ψ being the same for all gases of a given type. If we have two such gases 1 and 2, we have

$$B_{11} = v_{11}^*\psi(T/T_{11}^*), \quad B_{22} = v_{22}^*\psi(T/T_{22}^*), \tag{11·331·1}$$

while if the mutual potential energy of two molecules of gas 1 and gas 2 can be written in the same form, we also have

$$B_{12} = v_{12}^*\psi(T/T_{12}^*). \tag{11·331·2}$$

We may choose v_{11}^*, T_{11}^* and v_{22}^*, T_{22}^* as the critical volume and temperature of the gas 1 and the gas 2 respectively. It then remains to define v_{12}^*, T_{12}^*. The simplest possible assumption is that

$$\sigma_{12} = \tfrac{1}{2}(\sigma_{11}+\sigma_{22}), \quad E_{0,12} = (E_{0,11}\,E_{0,22})^{\frac{1}{2}}, \tag{11·331·3}$$

which is equivalent to

$$v_{12}^{*\frac{1}{3}} = \tfrac{1}{2}(v_{11}^{*\frac{1}{3}} + v_{22}^{*\frac{1}{3}}), \quad T_{12}^* = (T_{11}^*\,T_{22}^*)^{\frac{1}{2}}. \tag{11·331·4}$$

If we use the values of the critical constants given in table 8·1, p. 213, we find the values of v_{12}^* and T_{12}^* given in table 11·1 for a number of gas mixtures. (As usual, we cannot take T^* and v^* for hydrogen as the critical constants,

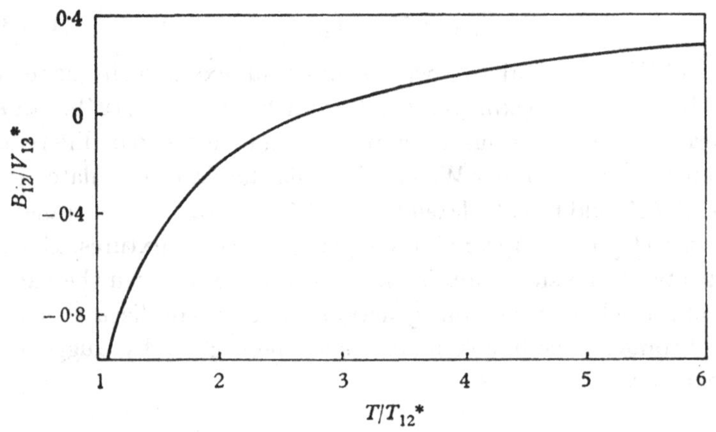

Fig. 11·3. The experimental curve for B_{12}/V_{12}^* as a function of T/T_{12}^*.

and we have to take the values already mentioned in § 7·422, namely, $T^* = 43·4°$ K., $v^* = 50$ cm.^{3}/gram molecule.) We may then plot B_{12}/v_{12}^* as a function of T/T_{12}^* for these mixtures, and it is found that they lie on a universal curve (fig. 11·3), which, except for the different units employed, is the same as the curve for single gases shown in fig. 8·9.

Table 11·1. *Values of T_{12}^* and v_{12}^* for gas mixtures*

	N_2/O_2	N_2/H_2	A/H_2	H_2/CO	CH_4/C_2H_6	$CH_4/$ n-C_4H_{10}
T_{12}^* in °K.	139·4	73·9	80·9	76·0	241·0	284·4
v_{12}^* in cm.^{3}/gram molecule	82·1	67·8	61·5	59·1	121·9	165·9

Chemical Reactions

11·4. The equations used in chemistry to describe chemical reactions can be supplemented so as to give concisely some of the thermodynamic features of the reactions. This can be done by indicating the state of the substances taking part in the reaction. For example, we may wish to consider the reaction in which $2n_1$ gram molecules of hydrogen at pressure P_1

and at temperature T_1 react with n_2 gram molecules of oxygen at pressure P_2 and temperature T_2 to give $2n_1$ gram molecules of water and n_2-n_1 gram molecules of oxygen at pressure P_3 and temperature T_3. However, in such a reaction neither the heat given out nor the external work done is in general determined by the initial and final states, and to calculate them we should have to know every detail of the course of the reaction. The thermodynamic functions, on the other hand, are completely determined by the initial and final states, but if we are only interested in these functions it is simpler to consider a reaction like that outlined above to consist of a succession of such changes as expansion, compression, heating, cooling and the chemical reaction proper. We shall therefore normally only consider chemical reactions between substances which are initially at the same temperature and pressure. If the initial and final temperatures are the same, there are essentially two different cases to consider according as the pressure or the volume is kept constant, the former being the more important. The heat of reaction is normally defined as the heat given out when the reaction takes place in a calorimeter, and is therefore $-(\Delta H)_P$ or $-(\Delta U)_V$ according as the pressure or the volume is kept constant. If the reaction can be made to proceed reversibly (it will then be isothermal and will involve the production of some kind of external work, such as electrical energy, other than the work of expansion), the maximum useful work will be obtained from the reaction and it will be either $-(\Delta G)_{T,P}$ or $-(\Delta F)_{T,V}$. The difference between ΔG and ΔH (or between ΔF and ΔU) is due to the heat absorbed or emitted in maintaining the temperature constant, and the change in the free energy can be greater or less than the heat of reaction.

The direction in which any chemical reaction proceeds can, ideally, be reversed by carrying it out in appropriate conditions, and it is therefore necessary to decide, with a certain degree of arbitrariness, upon one particular direction as being the direction of the 'forward reaction'. It is then convenient to write the equation of a chemical reaction, in which there are r' reactants $\mathscr{C}_1, \ldots, \mathscr{C}_{r'}$ and $r-r'$ products $\mathscr{C}_{r'+1}, \ldots, \mathscr{C}_r$, in the form

$$\sum_{i=1}^{r} \nu_i \mathscr{C}_i = 0, \tag{11·4·1}$$

where $\nu_1, \ldots, \nu_{r'}$ are negative and $\nu_{r'+1}, \ldots, \nu_r$ are positive. As usual, we measure the quantity of each constituent in gram molecules, and the stoichiometric numbers $-\nu_1, \ldots, -\nu_{r'}, \nu_{r'+1}, \ldots, \nu_r$, which are characteristic of the reaction, must be fixed once and for all, it not being sufficient here merely to fix their ratios. With these conventions, ΔH and ΔG are given by

$$\Delta H = \sum_1^r \nu_i H_i(T,P), \quad \Delta G = \sum_1^r \nu_i G_i(T,P), \tag{11·4·2}$$

and, for example, the equation of the hydrogen-oxygen reaction at 100° C. is written as

$$-2H_2\,(\text{gas}) - O_2\,(\text{gas}) + 2H_2O\,(\text{gas}) = 0;$$

$$-\Delta H = -116\,\text{kcal.}, \quad T = 100°\,\text{C.}, \quad P = 1\,\text{atmosphere}.$$

These equations make clear what is meant by ΔH or ΔG, namely, the algebraic sum of the heat functions or the free energies of the appropriate number of gram molecules of the various pure constituents. For gases, however, ΔH (and ΔG) is normally measured under conditions which are not those specified, and this is liable to lead to some confusion of thought. To measure ΔH for the hydrogen-oxygen reaction, for example, we could mix 2 gram molecules of hydrogen and 1 of oxygen in a calorimeter at temperature T and pressure P, and, having made arrangements to keep the pressure constant, we could then initiate the reaction by a spark and measure the heat evolved in the reaction and in the subsequent cooling of the calorimeter down to the original temperature T. This would give the correct value for ΔH, since it makes no difference whether we start with hydrogen and oxygen separated from one another or whether we start with them mixed, there being no heat of mixing for perfect gases. Also, the presence of one constituent in excess would make no difference to the result. For imperfect gases, however, the situation is different, since there is then a heat of mixing, but in practice the heat of mixing is insignificant compared with the heat of reaction, so that for all practical purposes we may start with a mixture of gases in a calorimeter to measure ΔH. For the free energy ΔG, on the other hand, it is not always unimportant whether we start with the separate reactants or with a mixture of them, since the free energy of mixing is not zero. The correct ΔG is the difference between the free energies of the separated products and of the separated reactants.

Since the quantities $H_i(T, P)$ and $G_i(T, P)$ refer to the pure substances and only depend upon T and P, the equations (11·4·2) are linear equations in the H_i's and G_i's. If, therefore, we have two reactions involving some of the same constituents we can combine the equations for ΔH and ΔG by the ordinary algebraic rules, provided that the temperature and pressure are the same in both reactions. In this way we can obtain ΔH and ΔG for other reactions. For example, from the reactions

$$\left.\begin{array}{c} -C\,(\text{solid}) - O_2\,(\text{gas}) + CO_2\,(\text{gas}) = 0, \\ T = 298{\cdot}16°\,\text{K.}, \quad P = 1\,\text{atmosphere}, \quad \Delta H = -94{,}052\,\text{cal.} \end{array}\right\} \qquad (11{\cdot}4{\cdot}3)$$

and

$$\left.\begin{array}{c} -CO\,(\text{gas}) - \tfrac{1}{2}O_2\,(\text{gas}) + CO_2\,(\text{gas}) = 0, \\ T = 298{\cdot}16°\,\text{K.}, \quad P = 1\,\text{atmosphere}, \quad \Delta H = -67{,}636\,\text{cal.}, \end{array}\right\} \qquad (11{\cdot}4{\cdot}4)$$

we can deduce

$$\left.\begin{array}{c} -C\,(\text{solid}) - \tfrac{1}{2}O_2\,(\text{gas}) + CO\,(\text{gas}) = 0, \\ T = 298{\cdot}16°\,\text{K.}, \quad P = 1\,\text{atmosphere}, \quad \Delta H = -26{,}416\,\text{cal.} \end{array}\right\} \qquad (11{\cdot}4{\cdot}5)$$

It would be difficult to obtain a reliable result for the heat of reaction of (11·4·5) from direct measurements.

11·41. *The equilibrium condition.* In considering reactions we must take into account both the forward and the back reactions, since in theory no reaction ever goes to completion, and an equilibrium is attained at which the forward and back reactions are of equal importance. According to the general theory given in § 3·6, the condition for equilibrium is that the internal energy shall be a minimum for given values of the entropy and of the volume, but here it is more convenient to use the equivalent condition $(dG)_{T,P}=0$. The previous discussion, however, assumed that all the constituents were present in fixed amounts, and to include chemical reactions we must extend the discussion to cover the interconvertibility of the constituents.

If we have a single chemical reaction characterized by the equation

$$\sum_{i=1}^{r} \nu_i \mathscr{C}_i = 0, \tag{11·41·1}$$

then, if the number of gram molecules of the constituent $\mathscr{C}_i$ is increased by dn_i, the meaning of the above equation is that $dn_i = \gamma\nu_i$, where γ is independent of i. For equilibrium we have therefore

$$dG(T, P, n_i)_{T,P} = 0, \tag{11·41·2}$$

subject to the conditions

$$\frac{dn_1}{\nu_1} = \frac{dn_2}{\nu_2} = \ldots = \frac{dn_r}{\nu_r}. \tag{11·41·3}$$

Now

$$(dG)_{T,P} = \Sigma\mu_i dn_i, \tag{11·41·4}$$

and hence (11·41·1) and (11·41·2) give

$$\Sigma\nu_i\mu_i = 0 \tag{11·41·5}$$

as the equilibrium condition.

If there are more chemical reactions than one, we have several chemical equations

$$\sum_{i=1}^{r} \nu_i^{(j)} \mathscr{C}_i = 0 \quad (j = 1, 2, \ldots), \tag{11·41·6}$$

where some of the $\nu_i^{(j)}$'s may be zero. The equilibrium condition is then given as the solution of the simultaneous equations

$$\sum_{i=1}^{r} \nu_i^{(j)} \mu_i = 0 \quad (j = 1, 2, \ldots). \tag{11·41·7}$$

11·411. *The affinity as defined by de Donder.* A somewhat different approach to the thermodynamics of chemical reactions has been given by de Donder. A variable ξ is introduced which defines 'the extent of the reaction', in the sense that, for a given value of ξ, the amount of the ith constituent present is $n_i = n_i^0 + \nu_i \xi$, where n_i^0 is the value of n_i in some initial state. We may then write, for example, the heat function and free energy of a mixture as $H(T, P, n_i^0 + \nu_i \xi)$ and $G(T, P, n_i^0 + \nu_i \xi)$. If we consider the n_i^0's to be constant, a variation in ξ means a variation in the amounts of the constituents present which is in accordance with the chemical reaction (11·41·1). De Donder normally writes the free energy simply as $G(T, P, \xi)$, and he then defines the affinity $\mathscr{A}$ of a chemical reaction by the relation

$$\mathscr{A} = -\partial G(T, P, \xi)/\partial \xi. \tag{11·411·1}$$

The reaction will proceed in the positive or negative directions according as $\mathscr{A}$ is positive or negative, while if $\mathscr{A}$ is zero the system is in equilibrium.

This method of developing the theory has certain advantages, but these can only be obtained by making substantial changes in the definitions of the physical quantities involved. In its simple form the theory is, moreover, less general than that given in the preceding sections, and if it is used in its general form it is, except for notation, identical with the usual theory.

The apparent simplicity of de Donder's method is best shown by writing dG in the form

$$dG = -S\,dT + V\,dP - \mathscr{A}\,d\xi, \tag{11·411·2}$$

which we may contrast with the more usual equation

$$dG = -S\,dT + V\,dP + \Sigma \mu_i\, dn_i. \tag{11·411·3}$$

The latter is the more general equation, and the former is the particular case obtained by putting $dn_i = \nu_i d\xi$. Hence

$$\mathscr{A} = -\Sigma \nu_i \mu_i, \tag{11·411·4}$$

and the μ_i's only occur explicitly in this combination. It must, however, be borne in mind that in the general case $\mathscr{A}$ is a function not merely of T, P and ξ, but of the n_i^0's as well, and, when the affinity of a reaction depends upon the initial amounts of the constituents present, the simplicity of the concept is lost. In the practically important case of a mixture of perfect gases, this complication does not arise. In the exposition given in § 11·41, the criterion concerning the direction in which a reaction would proceed is the sign of $\Delta G(T, P) \equiv \Sigma \nu_i G_i^0(T, P)$, where $G_i^0(T, P) \equiv \mu_i^0(T, P)$ is the free energy per gram molecule (which is identical with the thermodynamic potential) of the pure ith constituent. This criterion is therefore essentially different from that involving (11·411·1), since in the latter equation $\mu_i(T, P, n_j^0 + \nu_j \xi)$ refers to the thermodynamic potential of the ith constituent in a mixture containing all the reactants and products.

There is a similar difference in the definition of the heat of reaction, which, according to de Donder's theory is given by

$$-\frac{\partial H(T,P,\xi)}{\partial \xi} = -\Sigma \nu_i \left(\frac{\partial H(T,P,n_i)}{\partial n_i}\right)_{n_i=n_i^0+\nu_i\xi}. \qquad (11\cdot411\cdot5)$$

In general this is not the same as $-\Sigma \nu_i H_i^0$, where H_i^0 is the heat function per gram molecule of the pure ith component, though the two definitions agree for perfect gases.

De Donder defines the rate of reaction v by the relation

$$v = d\xi/dt, \qquad (11\cdot411\cdot6)$$

where ξ is supposed to be a function of the time t. It must, however, be borne in mind that the rate defined in this way has a very special meaning and is not necessarily related to the rate of a reaction as normally observed. If we consider two different values of ξ, say ξ_1 and ξ_2, the free energies $G(T,P,\xi_1)$ and $G(T,P,\xi_2)$ refer to two different equilibrium states of the mixture, which are only metastable and are in fact unstable if we allow the chemical reaction to take place between the various constituents. But, if the chemical reaction were allowed to occur, the mixture would not pass through a succession of equilibrium states connecting the states defined by ξ_1 and ξ_2 unless external constraints could be imposed upon the system to make it do so. The linear continuum of metastable states connecting ξ_1 and ξ_2 will therefore in general constitute only a virtual but not an actual 'reaction path'. The actual course of a reaction cannot be calculated without a knowledge of the kinetic rate constants.

The equilibrium theory of chemical reactions is, of course, the same whether de Donder's method or the normal method is adopted. The difference lies essentially in the definition and the treatment of such quantities as the heat of reaction, which refer to the properties of two non-equilibrium states. The reader will find a full exposition of de Donder's theory in the treatise by Prigogine & Defay (1954).

11·42. *Reactions between perfect gases.* When only perfect gases are present the equilibrium condition can be written down explicitly. For a perfect gas mixture we have, from (11·2·4),

$$\mu_i = -C_{P,i}^0 T\log T + \int_0^T C_{P,i}^1(T')\,dT' - T\int_0^T \frac{C_{P,i}^1(T')}{T'}\,dT' + RT\log\frac{n_i P}{n} + v_{0,i} - RTJ_i, \qquad (11\cdot42\cdot1)$$

where

$$RJ_i = s_{0,i} + R\log P^\dagger - C_{P,i}^0 - C_{P,i}^0 \log T^\dagger. \qquad (11\cdot42\cdot2)$$

If we introduce the partial pressures $p_i = n_i P/n$ and write

$$\mu_i = \mu_i^*(T) + RT\log p_i, \qquad (11\cdot42\cdot3)$$

then the equilibrium condition (11·41·5) can be put into the form

$$\prod_i p_i^{\nu_i} = K_P(T), \tag{11·42·4}$$

where K_P is a function of T only, defined by

$$\log K_P = -\sum_i \frac{\nu_i \mu_i^*}{RT} \tag{11·42·5}$$

$$= -\frac{\Delta H_0}{RT} + \frac{1}{R}\sum_i \nu_i C^0_{P,i} \log T$$

$$-\frac{1}{RT}\sum_i \nu_i \left[\int_0^T C^1_{P,i}(T')\,dT' - T\int_0^T \frac{C^1_{P,i}(T')}{T'}\,dT'\right] + J, \tag{11·42·6}$$

with

$$J = \Sigma \nu_i J_i, \quad \Delta H_0 = \Delta U_0 = \Sigma \nu_i U_{0,i}. \tag{11·42·7}$$

At constant temperature we have

$$\prod_i p_i^{\nu_i} = \text{constant}, \tag{11·42·8}$$

which is known as the law of mass action. The quantity K_P is called the equilibrium constant.

There are two other equilibrium constants that can be used. We have $p_i = x_i P$ and hence $\Pi p_i^{\nu_i} = P^{\Sigma \nu_i} \Pi x_i^{\nu_i}$. We may therefore replace (11·42·4) by the equation

$$\prod_i x_i^{\nu_i} = K_x, \tag{11·42·9}$$

where

$$K_x = K_P / P^{\Sigma \nu_i}. \tag{11·42·10}$$

We may also write

$$p_i = RTc_i / V, \tag{11·42·11}$$

and in this case (11·42·4) takes the form

$$\prod_i (c_i/V)^{\nu_i} = K_V, \tag{11·42·12}$$

where

$$K_V = K_P / (RT)^{\Sigma \nu_i}. \tag{11·42·13}$$

The equilibrium constant K_V is of less practical importance than K_P or K_x.

11·421. *The influence of temperature and pressure on the equilibrium constant.* By differentiating (11·42·6) we see that

$$RT^2 \frac{d}{dT}\log K_P = \Delta H_0 + \Sigma \nu_i \left(C^0_{P,i} T + \int_0^T C^1_{P,i}(T')\,dT'\right)$$

$$= \Sigma \nu_i \left(U_{0,i} + \int_0^T C_{P,i}(T')\,dT'\right) = \Sigma \nu_i H_i = \Delta H. \tag{11·421·1}$$

Hence

$$\frac{d \log K_P}{dT} = \frac{\Delta H}{RT^2}, \tag{11·421·2}$$

which gives a relation due to van't Hoff between the equilibrium constant and the heat of reaction. The equilibrium constant K_x satisfies the same equation. Similarly

$$\frac{d\log K_V}{dT}=\frac{d\log K_P}{dT}-\Sigma\nu_i\frac{d}{dT}\log RT$$

$$=\frac{\Sigma\nu_i(H_i-RT)}{RT^2}=\frac{\Delta U}{RT^2}, \qquad (11\cdot421\cdot3)$$

where $-\Delta U$ is the heat of reaction at constant volume.

Equation (11·421·2) (or equation (11·421·3)) can be used to find the heat of reaction, and this method of obtaining the heat of reaction is particularly valuable when the direct calorimetric measurements are difficult to carry out. Conversely, if we know K_P and ΔH at one particular temperature, and if the heat functions of the individual substances taking part in the reaction are known as functions of the temperature, K_P can be found over a range of temperatures.

If the forward direction of the reaction is defined so that the reaction is exothermic, i.e. so that $\Delta H < 0$, the temperature coefficient of K_P is negative and an increase in temperature will favour the reverse reaction. If, therefore, it is desired to carry out an exothermic reaction to as high a degree of completion as possible, it will be desirable to operate at low temperatures, whereas high temperatures are desirable for an endothermic reaction. But in practice, we have to take into account not only the equilibrium composition which is theoretically attainable but also the rate of reaction, since at low temperatures a reaction may proceed so slowly that for all practical purposes it is non-existent.

The equilibrium constant K_P is, from its definition, independent of pressure. K_x, on the other hand, depends strongly on the pressure unless $\Sigma\nu_i=0$, and in such a way that if there is a decrease in volume in the forward reaction an increase in pressure will favour the forward reaction.

We shall now discuss some simple examples of gas reactions to illustrate the principles outlined above.

11·422. *The dissociation of a molecule into two similar molecules.* If we denote the 'parent molecule' by $\mathscr{C}_2$ and its constituents by $\mathscr{C}$, the dissociation reaction can be written

$$-\mathscr{C}_2+2\mathscr{C}=0. \qquad (11\cdot422\cdot1)$$

Hence if p and p_2 are the partial pressures of $\mathscr{C}$ and $\mathscr{C}_2$, we have, by (11·42·4),

$$p^2/p_2=K_P(T). \qquad (11\cdot422\cdot2)$$

Alternatively, if x and x_2 are the mole fractions of single and double molecules, (11·42·9) and (11·42·10) give

$$x^2/x_2=K_P(T)/P. \qquad (11\cdot422\cdot3)$$

(Note that x is not the fraction of molecules dissociated. If r is the fraction dissociated, we have $x=2r/(1+r)$, and so $r=\frac{1}{2}x/(1-\frac{1}{2}x)$.) We may express (11·422·3) conveniently in terms of the density ρ of the gas. If ρ_0 is the density of the undissociated gas, then

$$\rho/\rho_0=x_2+\tfrac{1}{2}x, \tag{11·422·4}$$

and (11·422·3) becomes

$$\frac{4(1-\rho/\rho_0)^2}{2\rho/\rho_0-1}=\frac{K_P(T)}{P}. \tag{11·422·5}$$

In this particular type of reaction, the dissociation can be promoted by raising the temperature and by lowering the pressure.

The dissociation of nitrogen peroxide, N_2O_4, is one of the reactions to which thermodynamics was first applied, and it has therefore played a considerable part in the development of the subject. It is easily investigated at temperatures between 0 and 100° C., and the results can be well represented by the relation

$$\log_e K_P=21-6870/T, \tag{11·422·6}$$

if P is measured in atmospheres. This gives $\Delta H=13{,}600$ calories.

11·423. *The maximum extent of a reaction.* If the relative proportions of the reactants in any chemical reaction between perfect gases are varied, the maximum yield of the products is obtained when the reactants are initially present in the ratios of the stoichiometric numbers. To prove this, let n_i^0 $(i=1,\ldots,r')$ be the number of gram molecules present initially of the reactants, and let $n_i^0+\nu_i\xi$ $(i=1,\ldots,r')$ and $\nu_i\xi$ $(i=r'+1,\ldots,r)$ be the number of gram molecules of the reactants and products in the equilibrium state. Then (11·42·9) can be written as

$$\frac{\prod\limits_{i=1}^{r}(n_i^0+\nu_i\xi)^{\nu_i}}{\left[\sum\limits_{i=1}^{r}(n_i^0+\nu_i\xi)\right]^{\Sigma\nu_j}}=\frac{K_P(T)}{P^{\Sigma\nu_j}}, \tag{11·423·1}$$

where $n_{r'+1}^0,\ldots,n_r^0$ are all zero. In order to keep the initial pressure the same, the total number of molecules Σn_i^0 must be kept constant, and to find the maximum extent of the reaction we have to find the maximum value of ξ subject to the condition (11·423·1) and to the condition that Σn_i^0 is constant. If we introduce an undetermined multiplier λ, the conditions that ξ is a maximum may be stated in the form

$$\frac{\partial}{\partial\xi}\left(\log\frac{K_P}{P^{\Sigma\nu_j}}+\lambda\sum_{j=1}^{r'}n_j^0\right)=0,\quad \frac{\partial}{\partial n_i^0}\left(\log\frac{K_P}{P^{\Sigma\nu_j}}+\lambda\sum_{j=1}^{r'}n_j^0\right)=0\quad (i=1,\ldots,r'). \tag{11·423·2}$$

The condition arising from the differentiation with respect to n_i^0 is

$$\frac{\nu_i}{n_i^0+\nu_i\xi}-\frac{\sum_{j=1}^{r}\nu_j}{\sum_{j=1}^{r}(n_j^0+\nu_j\xi)}+\lambda=0 \quad (i=1,\ldots,r'),$$

which shows that n_i^0 must be proportional to ν_i for each of the reactants.

It should be noted that the above criterion for what has been called the maximum extent of the reaction is not always the one to be applied in practice. If only one of the reactants is considered to be economically important, we are usually interested in the conditions for the maximum conversion of this particular reactant and not in the most efficient utilization of all the reactants.

11·424. *Adiabatic reactions.* The discussion so far has been confined to chemical reactions in which the final temperature is the same as the initial temperature. At the opposite extreme we may consider reactions which take place adiabatically, so that the whole of the heat of the reaction is used in heating, or cooling, the gases. We have then two parameters to determine, namely, the extent of the reaction ξ and the final temperature T. If there are initially n_i^0 $(i=1,\ldots,r')$ gram molecules of the reactants present, the equilibrium condition (11·423·1) gives us one relation between ξ and T. To find a second relation we use the fact that, in an adiabatic change at constant pressure, the heat function of the system is unchanged. Hence, if T_0 is the initial temperature, we have

$$\sum_{i=1}^{r'} n_i^0 H_i(T_0)=\sum_{i=1}^{r'}(n_i^0+\nu_i\xi)H_i(T)+\sum_{i=r'+1}^{r}\nu_i\xi H_i(T). \qquad (11\cdot424\cdot1)$$

It is obvious without calculation that, if the reaction is exothermic (in which case K_p decreases as the temperature increases), the reaction will proceed to a smaller extent if the conditions are adiabatic than if they are isothermal.

11·425. *The synthesis of sulphur trioxide.* In modern sulphuric acid plants, sulphur trioxide is produced by the direct oxidation of sulphur dioxide. By burning sulphur or iron pyrites in air, a mixture of sulphur dioxide, oxygen and nitrogen is produced, and this 'burner gas' is passed through a large mass of a vanadium catalyst where the oxidation of the sulphur dioxide takes place. If we write the reaction as

$$\left.\begin{array}{c}-SO_2-\tfrac{1}{2}O_2+SO_3=0,\\ \Delta H=-95\cdot5\,\mathrm{kJ}\ (-22\cdot6\,\mathrm{kcal.}),\quad T=440^\circ\,\mathrm{C.},\quad P=1\ \mathrm{atmosphere},\end{array}\right\} \qquad (11\cdot425\cdot1)$$

and if n_1^0, n_2^0 and n are the numbers of gram molecules of sulphur dioxide, oxygen and nitrogen in the burner gas and if ξ is the number of gram molecules of sulphur trioxide produced, the equilibrium condition is

$$\frac{\xi(n_1^0+n_2^0+n-\tfrac{1}{2}\xi)^{\frac{1}{2}}}{(n_1^0-\xi)\,(n_2^0-\tfrac{1}{2}\xi)^{\frac{1}{2}}}=P^{\frac{1}{2}}K_P(T). \qquad (11·425·2)$$

The experimental values of $K_P(T)$ can be reasonably well represented by the relation

$$\log_e K_P = -10·755+11{,}400/T \qquad (11·425·3)$$

if P is measured in atmospheres. Vanadium catalysts do not function below about 400° C., so that the maximum feasible value of $\log K_P$ is about 6. With such a large value of K_P, the yield of the reaction is high.

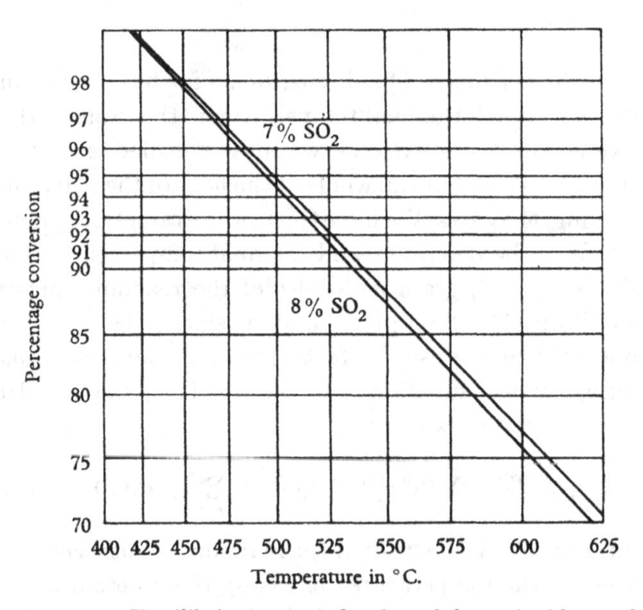

Fig. 11·4. Equilibrium curves for the sulphur trioxide synthesis.

In industrial plants the synthesis gas is at atmospheric pressure and it contains between 7 and 8 % of SO_2, and therefore between 14 and 13 % of O_2 and 79 % of N_2. The percentage conversion of SO_2 under such conditions is shown in fig. 11·4 as a function of the temperature (Gloag & Barritt, 1943), and it will be seen that the strength of the gas has only a minor effect upon the conversion. The actual yield during one pass through the catalyst bed is considerably less than that given by the curves in fig. 11·4. In the first place the reaction is not infinitely rapid, and equilibrium conditions are never reached. In the second place the reaction is not isothermal, and there is a considerable rise in the temperature of the gas during the reaction.

If we assume adiabatic instead of isothermal conditions we can calculate the conversion by the theory given in § 11·424. A typical result for a synthesis gas containing 8 % of SO_2 is that, if the inlet temperature is 440° C., the outlet temperature is 620° C. and the conversion is 71 %. (This is based upon taking the specific heats at constant pressure of N_2, O_2, SO_2 and SO_3 to be 7·5, 8·1, 13·5 and 12·8 cal./degree/gram molecule respectively. To obtain the results quoted, it is necessary to solve the equations (11·423·1) and (11·424·1) by trial and error.) If the conditions were isothermal, the conversion efficiency would be 98·5% at 440° C., while under the conditions occurring in a commercial plant the outlet temperature would be about 590° C. and the efficiency would be about 80 %. It will therefore be seen that there is a considerable loss in efficiency in conversion as compared with the conversion under isothermal conditions. To avoid this it is normal to use three or four converters and to cool the gas between the various stages of conversion.

11·43. *Reactions between imperfect gases.* If the gases taking part in a reaction are not perfect the preceding formulae require amendment. The appropriate changes are easily found, but since the effects of the departure from the perfect gas laws are in general small compared with the thermal effects caused by the reaction, the formulae are of little practical significance except for very high pressures.

As in § 11·3 we write

$$\mu_i(T, P, n_j) = \mu_i^*(T) + RT \log f_i, \tag{11·43·1}$$

where the fugacity f_i of the ith constituent has to satisfy the further condition

$$f_i \to n_i P/n \quad \text{as} \quad P \to 0. \tag{11·43·2}$$

The equilibrium condition is then

$$\prod_i f_i^{\nu_i} = K_P(T) = \exp\{-\Sigma \nu_i \mu_i^*(T)/RT\}, \tag{11·43·3}$$

where K_P is a function of the temperature only. It is clear from the definition that, since $\mu_i^*(T)$ is a function of T only, K_P is the same as the equilibrium constant when the pressure is low enough for the gases to be perfect. Hence

$$\frac{d}{dT} \log K_P = \frac{\Delta H^*}{RT^2}, \tag{11·43·4}$$

where $-\Delta H^*$ is not the heat of reaction as it actually takes place, but the heat of reaction at a sufficiently low pressure for the gases to behave as perfect gases. An alternative proof of this is as follows. We have

$$RT \log K_P = -\Sigma \nu_i \mu_i^*(T). \tag{11·43·5}$$

Then, if P^* is a sufficiently low pressure, the definitions (11·43·1) and (11·43·2) give

$$\mu_i^*(T) = \mu_i(P^*, T) - RT \log (n_i P^*/n). \tag{11·43·6}$$

Therefore

$$\frac{d}{dT}\frac{\mu_i^*(T)}{T} = \frac{\partial}{\partial T}\frac{\mu_i(T,P^*)}{T} = \frac{\partial}{\partial T}\frac{G_i(T,P^*)}{T} = -\frac{1}{T^2}\left(\frac{\partial H(T,P^*)}{\partial n_i}\right)_{T,P,n_j}. \tag{11·43·7}$$

But the heat function of a mixture of perfect gases is the sum of the heat functions of the individual gases, and the heat functions are functions of T only, i.e. $H(T,P^*) = \Sigma n_i H_i(T)$, so that

$$\frac{\partial}{\partial T}\log K_P = -\Sigma \nu_i \frac{\partial}{\partial T}\frac{\mu_i^*(T)}{RT} = \frac{\Sigma \nu_i H_i(T)}{RT^2} = \frac{\Delta H^*}{RT^2}. \tag{11·43·8}$$

To obtain the equilibrium condition in a form corresponding to (11·42·9) we write $f_i = x_i P \times f_i/p_i$, where p_i is the partial pressure defined by $p_i = x_i P$. Then (11·43·3) becomes

$$\prod_i x_i^{\nu_i} = K_P(T)/(K_\gamma P^{\Sigma \nu_i}), \tag{11·43·9}$$

where

$$K_\gamma = \prod_i (f_i/p_i)^{\nu_i}. \tag{11·43·10}$$

11·431. *The synthesis of ammonia.* The equilibrium constant for the reaction

$$-\tfrac{3}{2}H_2 - \tfrac{1}{2}N_2 + NH_3 = 0 \tag{11·431·1}$$

is given by the expression

$$\log_e K_P = -13{\cdot}55 + 6170/T. \tag{11·431·2}$$

The reaction can be made to take place at a practicable rate by using iron catalysts at a temperature of the order of 450° C., but at this temperature K_P is only 0·00654 and the conversion is very small. Since, however, the reaction takes place with a decrease in volume, it can be promoted by an increase in pressure, and it is customary in industrial practice to work at a pressure of 300 atmospheres.

According to § 11·423 we have to begin with a 3 : 1 ratio of hydrogen to nitrogen in order to achieve the maximum conversion of both gases (and the cost of making either gas is comparable with that of the other). We therefore assume that we have initially $3n_0$ gram molecules of hydrogen and n_0 of nitrogen, while when equilibrium is reached we have $3n_0 - 3\xi$ gram molecules of hydrogen, $n_0 - \xi$ gram molecules of nitrogen and 2ξ gram molecules of ammonia. The equilibrium condition is then

$$\frac{2\xi(4n_0 - 2\xi)}{(3n_0 - 3\xi)^{\frac{3}{2}}(n_0 - \xi)^{\frac{1}{2}}} = PK_P(T). \tag{11·431·3}$$

The mole fraction of the ammonia in the equilibrium mixture is $\xi/(2n_0 - \xi)$, and some values of this quantity, deduced from (11·431·2) and (11·431·3), are given in table 11·2.

Table 11·2. *Equilibrium mole fractions of ammonia at various temperatures and pressures*

Pressure in atmospheres	Temperature in °C. 400	500	600
1	0·0044	0·0013	0·0005
30	0·117	0·036	0·014
100	0·25	0·10	0·04

At higher pressures it is not permissible to treat the gases as perfect, and it is necessary to use equation (11·43·9) to describe the equilibrium instead of equation (11·42·9). The data from which to calculate K_γ accurately are

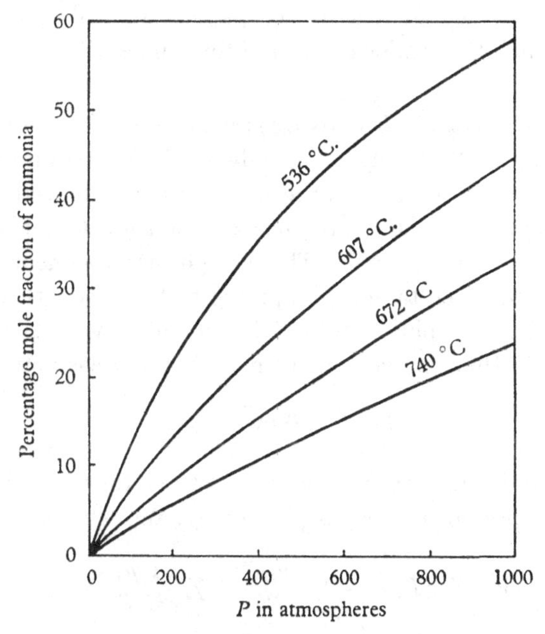

Fig. 11·5. The conversion of hydrogen and nitrogen to ammonia as a function of the pressure for various temperatures.

lacking, but a rough theory has been given by Larson (1924) assuming the validity of the relation (11·32·5) and calculating f_i^0 from the data given in fig. 8·11. The results for K_γ are given in table 11·13 and compared with the experimental results of Larson & Dodge. The agreement is unexpectedly good, and it will be seen that, with pressures of the order of 1000 atmospheres, the departure from the perfect gas laws increases the equilibrium constant by a factor 3. The mole fraction of the ammonia in the equilibrium mixture is shown in fig. 11·5 for various temperatures and pressures.

Table 11·3. *Calculated and observed values of K_γ for the ammonia synthesis; calculated values in brackets*

Pressure in atmospheres	Temperature in °C. 400	450	500
10	0·990 (0·986)	0·992 (0·988)	0·994 (0·992)
100	0·907 (0·895)	0·929 (0·905)	0·953 (0·914)
1000		0·285 (0·443)	0·387 (0·487)

A very detailed study of the effect of 'gas-imperfection' on the ammonia synthesis has been given in a series of papers by Gillespie & Beattie (1930*a*, *b*, *c*), to which the reader is referred for full details.

11·44. *Gas reactions in which solids are present.* Since the thermodynamic potential of a solid is effectively independent of the pressure and of the presence of other substances, the thermodynamic potential in the gas phase of any substance which is present as a solid must be considered as a function of the temperature only. Therefore in heterogeneous reactions, equation (11·42·4) must be taken as applying to those constituents which are present in the gas phase only, whereas for the others we have $\mu_i = \mu_i(T)_{\text{solid}}$. It is then convenient to modify the mass-action equation by writing

$$\prod_i{}' p_i^{\nu_i} = K'_P(T), \tag{11·44·1}$$

where the accent indicates that the product is to be taken over those components which are present in the gas phase only, and where

$$\log K'_P(T) = -\sum_{\text{gases}}{}' \frac{\nu_i \mu_i^*}{RT} - \sum_{\text{solids}}{}'' \frac{\nu_i \mu_i}{RT}. \tag{11·44·2}$$

Explicitly, by (9·3·1) and (11·2·4), we have, if we neglect $(Pv/RT)_{\text{solids}}$,

$$\begin{aligned}\log K'_P = &-\frac{\Delta H_0}{RT} + \frac{1}{R}\sum_{\text{gases}}{}' \nu_i C^0_{P,i} \log T \\ &- \frac{1}{RT}\sum_{\text{gases}}{}' \nu_i \left[\int_0^T C^1_{P,i}(T')\,dT' - T\int_0^T \frac{C^1_{P,i}(T')}{T'}\,dT'\right] \\ &+ \sum_{\text{solids}}{}'' \frac{\nu_i}{RT}\int_0^T dT' \int_0^{T'} \frac{C_{P,i}(T'')}{T''}\,dT'' + J,\end{aligned} \tag{11·44·3}$$

where

$$J = \sum_{\text{gases}}{}' \nu_i J_i + \sum_{\text{solids}}{}'' \nu_i s_{0,i}/R. \tag{11·44·4}$$

It is easily verified that
$$\frac{\partial \log K'_P}{\partial T} = \frac{\Delta H}{RT^2}. \tag{11·44·5}$$

It should, however, be borne in mind that K'_P is not strictly independent of the pressure, which should, therefore, be the pressure in the actual reaction.

11·441. *The producer-gas reaction.* In industrial practice, producer-gas, which ideally consists of carbon monoxide, nitrogen and carbon dioxide, is made by the partial oxidation of coke. The coke is in the form of a deep bed and air is led in at the bottom. In the lower part of the bed the coke burns to form carbon dioxide, and this is then reduced to carbon monoxide in the upper part of the bed. The producer-gas reaction is

$$\left.\begin{aligned} &-\mathrm{C}\,(\text{solid}) - \mathrm{CO_2}\,(\text{gas}) + 2\mathrm{CO}\,(\text{gas}) = 0, \\ &\Delta H = 172\cdot5\,\text{kJ.}\ (41\cdot25\,\text{kcal.}),\quad T = 25^\circ\,\text{C.},\quad P = 1\,\text{atmosphere,} \end{aligned}\right\} \tag{11·441·1}$$

and since it is endothermic it is favoured by a high temperature. Also it is adversely affected by pressure, the equilibrium condition being

$$\frac{\xi^2}{(n_0 + n - \frac{1}{2}\xi)(n_0 - \frac{1}{2}\xi)} = \frac{K'_P(T)}{P}, \tag{11·441·2}$$

where n_0 and n are the original numbers of gram molecules of CO_2 and N_2, and where ξ is the number of gram molecules of CO in the equilibrium mixture.

Some values of K'_P, with the pressures measured in atmospheres, are given in table 11·4. Since K'_P is very small at low temperatures and large at high temperatures, the equilibrium gas will contain almost wholly carbon dioxide at low temperatures and carbon monoxide at high temperatures, the change-over between the two extremes taking place largely between 700 and 800° C. where K'_P changes from a small value to a value greater than unity.

Table 11·4. *The equilibrium constant for the producer-gas reaction using average values relating to coke, charcoal and amorphous carbon*

T in °C. ...	400	500	600	700	800
K'_P	$7\cdot9 \times 10^{-5}$	$3\cdot3 \times 10^{-3}$	$6\cdot2 \times 10^{-2}$	$6\cdot2 \times 10^{-1}$	3·9
T in °C. ...	900	1000	1100	1200	1300
K'_P	18	65	$1\cdot9 \times 10^{2}$	$4\cdot9 \times 10^{2}$	$1\cdot1 \times 10^{3}$
T in °C. ...	1400	1500	1600	1700	
K'_P	$2\cdot2 \times 10^{3}$	$4\cdot3 \times 10^{3}$	$7\cdot1 \times 10^{3}$	$1\cdot2 \times 10^{4}$	

The Statistical Mechanics of Chemical Equilibrium

11·5. Since we have already obtained a theoretical expression for the thermodynamic potential of a perfect gas it is only necessary to substitute this expression for μ_i into the equilibrium condition (11·41·5). For completeness, however, we ought to verify that the previous calculations result in the correct dependence of the thermodynamic potential upon the number of systems present, and we therefore extend the theory given in §§ 5·32–5·34 to include cases where chemical reactions can occur.

For simplicity in notation we consider only the 'dissociation' reaction

$$-\mathscr{C}_{12}+\mathscr{C}_1+\mathscr{C}_2=0. \tag{11·5·1}$$

We then have to repeat the calculations of § 5·3 with the appropriate changes. If $\mathscr{N}^{(1)}$ is the total number of $\mathscr{C}_1$ atoms, of which $N^{(1)}$ exist as free atoms while $N^{(12)}$ are in the $\mathscr{C}_{12}$ molecules, and if $\mathscr{N}^{(2)}$, $N^{(2)}$, $N^{(12)}$ are the corresponding numbers of $\mathscr{C}_2$ atoms, then, with an obvious extension of the notation of § 5·3, the number of complexions of the assembly is given by

$$W=\prod_r \frac{M_r^{(1)N_r^{(1)}}}{N_r^{(1)}!}\prod_s \frac{M_s^{(2)N_s^{(2)}}}{N_s^{(2)}!}\prod_t \frac{M_t^{(12)N_t^{(12)}}}{N_t^{(12)}!}. \tag{11·5·2}$$

The total energy of the assembly is

$$U=\Sigma(N_r^{(1)}E_r^{(1)}+N_s^{(2)}E_s^{(2)}+N_t^{(12)}E_t^{(12)}), \tag{11·5·3}$$

where the energy zero is taken to be the same for each of the three sets of systems, and the conditions for the constancy of the total numbers of $\mathscr{C}_1$ and $\mathscr{C}_2$ atoms are

$$\Sigma(N_r^{(1)}+N_t^{(12)})=\mathscr{N}^{(1)},\quad \Sigma(N_s^{(2)}+N_t^{(12)})=\mathscr{N}^{(2)}. \tag{11·5·4}$$

If we now maximize $\log W-\vartheta U+\lambda_1\mathscr{N}^{(1)}+\lambda_2\mathscr{N}^{(2)}$, we find that we have to replace (5·3·5) by

$$\left.\begin{aligned} N_r^{(1)}=M_r^{(1)}\exp(\lambda_1-\vartheta E_r^{(1)}),\quad N_s^{(2)}=M_s^{(2)}\exp(\lambda_2-\vartheta E_s^{(2)}),\\ N_t^{(12)}=M_t^{(12)}\exp(\lambda_1+\lambda_2-\vartheta E_t^{(12)}).\end{aligned}\right\} \tag{11·5·5}$$

Hence $\qquad e^{\lambda_1}=N^{(1)}/Z_1,\quad e^{\lambda_2}=N^{(2)}/Z_2,\quad e^{\lambda_1+\lambda_2}=N^{(12)}/Z_{12},\qquad$ (11·5·6)

where

$$Z_1=\Sigma\exp(-\vartheta E_r^{(1)}),\quad Z_2=\Sigma\exp(-\vartheta E_s^{(2)}),\quad Z_{12}=\Sigma\exp(-\vartheta E_t^{(12)}), \tag{11·5·7}$$

and the parameters ϑ, λ_1 and λ_2 are determined by the equations

$$\begin{aligned} U&=-e^{\lambda_1}\frac{\partial Z_1}{\partial\vartheta}-e^{\lambda_2}\frac{\partial Z_2}{\partial\vartheta}-e^{\lambda_1+\lambda_2}\frac{\partial Z_{12}}{\partial\vartheta}\\ &=-N^{(1)}\frac{\partial\log Z_1}{\partial\vartheta}-N^{(2)}\frac{\partial\log Z_2}{\partial\vartheta}-N^{(12)}\frac{\partial\log Z_{12}}{\partial\vartheta},\end{aligned} \tag{11·5·8}$$

$$\mathscr{N}^{(1)}=e^{\lambda_1}Z_1+e^{\lambda_1+\lambda_2}Z_{12},\quad \mathscr{N}^{(2)}=e^{\lambda_2}Z_2+e^{\lambda_1+\lambda_2}Z_{12}. \tag{11·5·9}$$

We now define as before the generalized force A_i corresponding to the generalized coordinate a_i and we repeat the calculations of §§ 5·32–5·34, but we consider a variation in which ϑ, x_i, $N^{(1)}$, $N^{(2)}$ and $N^{(12)}$ change while $\mathcal{N}^{(1)}$ and $\mathcal{N}^{(2)}$ remain constant. Then the linear differential form for $dU + \Sigma A_i da_i$ has the integrating factor ϑ as before, and

$$\begin{aligned}\vartheta(dU + \Sigma A_i da_i) &= d(\vartheta U) - U d\vartheta + \vartheta \Sigma A_i da_i \\ &= d(\vartheta U + e^{\lambda_1} Z_1 + e^{\lambda_2} Z_2 + e^{\lambda_1+\lambda_2} Z_{12}) \\ &\quad - Z_1 e^{\lambda_1} d\lambda_1 - Z_2 e^{\lambda_2} d\lambda_2 - Z_{12} e^{\lambda_1+\lambda_2}(d\lambda_1 + d\lambda_2).\end{aligned}$$

The last three terms can be written as

$$\begin{aligned}&-N^{(1)} d\lambda_1 - N^{(2)} d\lambda_2 - N^{(12)}(d\lambda_1 + d\lambda_2) \\ &\quad = -d\{N^{(1)}\lambda_1 + N^{(2)}\lambda_2 + N^{(12)}(\lambda_1 + \lambda_2)\} + \lambda_1 d\mathcal{N}^{(1)} + \lambda_2 d\mathcal{N}^{(2)},\end{aligned}$$

and the last two terms of this are zero since $\mathcal{N}^{(1)}$ and $\mathcal{N}^{(2)}$ are constant. Hence, putting $\vartheta = 1/kT$, we have

$$dU + \Sigma A_i da_i = T(dS_1 + dS_2 + dS_{12}) + \mu_1 dN^{(1)} + \mu_2 dN^{(2)} + \mu_{12} dN^{(12)}, \quad (11\cdot5\cdot10)$$

where $\quad S_i = U_i/T + k(N^{(i)} \log Z_i - N^{(i)} \log N^{(i)} + N^{(i)}) \quad (i = 1, 2, 12) \quad (11\cdot5\cdot11)$

and $\quad \mu_1 = kT\lambda_1, \quad \mu_2 = kT\lambda_2, \quad \mu_{12} = kT(\lambda_1 + \lambda_2). \quad (11\cdot5\cdot12)$

We therefore see that, with S_i given by (11·5·11), which is the same relation as (5·33·3), we have

$$-\mu_{12} + \mu_1 + \mu_2 = 0, \quad (11\cdot5\cdot13)$$

which is the condition for thermodynamic equilibrium. Further, by (11·5·6) we have

$$\frac{N^{(12)}}{N^{(1)} N^{(2)}} = \frac{Z_{12}}{Z_1 Z_2}, \quad (11\cdot5\cdot14)$$

which is one form of the law of mass action.

The generalization of these results to chemical reactions involving any number of components is straightforward, and the details are left to the reader. The equilibrium condition is $\Sigma \nu_i \mu_i = 0$, and μ_i is given by

$$\mu_i = kT\lambda_i = kT \log (N^{(i)}/Z_i).$$

11·51. The translational partition function for a perfect gas has been evaluated in § 5·4. In the present problem we have, if the suffix i is used to denote the three species 1, 2 and 12,

$$Z_i(\vartheta) = B_i V \vartheta^{-\frac{3}{2}}, \quad (11\cdot51\cdot1)$$

where $\quad B_i = \varpi_i (2\pi m_i)^{\frac{3}{2}} h^{-3} \exp(-\vartheta E_0^{(i)}), \quad (11\cdot51\cdot2)$

$E_0^{(i)}$ being the lowest energy of the species i. Then

$$\mu_i = kT\lambda_i = kT \log (N^{(i)}/Z_i) = kT \log \{N^{(i)}/(B_i V k^{\frac{3}{2}} T^{\frac{3}{2}})\}. \quad (11\cdot51\cdot3)$$

If we write this in terms of the partial pressure $p_i = N^{(i)}kT/V$, we have

$$\mu_i = kT \log \frac{p_i}{B_i k^{\frac{5}{2}} T^{\frac{5}{2}}} = E_0^{(i)} + kT \log \left(\frac{p_i}{T^{\frac{5}{2}}} \frac{h^3}{\varpi_i (2\pi m_i)^{\frac{3}{2}} k^{\frac{5}{2}}} \right). \quad (11·51·4)$$

The equilibrium condition $\Sigma \nu_i \mu_i = 0$ therefore becomes

$$\Pi p_i^{\nu_i} = K_P \quad (11·51·5)$$

with

$$\log K_P = -\frac{\Sigma \nu_i E_0^{(i)}}{kT} + \tfrac{5}{2} \Sigma \nu_i \log T + J, \quad (11·51·6)$$

where J is explicitly given by

$$J = \Sigma \nu_i \log \{ \varpi_i (2\pi m_i)^{\frac{3}{2}} k^{\frac{5}{2}} / h^3 \}. \quad (11·51·7)$$

We can extend this result at once to diatomic and polyatomic molecules, by including the internal partition function and writing

$$Z = Z_{\text{trans.}} Z_{\text{rot.}} Z_{\text{vib.}} Z_{\text{el.}}.$$

This gives

$$\mu_i = E_0^{(i)} + kT \log \left(\frac{p_i}{(kT)^{\frac{5}{2}}} \frac{h^3}{\varpi_i (2\pi m_i)^{\frac{3}{2}}} \right) - kT \log Z_{\text{rot.}} - kT \log (Z_{\text{vib.}} Z_{\text{el.}}), \quad (11·51·8)$$

where we have separated out the rotational partition function since, except possibly for hydrogen at low temperatures, the rotational partition function invariably has its classical value. If we now denote the vibrational and electronic specific heat by $C_{P,i}^1(T)$, we see, by comparing equations (3·52·4), (3·52·5), (6·17·2) and (6·17·4) that we can write

$$-kT \log (Z_{\text{vib.}} Z_{\text{el.}}) = -T \int_0^T \frac{dT'}{T'^2} \int_0^{T'} C_{P,i}^1(T'')\, dT'' = \int_0^T \left(1 - \frac{T}{T'}\right) C_{P,i}^1(T')\, dT'. \quad (11·51·9)$$

Hence, since $Z_{\text{rot.}}$ is given by (6·12·6) or (6·13·6), we obtain equation (11·42·1) with

$$C_{P,i}^0 = \tfrac{7}{2}k, \quad J_i = \log \left(\frac{\varpi_i (2\pi m_i)^{\frac{3}{2}} k^{\frac{5}{2}}}{h^3} \frac{8\pi^2 k I^{(i)}}{h^2 \sigma_i} \right) \quad (11·51·10)$$

for linear molecules, and

$$C_{P,i}^0 = 4k, \quad J_i = \log \left(\frac{\varpi_i (2\pi m_i)^{\frac{3}{2}} k^{\frac{5}{2}}}{h^3} \frac{8\pi^2 (2\pi k)^{\frac{3}{2}} (I_1^{(i)} I_2^{(i)} I_3^{(i)})^{\frac{1}{2}}}{h^3 \sigma_i} \right) \quad (11·51·11)$$

for non-linear molecules, where the nuclear weights have been omitted.

If the energy levels of the molecules are known from an analysis of their spectra, we have all the data, with the exception of Δ_{H_0}, from which to calculate K_P as a function of the temperature. A measurement of the heat of reaction at any temperature, combined with a knowledge of the moments of inertia, the vibrational frequencies and the weights of the electronic ground states (or in exceptional cases, the electronic partition functions), is therefore sufficient to determine the equilibrium constant of any homogeneous gas reaction. In principle we can determine ΔH_0 for any gas reaction from the spectroscopic data concerning the dissociation energies of the molecules. In practice, however, the spectroscopic dissociation energies are not accurately known except for relatively few molecules, and ΔH_0 has in general to be determined by calorimetric methods.

The Application of Nernst's Principle to Chemical Reactions

11·6. If we know the equilibrium constant K_P as a function of the temperature, then, by equation (11·421·2), the heat of reaction can be obtained by the differentiation of K_P. An important problem, which was much discussed during the first thirty years of the present century, is whether conversely the equilibrium constant can be determined solely by measurements of the heat of reaction and of the heat capacities of the substances taking part in the reaction. A comparison of equations (7·1·1) and (11·421·2) shows that this problem is a special case of the more general question considered in Chapter 7, and therefore that the answer is yes if Nernst's principle applies to the substances taking part in the reaction. The only quantity in equation (11·42·6) which has not an obvious physical significance is J, which is called the chemical constant. Now, by (11·42·2), J is determined by the entropy constants of the various gases, and, if these can be found by calorimetric measurements, K_P can be obtained without carrying out any measurements of the partial pressures of the gases when the reaction is in equilibrium. The discussion of the calorimetric and of the spectroscopic entropy on §§ 7·4 and 7·5 shows that in general we can determine the entropy constant of a gas by calorimetric measurements of the heat capacity of the corresponding solid and liquid and of the heats of fusion and vaporization. We can therefore obtain K_P from the same measurements together with a measurement of ΔH_0.

By comparing equations (9·3·3), (11·42·2) and (11·42·7) we see that

$$J = \Sigma\nu_i i_i + \Sigma\nu_i s^{(i)}_{S,0}, \tag{11·6·1}$$

and since, according to Nernst's principle,

$$\Sigma\nu_i s^{(i)}_{S,0} \equiv 0, \tag{11·6·2}$$

we have

$$J = \Sigma\nu_i i_i. \tag{11·6·3}$$

This equation may, however, break down when any of the solids, whose heat capacities have to be measured in order to determine the i's, are in non-equilibrium states, since equation (11·6·2) will not then in general be valid. In such cases we could, of course, correct the vapour-pressure constants to take account of the residual entropy of the solids. But since this can only be done if we know the theoretical values of the entropy constants, we are led back to using the theoretical values of the chemical constants given in the preceding section, and the measurements of the vapour pressures are superfluous.

11·61. *The chemical constant of hydrogen.* We have already discussed the entropy constant and the vapour-pressure constant of hydrogen in §§ 7·52 and 9·311. The results are different according as the rotations are classical or not, but for gas reactions we can normally confine our attention to high temperatures. At temperatures greater than about 300° K. hydrogen behaves as a diatomic molecule with weight unity, excluding the nuclear spin weights, and with symmetry number $\sigma = 2$. The vapour-pressure constant, however, depends upon the residual entropy of solid hydrogen, which is either $\frac{3}{4}k \log 3$ or zero according to the definition of the specific heat of solid hydrogen below 12° K. for the purposes of calculating the vapour-pressure constant i. If therefore we take the normal definition of i, which depends upon extrapolating the specific heat to zero from 12° K., we must have $J = i + s_{S,0}/k$, where i is the high-temperature value of the vapour-pressure constant, and where $s_{S,0} = \frac{3}{4}k \log 3$ for hydrogen, while $s_{S,0}$ is $3k \log 3$ for deuterium and zero for HD. Thus the values J^* of J in practical units derived from the values of i^* given in § 9·311 are $-3·37$ for H_2, $-2·61$ for D_2 and $-2·68$ for HD. These are in close agreement with the theoretical values obtained from (11·51·10).

The Calculation of the Equilibrium Constants of Some Typical Gas Reactions

11·7. Measurements of the chemical constants of gas reactions cannot compare in accuracy with measurements of the calorimetric entropies of gases. Since the relations between the calorimetric entropy, the vapour-pressure constant and the chemical constant of a gas are now well known. investigations of the chemical constants themselves are less important today than they were twenty-five years ago. Also, the measured values of the chemical constants are best compared with the theoretical values (11·51·9), (11·51·10) and (11·51·11) which involve only the properties of the substances concerned with the gaseous state. It is, therefore, advantageous to go one stage further and compare the measured and calculated values of the equilibrium constants themselves as functions of the temperature rather than compare merely the chemical constants.

11·71. As already mentioned in § 11·51, the equilibrium constant of any perfect-gas reaction can be obtained from a knowledge of the energy levels of the various molecules together with a measured value of the heat of reaction at one temperature. The calculation is essentially the same as that of the spectroscopic entropy which was discussed in §7·411, except that here we are primarily interested in $\mu_i(T,P)$ and not in the entropy. If, as in (11·42·3), we write $\mu_i(T,P)=\mu_i^*(T)+RT\log p_i$, we can obtain $\mu_i^*(T)-E_0^{(i)}$ from (11·51·8) and from the expressions given in §§6·12, 6·13, 6·14 and 6·16 for the internal partition functions.

The quantity $\mu_i^*(T)-E_0^{(i)}$ has been calculated for a very large number of substances, but it is only possible to refer here to a few simple cases. We give in table 11·5 some values of $\{H_0-\mu_i^*(T)\}/T$ for a number of simple gases, whose molecular constants will be found in tables 6·2 and 6·3. The thermodynamic functions of these gases have been calculated by a number of authors, the most accurate results being those due to Wagman, Kilpatrick, Taylor, Pitzer & Rossini (1945).

Table 11·5. *Values of* $\{H_0-\mu_i^*(T)\}/T$ *for some simple gases in calories/degree/gram molecule*

T in °K. ...	298·16	400	600	800	1000	1200	1400
H_2	24·423	26·422	29·203	31·186	32·738	34·012	35·098
N_2	38·817	40·861	43·688	45·711	47·306	48·629	49·768
O_2	42·061	44·112	46·968	49·044	50·697	52·077	53·272
H_2O	37·172	39·508	42·768	45·131	47·018	48·605	49·989
CO	40·350	42·393	45·222	47·254	48·860	50·196	51·345
CO_2	43·555	45·828	49·238	51·895	54·109	56·019	57·706
CH_4	36·46	38·86	42·39	45·21	47·65	49·86	51·88

To apply these results to particular gas reactions we have to obtain ΔH_0 for the reactions concerned, but it is sufficient to determine $\Delta H(T)$ at any convenient temperature, since ΔH_0 may be obtained from $\Delta H(T)$ by using the calculated partition functions. $\Delta H(T)$ can be found by the direct calorimetric measurements of the heat of reaction, or, as explained in §11·4, by the combination of the heats of reaction of two or more related reactions. The equilibrium constants of a number of simple gas reactions obtained in this way are shown in fig. 11·6. The particular gas reactions considered are the following:

I. $-CO-\tfrac{1}{2}O_2+CO_2=0,\ \Delta H_0=-279{,}300$ joules ($-66{,}767$ cal.).

II. $-CO-H_2O+CO_2+H_2=0,\ \Delta H_0=-\ 40{,}400$ joules ($-\ 9{,}662$ cal.).

III. $-CH_4-CO_2+2CO+2H_2=0,\ \Delta H_0=\ 232{,}400$ joules ($55{,}552$ cal.).

IV. $-CH_4-H_2O+CO+3H_2=0,\ \Delta H_0=\ 192{,}000$ joules ($45{,}890$ cal.).

V. $-CH_4-2H_2O+CO_2+4H_2=0,\ \Delta H_0=\ 151{,}600$ joules ($36{,}230$ cal.).

Not all of these reactions are independent, and we have II + III = IV and 3IV − II − 2III = V.

11·72. Corresponding results for heterogeneous reactions can be obtained if we can evaluate $\mu_i(T) - E_0^{(i)}$ for the solids present. Now for a solid, if we neglect its compressibility, we have

$$\mu(T) - H_0 = \int_0^T \left(1 - \frac{T}{T'}\right) C_P(T')\, dT',$$

which can be calculated if the specific heat is known. The results for graphite are shown in table 11·6.

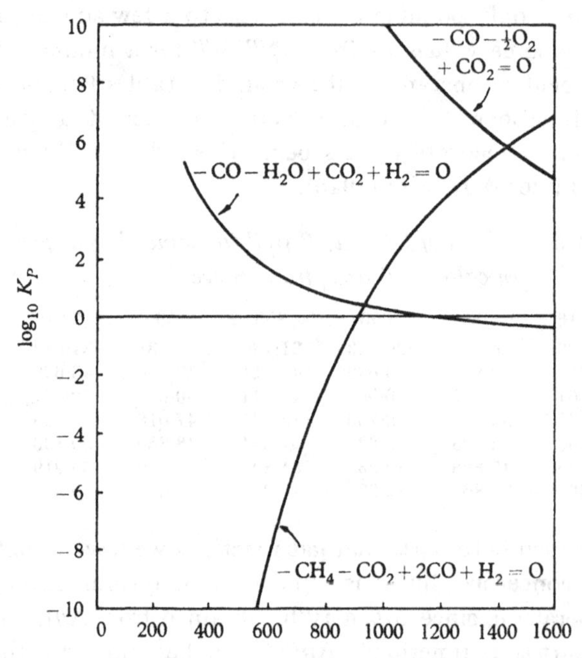

Fig. 11·6. The equilibrium constants of certain gas reactions.

Table 11·6. *Values of* $\{H_0 - \mu(T)\}/T$ *for graphite in calories/degree/gram molecule*

T in °K. ...	298·16	400	600	800	1000	1200
	0·5172	0·8245	1·477	2·138	2·771	3·365

The data in tables 11·5 and 11·6 enable us to calculate the equilibrium constants for the following reactions:

VI. $-C - CO_2 + 2CO = 0$, $\Delta H_0 = 165{,}500$ joules (39,565 cal.).

VII. $-C - H_2O + CO + H_2 = 0$, $\Delta H_0 = 125{,}100$ joules (29,902 cal.).

VIII. $-C - 2H_2 + CH_4 = 0$, $\Delta H_0 = -66{,}900$ joules ($-15{,}987$ cal.).

The results are shown in fig. 11·7.

11·73. *Isomerization reactions.* Since the molecules taking part have very similar structures, isomerization reactions are of a relatively simple type. Further, they provide examples of gas reactions in which alternative modes of reaction are of comparable importance. (Compare the corresponding discussion of reactions in solutions given in § 13·31.)

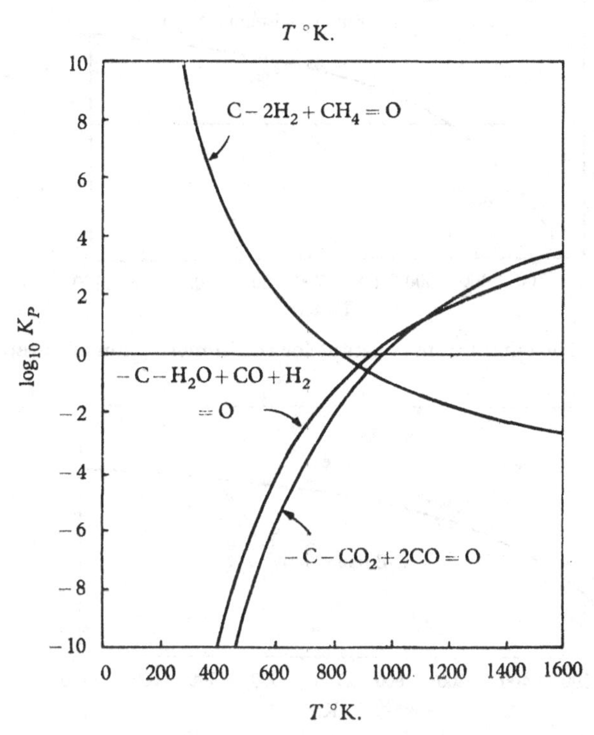

Fig. 11·7. The equilibrium constants of certain heterogeneous reactions.

The higher paraffins, with the formulae C_nH_{2n+2} $(n \geqslant 4)$ can exist in various isomeric forms. For example, butane can have the structure either of normal butane, $CH_3.CH_2.CH_2.CH_3$, or of isobutane, $CH(CH_3)_3$, while pentane has the three structures, normal pentane, $CH_3.CH_2.CH_2.CH_2.CH_3$, isopentane, $CH_3.CH_2.CH.(CH_3)_2$, and tetramethylmethane, $C(CH_3)_4$. If we denote the two butanes by A_1 and A_2, and the three pentanes by B_1, B_2 and B_3, we can write the isomerization reactions as

$$-A_1 + A_2 = 0, \tag{11·73·1}$$

$$-B_1 + B_2 = 0, \quad -B_1 + B_3 = 0, \quad -B_2 + B_3 = 0, \tag{11·73·2}$$

only two of the three reactions (11·73·2) being independent.

The partition functions of the paraffins can be calculated by standard methods (Pitzer, 1940*a*, *b*; Rossini, Prosen & Pitzer, 1941), and we can therefore calculate the differences between $\{\mu^*(T) - H_0\}/T$ for the various isomers. A knowledge of the heat of isomerization, which must be determined experimentally from the heats of combustion of the isomers, then enables us to find the equilibrium mole fractions of the various isomers,

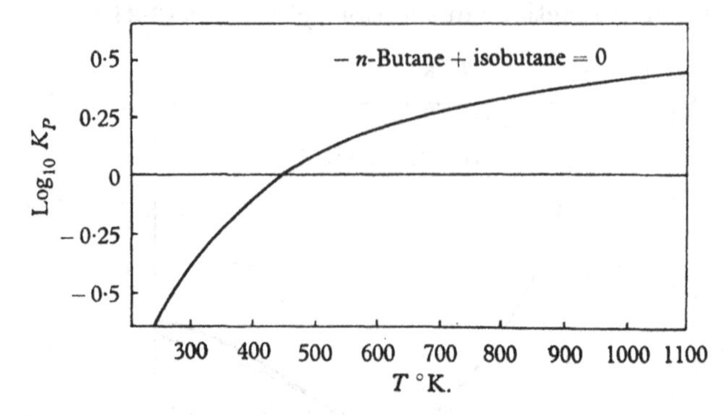

Fig. 11·8. The equilibrium constant for the isomerization of butane.

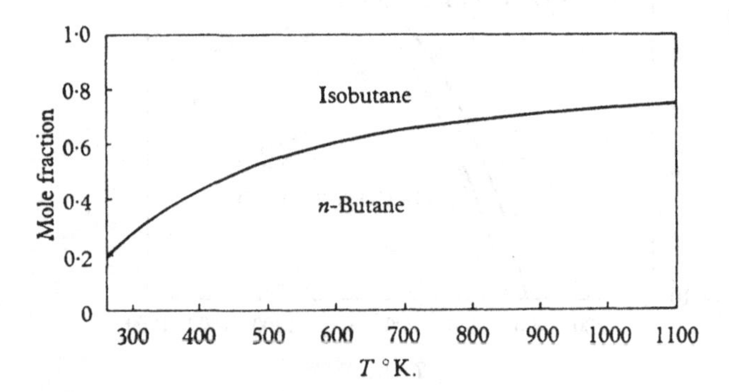

Fig. 11·9. The mole fractions of n-butane and isobutane.

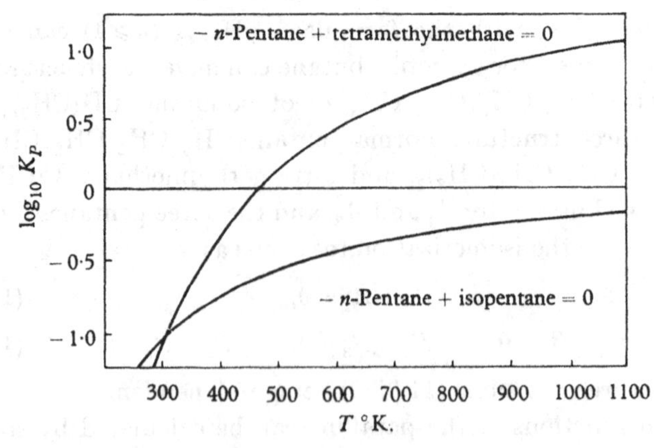

Fig. 11·10. The equilibrium constants for the isomerization of pentane.

the results for butane being given in figs. 11·8 and 11·9, and those for pentane being given in figs. 11·10 and 11·11. It will be seen that the relative stability of the isomers varies considerably over the temperature range.

There is a certain amount of experimental data with which to compare the theory, but the accuracy of the measurements is not high and the

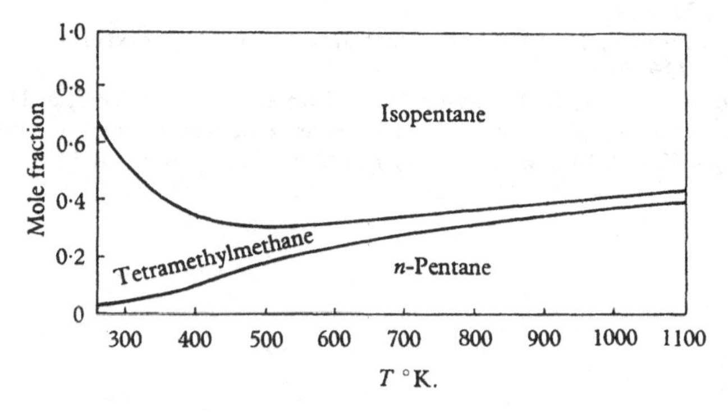

Fig. 11·11. The mole fractions of n-pentane, isopentane and tetramethylmethane.

probable error of the calculations is also appreciable. The experimental evidence is examined critically by Rossini, Prosen & Pitzer, and they conclude that there is reasonable agreement with the theory within, but on the verge of, the limits of accuracy at present obtainable.

REFERENCES

Beattie, J. A. (1949). Computation of the thermodynamic properties of real gases and mixtures of real gases. *Chem. Rev.* **44**, 141.

Beattie, J. A. & Stockmayer, W. H. (1940). Equations of state. *Rep. Progr. Phys.* **7**, 195.

Gibby, C. W., Tanner, C. C. & Masson, I. (1929). The pressures of gaseous mixtures. II. Helium and hydrogen and their intermolecular forces. *Proc. Roy. Soc.* A, **122**, 283.

Gillespie, L. J. & Beattie, J. A. (1930). The thermodynamic treatment of chemical equilibria in systems composed of real gases. *Phys. Rev.* **36**, 743, 1008; *J. Amer. Chem. Soc.* **52**, 4239.

Gloag, V. F. & Barritt, R. J. (1943). The manufacture of sulphuric acid in contact plants. *Trans. Instn Chem. Engrs, Lond.*, **21**, 31.

Guggenheim, E. A. & McGlashan, M. L. (1951). Corresponding states in mixtures of slightly imperfect gases. *Proc. Roy. Soc.* A, **206**, 448.

Larson, A. T. (1924). The ammonia equilibrium at high pressures. *J. Amer. Chem. Soc.* **46**, 367.

Larson, A. T. & Dodge, R. L. (1923). The ammonia equilibrium. *J. Amer. Chem. Soc.* **45**, 2918.

Masson, I. & Dolley, L. G. F. (1923). The pressures of gaseous mixtures. *Proc. Roy. Soc.* A, **103**, 524.

Pitzer, K. S. (1940*a*). Chemical equilibria, free energies and heat contents for gaseous hydrocarbons. *Chem. Rev.* **27**, 39.

Pitzer, K. S. (1940*b*). The vibration frequencies and thermodynamic functions of long chain hydrocarbons. *J. Chem. Phys.* **8**, 711.

Prigogine, I. & Defay, R. (1954). *Chemical thermodynamics*. London.

Rossini, F. D., Prosen, E. J. R. & Pitzer, K. S. (1941). Free energies and equilibria of isomerisation of the butanes, pentanes, hexanes and heptanes. *J. Res. Nat. Bur. Stand.* **27**, 529.

Tanner, G. C. & Masson, I. (1930). The pressures of gaseous mixtures. III. *Proc. Roy. Soc.* A, **126**, 268.

Wagman, D. D., Kilpatrick, J. E., Taylor, W. J., Pitzer, K. S. & Rossini, F. D. (1945). Heats, free energies and equilibrium constants of some reactions involving O_2, H_2, H_2O, C, CO, CO_2 and CH_4. *J. Res. Nat. Bur. Stand.* **34**, 143.

Chapter 12

SOLUTIONS

INTRODUCTION

12·1. In order to determine the properties of a solution it is necessary to know the free energy as a function of the temperature and the pressure (or the volume) and also as a function of the composition. If we are concerned with the heterogeneous equilibrium of more than one phase, we can deduce all the equilibrium phenomena from the condition that the thermodynamic potential μ_i of the ith constituent should be the same for all the phases in which the ith constituent occurs. It is the object of the thermodynamic theory of solutions to show how the various quantities which occur in the relevant formulae are interrelated and how they can be measured. Further, it is possible to find differential equations describing the phenomena which are more easily applied to the partial elucidation of the experimental data over a restricted range than the integrated equations, particularly when the latter involve more accurate experimental evidence than is available. In all cases, the thermodynamic relations are essentially identities, and in order to determine how the μ's, for example, depend upon the temperature, the pressure and the concentrations, the thermodynamic theory must be supplemented either by experimental data or by information derived from the statistical mechanics of particular theoretical models.

RÉSUMÉ OF THE THERMODYNAMIC EQUATIONS

12·2. The thermodynamic potentials are most simply defined by the relation (3·41·1), namely,

$$dG = -S\,dT + V\,dP + \Sigma\mu_i\,dn_i, \tag{12·2·1}$$

or the equivalent relation

$$d\frac{G}{T} = -\frac{H}{T^2}dT + \frac{V}{T}dP + \Sigma\frac{\mu_i}{T}dn_i. \tag{12·2·2}$$

These define the dependence of μ_i upon T and P, which is given by the equations

$$\left(\frac{\partial}{\partial T}\frac{\mu_i}{T}\right)_{P,n_j} = -\frac{1}{T^2}\left(\frac{\partial H}{\partial n_i}\right)_{T,P,n_j} \tag{12·2·3}$$

and

$$\left(\frac{\partial \mu_i}{\partial P}\right)_{T,n_j} = \left(\frac{\partial V}{\partial n_i}\right)_{T,P,n_j} \tag{12·2·4}$$

If, therefore, we can measure the differential heat function $\partial H/\partial n_i$ and the differential specific volume $\partial V/\partial n_i$ over a range of temperatures and pressures for a given composition, we can obtain μ_i over the same range. In general, if the temperature range is not too great, we may treat $\partial H/\partial n_i$ as a linear function of T, and for all normal pressures $\partial V/\partial n_i$ can be taken as being independent of P. The dependence of the μ's on the n's is, however, more difficult to determine.

The equations (12·2·3) and (12·2·4) can be supplemented by the $P(T, \mu_i)$ equation (§ 3·42), which in differential form is

$$V\,dP = S\,dT + \Sigma n_i\,d\mu_i. \tag{12·2·5}$$

We also have the consistency relations

$$\partial\mu_i/\partial n_j = \partial\mu_j/\partial n_i \tag{12·2·6}$$

and the stability conditions

$$(\partial\mu_i/\partial n_i)_{T,P,n_j} > 0, \tag{12·2·7}$$

which were derived in § 3·71.

Since the properties of any single phase depend upon the concentrations of the various constituents and not upon the total masses present, it would be more convenient to use the mole fractions x_i as the variables defining the composition rather than the n_i's themselves, were it not for the fact that the x_i's are not linearly independent but satisfy the relation $\Sigma x_i = 1$. It is therefore usually simpler to carry out the calculations with the n_i's and to change over to the x_i's after all the differentiations have been carried out. It is of course possible to use the x_i's throughout the calculation, but it is then necessary to eliminate one of the x_i's, and the equations become unsymmetrical. We can, for example, write

$$N = \Sigma n_i, \quad x_1 = 1 - \sum_{i\geq 2} x_i,$$

and use N and x_i $(i \geq 2)$ as the independent variables, in which case (12·2·1) becomes

$$dG = -S\,dT + V\,dP + N\sum_{i\geq 2}(\mu_i - \mu_1)\,dx_i + \sum_{i\geq 1} x_i\mu_i\,dN. \tag{12·2·8}$$

Hence

$$\sum_{i\geq 1} x_i\mu_i = \left(\frac{\partial G}{\partial N}\right)_{T,P,xi}, \quad \mu_i - \mu_1 = \frac{1}{N}\left(\frac{\partial G}{\partial x_i}\right)_{T,P,x_j} \quad (i \geq 2), \tag{12·2·9}$$

which give

$$\mu_1 = \frac{\partial G}{\partial N} - \frac{1}{N}\sum_{i\geq 2} x_i\frac{\partial G}{\partial x_i}, \quad \mu_i = \frac{\partial G}{\partial N} - \frac{1}{N}\sum_{j\geq 2} x_j\frac{\partial G}{\partial x_j} + \frac{1}{N}\frac{\partial G}{\partial x_i} \quad (i \geq 2). \tag{12·2·10}$$

It is often convenient, but never necessary, to designate one component of a solution as the solvent and the other components as solutes. We shall use or disregard this nomenclature as expediency dictates.

12·21. For binary solutions at constant temperature and pressure, equation (12·2·5) reduces to

$$n_1 \frac{\partial \mu_1}{\partial n_1} + n_2 \frac{\partial \mu_2}{\partial n_1} = 0, \tag{12·21·1}$$

a relation which is often known as the Gibbs-Duhem equation. In this case the stability condition (12·2·7) is equivalent to

$$(\partial \mu_2/\partial n_1)_{T,P,n_2} = (\partial \mu_1/\partial n_2)_{T,P,n_1} < 0. \tag{12·21·2}$$

These relations give a considerable amount of information concerning binary solutions.

If we treat μ_1 and μ_2 as functions either of x_1 or of x_2 ($x_1 + x_2 = 1$), (12·21·1) is equivalent to either of

$$x_1 \frac{\partial \mu_1}{\partial x_1} + x_2 \frac{\partial \mu_2}{\partial x_1} = 0, \quad x_1 \frac{\partial \mu_1}{\partial x_2} + x_2 \frac{\partial \mu_2}{\partial x_2} = 0. \tag{12·21·3}$$

Also
$$\frac{\partial \mu_1}{\partial n_1} = \frac{x_2}{N}\frac{\partial \mu_1}{\partial x_1} = -\frac{x_2}{N}\frac{\partial \mu_1}{\partial x_2}, \quad \frac{\partial \mu_2}{\partial n_2} = -\frac{x_1}{N}\frac{\partial \mu_2}{\partial x_1} = \frac{x_1}{N}\frac{\partial \mu_2}{\partial x_2} \tag{12·21·4}$$

and
$$\frac{\partial \mu_1}{\partial n_2} = -\frac{x_1}{N}\frac{\partial \mu_1}{\partial x_1} = \frac{x_1}{N}\frac{\partial \mu_1}{\partial x_2}, \quad \frac{\partial \mu_2}{\partial n_1} = \frac{x_2}{N}\frac{\partial \mu_2}{\partial x_1} = -\frac{x_2}{N}\frac{\partial \mu_2}{\partial x_2}, \tag{12·21·5}$$

since $\partial x_1/\partial n_1 = x_2/N$. Hence (12·21·1) is equivalent to

$$x_1 \partial \mu_1/\partial x_2 = x_2 \partial \mu_2/\partial x_1, \tag{12·21·6}$$

and (12·21·2) is equivalent to

$$\partial \mu_1/\partial x_1 > 0, \quad \partial \mu_2/\partial x_2 > 0. \tag{12·21·7}$$

In the general case of an arbitrary number of constituents, we can deduce, just as for binary solutions,

$$\lim_{n_j=0} (\partial \mu_1/\partial n_i)_{T,P,n_j} < 0 \quad (i \neq 1), \tag{12·21·8}$$

but this inequality is no longer necessarily true for non-zero values of the n_j's. In general, therefore, the equation

$$\Sigma n_i d\mu_i = 0,$$

taken in conjunction with (12·2·7), gives no information about the signs of any of the quantities $(\partial \mu_i/\partial n_j)_{T,P}$ when $i \neq j$.

Dilute and Ideal Solutions

12·3. It is necessary at some stage to know how the thermodynamic potentials depend upon the concentrations, and although many transformations of general validity can be derived which are independent of any particular functional form of the μ's, the exposition can be shortened by considering the explicit dependence of the μ's on the constitution at this point. In some ways the most informative approach is the following, though the choice is largely a matter of taste.

If we consider a binary mixture and keep the temperature and pressure constant, we have

$$n_1 d\mu_1 + n_2 d\mu_2 = 0. \qquad (12\cdot3\cdot1)$$

In the limit of $n_2 = 0$, we therefore have either $\partial\mu_1/\partial n_2 = 0$ or $\partial\mu_2/\partial n_2 = \infty$. Now there seems no *a priori* reason why $\partial\mu_1/\partial n_2$ should be zero for $n_2 = 0$, since this would mean that the free energy of a substance was unaltered to the first order by the addition of a second substance. It would be more reasonable to expect the limiting value of $\partial\mu_1/\partial n_2$ to be a non-zero constant, and in this case we should have

$$\partial\mu_2/\partial n_2 = A/n_2 \quad (n_2 \ll n_1), \qquad (12\cdot3\cdot2)$$

where, by (12·2·7), A is necessarily positive. We therefore have, for sufficiently small values of $x = n_2/(n_1 + n_2)$,

$$\mu_1 = \mu_1^0(T, P) - Ax, \quad \mu_2 = \mu_2^*(T, P) + A \log x. \qquad (12\cdot3\cdot3)$$

The above conclusion can only be considered to be a tentative one, since the initial assumption that $\partial\mu_1/\partial n_2 \neq 0$ for $n_2 = 0$ might be contested, and the argument gives no indication as to what the quantity A should be. We therefore require to appeal to experiment to determine A and to confirm the argument.

If we assume that the relations (12·3·3) are of general validity and apply to all substances, with a universal value of A, then the theory of perfect gas mixtures given in Chapter 11 shows that $A = RT$. We can avoid making this hypothesis, and can derive it instead, by invoking the experimental fact known as Henry's law, which states that, for a dilute solution, the partial pressure of a solute in the gaseous phase is proportional to the concentration of the solute in the solution. Since μ for the gas is given by

$$\mu^*(T) + RT \log p,$$

where p is the partial pressure, and since $p \propto x$, we see that (12·3·3) must in fact be correct for any solution, with $A = RT$, provided that we ascribe the same molecular weight to the solute in the solution as in the gas phase. (It is not always possible to do this, since the degree of association of the molecules may be different in the two phases. See § 13·1.)

We may generalize (12·3·3) to apply to a number of solutes in a dilute solution. By considering each solute in turn with the mole fractions of the others equal to zero, we have

$$\mu_i = \mu_i^*(T, P) + RT \log x_i \quad (i \geqslant 2). \qquad (12\cdot3\cdot4)$$

The consistency relations (12·2·6) then show that

$$\mu_1 = \mu_1^0(T, P) + RT \log (1 - \sum_{i \geqslant 2} x_i). \tag{12·3·5}$$

The quantity $\mu_1^0(T, P)$ is the thermodynamic potential of the pure solvent, but in general the quantities $\mu_i^*(T, P)$ have no such simple interpretation. The free-energy function from which (12·3·4) and (12·3·5) are derived is

$$G(T, P, n_i) = n_1 \mu_1^0(T, P) + \sum_{i \geqslant 2} n_i \mu_i^*(T, P) + RT(\sum_{i \geqslant 1} n_i \log n_i - N \log N), \tag{12·3·6}$$

where $n_1 \gg n_i$ ($i \geqslant 2$).

The equations (12·3·4)–(12·3·6) have been derived for dilute solutions from very general principles. We may expect them to hold more or less accurately for certain solutions for all values of the concentrations. Such solutions are called ideal solutions and they are the analogues of perfect gas mixtures. For such solutions $\mu_i^*(T, P)$ is the thermodynamic potential of the pure substance i since equation (12·3·4) holds up to and including $x_i = 1$. We may therefore write

$$\mu_i(T, P, n_j) = \mu_i^0(T, P) + RT \log x_i \quad (\text{all } i) \tag{12·3·7}$$

as the equations defining an ideal solution, where $\mu_i^0(T, P)$ is the thermodynamic potential of the ith component in the pure state. Correspondingly, the free-energy function G of an ideal solution may be written as

$$G(T, P, n_i) = \sum_{i \geqslant 1} n_i \mu_i^0(T, P) + RT(\sum_{i \geqslant 1} n_i \log n_i - N \log N). \tag{12·3·8}$$

This is identical in form with the expression (11·2·1) for the free energy of a mixture of perfect gases.

12·31. *The heat of mixing.* The heat function of an ideal solution is given by

$$H(T, P, n_i) \equiv -T^2 \frac{\partial}{\partial T} \frac{G}{T} = -T^2 \sum_{i \geqslant 1} n_i \frac{\partial}{\partial T} \frac{\mu_i^0(T, P)}{T} = \sum_{i \geqslant 1} n_i H_i^0(T, P). \tag{12·31·1}$$

The heat function of the mixture is therefore equal to the sum of the heat functions of the separate constituents of the mixture, and the heat of mixing of an ideal solution is zero. The heat of mixing must not be confused with the heat of solution, the latter term usually referring to the heat absorbed when a solute in any state of aggregation is dissolved in a liquid solvent. The heat of mixing, on the other hand, refers to the excess of the heat function of the solution over the sum of the heat functions of the constituents in the same state of aggregation as the solution, i.e. for a liquid solution all the constituents must be taken to be in the liquid state at the same temperature and pressure as the solution. In many cases this will mean extrapolating the properties of the constituents into regions in which they are metastable. The heat of solution of a solid which forms a perfect solution with a liquid solvent is positive and is equal to the latent heat of fusion of the solid. In similar circumstances the heat of solution of a gas in a liquid is negative and equal to minus the latent heat of vaporization.

The heat function of a dilute solution is given by

$$H(T,P,n_i)=n_1H_1^0(T,P)-T^2\sum_{i\geq 2}n_i\frac{\partial}{\partial T}\frac{\mu_i^*(T,P)}{T}. \qquad (12\cdot31\cdot2)$$

In this case the heat of mixing is not necessarily zero, since $\mu_i^*(T,P)$ is in general different from $\mu_i^0(T,P)$. However, H is a linear function of the n's, and so if two dilute solutions in the same solvent are mixed together, the heat of mixing is zero.

12·32. *The entropy of mixing.* The entropy of an ideal solution is given by

$$S(T,P,n_i)\equiv-\frac{\partial G}{\partial T}=\sum_{i\geq 1}n_is_i^0(T,P)-NR\sum_{i\geq 1}x_i\log x_i. \qquad (12\cdot32\cdot1)$$

The entropy of mixing is therefore $-R\Sigma x_i\log x_i$, just as for a mixture of perfect gases. This is not true for dilute solutions which are not ideal, since $-\partial\mu_i^*(T,P)/\partial T$ is not equal to the entropy of the pure component, which is given by $-\partial\mu_i^0(T,P)/\partial T$.

Heterogeneous Equilibrium when one Phase Consists of a Single Component

12·4. The discussion of the heterogeneous equilibrium of solutions is much simplified when one of the phases (either solid or gaseous) consists of a pure substance. We therefore begin by discussing this case and derive explicit formulae for ideal solutions. In all cases where only the thermodynamic potential of the solvent appears explicitly, the formulae for dilute solutions are limiting cases of those for ideal solutions, and are therefore not given separately. In the general case where no assumptions are made concerning the nature of the solution, it is usually necessary to give the formulae in differential form. We use the suffix 1 to refer to the solvent and the suffixes 2, 3, ... to refer to the solutes, while the suffixes S, L and G refer to the solid, liquid and gaseous phases. An index or a suffix zero indicates that the quantity concerned refers to a pure component.

12·41. *The freezing-point of a solution when the solvent separates out.* In this case the equilibrium condition is

$$\mu_{1,S}(T,P)=\mu_{1,L}(T,P,n_i), \qquad (12\cdot41\cdot1)$$

which, for an ideal solution, is

$$RT\log(1-\sum_{i\geqslant 2}x_i)=\mu^0_{1,S}(T,P)-\mu^0_{1,L}(T,P). \qquad (12\cdot41\cdot2)$$

If T_0 is the freezing-point of the pure solvent, then $\mu^0_{1,L}(T_0,P)=\mu^0_{1,S}(T_0,P)$, and by using equation (12·2·3), the relation (12·41·2) can be expressed in terms of the heat functions of the solvent in the solid and supercooled liquid states. (These heat functions are directly measurable quantities, though that for the supercooled liquid has to be obtained by extrapolation from above the normal melting-point.) The result is that for an ideal solution

$$\log(1-\sum_{i\geqslant 2}x_i)=\log x_1=\int_{T_0}^{T}\{H_{1,L}(T',P)-H_{1,S}(T',P)\}\frac{dT'}{RT'^2}, \qquad (12\cdot41\cdot3)$$

and, since $H_{1,L}>H_{1,S}$, the freezing-point is lowered by the presence of the solutes. For a dilute solution, this reduces to

$$T_0-T=RT_0^2\sum_{i\geqslant 2}x_i/L, \qquad (12\cdot41\cdot4)$$

where L is the latent heat of fusion of the pure solvent.

In the general case we can transform (12·41·1) into a differential equation as follows. If P and n_1 are constant we have

$$d\frac{\mu_{1,S}(T,P)}{T}=\frac{\partial}{\partial T}\frac{\mu_{1,S}}{T}dT=d\frac{\mu_{1,L}(T,P,n_i)}{T}=\frac{\partial}{\partial T}\frac{\mu_{1,L}}{T}dT+\frac{1}{T}\sum_{i\geqslant 2}\frac{\partial\mu_{1,L}}{\partial n_i}dn_i, \qquad (12\cdot41\cdot5)$$

which, by (12·2·3), can be written as

$$dT=T\sum_{i\geqslant 2}\frac{\partial\mu_{1,L}}{\partial n_i}dn_i\Big/\left(\frac{\partial H_L}{\partial n_1}-\frac{\partial H_S}{\partial n_1}\right)_{T,P,n_j}. \qquad (12\cdot41\cdot6)$$

Here $\partial H_S/\partial n_1$ is the heat function per unit mass of the pure solvent in the solid state, while $\partial H_L/\partial n_1$ is the differential heat function per unit mass of the solvent in the liquid solution.

12·411. Measurements of the freezing-points of solutions are often made the basis for the experimental determination of the thermodynamic potential of the solvent. It is then convenient to introduce the quantity

$$\mu_{1,L}(T,P,n_i)-\mu^0_{1,L}(T,P) \quad (T<T_0),$$

which by (12·41·1) is equal to $\mu^0_{1,S}(T,P)-\mu^0_{1,L}(T,P)$. This latter quantity is determined by the latent heat of fusion of the pure solid solvent, extrapolated below the normal melting-point. If ΔH_0 is the latent heat of fusion at the normal melting-point T_0, and if $\Delta C_P=C_{P,L}-C_{P,S}$ is the excess of the specific heat of the liquid over that of the solid, we have, by (12·2·3),

$$\frac{\mu^0_{1,L}(T,P)}{T}-\frac{\mu^0_{1,S}(T,P)}{T}=-\int_{T_0}^{T}\left[\Delta H_0+\int_{T_0}^{T'}\Delta C_P(T'')\,dT''\right]\frac{dT'}{T'^2}, \qquad (12\cdot411\cdot1)$$

since $$\mu^0_{1,L}(T_0, P) = \mu^0_{1,S}(T_0, P).$$

If we treat ΔC_P as a constant and consider a restricted temperature range such that $T_0 \gg T_0 - T$, (12·411·1) becomes

$$\mu^0_{1,L}(T, P) - \mu^0_{1,S}(T, P) = -\frac{\Delta H_0}{T_0}(T_0 - T) - \frac{\Delta C_P}{2T_0}(T_0 - T)^2 + \dots \quad (12\cdot411\cdot2)$$

For water, $\Delta H_0 = 1438$ calories per gram molecule, and $\Delta C_P = 9$ calories per gram molecule for not too large temperature ranges. If, therefore, T is not lower than about $-30°$ C., we have

$$\mu_{1,L}(T, P, n_i) - \mu^0_{1,L}(T, P) = -5{\cdot}262(T_0 - T) + 0{\cdot}0165(T_0 - T^2) + \dots \quad (12\cdot411\cdot3)$$

The value of T in this equation is the freezing-point of the solution corresponding to given n's. To determine $\mu_{1,L}$ at other temperatures, we require to integrate equation (12·2·3) or its equivalent.

12·42. *The boiling-point and vapour pressure of a solvent containing a non-volatile solute.* In this case the equilibrium condition is

$$\mu_{1,L}(T, P, n_i) = \mu_{1,G}(T, P), \quad (12\cdot42\cdot1)$$

which only differs from (12·41·1) by the replacing of the solid phase by the gaseous phase. The boiling-point is the analogue of the freezing-point, the pressure being constant, the only difference being that the boiling-point is raised by the presence of the solutes while the freezing-point is lowered. Since the formulae for the boiling-point are otherwise the same, they need not be repeated.

To find the effect on the vapour pressure we must keep the temperature constant. For an ideal solution we have

$$RT \log\left(1 - \sum_{i \geqslant 2} x_i\right) = \mu^0_{1,G}(T, P) - \mu^0_{1,L}(T, P), \quad (12\cdot42\cdot2)$$

and, if P_0 is the vapour pressure of the pure solvent at temperature T, then $\mu^0_{1,L}(T, P_0) = \mu^0_{1,G}(T, P_0)$. Also, by (12·2·4),

$$\mu^0_{1,G}(T, P) - \mu^0_{1,L}(T, P) = \int_{P_0}^{P} \{v_{1,G}(T, P') - v_{1,L}(T, P')\}\, dP', \quad (12\cdot42\cdot3)$$

and so

$$RT \log\left(1 - \sum_{i \geqslant 2} x_i\right) = RT \log x_1 = \int_{P_0}^{P} \{v_{1,G}(T, P') - v_{1,L}(T, P')\}\, dP'. \quad (12\cdot42\cdot4)$$

If we treat the liquid as incompressible and the vapour as a perfect gas, this becomes

$$RT \log\left(1 - \sum_{i \geqslant 2} x_i\right) = RT \log x_1 = RT \log(P/P_0) - (P - P_0)\, v_{1,L}. \quad (12\cdot42\cdot5)$$

If, further, we neglect $v_{1,L}$, we have

$$P = x_1 P_0. \tag{12·42·6}$$

In practice this could only apply to a dilute solution, since, if the solute is non-volatile and the solvent is volatile, the solution cannot be a perfect one.

In the general case, by differentiating (12·42·1), keeping T and n_1 constant and using (12·2·4), we have

$$dP = \sum_{i\geqslant 2} \frac{\partial \mu_{1,L}}{\partial n_i} dn_i \bigg/ \left(\frac{\partial V_G}{\partial n_1} - \frac{\partial V_L}{\partial n_1}\right). \tag{12·42·7}$$

Since $\partial V_G/\partial n_1$ is the specific volume of the vapour while $\partial V_L/\partial n_1$ is the differential specific volume of the solvent in the solution, which is small compared with the specific volume of the vapour, the denominator in (12·42·7) is positive. The vapour pressure is therefore lowered by the presence of a non-volatile solute.

12·43. *The osmotic pressure of a solution.* If a solution is separated from a pure solvent by a rigid wall which is permeable only to the solvent, equilibrium is impossible unless the pressure on the solution exceeds that on the solvent. The excess pressure Π which must be exerted on the solution to maintain equilibrium is called the osmotic pressure. Π is determined by the equilibrium condition

$$\mu_{1,L}(T, P+\Pi, n_i) = \mu_{1,L}(T, P, 0), \tag{12·43·1}$$

which, for an ideal solution, takes the form

$$RT \log (1 - \sum_{i\geqslant 2} x_i) = RT \log x_1 = \mu^0_{1,L}(T, P) - \mu^0_{1,L}(T, P+\Pi) \tag{12·43·2}$$

$$= -\int_P^{P+\Pi} v_{1,L}(T, P')\, dP', \tag{12·43·3}$$

by (12·2·4). If the liquid is considered to be incompressible, this becomes

$$\Pi v_{1,L}(T) = -RT \log (1 - \sum_{i\geqslant 2} x_i) = -RT \log x_1. \tag{12·43·4}$$

Hence, for a dilute solution, we have

$$\Pi v_{1,L}(T) = RT \sum_{i\geqslant 2} x_i. \tag{12·43·5}$$

This equation is formally analogous to the equation of state of a perfect gas, and for this reason it is often stated that the solute molecules in a dilute solution behave as if they formed a perfect gas. The analogy is, however, an imperfect one and is apt to be very misleading.

The osmotic pressure may reach very high values. If, for example, we have an aqueous solution at 15° C. in which the mole fraction of the solute is 0·01, the osmotic pressure is 13 atmospheres.

12·44. *The effect of temperature on the solubility of a solute.* It is sufficient to consider one solute, which we take to be solid, though the equations apply, with the appropriate changes, to gaseous or liquid solutes. If we denote the solute by the suffix 2, the condition for the solution to be saturated is

$$\mu_{2,S}(T,P)=\mu_{2,L}(T,P,n_i). \tag{12·44·1}$$

For an ideal solution this becomes

$$RT\log x_2=\mu^0_{2,S}(T,P)-\mu^0_{2,L}(T,P), \tag{12·44·2}$$

i.e.

$$\log x_2=\int_{T_0}^{T}\{H^0_{2,L}(T',P)-H^0_{2,S}(T',P)\}\frac{dT'}{RT'^2}, \tag{12·44·3}$$

by (12·2·3), where T_0 is the melting-point of the solute. These equations cannot apply to a dilute solution containing a solid solute, since when a dilute solution is cooled it is the solvent and not the solute which first separates out.

12·45. *A non-volatile solvent and a single gaseous solute.* In this case the equilibrium condition is

$$\mu_{2,L}(T,P,n_i)=\mu_{2,G}(T,P). \tag{12·45·1}$$

If the solute is a perfect gas in the gas phase, we have

$$\mu_{2,G}(T,P)=\mu_{2,G}(T,P^\dagger)+RT\log(P/P^\dagger)=\mu^*_{2,G}(T)+RT\log P, \tag{12·45·2}$$

where as usual $P^\dagger$ is a fixed standard pressure, and where, by (12·2·3),

$$\frac{\partial}{\partial T}\frac{\mu_{2,G}(T,P^\dagger)}{T}=-\frac{H_{2,G}(T)}{T^2}. \tag{12·45·3}$$

If the solution is dilute, we have

$$\mu_{2,L}(T,P,n_i)=\mu^*_{2,L}(T,P)+RT\log x_2, \tag{12·45·4}$$

and, on inserting (12·45·2) and (12·45·4) into (12·45·1), we see that

$$P=hx_2, \tag{12·45·5}$$

where

$$\log h=\log P^\dagger+\frac{\mu^*_{2,L}(T,P)-\mu_{2,G}(T,P^\dagger)}{RT}=\frac{\mu^*_{2,L}(T,P)-\mu^*_{2,G}(T)}{RT}. \tag{12·45·6}$$

This is Henry's law, with an explicit expression for the proportionality factor h, which is practically independent of P. Since the quantity $\mu_{2,G}(T,P^\dagger)$ (or $\mu^*_{2,G}(T)$) depends explicitly on the entropy constant of the gas, h cannot be determined from calorimetric measurements alone. The temperature coefficient of h, however, does not depend upon the entropy constant, and, since

$$\frac{\partial}{\partial T}\frac{\mu^*_{2,L}}{T}=\frac{\partial}{\partial T}\frac{\mu_{2,L}}{T}=-\frac{1}{T^2}\frac{\partial H_{2,L}}{\partial n_2},$$

it is given by $$RT^2\,\partial \log h/\partial T = \mathrm{H}_{2,G} - \partial H_{2,L}/\partial n_2. \qquad (12\cdot45\cdot7)$$

The quantity on the right is the differential heat of evaporation of the solute.

12·46. The formulae derived in the preceding section may be used in a variety of ways. In the first place, since all the effects depend upon the molar concentrations of the solutes, we may use them to determine the molecular weights of the solutes. For example, if we have a dilute binary solution in which the molecular weight of the solute is M_2^*, and if the mass in grams of the solvent is m_1 while that of the solute is m_2, equation (12·41·4) can be written as

$$\frac{T_0-T}{m_2/m_1}=\frac{RT_0^2}{M_2^*L^*}, \qquad (12\cdot46\cdot1)$$

where L^* is the latent heat of fusion of the solvent per gram. Thus M_2^* can be determined by measuring the depression in the freezing-point of a dilute solution.

Secondly, the thermodynamic potential $\mu_{1,L}$ of the solvent in the solution is equal to the value of $\mu_{1,L}$ in any pure phase in equilibrium with the solution, and we can therefore find the value of $\mu_{1,L}$ in the solution from a knowledge of the thermodynamic properties of the pure solvent. As already described in § 12·411, we can, for example, use equation (12·41·1) to determine $\mu_{1,L}(T,P,n_i)$ at the freezing-point $T(n_i)$ of the solution in terms of $\mu_{1,S}(T,P)$. The value of $\mu_{1,L}(T,P,n_i)$ at any other temperature (or pressure) can then be found from equation (12·2·3) (or equation (12·2·4)) which only involves measurements at constant composition. Any of the phenomena described may be used to determine $\mu_{1\,L}$ in this way, but in general the determination of the freezing-point of the solution is the most reliable method for dilute solutions, while for concentrated solutions of a non-volatile solute it is simpler to measure the vapour pressure of the solution. Correspondingly, if the solute is volatile its thermodynamic potential can be found from its partial pressure in the vapour phase.

Finally, if the thermodynamic potential of one constituent of a binary solution has been determined by one of the above methods, that of the second constituent can be found by integrating equation (12·21·1). On account of the logarithmic singularities in the thermodynamic potentials, the integration is most easily effected by writing

$$\mu_1=\mu_1^0(T,P)+RT\log\{f_1(1-x_2)\},\quad \mu_2=\mu_2^*(T,P)+RT\log(f_2x_2), \qquad (12\cdot46\cdot2)$$

where f_1 and f_2 tend to unity as x_2 tends to zero. The general relation (12·21·6) then becomes

$$x_2\frac{\partial \log f_2}{\partial x_2}=-(1-x_2)\frac{\partial \log f_1}{\partial x_2}, \qquad (12\cdot46\cdot3)$$

of which the integrated form is

$$\log f_2 = -\int_0^{x_2} \frac{1-x_2}{x_2}\frac{\partial \log f_1}{\partial x_2}\,dx_2. \qquad (12\cdot46\cdot4)$$

This formula may be illustrated by considering the case in which $\log f_1$ can be written in the form

$$\log f_1 = -\frac{\alpha}{\{1+\beta(1-x_2)/x_2\}^2}, \qquad (12\cdot46\cdot5)$$

an expression which fits a number of experimental results. If we put $t=(1-x_2)/x_2$, then (12·46·4) becomes

$$\log f_2 = -\int_\infty^t t\frac{\partial \log f_1}{\partial t}\,dt = -t\log f_1 + \int_\infty^t \log f_1\,dt$$

$$= \frac{\alpha(1+2\beta t)}{\beta(1+\beta t)^2} = \frac{\alpha}{\beta}\frac{1+2\beta(1-x_2)/x_2}{\{1+\beta(1-x_2)/x^2\}^2}. \qquad (12\cdot46\cdot6)$$

Heterogeneous Equilibrium when both Phases are Mixtures

12·5. If two phases are in contact with one another and if neither of them are pure substances, there is an equilibrium condition for each component, and the problem is much more complex than that treated in § 12·4. We therefore restrict the discussion to binary solutions.

If we have two phases in contact, which for definiteness we take as the liquid and gaseous phases, there are four parameters defining the state which can be taken as T, P and the compositions of the two phases. There are two equilibrium conditions to be satisfied, and the system has therefore two degrees of freedom. Thus, for a given value of the pressure, the temperature defines the composition of both the liquid and the gaseous phases.

The equilibrium conditions are

$$\mu_{1,L} = \mu_{1,G} = \mu_1, \quad \mu_{2,L} = \mu_{2,G} = \mu_2, \qquad (12\cdot5\cdot1)$$

which can be written in differential form as follows. The $P(T,\mu_i)$ equations for the two phases are

$$\left.\begin{aligned} S_L dT - V_L dP + n_{1,L} d\mu_1 + n_{2,L} d\mu_2 = 0,\\ S_G dT - V_G dP + n_{1,G} d\mu_1 + n_{2,G} d\mu_2 = 0,\end{aligned}\right\} \qquad (12\cdot5\cdot2)$$

and so $$\left(\frac{S_G}{n_{1,G}} - \frac{S_L}{n_{1,L}}\right)dT - \left(\frac{V_G}{n_{1,G}} - \frac{V_L}{n_{1,L}}\right)dP + \left(\frac{n_{2,G}}{n_{1,G}} - \frac{n_{2,L}}{n_{1,L}}\right)d\mu_2 = 0. \qquad (12\cdot5\cdot3)$$

Therefore, if P is constant, T, considered as a function of the composition of either phase, is a maximum or a minimum when $x_{2,G} = x_{2,L}$, i.e. when the compositions of the two phases are identical. Conversely, when the compositions of the two phases are identical, T must be a maximum or a minimum. The corresponding statements hold for P if T is constant. These results are known as the Gibbs-Konowalow rules, and they enable us to classify the types of equilibrium diagram that are to be expected.

12·51. *Equilibrium diagrams.* When we deal with a solution from which a pure solid phase separates out on cooling, there are only two possibilities. If the solution is dilute the solvent separates out on cooling and the solution becomes more concentrated. As the temperature is lowered, more and more solvent separates out until the solution solidifies as a whole and forms a mixture of the separate crystals of the two components. This mixture is

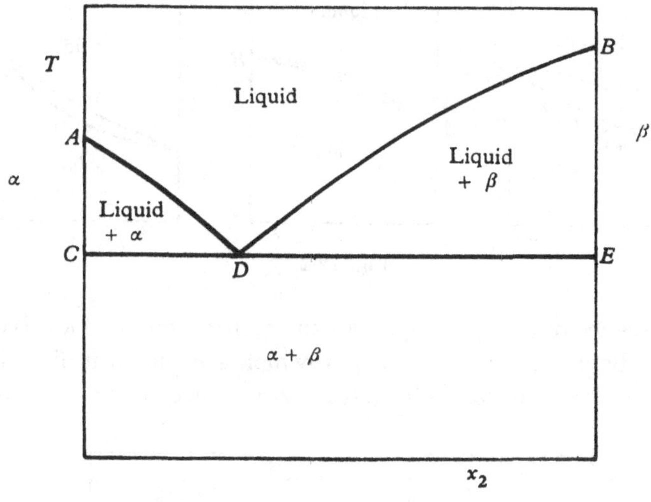

Fig. 12·1

called a eutectic mixture, and the temperature of solidification is called the eutectic temperature. If, on the other hand, the solution is a concentrated one, it is the solute that will separate out when the solution is cooled, and the solution will become less and less concentrated as the temperature falls, until the eutectic temperature is reached when the solution solidifies as a whole. This behaviour is illustrated in fig. 12·1, which is the equilibrium diagram for the liquid and solid phases of two substances α and β which are miscible in all proportions in the liquid phase, but which do not mix at all in the solid phase. The points A and B are the melting-points of the two pure components, while AD and BD are the temperature-concentration curves for the equilibrium of the liquid in contact with pure solid α and pure solid β respectively. These curves meet at D, which is the eutectic point defined by

$$\mu_{\alpha,L}(T,x)=\mu_{\alpha,S}(T), \quad \mu_{\beta,L}(T,x)=\mu_{\beta,S}(T).$$

(Since we are dealing with liquid and solid phases, the pressure may be omitted.) Any point in the diagram above the curves AD and BD refers to a homogeneous liquid state. A point in the region ACD (or BED) refers to a heterogeneous mixture of liquid plus pure solid α (or β), the concentration of the liquid being determined by the point on AD (or BD) with the same value of the temperature, while a point in the region below CDE refers to a heterogeneous mixture of pure crystals of α and pure crystals of β.

12·511. If the components in a binary solution are miscible in all proportions in the solid and in the liquid phases, it is no longer necessarily true

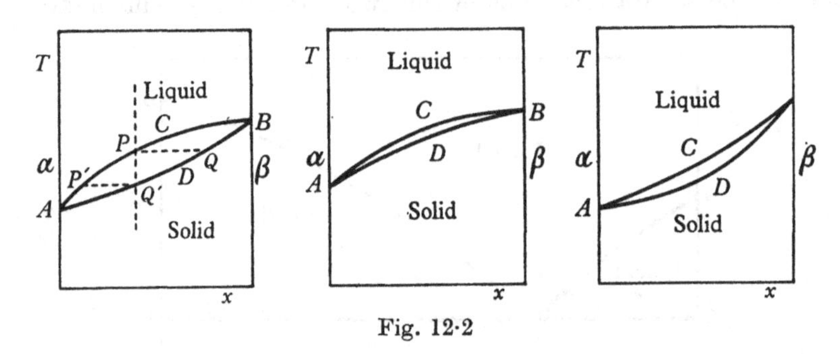

Fig. 12·2

that the presence of a solute depresses the melting-point of a solvent. The equilibrium diagrams are of five types, which are shown in figs. 12·2 and 12·3. There are two curves ACB and ADB connecting the melting-points

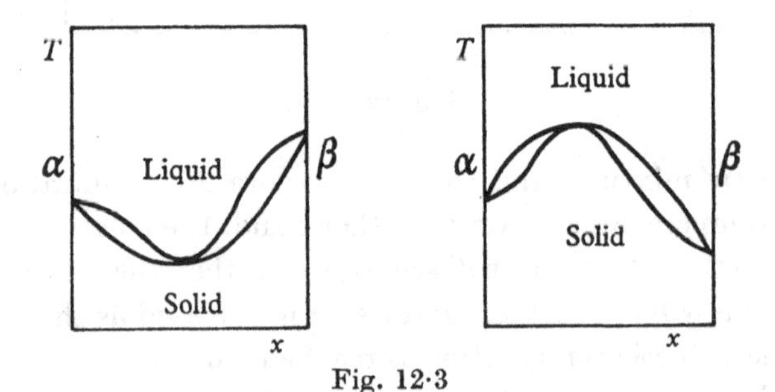

Fig. 12·3

of the two pure solids. Above ACB and below ADB the solutions are homogeneous (being liquid and solid respectively) while the region inside $ADBC$ refers to heterogeneous mixtures, and the intersections of a line of constant temperature with ACB and ADB define the concentrations of the liquid and solid phases which are in equilibrium with one another at that temperature. The curve ACB is usually known as the liquidus curve, while ADB is known as the solidus curve.

The equilibrium diagrams shown in fig. 12·2 occur when the two components have very similar molecules, examples of the three types of diagram being given by the nickel-copper, silver-gold and nickel-cobalt alloys respectively. If a solution of composition x_L is cooled until it reaches the point P on the liquidus curve, a solid of composition x_S (corresponding to the point Q) separates out. If the solution is further cooled and kept in contact with the solid, the concentrations of the solute in both the liquid and the solid diminish (the total amount of solid present increasing and the total amount of liquid present diminishing) until the point P' on the liquidus curve and the point Q' on the solidus curve are reached, when the heterogeneous mixture consists of a solid solution of composition $x'_S = x_L$ in contact with a vanishingly small quantity of liquid of compositon x'_L corresponding to the point P'. On further cooling, the mass becomes a homogeneous solid solution of composition $x'_S = x_L$. The cooling may, however, be carried out in a different fashion, by removing the solid at every stage as soon as it is formed. In this case we have to consider the equilibrium of the liquid in contact with a vanishingly small amount of solid. The solids removed will then have concentrations varying continuously from x_S to $x = 0$.

If both solutions are perfect, the equilibrium conditions are

$$\mu^0_{1,S}(T, P) + RT \log(1 - x_{2,S}) = \mu^0_{1,L}(T, P) + RT \log(1 - x_{2,L}), \quad (12·511·1)$$

$$\mu^0_{2,S}(T, P) + RT \log x_{2,S} = \mu^0_{2,L}(T, P) + RT \log x_{2,L}, \quad (12·511·2)$$

of which the solutions are

$$\left.\begin{aligned} x_{2,L} &= \frac{\exp\{(\mu^0_{1,L} - \mu^0_{1,S})/RT\} - 1}{\exp\{(\mu^0_{1,L} - \mu^0_{1,S})/RT\} - \exp\{(\mu^0_{2,L} - \mu^0_{2,S})/RT\}}, \\ x_{2,S} &= x_{2,L} \exp\{(\mu^0_{2,L} - \mu^0_{2,S})/RT\}. \end{aligned}\right\} \quad (12·511·3)$$

If the melting-points of the two substances are T_1 and T_2 ($T_1 < T_2$), and if the temperature range is sufficiently small for the μ's to be considered as linear functions of T, we can write

$$(\mu^0_{1,L} - \mu^0_{1,S})/RT = -\alpha(T - T_1)/T, \quad (\mu^0_{2,L} - \mu^0_{2,S})/RT = -\beta(T - T_2)/T, \quad (12·511·4)$$

and the equations (12·511·3) become

$$\left.\begin{aligned} x_{2,L} &= \frac{\exp[-\alpha(T - T_1)/T] - 1}{\exp[-\alpha(T - T_1)/T] - \exp[-\beta(T - T_2)/T]}, \\ x_{2,S} &= \frac{1 - \exp[\alpha(T - T_1)/T]}{\exp[\beta(T - T_2)/T] - \exp[\alpha(T - T_1)/T]}. \end{aligned}\right\} \quad (12·511·5)$$

The liquidus and solidus curves may have any of the forms shown in fig. 12·2, depending upon the relative values of the four constants α, β, T_1, T_2.

Equilibrium diagrams of the form shown in fig. 12·3 are of frequent occurrence. If the temperature-concentration curves have a maximum or a minimum, it is necessary, as shown in § 12·5, for the liquid and the solid to have the same composition at the maximum or minimum, and hence the liquidus and solidus curves must touch there. This property of the liquidus and solidus curves always touching at a maximum or a minimum shows that the gradients of the two curves at any given concentration must have the same sign. Typical examples of temperature-concentration curves which have a maximum or a minimum are provided by the acetone-chloroform system and the manganese-cobalt system respectively.

12·512. All the general results concerning the equilibrium between solid and liquid solutions apply equally to the equilibrium between liquid solutions and their vapours. Corresponding to the liquidus and solidus

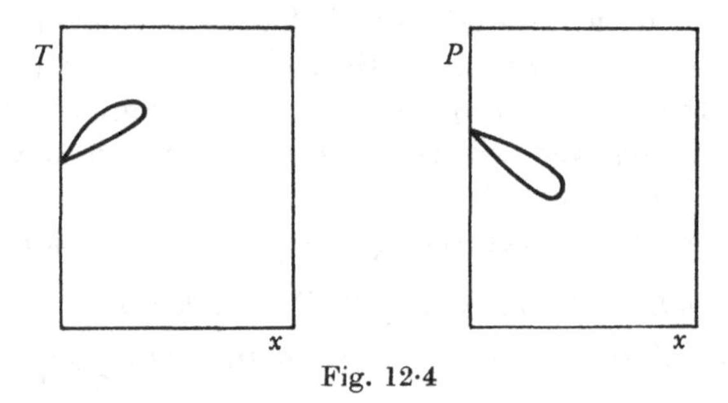

Fig. 12·4

curves we have condensation and vaporization curves, the only essential difference being that if vapours are present it is necessary to specify the pressure and to consider separate equilibrium diagrams according as the pressure or the temperature is kept constant (or to consider a three dimensional T, P, x diagram). The T, x equilibrium diagrams (i.e. with P constant) are similar in general form to those shown in figs. 12·2 and 12·3. The P, x equilibrium diagrams (i.e. with T constant) are also similar in appearance with the exception that, if a $T(x)$ curve has a positive gradient, the corresponding $P(x)$ curve will have a negative gradient since the liquid with the lower boiling-point will have the higher vapour pressure. It is, however, possible for the T, x and P, x equilibrium diagrams to have the forms shown in fig. 12·4. This occurs in the P, x diagram when the temperature lies between the critical temperatures of the two constituents. Similarly, the phenomenon must occur in the T, x diagram when the pressure lies between the critical pressures of the constituents.

The behaviour of a binary mixture in the critical region can perhaps best be illustrated by drawing in the P, T diagram the family of vaporization and condensation curves for various values of the composition. In fig. 12·5 the curves 1 and 6 are the liquid-vapour curves of the pure components, ending at the critical points C_α and C_β respectively. For a mixture with a fixed composition, the liquid-vapour curve forms a loop, one portion of which is the vapour branch and the other is the liquid branch, the two branches joining at the critical point relating to the given composition. If we consider all possible compositions, the loops fill in a region bounded by the curves 1 and 6 and a curve $C_\alpha C_\beta$, which is the envelope of the loops and is the locus of the critical points. The intersection of any vapour curve and any liquid curve corresponds to two phases which have different compositions but the same temperature and pressure. The phases can therefore exist together in equilibrium. It should be noted that the critical temperature of a gas mixture of a given composition, unlike that of a single gas, is not necessarily the maximum temperature on the liquid-vapour curve, nor is the critical pressure the maximum pressure on the same curve.

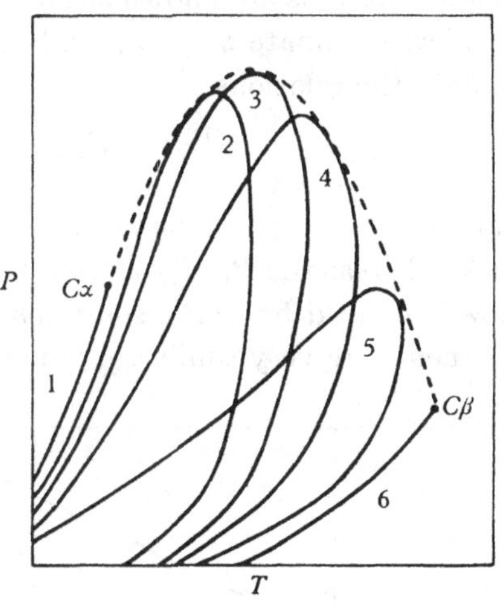

Fig. 12·5

12·513. *Raoult's law.* The conditions for the equilibrium of a perfect solution and its vapour (the vapour being treated as a perfect gas mixture) are

$$\mu^0_{1,L}(T,P)+RT\log x_{1,L}=\mu_{1,G}(T,P^\dagger)+RT\log(x_{1,G}P/P^\dagger), \quad (12·513·1)$$

$$\mu^0_{2,L}(T,P)+RT\log x_{2,L}=\mu_{2,G}(T,P^\dagger)+RT\log(x_{2,G}P/P^\dagger), \quad (12·513·2)$$

where, as usual, $P^\dagger$ is a standard pressure. (Alternatively, we may write the expressions for the thermodynamic potentials in the gas phase in the form $\mu^*_{i,G}(T)+RT\log(x_{i,G}P)$ and avoid the appearance of the standard pressure $P^\dagger$.) Also, if $P^0_1(T)$ and $P^0_2(T)$ are the natural vapour pressures of the pure liquids, we have

$$\mu^0_{1,L}(T,P^0_1)=\mu_{1,G}(T,P^\dagger)+RT\log(P^0_1/P^\dagger), \quad (12·513·3)$$

$$\mu^0_{2,L}(T,P^0_2)=\mu_{2,G}(T,P^\dagger)+RT\log(P^0_2/P^\dagger). \quad (12·513·4)$$

Hence, if we neglect the effect of pressure on the liquid phases,

$$x_{1,G}=x_{1,L}P^0_1/P, \quad x_{2,G}=x_{2,L}P^0_2/P. \quad (12·513·5)$$

These relations are more usually expressed in terms of the partial pressures $p_1 = x_{1,G}P$, $p_2 = x_{2,G}P$, and they can then be written as

$$p_1 = x_{1,L}P_1^0, \quad p_2 = x_{2,L}P_2^0. \qquad (12{\cdot}513{\cdot}6)$$

These relations are known as Raoult's law.

If we eliminate $x_{1,G}$, $x_{2,G}$ and $x_{1,L}$, $x_{2,L}$ respectively from (12·513·5), we obtain the relations

$$P = x_{1,L}P_1^0 + x_{2,L}P_2^0, \quad \frac{1}{P} = \frac{x_{1,G}}{P_1^0} + \frac{x_{2,G}}{P_2^0}. \qquad (12{\cdot}513{\cdot}7)$$

The P, $x_{2,L}$ curve is therefore a straight line in accordance with Raoult's law, whereas the P, $x_{2,G}$ curve is the hyperbola shown in fig. 12·6. Raoult's law is obeyed by many solutions in which the molecules of the two constituents are very similar. A typical example is shown in fig. 12·7.

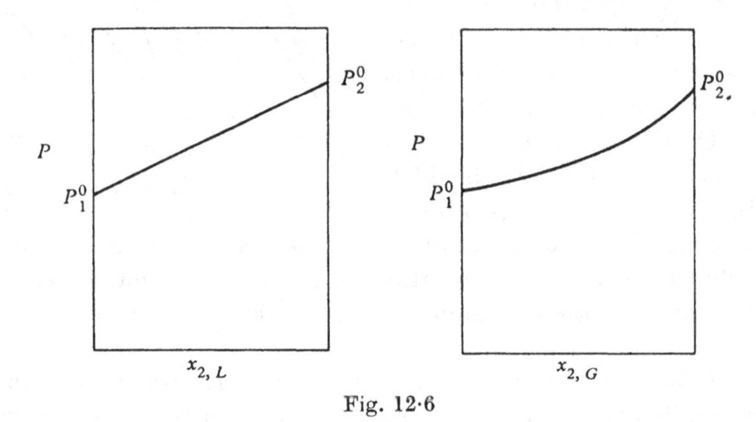

Fig. 12·6

If we consider a dilute solution in which the solvent is denoted by the suffix 1, the equations for the solvent are the same as those given above, while we must replace equation (12·513·2) for the solute by the relation

$$\mu_{2,L}^*(T,P) + RT\log x_{2,L} = \mu_{2,G}(T,P^\dagger) + RT\log(x_{2,G}P/P^\dagger). \qquad (12{\cdot}513{\cdot}8)$$

We therefore still have the relations

$$x_{1,G} = x_{1,L}P_1^0/P, \quad p_1 = x_{1,L}P_1^0, \qquad (12{\cdot}513{\cdot}9)$$

while the equations for the solute are

$$x_{2,G} = hx_{2,L}P_2^0/P, \quad p_2 = hx_{2,L}P_2^0, \qquad (12{\cdot}513{\cdot}10)$$

where

$$h = \exp\{(\mu_{2,L}^* - \mu_{2,L}^0)/RT\}. \qquad (12{\cdot}513{\cdot}11)$$

This is Henry's law, and in general $h \neq 1$. A typical example of Henry's law, with the deviations from it in concentrated solutions, is shown in fig. 12·8, which relates to acetic acid and toluene mixtures. The gradient of the vapour-pressure curve for toluene is reasonably constant up to $x_{2,L}=0{\cdot}2$, but it has a greater value than that given by Raoult's law. (If Raoult's law held, the vapour-pressure curves would be straight lines connecting the end-points of the curves for $x_{2,L}=0$ and $x_{2,L}=1$.) For $x_{2,L}>0{\cdot}2$ the gradient of the toluene vapour-pressure curve steadily diminishes.

12·514. If there is incomplete mixing in one or more of the phases, which are then necessarily liquid or solid phases, the equilibrium diagrams become more complicated. If we consider a single (liquid or solid) phase,

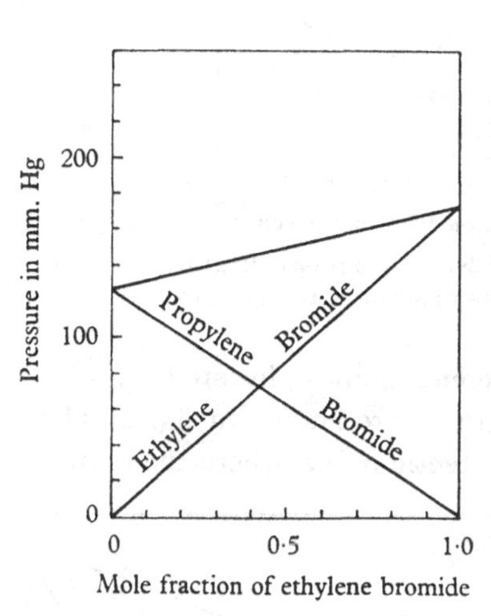

Fig. 12·7. The partial and total vapour pressures of ethylene bromide and propylene bromide at 85° C. as functions of the mole fraction in the liquid phase.

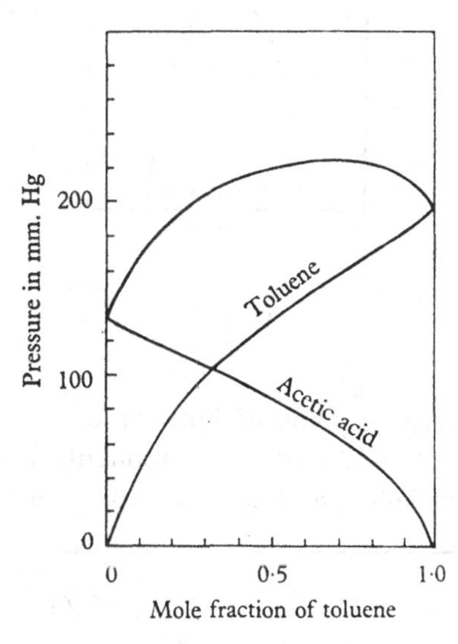

Fig. 12·8. The partial and total vapour pressures of toluene and acetic acid at 70° C. as functions of the mole fraction in the liquid phase.

then if there is incomplete mixing, the equilibrium diagram will have the general form shown in fig. 12·9. At high temperatures the components will be miscible in all proportions, but there is a critical temperature below which, for a certain range of concentrations, the solution splits up into two separate (liquid or solid) phases. An example is provided by the water-phenol system. Similarly, it is possible for there to be a lower critical temperature. For example, water and diethylamine form homogeneous mixtures below 143° C. and heterogeneous mixtures above. Further, it sometimes happens that there is a lower critical temperature as well, so that the region of heterogeneous phases is bounded by a closed curve in the T, x diagram, a well-known example being provided by the nicotine-water system (fig. 12·10).

If next we consider two phases which we may take to be liquid and solid, in the less stable of which (the liquid) there is complete mixing, while in the more stable phase (the solid) there is only limited solubility of each component in the other, the equilibrium diagram is of two types according as

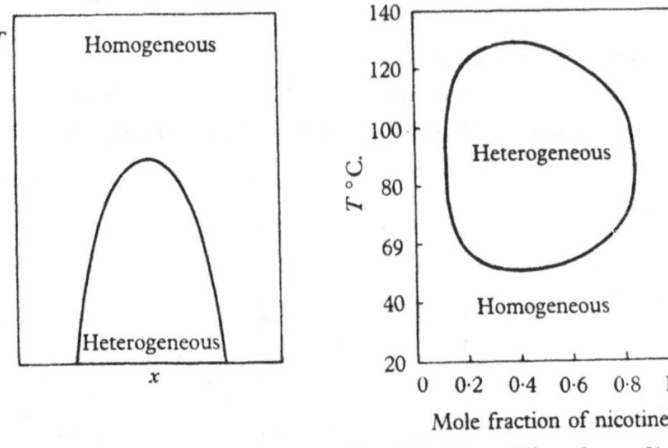

Fig. 12·9

Fig. 12·10. The phase diagram for the nicotine-water system.

the freezing-point of both or of only one component is lowered by the presence of the other component. The diagrams are shown in fig. 12·11, and in each case there is a triple point, the eutectic point, where the liquid

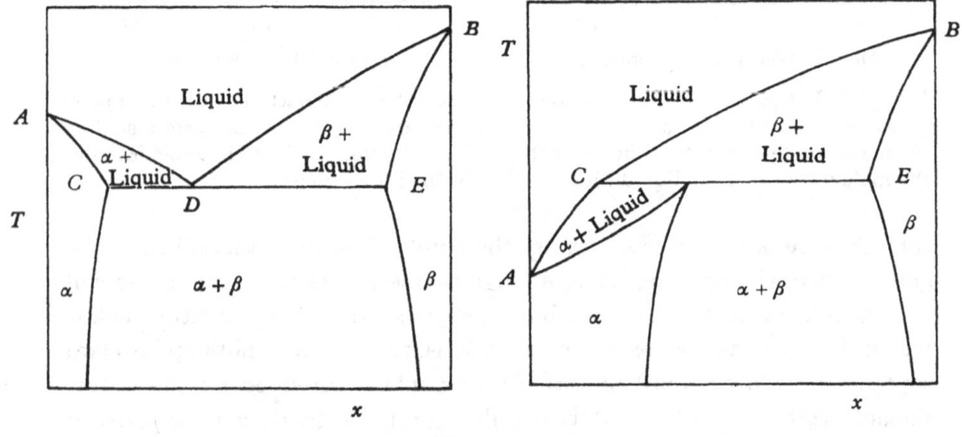

Fig. 12·11

with a composition corresponding to the point D is in equilibrium with two solid solutions with concentrations corresponding to the points C and E. An example of the first type is provided by the lead-tin alloys, and of the second type by the magnesium-cadmium alloys. The extreme case of this type occurs when the components are completely immiscible in the solid phase, a situation which we have already discussed in § 12·51.

12·515. In many alloys the components are immiscible in the solid phase or nearly so, but form a chemical compound $\alpha_m\beta_n$. In this case the equilibrium diagram can be considered to consist of two separate parts, the equilibrium diagram for phases consisting of homogeneous or heterogeneous mixtures of α and $\alpha_m\beta_n$, and the equilibrium diagram for phases relating to mixtures of $\alpha_m\beta_n$ and β. The equilibrium diagram therefore consists of two diagrams of the type illustrated in fig. 12·1, and there are two triple or

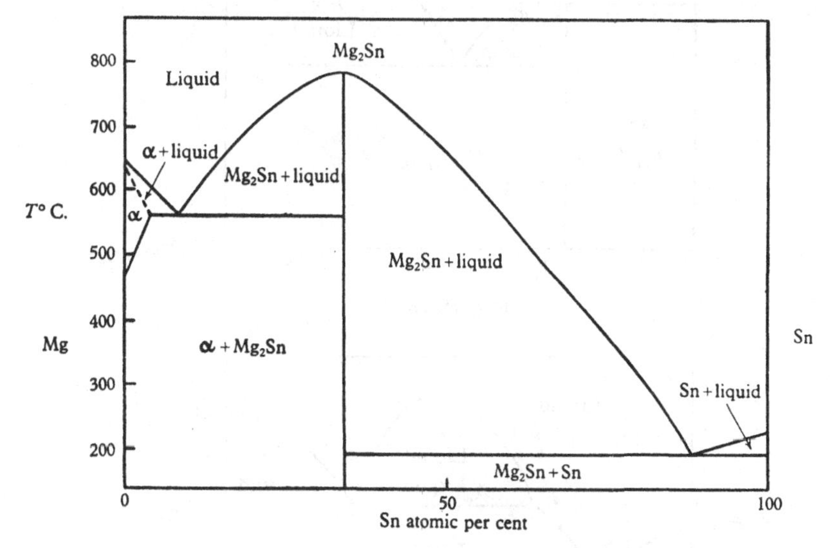

Fig. 12·12. The phase diagram of the magnesium-tin alloys.

eutectic points. The equilibrium diagram of the magnesium-tin alloys, which is of this type, is shown in fig. 12·12. In some alloys the compound is not stable up to the melting-point, and in this case the equilibrium diagram is as shown in fig. 12·13.

If two substances are completely immiscible in the solid phase, it frequently happens that they are only partly miscible in the liquid phase. The equilibrium diagram is then of the type shown in fig. 12·14. In addition to the usual features there is a region *FGI* which corresponds to a heterogeneous mixture of two liquids with different compositions, and there is a critical point *I*. There are also two triple points *D* and *G*. At *D* the liquid phase is in equilibrium with the two pure solids, while at *G* the solid β is in equilibrium with the liquids whose compositions correspond to the points *F* and *G*.

NON-IDEAL SOLUTIONS

12·6. In determining the thermodynamic potentials experimentally from freezing-point, vapour pressure or osmotic data, it is customary to

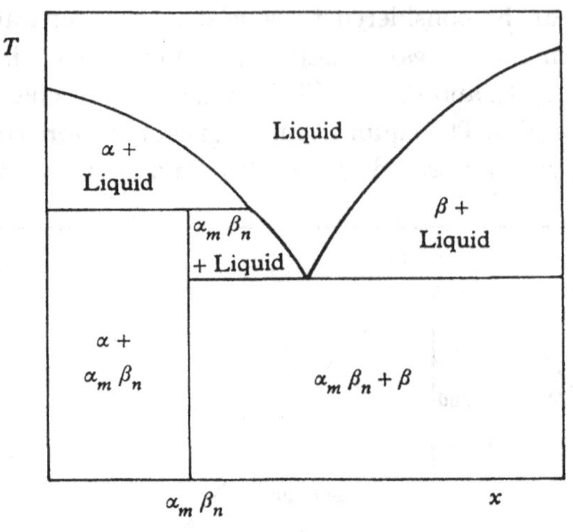

Fig. 12·13

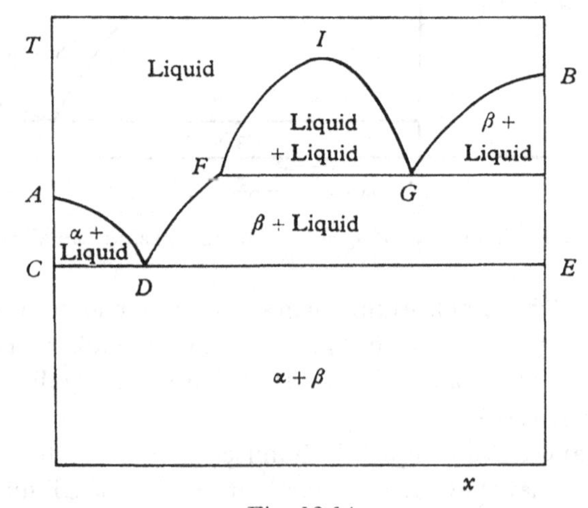

Fig. 12·14

express the departure from ideality of a solution in one of three ways. First, we may write

$$\mu_i(T, P, n_j) = \mu_i^0(T, P) + RT \log (f_i x_i), \tag{12·6·1}$$

which is most appropriate when we are interested in the vapour pressure of a solution over the whole range of concentrations. If the vapour phase can be treated as a perfect gas mixture, the argument given in § 12·513 shows that the partial pressure of the ith component is given by

$$p_i = f_i x_{i,L} P_i^0, \tag{12·6·2}$$

which replaces Raoult's law. If, however, the thermodynamic potential of the solvent is determined by measurements of the freezing-point of the osmotic pressure, it is simpler to write

$$\mu_i(T, P, n_j) = \mu_i^0(T, P) + g_i RT \log x_i, \tag{12·6·3}$$

since (12·41·3), (12·41·4) and (12·43·3) are then replaced by

$$g_1 \log(1 - \sum_{i \geq 2} x_i) = \int_{T_0}^{T} \{H_{1,L}(T', P_0) - H_{1,S}(T', P_0)\} \frac{dT'}{RT'^2}, \tag{12·6·4}$$

by

$$T_0 - T = RT_0^2 g_1 \sum_{i \geq 2} x_i / L \quad (x_i \ll 1), \tag{12·6·5}$$

and by

$$\int_P^{P+\Pi} V_{1,L}(T, P')\, dP' = -g_1 RT \log(1 - \sum_{i \geq 2} x_i) \tag{12·6·6}$$

respectively. The quantity g_1 is usually called the osmotic coefficient. Thirdly, if we are only concerned with relatively dilute solutions which are such that Raoult's law does not apply to the solutes, it is preferable to employ either (12·6·1) or (12·6·3) for the solvent, i.e. for $i = 1$, but to replace (12·6·1) for $i \geq 2$ by

$$\mu_i = \mu_i^*(T, P) + RT \log(\gamma_i x_i), \quad \lim_{x_i \to 0} \gamma_i = 1, \tag{12·6·7}$$

since in this form the departure from Henry's law is most apparent.

It will be noted that

$$\log f_i = (g_i - 1) \log x_i, \tag{12·6·8}$$

but in general γ_i can only be determined from f_i or g_i if $\mu_i^*(T, P)$ is known. For a binary mixture, however, the relation

$$(1 - x_2) \frac{\partial \mu_1}{\partial x_2} + x_2 \frac{\partial \mu_2}{\partial x_2} = 0 \tag{12·6·9}$$

gives

$$(1 - x_2) \left(\frac{\partial g_1}{\partial x_2} \log(1 - x_2) - \frac{g_1}{1 - x_2} \right) + x_2 \left(\frac{1}{\gamma_2} \frac{\partial \gamma_2}{\partial x_2} + \frac{1}{x_2} \right) = 0, \tag{12·6·10}$$

which integrates to

$$\log \gamma_2 = -\int_0^{x_2} \frac{1 - x_2}{x_2} \log(1 - x_2) \frac{\partial g_1}{\partial x_2}\, dx_2 - \int_0^{x_2} \frac{1 - g_1}{x_2}\, dx_2. \tag{12·6·11}$$

In order to obtain γ_2 from the formula it is necessary to know how $g_1 - 1$ behaves for small values of x_2. If we assume

$$g_1 = 1 - a x_2^r + O(x_2^{2r}) \quad (0 < r \leq 1), \tag{12·6·12}$$

then

$$\log \gamma_2 = -(1 + 1/r)\, a x_2^r + O(x_2^{2r}). \tag{12·6·13}$$

For ordinary solutes, $r = 1$ and

$$\log \gamma_2 = -2(1 - g_1) + O(x_2^2) \quad (r = 1). \tag{12·6·14}$$

For electrolytes, on the other hand, $r = \frac{1}{2}$ (see § 13.21) and

$$\log \gamma_2 = -3(1 - g_1) + O(x_2) \quad (r = \tfrac{1}{2}). \tag{12·6·15}$$

12·61. If the vapour phase in equilibrium with a solution can be treated as a perfect gas mixture, we have

$$\mu_{1,L} = \mu_{1,G}(T, P^\dagger) + RT \log (p_1/P^\dagger), \quad \mu_{2,L} = \mu_{2,G}(T, P^\dagger) + RT \log (p_2/P^\dagger). \tag{12·61·1}$$

where p_1 and p_2 are the partial pressures of the two components. We may therefore write the general equation

$$x_{1,L} \frac{\partial \mu_{1,L}}{\partial x_{2,L}} + x_{2,L} \frac{\partial \mu_{2,L}}{\partial x_{2,L}} = 0 \tag{12·61·2}$$

in the form
$$(1 - x_{2,L}) \frac{\partial \log p_1}{\partial x_{2,L}} + x_{2,L} \frac{\partial \log p_2}{\partial x_{2,L}} = 0, \tag{12·61·3}$$

which is known as the Duhem-Margules equation. If we know the total vapour pressure $P = p_1 + p_2$ as a function of $x_{2,L}$, equation (12·61·3) can be written in the form

$$\frac{\partial p_1}{\partial x_{2,L}} = \frac{x_{2,L} p_1}{p_1 - (1 - x_{2,L}) P} \frac{\partial P}{\partial x_{2,L}}, \tag{12·61·4}$$

which can be integrated numerically to give p_1, and therefore p_2, as a function of $x_{2,L}$.

Some examples of the behaviour of the vapour pressures of non-ideal solutions are given in fig. 12·15.

12·62. *Regular solutions.* The formulae given in the preceding section are of no theoretical importance and are merely convenient ways of expressing the experimental results. If for any solution we have derived a theoretical formula for the free energy G, all the properties of the solution can be obtained directly from G.

An obvious extension of the theory is to solutions which have non-zero heats of mixing but for which the general expression $-R\Sigma x_i \log x_i$ for the entropy of mixing still holds. The free energy can then only differ from the free energy G_{ideal} of a perfect solution by an expression which is a homogeneous function of the first degree in the n_i's and which is independent of T and P. Such solutions are known as regular solutions (Hildebrand, 1929, 1936), and we can expect solutions in which the molecules have similar shapes, but for which the interaction energies are appreciable, to behave approximately as regular solutions.

Binary regular solutions are characterized by the free-energy function

$$G(T, P, n_i) = G(T, P, n_i)_{\text{ideal}} + wn_1n_2/(n_1+n_2), \qquad (12\cdot62\cdot1)$$

where w is independent of T, P and of the composition. Hence

$$\mu_1 = \mu_1^0(T, P) + RT\log x_1 + wx_2^2, \quad \mu_2 = \mu_2^0(T, P) + RT\log x_2 + wx_1^2. \qquad (12\cdot62\cdot2)$$

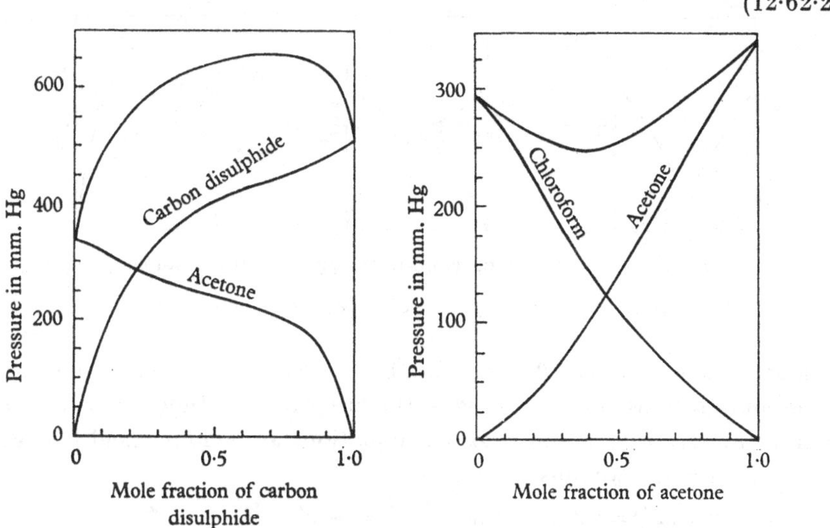

Fig. 12·15. The partial and total pressures of the systems acetone-carbon bisulphide and acetone-chloroform at 35° C. as functions of the mole fraction in the liquid phase.

If the gas phase is treated as perfect, we see by comparison with (12·313·1) and (12·513·2) that Raoult's law is replaced by

$$p_1 = x_{1,L}\exp(wx_{2,L}^2/RT)\,P_1^0, \quad p_2 = x_{2,L}\exp(wx_{1,L}^2/RT)\,P_2^0, \qquad (12\cdot62\cdot3)$$

and that Henry's law is explicitly

$$p_1 = x_{1,L}P_1^0, \quad p_2 = x_{2,L}\,e^{w/RT}\,P_2^0 \quad (x_{2,L} \ll 1), \qquad (12\cdot62\cdot4)$$

$$p_2 = x_{2,L}P_2^0, \quad p_1 = x_{1,L}\,e^{w/RT}\,P_1^0 \quad (x_{1,L} \ll 1). \qquad (12\cdot62\cdot5)$$

The formulae are, of course, symmetrical in the species 1 and 2, but if we wish to consider the liquid 1 as the solvent, the following are the expressions for g_1, γ_2 and $\mu_2^*(P, T)$:

$$\left.\begin{aligned} g_1 &= 1 + \frac{w(1-x_1^2)}{RT\log x_1}, \\ \mu_2^*(T, P) &= \mu_2^0(T, P) + w, \quad \log\gamma_2 = -2wx_2(2-x_2)/RT. \end{aligned}\right\} \qquad (12\cdot62\cdot6)$$

If $w < 0$ the vapour-pressure concentration curves lie below those given by Raoult's law, whereas if $w > 0$ the curves lie above those for an ideal solution (see fig. 12·16). In the latter case, moreover, the behaviour can be drastically different from that of a perfect solution since, if T is sufficiently small, incomplete mixing can take place and there is a two-phase region for certain concentration ranges.

If μ_1 as a function of x_1 has a maximum and a minimum, the $\mu_1(x_1)$ curve has three intersections with the line $\mu_1 = \mu_1^*$ if μ_1^* lies between the maximum and minimum values of μ_1. The middle intersection cannot correspond to

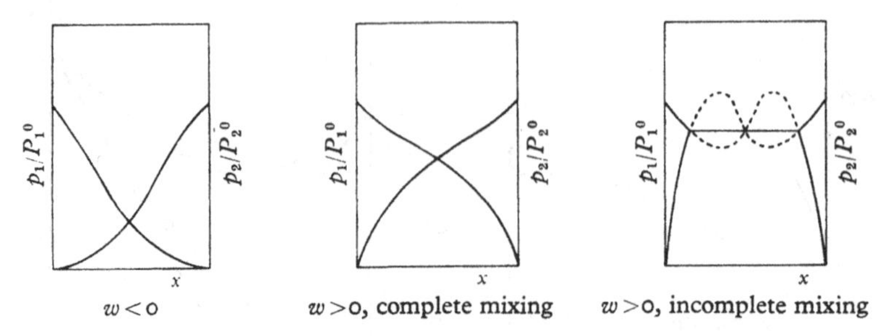

Fig. 12·16. The vapour-pressure curves for regular mixtures.

a real state since $\partial\mu_1/\partial x_1 < 0$ there, and, by (12·2·7), such a state is unstable. If the compositions corresponding to the other intersections are x_1' and x_1'', we have $\mu_1(x_1') = \mu_1(x_1'')$, and since the equations are symmetrical between the species 1 and 2 we must also have

$$\mu_2(x_2') = \mu_2(x_2''), \quad \text{where} \quad x_2' = 1 - x_1', \quad x_2'' = 1 - x_1''.$$

By symmetry we have in addition $x_1' = x_2''$ and $x_1'' = x_2'$, so that

$$x_1' + x_1'' = x_2' + x_2'' = 1.$$

There are therefore two coexistent liquid phases with compositions x_1', x_2' and x_1'', x_2''.

The two-phase region occurs when the $\mu_1(x_1)$ curve has a real maximum and a real minimum, which occur at the composition

$$x_1 = \tfrac{1}{2} \pm \sqrt{(\tfrac{1}{4} \pm \tfrac{1}{2}RT/w)}.$$

These turning points are real if $RT < \frac{1}{2}w$, and the minus sign refers to the maximum and the plus sign to the minimum of μ_1. Hence, if $RT < \frac{1}{2}w$, μ_1 (and μ_2) has a real maximum and minimum, and incomplete mixing occurs. If $RT = \frac{1}{2}w$, the maximum and the minimum of μ_1 coincide at $x_1 = \frac{1}{2}$, which corresponds to a critical point, while, if $RT > \frac{1}{2}w$, mixing is always complete. The vapour-pressure curves are shown in fig. 12·16, including the unstable portions of the isotherms when mixing is incomplete.

REFERENCES

Hildebrand, J. H. (1929). Regular solutions. *J. Amer. Chem. Soc.* **51**, 66.
Hildebrand, J. H. (1936). *The solubility of non-electrolytes*. New York.

Chapter 13

SOLUTIONS OF ELECTROLYTES AND ELECTROCHEMICAL SYSTEMS

SOLUTIONS OF ELECTROLYTES

13·1. *Introduction.* When a polar substance is dissolved in water, the electrical forces between the ions forming the molecule are weakened owing to the very high dielectric constant of water (88 e.s.u.), and the molecules are dissociated to a greater or less extent. If every molecule is completely dissociated into ν ions, we may use all the formulae derived in Chapter 12 provided either that we consider each species of ion to be a separate constituent in the solution (with a constant ratio of their concentrations in accordance with the chemical formula of the molecule), or that we consider the molecular weight of the solute to be M^*/ν, where M^* is the molecular weight of the undissociated substance. Thus if x denotes the mole fraction of the solute in a dilute solution, we have

$$\mu_{1,L} = \mu^0_{1,L}(T,P) - \nu RTx, \quad \mu_{2,L} = \mu^*_{2,L}(T,P) + \nu RT \log x, \quad (13\cdot1\cdot1)$$

so that equation (12·41·4) for the depression of the freezing-point becomes

$$T_0 - T = \nu RT_0^2 x/L, \quad (13\cdot1\cdot2)$$

while formula (12·43·5) for the osmotic pressure becomes

$$\Pi v_{1,L}(T) = \nu RTx. \quad (13\cdot1\cdot3)$$

13·11. *The degree of dissociation.* The foregoing theory, which is due to Arrhenius, only applies at infinite dilution where dissociation is complete. To generalize the theory we suppose that each molecule dissociates into ν_+ positive ions A_+ and ν_- negative ions B_-, where $\nu = \nu_+ + \nu_-$. If $z_+\epsilon$ is the charge on an A_+ ion and $z_-\epsilon$ that on a B_- ion, the condition for the solution to be electrically neutral is

$$\nu_+ z_+ + \nu_- z_- = 0, \quad (13\cdot11\cdot1)$$

and the dissociation is characterized by the equation

$$-(A_+)_{\nu_+}(B_-)_{\nu_-} + \nu_+ A_+ + \nu_- B_- = 0. \quad (13\cdot11\cdot2)$$

Now, if we denote the mole fractions of the neutral molecules, the positive ions and the negative ions by x_n, x_+ and x_- respectively, we may write

$$\left.\begin{aligned} \mu_1 &= \mu_1^0(T,P) - RT\log(1 - x_n - x_+ - x_-), \\ \mu_i &= \mu_i^*(T,P) + RT\log x_i \;\; (i = n, +, -) \end{aligned}\right\} \quad (13\cdot11\cdot3)$$

for sufficiently small values of x_i, and equation (13·11·3) then gives

$$-\mu_n + \nu_+\mu_+ + \nu_-\mu_- = 0, \tag{13·11·4}$$

i.e.

$$x_+^{\nu_+} x_-^{\nu_-}/x_n = K_x(T), \tag{13·11·5}$$

where the equilibrium constant $K_x(T)$ is a function of T only if, as usual, we neglect the effect of pressure.

If n_n, n_+ and n_- are the numbers of gram molecules present of neutral molecules, positive ions and negative ions of the solute, and if n and n_0 are the total number of gram molecules of the solute, calculated as if undissociated, and of the solvent, then

$$n_n = (1-\alpha)\,n, \quad n_+ = \nu_+\alpha n, \quad n_- = \nu_-\alpha n, \tag{13·11·6}$$

$$\left.\begin{aligned} x_n &= \frac{(1-\alpha)\,n}{n_0 + \{1+(\nu-1)\,\alpha\}\,n}, \\ x_+ &= \frac{\nu_+\alpha n}{n_0 + \{1+(\nu-1)\,\alpha\}\,n}, \\ x_- &= \frac{\nu_-\alpha n}{n_0 + \{1+(\nu-1)\,\alpha\}\,n}, \end{aligned}\right\} \tag{13·11·7}$$

where α is the fraction of solute molecules dissociated. Hence (13·11·5) becomes

$$\frac{\alpha^\nu}{1-\alpha}\left(\frac{n}{n_0 + \{1+(\nu-1)\,\alpha\}\,n}\right)^{\nu-1} = \frac{K_x(T)}{\nu_+^{\nu_+}\nu_-^{\nu_-}}. \tag{13·11·8}$$

In practical work it is usual to express the composition of a solution in terms of the number of gram molecules of the solute per litre of the solution. If we write $c_i = n_i/V$, then $x_i = c_i V/\Sigma n_i$, where Σn_i is the total number of gram molecules (of solvent, solute molecules and solute ions) in the volume V. We then have

$$c_+^{\nu_+} c_-^{\nu_-}/c_n = K_c, \tag{13·11·9}$$

and

$$\frac{\alpha^\nu}{1-\alpha}\left(\frac{n}{V}\right)^{\nu-1} = \frac{K_c}{\nu_+^{\nu_+}\nu_-^{\nu_-}}, \tag{13·11·10}$$

where

$$K_c = K_x(T)\,(\Sigma n_i/V)^{\nu-1}. \tag{13·11·11}$$

We may use whichever of $K_x(T)$ or K_c is the more convenient, but it must be borne in mind when considering numerical values that the units are different in the two cases. (K_c is not strictly a function of T only, since the total number of gram molecules is not constant. In practice, however, we are only interested in dilute solutions, and $V/\Sigma n$ is then constant and equal to the specific volume of the solvent.)

13·12. *The conductivity of solutions of electrolytes.* Unless we can measure the equilibrium constant $K_x(T)$, equations (13·11·5) and (13·11·8) are purely formal relations between the concentrations of the ions. In order to make any further progress we therefore have to relate $K_x(T)$ to some directly measurable property of the solution. We now show that, with certain assumptions, $K_x(T)$ can be found from the electrical conductivity. We denote by the symbol Λ the molal conductivity, i.e. the specific conductivity divided by the number of gram molecules per unit volume of the solute, calculated on the molecular weight of the undissociated substance. Then

$$\Lambda = (n_+ \sigma_+ + n_- \sigma_-)/n, \tag{13·12·1}$$

where σ_+ and σ_- are characteristic of the positive and negative ions respectively and are assumed to be independent of the concentrations. The limiting value Λ_∞ of Λ for infinite dilution is $\nu_+ \sigma_+ + \nu_- \sigma_-$, and

$$\Lambda/\Lambda_\infty = \alpha. \tag{13·12·2}$$

We may therefore write (13·11·8) as

$$\frac{(\Lambda/\Lambda_\infty)^\nu}{1-\Lambda/\Lambda_\infty}\left(\frac{n}{n_0+\{1+(\nu-1)\,\Lambda/\Lambda_\infty\}\,n}\right)^{\nu-1} = \frac{K_x(T)}{\nu_+^{\nu_+}\nu_-^{\nu_-}}, \tag{13·12·3}$$

or, by (13·11·10), as

$$\frac{(\Lambda/\Lambda_\infty)^\nu}{1-\Lambda/\Lambda_\infty}\left(\frac{n}{V}\right)^{\nu-1} = \frac{K_c}{\nu_+^{\nu_+}\nu_-^{\nu_-}}. \tag{13·12·4}$$

For a binary electrolyte this becomes

$$\frac{(\Lambda/\Lambda_\infty)^2}{1-\Lambda/\Lambda_\infty}\frac{n}{V} = K_c, \tag{13·12·5}$$

a relation which is usually known as Ostwald's dilution law. If the assumptions made above are valid, we can determine K_x and K_c by measurements of the electrical conductivity as a function of the composition.

Table 13·1. *The molar conductivity of dilute acetic acid solutions at* 25° C. $\Lambda_\infty = 392$

n/V in gram molecules/litre	Λ in ohm^{-1} cm.^{2}/gram molecule	$K_c \times 10^5$
0·07369	6·086	1·835
0·01842	12·091	1·818
0·004606	23·81	1·811
0·001151	46·13	1·806
0·0002879	86·71	1·807

The equation (13·12·5) is well obeyed by a large number of substances known as weak electrolytes, of which a typical example is acetic acid CH_3COOH, which reacts with water to give the ions OH_3^+ and CH_3COO^-. According to the data given in table 13·1, the dissociation constant of acetic acid at 25° C. in moderately dilute solutions is $1{\cdot}8 \times 10^{-5}$ gram molecule/litre. K_c is a function of the temperature and increases with increasing T. This means that the dissociation of acetic acid in a dilute aqueous solution is an endothermic reaction.

Strong Electrolytes

13·2. According to equation (12·41·4), the lowering of the freezing-point of a dilute solution of an electrolyte is given by

$$T_0 - T = RT_0^2 \sum_{i\geqslant 2} x_i/L, \tag{13·2·1}$$

where, by (13·11·7),

$$\Sigma x_i = x_n + x_+ + x_- = \{1 + (\nu - 1)\alpha\} n/n_0. \tag{13·2·2}$$

If, in accordance with (12·6·5), we write (13·2·1) as

$$T_0 - T = \frac{RT_0^2}{L} g_1 \frac{\nu n}{n_0}, \tag{13·2·3}$$

then $g_1 = 1$ if the solute is completely dissociated; and when the dissociation is incomplete we have

$$1 - g_1 = (1-\alpha)(\nu - 1)/\nu, \tag{13·2·4}$$

which, by (13·11·9), can be written as

$$1 - g_1 = \frac{\nu - 1}{\nu}\left(\frac{n}{V}\right)^{\nu-1} \frac{\nu_+^{\nu_+}\nu_-^{\nu_-}}{K_c} \tag{13·2·5}$$

if α is very near to unity. If, therefore, we plot $1 - g_1$ as a function of n/V, the curve should have a finite gradient at the origin (a zero gradient if $\nu > 2$). It is found experimentally that while this rule is correct for weak electrolytes it fails completely for the class of substances known as strong electrolytes, which includes many inorganic acids, bases and salts. For strong electrolytes, (13·2·5) has to be replaced by

$$1 - g = a'(n/V)^{\frac{1}{2}}, \tag{13·2·6}$$

where a' is a function of the temperature. This anomalous behaviour of strong electrolytes can also be demonstrated by calculating the values of K_c for very dilute solutions. As shown by the results for sodium chloride which are given in table 13·2, K_c shows no signs of tending to a definite limit as $n/V \to 0$.

Table 13·2. *Values of K_c for aqueous solutions of* NaCl *at* 18° C.

n/V in gram molecules/litre	Λ/Λ_∞	K_c
10^{-1}	0·8444	0·4584
10^{-2}	0·9355	0·1358
10^{-3}	0·9772	0·0419
10^{-4}	0·9921	0·0123

13·21. The apparent values of K_c for strong electrolytes are so large that the molecules must be completely dissociated even in solutions which would not normally be classed as very dilute. The anomalous behaviour of strong electrolytes is therefore not due to an anomalous variation of the degree of dissociation with concentration, but to the effect of the electrostatic forces between the ions.

The calculation of the contribution of the long-range electrostatic forces to the free energy of an electrolyte solution is one of the most complex problems in statistical mechanics, major contributions to which have been made by Milner (1912) and by Debye & Hückel (1923). (See, for example, Falkenhagen (1934) or Fowler (1936).) If certain simplying assumptions are made, it can be shown that, for extremely dilute solutions,

$$G(T,P,n_i) = n_1\mu_1^0(T,P) + \sum_{i\geqslant 2} n_i\mu_i^*(T,P) + RT\Big(\sum_{i\geqslant 1} n_i \log n_i - N\log N\Big)$$
$$-\frac{2\pi^{\frac{1}{2}}\mathscr{L}^2}{3V^{\frac{1}{2}}}RT\left(\frac{\epsilon^2}{DRT}\sum_{i\geqslant 2} n_i z_i^2\right)^{\frac{3}{2}} \quad \Big(N = \sum_{i\geqslant 1} n_i\Big), \quad (13·21·1)$$

where $\mathscr{L}$ is Avogadro's number, D is the dielectric constant of the solvent and where $z_i\epsilon$ is the charge on the ion i. To determine the thermodynamic potential of the solvent we differentiate with respect to n_1, taking account of the fact that V is a function of the n_i's. Since the solution is dilute, we have $V = n_1/\rho$ and $\partial V/\partial n_1 = 1/\rho$, where ρ is the density of the solvent, and hence

$$\mu_1 = \mu_1^0 + RT\log\Big(1 - \sum_{i\geqslant 2} x_i\Big) + \tfrac{1}{3}\mathscr{L}^2(\pi\rho)^{\frac{1}{2}}RT\left(\frac{\epsilon^2}{DRT}\sum_{i\geqslant 2} z_i^2 x_i\right)^{\frac{3}{2}}. \quad (13·21·2)$$

We therefore see that the osmotic coefficient g_1, defined by (12·6·3), is given by

$$g_1 = 1 - \tfrac{1}{3}\mathscr{L}^2(\pi\rho)^{\frac{1}{2}}\left(\frac{\epsilon^2}{DRT}\sum_{i\geqslant 2} z_i^2 x_i\right)^{\frac{3}{2}} \Big/ \sum_{i\geqslant 2} x_i. \quad (13·21·3)$$

For a single electrolyte which splits up into ν_+ positive ions and ν_- negative ions, we have, for electrical neutrality, $\nu_+z_+ + \nu_-z_- = 0$, and, if x is the mole fraction of the solute calculated on its molecular weight in the undissociated state, $x_+ = \nu_+x$, $x_- = \nu_-x$, provided that dissociation is complete (equation (13·11·7) with $\alpha = 1$). We may then write

$$g_1 = 1 - ax^{\frac{1}{2}}, \quad (13·21·4)$$

in agreement with the empirical law (12·2·6), where

$$a = \tfrac{1}{3}\mathscr{L}^2(\pi\rho)^{\frac{1}{2}}\left(\frac{\epsilon^2}{DRT}\right)^{\frac{3}{2}}\frac{(\nu_+z_+^2 + \nu_-z_-^2)^{\frac{3}{2}}}{\nu_+ + \nu_-} = \alpha\frac{(\nu_+z_+^2 + \nu_-z_-^2)^{\frac{3}{2}}}{\nu_+ + \nu_-}. \quad (13·21·5)$$

For water at 0° C. we have $D = 88$ e.s.u. and $\rho = 1$ gram/cm.$^3 = \frac{1}{18}$ gram molecule/cm.^{3}, and so $a = 2$.

13·22. It is found experimentally that the equations of the preceding section have an extremely small range of validity, and therefore that it is desirable to extend the theory to apply to less dilute solutions in which the concentrations of the ions are of the order of 0·1 gram ions/litre. This, however, is a most difficult problem, and it has only been partially solved (and even then by making distinctly crude approximations).

Debye & Hückel's theory of the effect of the electrostatic forces between the ions is based upon a model in which the diameter of an ion is d, and the expression (13·21·1) is the limiting form of the free energy when d tends to zero, i.e. when the ions are treated as point charges. A more general theory is obtained by considering a non-zero value of d, but further correction terms are necessary which have been discussed by a number of authors, in particular by Guggenheim (1935). If both these generalizations are included, the free energy can be written as

$$G(T,P,n_i)=n_1\mu_1^0(T,P)+\sum_{i\geq 2} n_i\mu_i^*(T,P)+RT\Big(\sum_{i\geq 1} n_i\log n_i-N\log N\Big)-\frac{\mathscr{L}\epsilon^2}{3D}\sum_{i\geq 2} n_i z_i^2\frac{\xi}{d}\tau(\xi)+\frac{RT}{V}\sum_{i,j} n_i n_j\phi_{i,j}, \qquad (13\cdot22\cdot1)$$

where

$$\left.\begin{aligned}\xi=\kappa d, \quad \kappa&=2\mathscr{L}\epsilon\Big\{\pi\sum_{i\geq 2} n_i z_i^2/(VDRT)\Big\}^{\frac{1}{2}},\\ \tau(\xi)&=\frac{3}{\xi^3}\{\log(1+\xi)-\xi+\tfrac{1}{2}\xi^2\}.\end{aligned}\right\} \qquad (13\cdot22\cdot2)$$

The $\phi_{i;j}$'s are parameters which do not occur in the Debye-Hückel theory, and which are independent of V and the n_i's. Then

$$\mu_1=\mu_1^0+RT\log\Big(1-\sum_{i\geq 2} x_i\Big)+\frac{\mathscr{L}\epsilon^2}{6D}\frac{\xi}{d}\sigma(\xi)\sum_{i\geq 2} x_i z_i^2-RT\rho\sum_{i,j} x_i x_j\phi_{ij}, \qquad (13\cdot22\cdot3)$$

where $$\sigma(\xi)=\frac{3}{\xi^3}\left\{1+\xi-\frac{1}{1+\xi}-2\log(1+\xi)\right\}=1-\tfrac{3}{2}\xi+\tfrac{9}{5}\xi^2-\dots, \qquad (13\cdot22\cdot4)$$

and the osmotic coefficient g_1 is given by

$$g_1=1-\frac{\mathscr{L}\epsilon^2}{6D}\frac{\xi}{d}\sigma(\xi)\frac{\sum_{i\geq 2} x_i z_i^2}{\sum_{i\geq 2} x_i}+\rho\frac{\sum_{i\geq 2} x_i x_j\phi_{ij}}{\sum_{i\geq 2} x_i}. \qquad (13\cdot22\cdot5)$$

It is found that the only quantities ϕ_{ij} which need to be taken as different from zero are those relating to pairs of ions of different charges.

13·23. *Comparison with experiment.* In practical work it is customary to use the concentrations c_i in gram ions/litre and to introduce a quantity I, known as the ionic strength, which is defined by

$$I=\tfrac{1}{2}\Sigma c_i z_i^2. \qquad (13\cdot23\cdot1)$$

For a binary electrolyte, I reduces to c, the concentration of the electrolyte calculated on the molecular weight of the undissociated salt. If there is only one electrolyte characterized by the integers ν_+, ν_-, z_+, z_- which satisfy (13·11·1), we have $c_+ = \nu_+ c$, $c_- = \nu_- c$ and the numerical values of g_1 for water at 0° C. are given by

$$1 - g_1 = 0{\cdot}374\nu_+\nu_- I^{\frac{1}{2}}\sigma\left(\frac{I^{\frac{1}{2}}d}{3{\cdot}08\times 10^{-8}}\right) - \bar{\nu}c\phi, \qquad (13{\cdot}23{\cdot}2)$$

where $\bar{\nu} = 2\nu_+\nu_-/(\nu_+ + \nu_-)$.

The formula (13·23·2) contains two adjustable parameters, but for simplicity d is often taken to be fixed and equal to $3{\cdot}08\times 10^{-8}$ cm., since it is found that the presence of single parameter ϕ is sufficient to enable (13·23·2) to be fitted to the simple experimental data. Some curves of $(1-g_1)/\nu_+\nu_-$ as a function of $I^{\frac{1}{2}}$ are shown in fig. 13·1, and the values of $\bar{\nu}\phi$ required to reproduce them are given in table 13·3.

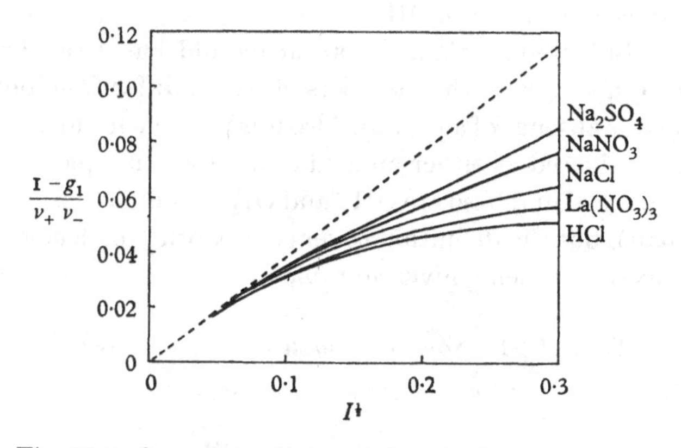

Fig. 13·1. Osmotic coefficients of single electrolytes at 0° C. as functions of the ionic strength.

Table 13·3. *Values of $\bar{\nu}\phi$ for single electrolytes in water at* 0° C.

	HCl	NaCl	$NaNO_3$	Na_2SO_4	$La(NO_3)_3$
$\bar{\nu}$	1	1	1	$\frac{4}{3}$	$\frac{3}{2}$
$\bar{\nu}\phi$	0·275	0·135	0	−0·60	3·9

Acids and Bases

13·3. *Definitions.* The classical definition of acids and bases as substances which can neutralize one another to produce a salt and water is unnecessarily restrictive and places too much emphasis upon aqueous solutions. A wider definition, due to Brönsted (1923), is that an acid is any molecule or ion capable of losing a proton and that a base is any molecule or ion capable of capturing a proton. It is then necessary to describe the chemical action between acids and bases in a manner which differs somewhat from the classical description. When an aqueous solution of hydrochloric acid is neutralized by an aqueous solution of caustic soda, the sodium chloride, which according to convention is formed, does not exist as individual NaCl molecules but as separate Na^+ and Cl^- ions. (This is true even in the pure solid state.) It is therefore simpler and better to say that the hydrochloric acid consists of H^+ and Cl^- ions, while the caustic soda consists of Na^+ and OH^-, and that in the process of neutralization the H^+ and OH^- ions combine to give neutral H_2O molecules, the Na^+ and Cl^- ions playing no part in the reaction. Further, it is probable that the hydrogen ions do not exist as separate entities but attach themselves to water molecules to form OH_3^+ ions. We may then say that the acidic character of an aqueous solution of hydrochloric acid is due to the presence of the acid OH_3^+, while the basic character of an aqueous solution of caustic soda is due to the presence of the base OH^-.

If we adopt Brönsted's definition, all acids and bases can be grouped together in conjugate pairs, the members of which differ from one another by the presence or absence of a proton. Also it is possible for many molecules or ions to be considered as either an acid or a base. In aqueous solutions, the fundamental acid and base are OH_3^+ and OH^- (though these do not form a conjugate pair), but the definition covers many other molecules and ions, some typical examples being given in table 13·4.

Table 13·4. *Some conjugate acids and bases*

Acid	Base
OH_3^+	OH_2
OH_2	OH^-
H_2SO_4	HSO_4
HSO_4^-	SO_4^{2-}
NH_4^+	NH_3
HCl	Cl^-
H_2CO_3	HCO_3^-
HCO_3^-	CO_3^{2-}
CH_3COOH	CH_3COO^-
$NH_3^+CH_2COOH$	$NH_3^+CH_2COO^-$
$NH_3^+CH_2COO^-$	$NH_2CH_2COO^-$

13·31. *Acid-base reactions.* If A, B and A', B' denote two conjugate pairs of acids and bases, every acid-base reaction is of the type

$$A+B' \rightleftharpoons B+A'. \tag{13·31·1}$$

We have, for example,

$$CH_3COOH+H_2O \rightleftharpoons CH_3COO^- + OH_3^+, \tag{13·31·2}$$

which is the modern way of representing the dissociation of acetic acid in water. If the solution is dilute, the law of mass action (13·11·9) gives

$$c_{Ac^-}c_{OH_3^+}/c_{HAc} = K_c(T), \tag{13·31·3}$$

where we use the symbol HAc to denote CH_3COOH and Ac^- to denote CH_3COO^-. More generally, if the solution is not extremely dilute, we write with the notation of § 12·6 for non-ideal solutions,

$$\mu_{H_2O} = \mu^0_{H_2O}(T) + RT \log (f_{H_2O} x_{H_2O}), \tag{13·31·4}$$

where $f_{H_2O} \to 1$ as $x_{H_2O} \to 1$, and

$$\mu_A = \mu^*_A(T) + RT \log (\gamma_A x_A) \tag{13·31·5}$$

for $A = \text{HAc}, \text{Ac}^-, \text{OH}_3^+$, where $\gamma_A \to 1$ as $x_A \to 0$. The law of mass action is then

$$\frac{\gamma_{Ac^-} x_{Ac^-} \gamma_{OH_3^+} x_{OH_3^+}}{f_{H_2O} x_{H_2O} \gamma_{HAc} x_{HAc}} = \exp [(\mu^0_{H_2O} + \mu^*_{HAc} - \mu_{Ac^-} - \mu_{OH_3^+})/RT] = K_x(T). \tag{13·31·6}$$

Except in concentrated solutions we may omit the factors $f_{H_2O} x_{H_2O}$ and absorb them into $K_x(T)$. Alternatively, if we wish to use the concentrations $c_i = n_i/V$ and generalize (13·31·3) to apply when the γ's differ appreciably from unity, we may define the activities‡ of the various constituents by the symbols

$$\{\text{HAc}\} = \gamma_{HAc} c_{HAc}, \quad \{\text{Ac}^-\} = \gamma_{Ac^-} c_{Ac^-}, \quad \{\text{OH}_3^+\} = \gamma_{OH_3^+} c_{OH_3^+}. \tag{13·31·7}$$

We then have
$$\{\text{Ac}^-\}\{\text{OH}_3^+\}/\{\text{HAc}\} = K_c(T), \tag{13·31·8}$$

where we have absorbed the factor relating to the water into $K_c(T)$, since the formula does not apply in any case to concentrated solutions. Every acid-base reaction has an equilibrium constant $K_c(T)$ which, as we have already seen in § 13·12, can be found by measuring the electrical conductivity. According to the data given in table 13·1 $K_{HAc}(T) = 1·8 \times 10^{-5}$ gram molecule/litre at 25° C., and it increases with the temperature. The corresponding equilibrium constant $K_{NH_4^+}(T)$ for the reaction

$$NH_4^+ + H_2O \rightleftharpoons NH_3 + OH_3^+$$

is 6×10^{-10} gram molecule/litre at 25° C.

13·32. *The equilibrium constant of water.* However carefully water is purified, it is always found to have a measurable electrical conductivity, the values for 'ideally pure water' being as shown in table 13·5. The conductivity is due to the presence of OH_3^+ and OH^- ions formed by the reaction

$$H_2O + H_2O \rightleftharpoons OH_3^+ + OH^-, \tag{13·32·1}$$

‡ The activities, which are much used in practical work, but which have no theoretical significance, are generalized concentrations, or ratios of concentrations, which can be defined in various ways to suit the particular problem under consideration. Unless the various γ's, or their equivalents, can be measured or calculated, the introduction into any formula of the activities instead of the concentrations or the mole fractions is a purely formal generalization.

water acting as both an acid and as base. The law of mass action gives

$$\{OH_3^+\}\{OH^-\} = K_W(T), \qquad (13\cdot32\cdot2)$$

but in this case $K_W(T)$ cannot be found by measuring the electrical conductivity and applying Ostwald's dilution formula (13·12·5). It can, however, be found by certain indirect methods, which are discussed in books on physical chemistry, provided that certain assumptions are made. Its values at 0, 20 and 25° C. are usually taken as $0{\cdot}115 \times 10^{-14}$, $0{\cdot}68 \times 10^{-14}$ and $1{\cdot}01 \times 10^{-14}$ gram molecule/(litre)^2. In pure water we necessarily have $\{OH_3^+\} = \{OH^-\}$, and so $\{OH_3^+\} = \{OH^-\} = 10^{-7}$ gram ion/litre at 25° C.

Table 13·5. *The conductivity of pure water at various temperatures*

T in °C. ...	0	18	25	34	50
$\sigma \times 10^7$ in ohm^{-1} cm.^{2}	0·15	0·43	0·62	0·95	1·87

In view of (13·32·2) we may write any equilibrium condition in an aqueous solution in more than one way. For example, we may write

$$\{NH_3\}\{OH_3^+\}/\{NH_4^+\} = K_{NH_4^+}, \quad \{NH_4^+\}\{OH^-\}/\{NH_3\} = K_{NH_3} = K_W/K_{NH_4^+}, \qquad (13\cdot32\cdot3)$$

the latter corresponding to the reaction

$$NH_3 + H_2O \rightleftharpoons NH_4^+ + OH^-.$$

13·321. *The* pH *of a solution.* The acidity or alkalinity of a solution is usually defined in terms of the concentration of the hydrogen ions (or of the OH_3^+ ions) in the solution. Strictly speaking it is the activity and not the concentration of the ions that is measured, and we define the pH of the solution by the relation

$$pH = -\log_{10}\{OH_3^+\}, \qquad (13\cdot321\cdot1)$$

where $\{OH_3^+\}$ is expressed in gram ions/litre. If we define a neutral solution as one in which the concentration of OH_3^+ ions is the same as in pure water, the pH of a neutral solution is 7, while if the pH is less than 7 the solution is acidic, and if the pH is greater than 7 the solution is alkaline. For the theory of the methods of measuring the pH, the reader is referred to books on electrochemistry (e.g. Glasstone, 1945).

13·33. *Neutralization curves.* We now consider how the pH of an aqueous solution varies with the addition of acids or bases. Near the equivalence point we must take into account the intrinsic ionization of the water itself, and we therefore have to deal with two chemical equilibria simultaneously. There are a large number of special cases, and we only deal with a few of these to show the principles involved. For the sake of simplicity we shall assume that the solutions are such that the activities can be replaced by the concentrations, even for strong electrolytes. We shall also assume that all strong electrolytes are completely dissociated.

A strong acid and a strong base. Let n_a and n_b be the number of gram molecules of the acid and of the base. Then since the solution is electrically neutral, there must be equal numbers of positive and negative ions. But the acid provides n_a negative ions and the base provides n_b positive ions, so that we must have

$$[\mathrm{OH}^-]+n_a/V=[\mathrm{OH}_3^+]+n_b/V, \tag{13·33·1}$$

where V is the volume of the solution and where we use square brackets to denote the molar concentrations. We also have the relation

$$[\mathrm{OH}^-][\mathrm{OH}_3^+]=K_W,$$

and hence $[\mathrm{OH}_3^+]$ is the positive root of the equation

$$[\mathrm{OH}_3^+]([\mathrm{OH}_3^+]-n/V)=K_W,$$

where $n=n_a-n_b$. Therefore

$$[\mathrm{OH}_3^+]=\frac{1}{2}\frac{n}{V}+\sqrt{\left(\frac{1}{4}\frac{n^2}{V^2}+K_W\right)}, \tag{13·33·2}$$

and this applies whether n is positive or negative. The general behaviour of the pH as a function of n/V is shown in fig. 13·2, and it will be seen that there is a very rapid variation in the neighbourhood of the equivalence point.

A weak acid and a strong base. Since the number of gram molecules of the undissociated acid HA plus the number of gram ions of the acid ions A^- must be equal to n_a the total number of gram molecules of the acid, we have

$$[\mathrm{HA}]+[\mathrm{A}^-]=n_a/V. \tag{13·33·3}$$

Also, since the solution must be electrically neutral, the number of negative charges must be equal to the number of positive charges. Therefore since the base provides n_b positive ions, we have

$$[\mathrm{A}^-]+[\mathrm{OH}^-]=[\mathrm{OH}_3^+]+n_b/V. \tag{13·33·4}$$

In addition, we have the two equilibrium conditions

$$[\mathrm{A}^-][\mathrm{OH}_3^+]/[\mathrm{HA}]=K_{\mathrm{HA}}, \quad [\mathrm{OH}_3^+][\mathrm{OH}^-]=K_W, \tag{13·33·5}$$

and by combining these equations we obtain the relation

$$[\mathrm{OH}_3^+]=\frac{(n_a-n_b)/V-[\mathrm{OH}_3^+]+K_W/[\mathrm{OH}_3^+]}{n_b/V+[\mathrm{OH}_3^+]-K_W/[\mathrm{OH}_3^+]}K_{\mathrm{HA}}, \tag{13·33·6}$$

which is a cubic equation for $[\mathrm{OH}_3^+]$. Various approximate forms of this equation can be given, depending upon the relative magnitudes of n_a/V, n_b/V and K_W. The general behaviour of the pH curve as a function of

$(n_b - n_a)/V$ for $n_a/V = 0{\cdot}1$, $K_{HA} = 1{\cdot}8 \times 10^{-5}$ is shown in fig. 13·3. It will be seen that when $n_a = n_b$ we do not have a pH of 7 but a pH of $9 - \frac{1}{2}\log_{10} 1{\cdot}8$. In other words, the solution has an alkaline reaction when $n_a = n_b$ due to the presence of some undissociated acid.

The case of a weak base and a strong acid can be treated along exactly similar lines, the roles of the OH_3^+ and OH^- ions being interchanged.

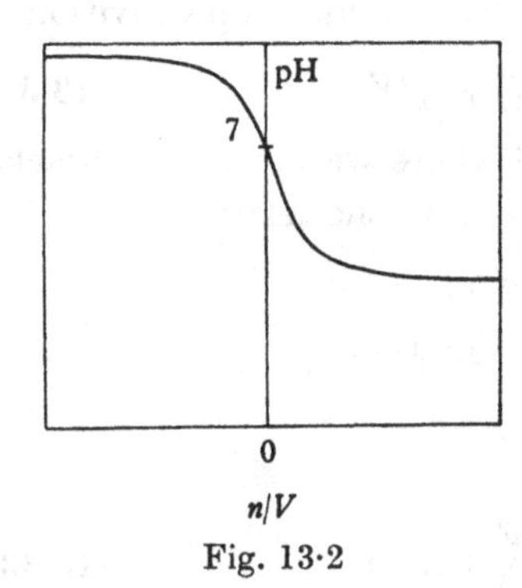

Fig. 13·2

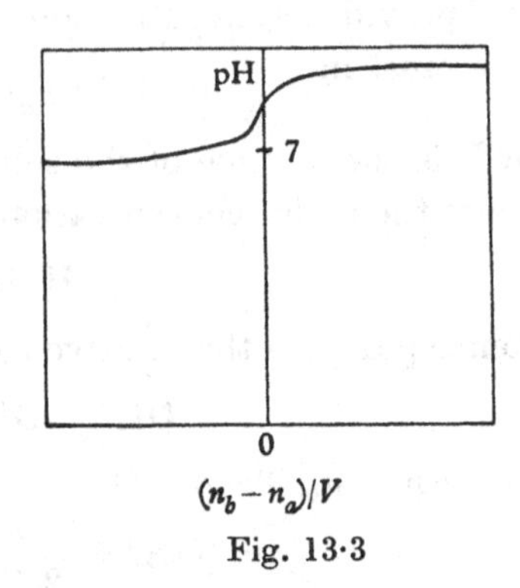

Fig. 13·3

A weak acid and a weak base. In this case the reactions may be written

$$HA + H_2O \rightleftharpoons A^- + OH_3^+, \quad B + H_2O \rightleftharpoons BH^+ + OH^-. \qquad (13{\cdot}33{\cdot}7)$$

Since the total numbers of gram molecules of acid either as HA or as A^- is n_a, and of base either as B or as BH^+ is n_b, we have

$$[HA] + [A^-] = n_a/V, \quad [B] + [BH^+] = n_b/V. \qquad (13{\cdot}33{\cdot}8)$$

Also, the condition for the electrical neutrality of the solution is

$$[A^-] + [OH^-] = [BH^+] + [OH_3^+], \qquad (13{\cdot}33{\cdot}9)$$

and in addition we have the relations

$$[A^-][OH_3^+]/[HA] = K_{HA}, \quad [BH^+][OH^-]/[B] = K_B, \quad [OH_3^+][OH^-] = K_W. \qquad (13{\cdot}33{\cdot}10)$$

If we eliminate [HA] from the first pair of the equations (13·33·8) and (13·33·10) and eliminate [B] from the second pair, we obtain expressions for $[A^-]$ and $[BH^+]$ which when substituted into (13·33·9) give the following quartic for $[OH_3^+]$:

$$[OH_3^+] = \frac{n_a/V}{1 + [OH_3^+]/K_{HA}} - \frac{[OH_3^+]\, n_b/V}{[OH_3^+] + K_W/K_B} + \frac{K_W}{[OH_3^+]}. \qquad (13{\cdot}33{\cdot}11)$$

This reduces to (13·33·6) if we take $K_B = \infty$.

As a typical weak acid and weak base we may consider acetic acid with $K_{HA} = 1{\cdot}8 \times 10^{-5}$ and ammonia with $K_W/K_B = 6 \times 10^{-10}$. We then see that, with a negligible error, $[OH_3^+]=10^{-7}$ when $n_a=n_b$, so that the solution is neutral when there are equal quantities of acid and base. The general shape of the pH curve is shown in fig. 13·4 for $n_b/V=0{\cdot}1$.

A weak dibasic acid and a strong base. We write the two reactions of the acid as

$$H_2A+H_2O\rightleftharpoons HA^-+OH_3^+,\quad HA^-+H_2O\rightleftharpoons A^{2-}+OH_3^+, \tag{13·33·12}$$

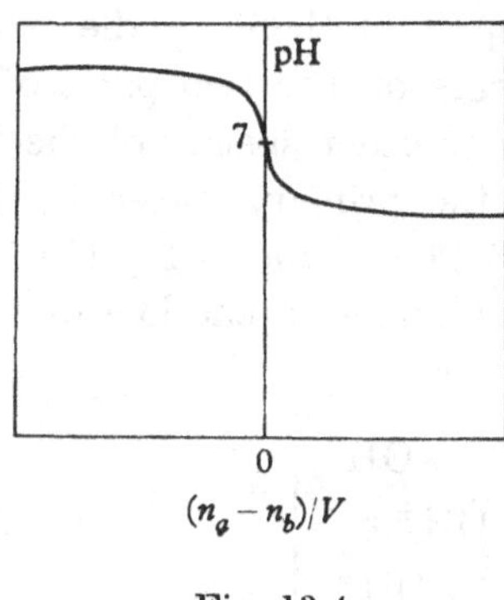

Fig. 13·4

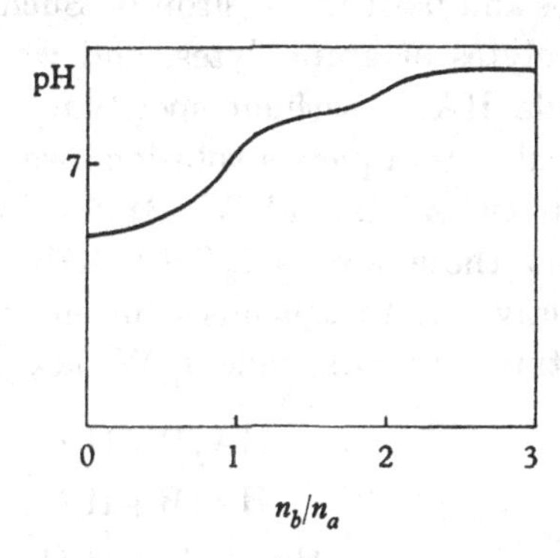

Fig. 13·5

and the equilibrium conditions as

$$[HA^-][OH_3^+]/[H_2A]=K_{H_2A},\quad [A^{2-}][OH_3^+]/[AH^-]=K_{HA^-}. \tag{13·33·13}$$

Then

$$[H_2A]+[HA^-]+[A^{2-}]=n_a/V, \tag{13·33·14}$$

while the condition for the electrical neutrality of the solution is

$$[OH^-]+[HA^-]+2[A^{2-}]=[OH_3^+]+n_b/V. \tag{13·33·15}$$

By combining all these equations we obtain the following quartic equation for $[OH_3^+]$:

$$\frac{n_a}{V}\left(1+2\frac{K_{HA^-}}{[OH_3^+]}\right)=\left(\frac{n_b}{V}+[OH_3^+]-\frac{K_W}{[OH_3^+]}\right)\left(1+\frac{[OH_3^+]}{K_{H_2A}}+\frac{K_{HA^-}}{[OH_3^+]}\right). \tag{13·33·16}$$

In any practical example all the terms in the above expression will not have equal importance for any given values of n_a/V and n_b/V. The general shape of the pH curve is shown in fig. 13·5 as a function of the amount of base added. On physical grounds the curve should show inflexions for $n_b/n_a=1$ and for $n_b/n_a=2$, but in practice one or other of these may be masked. The range of variation of K_{H_2A}/K_{HA^-} is very large for the different dibasic acids, and the neutralization curves show a correspondingly large variability. (For carbonic acid at 25° C., $K_{H_2A}=3\times10^{-7}$, $K_{HA^-}=6\times10^{-11}$; for oxalic acid, $K_{H_2A}=6{\cdot}5\times10^{-2}$, $K_{HA^-}=6\times10^{-5}$, while sulphuric acid is a strong acid during its first stage of ionization, so that it can then be considered to be completely ionized. Its second dissociation constant is $K_{HA^-}=1{\cdot}2\times10^{-2}$.)

13·34. *Amphoteric substances.* Many organic compounds contain both basic and acidic grounds and may tend to lose a proton at one part of the molecule and capture a proton at another portion of the molecule. For example, glycine or amino-acetic acid, which in aqueous solutions is usually considered to have the formula $NH_3^+CH_2COO^-$, contains acidic NH_3^+ groups and basic COO^- groups. Such substances are known as amphoteric electrolytes or ampholytes, and we may represent them by the general formula HA.B without specifying the charges on the two parts of the molecule. In aqueous solutions we have to consider species of the type $HA.BH^+$, $A^-.B$ and $A^-.BH^+$ as well as the primary molecule. (For glycine these are $NH_3^+CH_2COOH$, $NH_2CH_2COO^-$ and NH_2CH_2COOH respectively.) In aqueous solution we have a number of reactions of which only three are independent. We take these to be

$$\left.\begin{aligned} HA.B + H_2O &\rightleftharpoons HA.BH^+ + OH^-, \\ HA.B + H_2O &\rightleftharpoons A^-.B + OH_3^+, \\ HA.BH^+ + H_2O &\rightleftharpoons A^-.BH^+ + OH_3^+. \end{aligned}\right\} \tag{13·34·1}$$

The equilibrium conditions are

$$\left.\begin{aligned} [HA.BH^+][OH^-]/[HA.B] = K_1, \quad [A^-.B][OH_3^+]/[HA.B] = K_2, \\ [A^-.BH^+][OH_3^+]/[HA.BH^+] = K_3, \end{aligned}\right\} \tag{13·34·2}$$

and, if the total number of gram molecules of HA.B in all forms is n_a, we have

$$n_a/V = [HA.B] + [HA.BH^+] + [A^-.B] + [A^-.BH^+]. \tag{13·34·3}$$

Also, the condition for the electrical neutrality of the solution is

$$[A^-.B] + [OH^-] = [OH_3^+] + [HA.BH^+]. \tag{13·34·4}$$

On combining these equations we find that

$$[OH_3^+]^2 = K_W + \frac{n_a}{V}\left(K_2 - \frac{K_1}{K_W}[OH_3^+]^2\right) \Big/ \left(1 + \frac{K_1}{K_W}[OH_3^+]^2 + \frac{K_1 K_3}{K_W} + \frac{K_2}{[OH_3]}\right). \tag{13·34·5}$$

Thus $[OH_3^+]$ is not equal to $\sqrt{K_W}$ unless $K_1 = K_2$.

The values of the various constants for glycine at 25° C. are

$$K_1 = 2{\cdot}1 \times 10^{-12}, \quad K_2 = 1{\cdot}9 \times 10^{-10}, \quad K_3 = 2 \times 10^{-8}.$$

Equation (13·34·5) then reduces approximately to

$$[OH_3^+]^2 = K_W + K_2 n_a/V = K_W + 1{\cdot}9 \times 10^{-10} n_a/V.$$

An aqueous solution of glycine is therefore slightly acidic.

In strongly acid solutions, ampholytes are positively charged, while in strongly alkaline solutions they are negatively charged. At some intermediate point, called the isoelectric point, the average charge on the ampholyte molecules will be zero. At the isoelectric point we must have $[HA.BH^+] = [A^-.B]$, and hence, by (13·34·2),

$$[OH_3^+]^2 = K_2 K_W / K_1. \qquad (13\cdot34\cdot6)$$

For glycine at 25° C., the isoelectric point has a pH of approximately 6.

Reversible Electric Cells

13·4. *Heterogeneous reactions between solids.* The problems raised by the consideration of heterogeneous chemical reactions between solids or immiscible liquids are somewhat different from those associated with gaseous equilibria. When the reaction is completely heterogeneous, so that all the reactants and products form separate phases, the thermodynamic potential of any particular phase is entirely independent of the amount of the other phases present, and there can therefore be no question of an equilibrium state existing in which the various constituents are all present in well-defined ratios. Such a heterogeneous reaction, at any particular temperature and pressure, will go to completion in the direction which produces a diminution in the free energy. While the heat of reaction can be determined by carrying out the reaction irreversibly in a calorimeter, there is no way of determining the free energy by purely calorimetric methods. Integration of the Gibbs-Helmholtz equation leaves the entropy constant undefined, and this cannot be found unless further principles are invoked which have been discussed in § 11·7. This difficulty can be overcome if it is possible to carry out the reaction in a reversible manner (see the discussion of the meaning of the free-energy function in § 3·1), and this can be achieved if the reaction can be made the basis for a reversible electric cell. (This applies whether the substances concerned are solids, liquids or gases.)

13·41. *Reversible and irreversible electric cells.* It is only rarely that an electric cell is fully reversible, but there are many cells which are completely irreversible while others can be considered as being reversible in ideal circumstances. If two electrodes, one of pure zinc and the other of pure copper, are placed without touching in dilute sulphuric acid, no chemical reaction takes place. If, however, the electrodes are joined externally by a conductor, the zinc dissolves, hydrogen is evolved from the copper electrode and a positive current will pass through the cell from the zinc to the copper. But if we have an electromotive force in the external circuit sufficient to reverse the direction of the current, the copper dissolves and hydrogen is given off at the zinc electrode. This cell is therefore essentially irreversible.

A simple example of a reversible cell is provided by the Daniell cell, which consists of a zinc electrode immersed in an aqueous solution of zinc sulphate separated by a diaphragm, which is porous to sulphate ions, from a solution of copper sulphate into which dips a copper electrode. When the electrodes are joined externally a positive current passes through the cell from the zinc to the copper, zinc dissolves from the zinc electrode while sulphate ions diffuse into the zinc compartment to preserve electrical neutrality, and at the same time copper is deposited on the copper electrode. If an external electromotive force is used to reverse the direction of the current, copper dissolves and zinc is deposited. The cell is therefore reversible and we may write the chemical action as

$$Zn + CuSO_4 \rightleftharpoons ZnSO_4 + Cu.$$

13·411. To every reversible electric cell there corresponds a chemical reaction, though not all of the components of the cell may appear explicitly in the reaction. A well-known reversible cell is one which consists of a lead electrode in contact with solid lead chloride, the other electrode being liquid mercury in contact with solid mercurous chloride, and the electrolyte being an aqueous solution of potassium chloride. The potassium chloride plays no part in the chemical reaction, its function being to allow the free passage of ions and to avoid polarization effects due to the accumulation of ions at the surfaces of the lead chloride and the mercurous chloride. The chemical reaction that takes place is the heterogeneous reaction

$$-\text{Pb (solid)} + \text{PbCl}_2\text{ (solid)} - 2\text{HgCl (solid)} + 2\text{Hg (liquid)} = 0. \quad (13\cdot411\cdot1)$$

To study the free energy of the reaction represented by equation (13·411·1) we suppose that the electrodes are joined externally to a condenser, and that the potential difference between the plates is adjusted so as to give zero current. The potential difference $\mathscr{E}$ is then defined to be the electromotive force of the cell. If the potential difference across the condenser is made slightly less than $\mathscr{E}$, a current will be produced by the cell and the reaction will take place in the forward direction. If, on the other hand, the potential difference is made slightly greater than $\mathscr{E}$, a current will flow in the reverse direction and the backward chemical reaction will take place. We shall write the chemical equations in such a way that there is a positive current through the cell from left to right, and also that 1 gram molecule of the primary reacting substance (in the above case lead) appears in the equation.

By Faraday's law of electrolysis, if δn gram molecules of a substance react in a cell, the amount of electricity produced is equal to $\delta n \mathscr{N} \mathscr{F}$, where $\mathscr{N}$ is the valency of the substance and $\mathscr{F}$ is Faraday's constant, 96,494 coulombs per gram molecule. Hence if $\mathscr{E}$ is the electromotive force, the external electrical work done is $\delta n \mathscr{N} \mathscr{F} \mathscr{E}$. If we keep the cell at constant temperature and pressure, the electrical work done must be equal to the maximum work obtainable over and above the necessary work of expansion, i.e. to minus the change in the free energy G. Hence if ΔG denotes the change in the free energy of the cell when 1 gram molecule of the primary reacting substance disappears, we have

$$\mathscr{N} \mathscr{F} \mathscr{E} = -(\Delta G)_{T,P}. \tag{13·411·2}$$

13·42. *The temperature coefficient of the electromotive force.* The electromotive force of a cell measures the change in the free energy due to the chemical reaction and not the change in the heat function. We can, however, readily obtain the latter quantity. For if we insert (13·411·2) into the Gibbs-Helmholtz equation $\Delta H = \Delta G - T(\partial \Delta G/\partial T)_P$ we find that

$$\mathscr{E} + \frac{\Delta H}{\mathscr{N} \mathscr{F}} = T\left(\frac{\partial \mathscr{E}}{\partial T}\right)_P. \tag{13·42·1}$$

Equations (13·411·2) and (13·42·1) give the simplest and most reliable method of determining ΔG and ΔH for any reaction, provided that it can be carried out reversibly in an electric cell. The term on the right of (13·42·1) multiplied by $\mathscr{N} \mathscr{F}$ is the (reversible) heat absorbed from the surroundings. (Note that if the cell is short-circuited so that the chemical reaction takes place irreversibly, the heat given out by the cell is $-\Delta H$.) ΔG can be either greater or less than ΔH.

In the reaction (13·411·1), we have $\mathscr{N} = 2$ and $\mathscr{E} = 0·5357$ V. at 25° C. This gives

$$-\Delta G_{298^\circ \text{K.}} = 103{,}330 \text{ joules } (24{,}720 \text{ calories}).$$

Also $\partial \mathscr{E}/\partial T = 0·000145$ V./degree, which gives

$$-\Delta H_{298^\circ \text{K.}} = 95{,}010 \text{ joules } (22{,}730 \text{ calories}).$$

In this case the free-energy change is greater than the heat function change by 8320 joules, which is the heat absorbed by the cell from its surroundings.

In the reversible cell consisting of a mercury electrode in contact with mercurous chloride immersed in an aqueous solution of potassium chloride, and with platinum as the second electrode in contact with chlorine gas, the reaction is

$$-\text{Hg (liquid)} - \tfrac{1}{2}\text{Cl}_2 \text{ (gas, 1 atmosphere)} + \text{HgCl (solid)} = 0. \tag{13·42·2}$$

Here we have $\mathcal{N}=1$, $\mathcal{E}=1{\cdot}0894$ V., $\partial\mathcal{E}/\partial T=-0{\cdot}00958$ V./degree at 25° C. These figures give

$$-\Delta G_{298^\circ \text{K.}} = 103{,}270 \text{ joules } (25{,}137 \text{ calories})$$

and

$$-\Delta H_{298^\circ \text{K.}} = 130{,}830 \text{ joules } (31{,}300 \text{ calories}).$$

The cell therefore gives up 27,560 joules to its surroundings.

13·43. *Cells containing solutions and gases.* The expression (13·411·2) for the electromotive force is more informative if written out in full in terms of the stoichiometric numbers ν_i and the thermodynamic potentials μ_i of the substances taking part in the reaction. It is convenient to consider separately the substances which are solids (or pure liquids), solutions and gases. We then have

$$-\mathcal{N}\mathcal{F}\mathcal{E} = \sum_{\text{solids}} \nu_i\mu_i + \sum_{\text{solutions}} \nu_i\mu_i + \sum_{\text{gases}} \nu_i\mu_i. \qquad (13{\cdot}43{\cdot}1)$$

For any solid or pure liquid, μ_i is a function of T only. In a solution it is a function of T and the composition, whereas for gases it is a function of T, P and the composition.

If the cell contains only solids and perfect gases we may write

$$-\mathcal{N}\mathcal{F}\mathcal{E} = \sum_{\text{solids}} \nu_i\mu_i + \sum_{\text{gases}} \nu_i\{\mu_i^*(T) + RT\log p_i\}, \qquad (13{\cdot}43{\cdot}2)$$

where p_i is the partial pressure of the ith gas. More generally, we may write for a gas $\mu_i = \mu_i^*(T) + RT\log f_i$, where f_i is the fugacity. We then have

$$\mathcal{E} = \frac{RT}{\mathcal{N}\mathcal{F}} K'_P(T) - \frac{RT}{\mathcal{N}\mathcal{F}} \log \Pi f_i^{\nu_i} = \mathcal{E}_0(T) - \frac{RT}{\mathcal{N}\mathcal{F}} \log \Pi f_i^{\nu_i}, \qquad (13{\cdot}43{\cdot}3)$$

where

$$\mathcal{E}_0(T) = \frac{RT}{\mathcal{N}\mathcal{F}} K'_P(T), \quad K'_P(T) = -\sum_{\text{gases}} \frac{\nu_i\mu_i^*}{RT} - \sum_{\text{solids}} \frac{\nu_i\mu_i}{RT}. \qquad (13{\cdot}43{\cdot}4)$$

If therefore we can measure $\mathcal{E}$ as a function of the composition we can determine the fugacities, and by finding $\mathcal{E}_0(T)$ we can determine the equilibrium constant $K'_P(T)$ of the reaction.

An example of a cell of this type is one which consists of hydrogen, mercurous chloride, mercury and aqueous hydrochloric acid, the reaction being

$$-H_2\,(\text{gas}) - 2HgCl\,(\text{solid}) + 2Hg\,(\text{liquid}) + 2HCl\,(\text{aqueous}) = 0. \qquad (13{\cdot}43{\cdot}5)$$

Here $\mathcal{N}=2$, and, if P is the pressure of the hydrogen, we have

$$\mathcal{E} = \text{constant} + \tfrac{1}{2}(RT/\mathcal{F})\log P. \qquad (13{\cdot}43{\cdot}6)$$

The verification of this relation is shown in table 13·5.

A reaction of a somewhat different type occurs if we have hydrogen at two different pressures P_1 and P_2 $(P_1 > P_2)$ in contact with two different platinum or iridium electrodes which dip into dilute sulphuric acid. The amount of hydrogen at the higher pressure diminishes while that at the lower pressure increases, and the reaction is

$$-H_2\,(\text{gas}, P_1) + H_2\,(\text{gas}, P_2) = 0. \qquad (13\cdot43\cdot7)$$

Hence

$$\mathscr{E} = -\frac{\Delta G}{2\mathscr{F}} = \frac{RT}{2\mathscr{F}} \log \frac{P_1}{P_2}. \qquad (13\cdot43\cdot8)$$

At 15° C. and with $P_1 = 2P_2$, this gives $\mathscr{E} = 8\cdot5 \times 10^{-3}$ V.

Table 13·5. *The effect of pressure on the electromotive force of the hydrogen, mercurous chloride cell at* 25° C.

P in cm. Hg	76	78·7	80·6	82·2
$\mathscr{E}$ observed in V.	0·40089	0·40134	0·40163	0·40190
$\mathscr{E}$ calculated in V.	—	0·40134	0·40165	0·40189

13·431. If the cell contains solutions we may write

$$\mu_i = \mu_i^*(T) + RT \log (\gamma_i x_i) \qquad (13\cdot431\cdot1)$$

for every ionic species i. We then have

$$-\mathscr{N}\mathscr{F}\mathscr{E} = \sum_{\text{solids}} \nu_i \mu_i + \sum_{\text{solutions}} \nu_i \{\mu_i^*(T) + RT \log (\gamma_i x_i)\}, \qquad (13\cdot431\cdot2)$$

i.e.

$$\mathscr{E} = \mathscr{E}_0(T) - \frac{RT}{\mathscr{N}\mathscr{F}} \sum_{\text{solutions}} \nu_i \log (\gamma_i x_i), \qquad (13\cdot431\cdot3)$$

where

$$\mathscr{E}_0(T) = \frac{RT}{\mathscr{N}\mathscr{F}} K_P(T), \quad K_P(T) = -\sum_{\text{solutions}} \frac{\nu_i \mu_i^*}{RT} - \sum_{\text{solids}} \frac{\nu_i \mu_i}{RT}. \qquad (13\cdot431\cdot4)$$

If $\mathscr{E}$ can be measured as a function of the composition we can determine $K_P(T)$ and the activity coefficients γ_i. This is one of the most valuable methods of measuring equilibrium constants and activities.

13·432. *Concentration cells.* The Clark cell, which has the reaction

$$-\text{Zn (solid)} + \text{ZnSO}_4 \text{ (aqueous)} - \text{HgSO}_4 \text{ (solid)} + \text{Hg (liquid)}, \qquad (13\cdot432\cdot1)$$

has an electromotive force which depends upon the concentration of the zinc sulphate solution as well as on the temperature, the electromotive force increasing as the concentration diminishes. Here we have $\mathscr{N} = 2$ and two ions Zn^{2+} and SO_4^{2-}. Therefore

$$\mathscr{E} = \mathscr{E}_0(T) - \tfrac{1}{2}(RT/\mathscr{F}) \log (\gamma_{Zn^{2+}} x_{Zn^{2+}} \gamma_{SO_4^{2-}} x_{SO_4^{2-}}). \qquad (13\cdot432\cdot2)$$

Since $x_{Zn^{2+}} = x_{SO_4^{2-}}$, this can be written as

$$\mathscr{E} = \mathscr{E}_0(T) - (RT/\mathscr{F}) \log (\gamma x_{Zn^{2+}}), \qquad (13\cdot432\cdot3)$$

where

$$\gamma^2 = \gamma_{Zn^{2+}} \gamma_{SO_4^{2-}}. \qquad (13\cdot432\cdot4)$$

If we have two Clark cells at the same temperature in opposition with different concentrations, we have a pure concentration cell and

$$\mathscr{E} = \mathscr{E}_1 - \mathscr{E}_2 = \frac{RT}{\mathscr{F}} \log \frac{\gamma^{(2)} x^{(2)}_{Zn^{2+}}}{\gamma^{(1)} x^{(1)}_{Zn^{2+}}}. \tag{13·432·5}$$

13·433. *Concentration cells with transference.* In the whole of the preceding discussion we have implicitly assumed that ions are transferred only from solids (or from gases) to solutions, and vice versa, and that no transfer of ions takes place between two solutions. This assumption is never strictly true, though it is often an extremely good approximation. For example, if we consider the reaction (13·411·1), both the lead chloride and the mercurous chloride are slightly soluble in the aqueous solution of potassium chloride, and there is a continuous change from a solution of potassium chloride and lead chloride near one electrode to a solution of potassium chloride and mercurous chloride near the other electrode. In a strict analysis of the reactions in the cell we ought therefore to take into account the possible movement of lead and mercurous ions. However, the concentrations of these ions are so small that the fraction of the current which they could carry is negligible, and our previous analysis is correct for all practical purposes. Cells in which the composition of the electrolyte solution effectively does not vary with position between the electrodes are called cells without transference (as are also cells with several intermediate electrodes, composed of units each of which is a cell without transference). In all other cells there is a marked variation in composition (either gradual or abrupt) in the electrolyte, so that we must consider the electrolytic solution as being composed of two or more different liquids in direct contact with one another. Such cells are called cells with transference, since we have to take explicitly into account the transference of ions directly from one solution to the next.

A cell which consists of two zinc electrodes dipping into two solutions of zinc sulphate of different compositions, which are separated by a porous membrane, is reversible if we neglect the spontaneous diffusion of the two solutions into one another. The positive current in the cell is from the less to the more concentrated solution. In this cell the two solutions are in contact with one another and the current is carried by both the Zn^{2+} and the SO_4^{2-} ions. We therefore need to know the fraction α of the current carried by the SO_4^{2-} in order to calculate the electromotive force. If $\mathscr{F}$ units of charge are produced in the cell, 1 gram equivalent of zinc dissolves from the negative electrode (a gram equivalent is equal to a gram molecule divided by the valency $\mathscr{N}$), but $1-\alpha$ gram equivalents of zinc are transferred from the cathode compartment to the anode compartment, so that the net increase in the zinc in the cathode compartment is α gram equivalents, while α gram equivalents of SO_4^{2-} ions are transferred in the opposite direction, and 1 gram equivalent of zinc is deposited on the anode. The chemical reaction may therefore be considered to be

$$\alpha\, Zn^{2+}\,(\text{mole fraction } x_1) + \alpha\ SO_4^{2-}\,(\text{mole fraction } x_1)$$
$$-\alpha\, Zn^{2+}\,(\text{mole fraction } x_2) - \alpha\ SO_4^{2-}\,(\text{mole fraction } x_2) = 0, \quad (13\cdot433\cdot1)$$

and the electromotive force is given by

$$\mathscr{E} = -\frac{\alpha RT}{2\mathscr{F}} \log \frac{\gamma^{(1)}_{Zn^{2+}} x^{(1)}_{Zn^{2+}} \gamma^{(1)}_{SO_4^{2-}} x^{(1)}_{SO_4^{2-}}}{\gamma^{(2)}_{Zn^{2+}} x^{(2)}_{Zn^{2+}} \gamma^{(2)}_{SO_4^{2-}} x^{(2)}_{SO_4^{2-}}} = \frac{\alpha RT}{\mathscr{F}} \log \frac{\gamma^{(2)} x^{(2)}_{Zn^{2+}}}{\gamma^{(1)} x^{(1)}_{Zn^{2+}}}, \quad (13\cdot433\cdot2)$$

since $x_{Zn^{2+}} = x_{SO_4^{2-}}$ and $\gamma^2 = \gamma_{Zn^{2+}} \gamma_{SO_4^{2-}}$. More complicated cells can be treated in the same way.

To utilize equation (13·433·2) we need to be able to measure or calculate all the quantities concerned, and the procedure will be different for different cells. We may, for example, be able to obtain the values of α and γ from other measurements, in which case we can calculate $\mathscr{E}$. Alternatively, if we can find the γ's from, say, the freezing-points of the solutions, measurements of $\mathscr{E}$ enable us to find α.

Résumé of the Methods of Determining the Free Energy

13·5. All the properties of a thermodynamic system can be obtained if we know the free-energy function $G(T, P, n_i)$ (or its generalization for systems characterized by other parameters than the pressure and the volume). Alternatively, since

$$G(T, P, n_i) = \sum_i n_i \mu_i(T, P, n_j), \quad (13\cdot5\cdot1)$$

it is sufficient if we know all the μ_i's. In the preceding chapters, various methods have been outlined by which we can obtain G or the μ_i's for solids, liquids and gases, and it is convenient at this point to summarize them briefly.

13·51. *Solids and pure liquids.* For solids and pure liquids μ is very nearly independent of the pressure and can usually be taken to be a function of the temperature only. The quantity $\mu(T) - H_0$ can be found from the specific heat or from the heat function by the formulae

$$\mu(T) - H_0 = \int_0^T \left(1 - \frac{T}{T'}\right) C_P(T')\, dT' = -T \int_0^T \frac{dT'}{T'^2} \int_0^{T'} C_P(T'')\, dT'' \quad (13\cdot51\cdot1)$$

$$= -T \int_0^T \{H(T') - H_0\} \frac{dT'}{T'^2}. \quad (13\cdot51\cdot2)$$

In accordance with the third law of thermodynamics, we have taken the entropy to be zero at $T=0$. For high pressures the dependence of μ upon P cannot be neglected, and we may then calculate $\mu(T,P)$ from the formula

$$\mu(T,P)=\mu(T,0)+\int_0^P v(T,P')\,dP'. \qquad (13\cdot51\cdot3)$$

The quantity $H(T,P)-H_0$ is directly measurable by calorimetric means, but the constant H_0 is arbitrary. We can, however, find $\Delta H_0=\Sigma\nu_i H_{0,i}$, the change in the heat functions of a number of solids at $T=0$ in a chemical reaction. To do this we have to measure $\Delta H(T,P)=\Sigma\nu_i H_i(T,P)$ and all the quantites $H_i(T,P)-H_{0,i}$, $\Delta H(T,P)$ being obtained either directly from the heat of reaction or indirectly from a combination of the heats of reactions of two or more associated reactions. The same arguments regarding the heat functions apply to gases and to solutions.

Finally, if we carry out a given reaction between solids and pure liquids in a reversible electric cell, we can find $\Sigma\nu_i\mu_i(T)$ from the relation

$$\Sigma\nu_i\mu_i(T)=-\mathcal{N}\mathcal{F}\mathcal{E}. \qquad (13\cdot51\cdot4)$$

13·52. *Gases.* We can obtain the free energy of a perfect gas from measurements of its specific heat by means of equation (3·52·4), which gives

$$\mu(T,P)_{\text{ideal}}=\int_0^{T} C_P(T')\,dT'-C_P^0 T\log\frac{T}{T^\dagger}-T\int_0^{T}\frac{C_P^1(T')}{T'}\,dT' + RT\log\frac{P}{P^\dagger}+U_0-TS_0. \qquad (13\cdot52\cdot1)$$

For a real gas we may then apply equation (8·5·3), namely,

$$\mu(T,P)=\mu(T,P)_{\text{ideal}}+\int_0^P\left(v(T,P')-\frac{RT}{P'}\right)dP', \qquad (13\cdot52\cdot2)$$

but these equations do not enable us to determine the entropy constant. To do this, we can measure the free energy of the solid and of the liquid up to the boiling-point, and this must be equal to the free energy of the gas for that particular temperature and pressure. In this way we can obtain $\mu(T,P)-H_0$, the entropy constant having been automatically chosen so as to satisfy the third law of thermodynamics. Alternatively, we can use the theoretical value of the entropy constant calculated from one of the formulae (5·4·13), (6·17·2) and (6·17·4).

We can adopt the same method for gas mixtures, basing the calculations essentially on equation (11·3·2), but in general our knowledge of the equation of state for imperfect gas mixtures is insufficient to enable the calculations to be carried out, so that most investigations refer to perfect gases. The thermodynamic potential μ_i of the ith constituent of a perfect gas mixture can be written in the form

$$\mu_i(T, P, n_j) = \mu_i^*(T) + RT \log p_i \tag{13·52·3}$$

(see equation (11·42·3)), where $p_i = n_i P/\Sigma n_j$ is the partial pressure of the ith gas, and $\mu_i^*(T) - H_{0,i}$ can be found from equation (11·2·4), since the entropy constant $s_{0,i}$ is the same as for the pure gas.

The quantity $\Sigma \nu_i \mu_i^*(T)$ for a given chemical reaction can be found from the equilibrium constant K_p (equation (11·42·5)) or from the electromotive force of an electric cell in which the reaction is allowed to take place.

In practical work the quantities $\mu_i^*(T)$ and $\Sigma \nu_i \mu_i^*(T)$ are often called the standard free energy and the standard free-energy change in a chemical reaction. This nomenclature is unilluminating and is best avoided.

13·53. *Solutions.* If we write the thermodynamic potential of the ith constituent in a solution in the form

$$\mu_i(T, n_j) = \mu_i^*(T) + RT \log \{\gamma_i(T, x_j)\, x_i\}, \tag{13·53·1}$$

we can determine the quantities

$$\Sigma \nu_i \mu_i^*(T), \quad \Pi \gamma_i^{\nu_i} \tag{13·53·2}$$

for a reaction between the constituents if we can measure the electromotive force of an electric cell, in which the reaction takes place, as a function of the concentrations (see equations (13·431·3) and (13·431·4)).

Other methods available for the determination of the free energy of a solution are the measurement of the freezing point, the boiling point, the vapour pressure and the osmotic pressure (see § 12·46).

REFERENCES

Brönsted, J. N. (1923). The concept of acids and bases. *Rec. Trav. chim. Pays-Bas*, **42**, 718.

Debye, P. & Hückel, E. (1923). The theory of electrolytes. *Phys. Z.* **24**, 185.

Falkenhagen, H. (1934). *Electrolytes.* Oxford.

Fowler, R. H. (1936). *Statistical mechanics*, 2nd ed. Cambridge.

Glasstone, S. (1945). *Electrochemistry of solutions*, 2nd ed. London.

Guggenheim, E. A. (1935). Specific thermodynamic properties of aqueous solutions of strong electrolytes. *Phil. Mag.* (7), **19**, 588.

Milner, S. R. (1912). The virial of a mixture of ions. *Phil. Mag.* (6), **23**, 551.

Chapter 14

FURTHER TOPICS IN SOLIDS

RUBBER

14·1. *The structure of rubber.* The thermal and elastic properties of rubber differ profoundly from those of normal solids. In the first place, the amount by which rubber can be stretched reversibly is extremely large, and in the second place the tension in a piece of stretched rubber, instead of decreasing as the temperature is raised, is roughly proportional to the absolute temperature provided that the extension is not too small.

Rubbers, either natural or synthetic, are high polymers of considerable complexity. Natural rubber consists of long chains formed of isoprene units C_5H_8, and a molecule of rubber can be denoted by

$$(\text{—C(CH}_3\text{)=CH—CH}_2\text{—CH}_2\text{—})_n,$$

where n is of the order of 100–5000. In such molecules, internal rotations are possible round each single carbon-carbon bond provided that the temperature is high enough for the kinetic energy of rotation to be sufficient to overcome the potential barriers to complete rotation (compare § 6·15). No such rotation, however, can occur round the double bonds, and the $\text{—C—C(CH}_3\text{)=CH—CH}_2\text{—}$ structure must be treated as a semi-rigid one in which the carbon atoms can only execute small oscillations about their equilibrium positions. (Note that the CH_3 group can still rotate as a rigid structure round the bond joining it to the main chain.) There are then two essentially different configurations of the chain according as the single carbon-carbon bonds adjacent to a double bond are in the *cis*-position ⟍═⟋ or in the *trans*-position ⟍⫽⟍. Both possibilities occur, Hevea rubber being composed of *cis*-molecules, while Balata rubber or gutta-percha is formed of *trans*-molecules (see fig. 14·1). In natural rubbers, the molecular configurations are either all *cis* or all *trans*. In synthetic materials, on the other hand, it is possible for the *cis* and *trans* configurations to be distributed at random over the length of a chain.

Since rotation can occur to a greater or less extent round every single carbon-carbon bond, the long rubber molecules can very readily take up highly coiled non-planar configurations unless prevented from doing so by constraints. In very pure rubber there are no restraining forces, and the rubber behaves like a highly viscous liquid. To obtain rubber in a useful form it is necessary to vulcanize it, a process which consists in attaching sulphur atoms to certain of the double bonds and forming cross-linkages between the various long chains. Ordinary commercial solid rubber is therefore a branched high polymer. The constraints introduced by the cross-links are sufficient to give the rubber some rigidity of shape and to help to straighten the molecules, but the molecules are still highly coiled, and they can be straightened still further by stretching the rubber. Further, since the coiling and uncoiling of the molecules take place by free rotation round the carbon-carbon bonds, they are accomplished with very little change in the internal energy. The entropy, on the other hand, is clearly less for straight chains than for coiled chains, and we may write the free energy of a strip of rubber of length l as $F = \text{constant} - TS(l, T)$, where $S(l, T)$ is a decreasing function of l. The tension is $\partial F/\partial l = -T\,\partial S/\partial l$, which therefore increases with T.

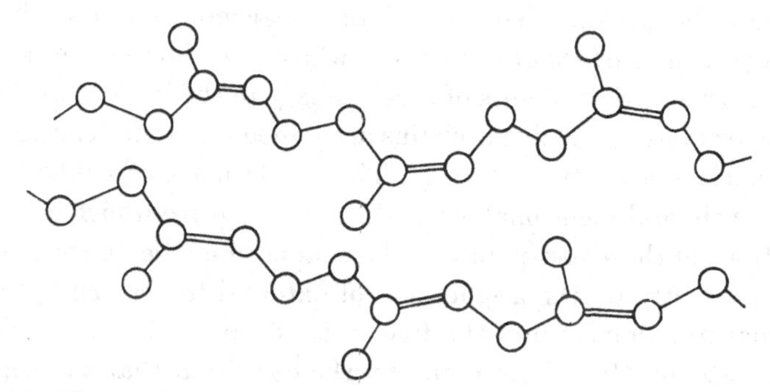

Fig. 14·1. The configurations of a rubber molecule in the *cis* and *trans* forms.

To make these statements more precise we have to calculate the partition function Z_{system} and the free energy of a given mass of rubber as functions of the deformation coordinates of the rubber, where

$$F = -kT \log Z_{\text{system}}, \quad Z_{\text{system}} = \sum_i W(U_i) \exp(-U_i/kT). \qquad (14\cdot1\cdot1)$$

If we assume that the internal energy is independent of the degree of coiling of the molecules, the problem is reduced to the calculation of the number of configurations W that a molecule can take up when its end-points are held in arbitrary fixed positions. This is a problem of extreme mathematical complexity, to which no rigorous solution has yet been found. There are, however, numerous approximate theories which lead to more or less the same end-results, whether their foundations are soundly or insecurely based.

The physical basis of the various mathematical theories of the elasticity of rubber was outlined by Meyer, von Susich & Valko (1932) and by Busse (1932), while the calculation of the probable configurations of a coiled chain was first considered by Kuhn (1934) and by Guth & Mark (1934). More elaborate theories have been given by Wall (1942, 1943 *b*), by James (1942), by James & Guth (1943, 1953), by Flory & Rehner (1943) and by Treloar (1943). Of these theories, the most soundly based is that due to James & Guth, which we adopt here. A simplified version of this theory has been given by James & Guth (1949). The account which follows is restricted to the most important theoretical aspects of the problem. For a detailed description of the physical properties of rubber and a more elaborate discussion of some of the theories proposed, the reader is referred to the monograph by Treloar (1949).

14·11. *The network model of rubber.* We consider a mass of rubber which, when in equilibrium, is in the form of a rectangular parallepiped and which is assumed to be isotropic. The long-chain linear molecules have initial and end-points on the surface of the rubber whose coordinates are denoted by $\mathbf{r}_i$, and there are various points of cross-linkage in the interior of the rubber where two or more of the linear chains are joined together. The coordinates of the junctions are denoted by $\boldsymbol{\xi}_i$. The whole mass of rubber therefore consists of a three-dimensional network, which runs from surface to surface. The positions of those end-points of the chains which lie on the surface can be altered directly by the application of external forces. The positions of the internal junctions and of the free ends of any chains which do not lie on the surface are then determined by the condition that the whole mass must be in equilibrium. In the present state of the theory we can give no detailed explanation of the rigidity of rubber. We must therefore introduce this as an *ad hoc* assumption.

If the chains are highly coiled, the number of configurations consistent with a given separation r between the initial and end-points of a chain will be very large. Also, if $\overline{r^2}$ is the mean of the square of the separation, the distribution function giving the number of configurations of the chain as a function of r will be of the Gaussian form. We therefore first consider the calculation of the distribution function for the configurations of a single chain.

14·12. *The configuration of a long-chain molecule.* The simplest long-chain molecule to consider is one which contains n carbon-carbon bonds, each of length l, about which free rotation is possible, the angle between adjacent bonds being fixed and equal to α (Eyring, 1932). Let the bonds be denoted by vectors $\mathbf{l}_1, \mathbf{l}_2, \ldots, \mathbf{l}_n$, and let $\mathbf{r}$ be the position vector of the end of the chain relative to the initial atom. Then $\mathbf{r} = \Sigma \mathbf{l}_i$, where $\mathbf{l}_i^2 = l^2$ and $\mathbf{l}_i \,.\, \mathbf{l}_{i+1} = l^2 \cos\alpha$. To determine the average value $\overline{r^2}$ of $\mathbf{r}^2$, we use the fact that the component of $\mathbf{l}_2$ in the direction of $\mathbf{l}_1$ is $l\cos\alpha$, while the average component of $\mathbf{l}_3$ in the direction of $\mathbf{l}_1$ is $l\cos^2\alpha$, since there is free rotation round every bond. Similarly, the average component of $\mathbf{l}_i$ in the direction of $\mathbf{l}_1$ is $l\cos^i\alpha$. Also the average values of $\mathbf{l}_i.\mathbf{l}_{i+j}$ $(i=1,\ldots,n-j;\, j=1,\ldots,n-2)$ are independent of i, and they are therefore equal to $l^2\cos^j\alpha$. But

$$r^2=\sum_i \mathbf{l}_i^2+2\sum_{k>i}\mathbf{l}_i.\mathbf{l}_k=\sum_{i=1}^{n}\mathbf{l}_i^2+2\sum_{j=1}^{n-1}\sum_{i=1}^{n-j}\mathbf{l}_i.\mathbf{l}_{i+j}, \qquad (14·12·1)$$

and so
$$\overline{r^2}=nl^2+2\sum_{j=1}^{n-1}\sum_{i=1}^{n-j}\overline{\mathbf{l}_i.\mathbf{l}_{i+j}}=nl^2+2l^2\sum_{j=1}^{n-1}(n-j)\cos^j\alpha. \qquad (14·12·2)$$

On summing the series we obtain the alternative expression

$$\overline{r^2}=\frac{1+\cos\alpha}{1-\cos\alpha}nl^2-2\frac{1-\cos^n\alpha}{1-\cos\alpha}l^2\cos\alpha, \qquad (14·12·3)$$

which, for large values of n, becomes

$$\frac{\overline{r^2}}{nl^2}=\frac{1+\cos\alpha}{1-\cos\alpha}+O\left(\frac{1}{n}\right). \qquad (14·12·4)$$

For a paraffin chain $\alpha=70°\,30'$, $\cos\alpha=\frac{1}{3}$ and so $\overline{r^2}=2nl^2$. (For paraffin chains $l=1·54\times10^{-8}$ cm.) If we assume that the distribution function for $\mathbf{r}$ is the normal Gaussian function, which will be true for sufficiently small values of r/nl (see also § 14·141), the probability that the position vector of the relative displacement of the ends of the chain will have a length lying between r and $r+dr$ is given by

$$4\pi r^2P(r)\,dr=\frac{4}{\sqrt{\pi}}\left(\frac{3}{2\overline{r^2}}\right)^{\frac{3}{2}}\exp\left(-\tfrac{3}{2}r^2/\overline{r^2}\right)r^2dr. \qquad (14·12·5)$$

14·121. *Molecules containing double bonds.* In the long-chain rubber molecules every fourth carbon-carbon bond is a double bond, and the equations of the preceding section need modification. The requisite formulae have been worked out by Wall (1943*a*), but, on account of their complexity, they are not reproduced here. There are two changes necessary. First, the interatomic distance l' is less for a double bond than for a single bond, and secondly the angle β between the direction of the double bond and the neighbouring single bond is less than α the angle between two single bonds. With the experimental values $l=1·54\times10^{-8}$ cm., $l'=1·34\times10^{-8}$ cm., $\alpha=68°\,30'$, $\beta=55°\,40'$, Wall finds that

$$\overline{r^2}/n=8·4\times10^{-16}\text{ cm.}^2 \qquad (14·121·1)$$

for a chain consisting of *trans*-molecules,

$$\overline{r^2}/n=4·1\times10^{-16}\text{ cm.}^2 \qquad (14·121·2)$$

for a chain consisting of *cis*-molecules, and

$$\overline{r^2}/n = 6 \times 10^{-16}\,\text{cm.}^2 \tag{14·121·3}$$

for a chain in which the *cis* and *trans* configurations are distributed at random over the chain.

14·13. *The partition function.* We suppose that the mass of rubber, which is a rectangular parallepiped of sides L_x^0, L_y^0, L_z^0 and volume $V_0 = L_x^0 L_y^0 L_z^0$ when under no forces, is stretched so that the sides are of lengths L_x, L_y, L_z and the volume is $V = L_x L_y L_z$. We can write down the partition at once if we assume that the number of configurations of the rubber can be expressed in terms of the coordinates of the surface end-points $\mathbf{r}_i$ of the chains by the obvious generalization of equation (14·12·5), namely,

$$W(\mathbf{r}_i) = \text{constant} \times \exp\left\{-\frac{3}{2l^2} \sum_{i>j} \frac{1}{n_{ij}} (\mathbf{r}_i - \mathbf{r}_j)^2\right\}, \tag{14·13·1}$$

where n_{ij} is the average number of links in the chains joining the points $\mathbf{r}_i$ and $\mathbf{r}_j$. (The quantity n_{ij} is, however, not well defined, since there must be many possible paths from $\mathbf{r}_i$ to $\mathbf{r}_j$, and we are probably not justified in distinguishing between n_{ij} and a common average value $\bar{n}$.) Now

$$x_i = x_i^0 L_x/L_x^0 = \alpha_x x_i^0, \quad y_i = y_i^0 L_y/L_y^0 = \alpha_y y_i^0, \quad z_i = z_i^0 L_z/L_z^0 = \alpha_z z_i^0, \tag{14·13·2}$$

and so (14·13·1) can be written as

$$W(\mathbf{r}_i) = \text{constant} \times \exp\left[-\tfrac{1}{2}K(\alpha_x^2 + \alpha_y^2 + \alpha_z^2)\right], \tag{14·13·3}$$

where

$$K = \frac{3}{l^2} \sum_{i>j} \frac{1}{n_{ij}} (x_i^0 - x_j^0)^2 = \frac{3}{l^2} \sum_{i>j} \frac{1}{n_{ij}} (y_i - y_i^0)^2 = \frac{3}{l^2} \sum_{i>j} \frac{1}{n_{ij}} (z_i - z_i^0)^2, \tag{14·13·4}$$

the three expressions being equal since the rubber is isotropic. We may write this as

$$K = \frac{N}{l^2} \frac{\overline{L_0^2}}{\bar{n}}, \tag{14·13·5}$$

where N is the number of chains, $\bar{n}$ is the average number of links in the chains and $\overline{L_0^2}$ is the average of the square of the distance apart of the initial and end-points of the chains lying on the surface of the rubber. None of these quantities is well defined, and the expression (14·13·5) is therefore merely a convenient alternative form of (14·13·4).

We can verify the form assumed above for $W(\mathbf{r}_i)$ in the following way. Let $W(\mathbf{r}, \boldsymbol{\xi})\,d\boldsymbol{\xi}$ be the number of configurations of a chain whose initial atom has the position vector $\mathbf{r}$ while the position vector of the junction with another chain lies in the volume $d\boldsymbol{\xi} = d\xi\, d\eta\, d\zeta$ centred round $\boldsymbol{\xi}$. Similarly, let $W(\xi, \xi')\,d\xi'$ be the corresponding number of configurations for the portion of a chain between two cross-links. Then

$$Z_{\text{system}} = \int \prod_{i,j,k} W(\mathbf{r}_i, \xi_j)\, W(\xi_j, \xi_k)\, d\xi_j d\xi_k \exp\left[-U(\mathbf{r}_i)/kT\right], \quad (14\cdot13\cdot6)$$

where $U(\mathbf{r}_i)$ depends only upon the positions of the external boundaries so long as there is no energy change in coiling and uncoiling a chain. For most purposes we can assume that U is a function of the volume V only. Now if the distribution function is the Gaussian function,

$$-\log\{\Pi\, W(\mathbf{r}_i, \xi_j)\, W(\xi_j, \xi_k)\}$$

is a homogeneous quadratic function in the r's and the ξ's, which by a linear transformation of the ξ's can be transformed into

$$\sum_l \gamma_l \xi_l'^2 + \sum_{i,j} \gamma_{ij} \mathbf{r}_i \cdot \mathbf{r}_j,$$

so that we may write (14·13·6) as

$$Z_{\text{system}} = \exp\left[-U(V)/kT\right] \int \exp\left[-\sum_{i,j} \gamma_{ij} \mathbf{r}_i \cdot \mathbf{r}_j - \sum_l \gamma_l \xi_l'^2\right] d\xi_1 d\xi_2 \ldots. \quad (14\cdot13\cdot7)$$

If we substitute for x_i, y_i and z_i from (14·13·2), this becomes

$$Z = \text{constant} \times \exp\left[-\tfrac{1}{2}K(\alpha_x^2 + \alpha_y^2 + \alpha_z^2) - U(V)/kT\right], \quad (14\cdot13\cdot8)$$

where
$$K = 2\sum_{i,j} \gamma_{ij} x_i^0 x_j^0 = 2\sum_{i,j} \gamma_{ij} y_i^0 y_j^0 = 2\sum_{i,j} \gamma_{ij} z_i^0 z_j^0, \quad (14\cdot13\cdot9)$$

the three expressions being equal since the rubber is isotropic. We have therefore shown that $W(\mathbf{r}_i)$ is of the form given in (14·13·3).

The variable part of the free energy is given by

$$F = U(V) + \tfrac{1}{2}K(\alpha_x^2 + \alpha_y^2 + \alpha_z^2)\, kT. \quad (14\cdot13\cdot10)$$

To determine the external normal forces X, Y, Z on the faces of the block, which balance the internal tensions of the chains, and which tend to increase L_x, L_y, L_z, we have

$$X = \partial F/\partial L_x, \quad Y = \partial F/\partial L_y, \quad Z = \partial F/\partial L_z, \quad (14\cdot13\cdot11)$$

where $V = L_x L_y L_z$. Hence

$$\left.\begin{aligned} X &= (\partial U/\partial V)\, L_y L_z + KkT L_x/L_x^{02}, \\ Y &= (\partial U/\partial V)\, L_z L_x + KkT L_y/L_y^{02}, \\ Z &= (\partial U/\partial V)\, L_x L_y + KkT L_z/L_z^{02}. \end{aligned}\right\} \quad (14\cdot13\cdot12)$$

Let V^* be the equilibrium volume of the rubber when $T = 0$. Then $U(V)$ is a minimum at $V = V^*$, and we may write

$$U(V) = U(V^*) + \tfrac{1}{2}(V - V^*)^2\, U''(V^*) + \ldots \quad (U''(V^*) > 0). \quad (14\cdot13\cdot13)$$

When $T \neq 0$, the stress-free state is obtained from (14·13·12) and (14·13·13) by putting $X = Y = Z = 0$, $\alpha_x = \alpha_y = \alpha_z = 1$, which gives

$$V_0(T) = V^* - KkT/\{V_0(T)\, U''(V^*)\}. \qquad (14\cdot 13\cdot 14)$$

Thus, owing to the tendency of the chains to contract as the temperature, and correspondingly the entropy, increases, the coefficient of expansion is negative. This contraction is superposed on the expansion due to the dependence of the internal energy on the temperature, an effect which is excluded from the present theory.

14·131. *Special types of strain.* If the deformation is a simple stretch in the x direction, we have $Y = Z = 0$. The last two of the equations (14·13·12) then give

$$\frac{\partial U}{\partial V} = -\frac{KkT}{V}\alpha_y^2 = -\frac{KkT}{V}\alpha_z^2 = -\frac{KkT}{V}\alpha_y\alpha_z, \qquad (14\cdot 131\cdot 1)$$

and the first of the equations becomes

$$\frac{X}{L_y^0 L_z^0} = \frac{KkT}{V_0}\left(\alpha_x - \frac{V}{V_0}\frac{1}{\alpha_x^2}\right), \qquad (14\cdot 131\cdot 2)$$

where $X/(L_y^0 L_z^0)$ is the force per unit area, referred to the unstretched state. The tension is therefore proportional to the absolute temperature.

If we treat the rubber as incompressible, we have $V = V_0$ and

$$\frac{X}{L_y^0 L_z^0} = \frac{KkT}{V_0}\left(\alpha_x - \frac{1}{\alpha_x^2}\right). \qquad (14\cdot 131\cdot 3)$$

We may derive this relation more directly from (14·13·10) by noting that, if the rubber is incompressible, we must have $\alpha_x\alpha_y\alpha_z = 1$ and therefore that $\alpha_y = \alpha_z = 1/\alpha_x^{\frac{1}{2}}$ in a simple stretch. Then (14·13·10) becomes

$$F = U(V_0) + \tfrac{1}{2}K(\alpha_x^2 + 2/\alpha_x)\,kT, \qquad (14\cdot 131\cdot 4)$$

from which (14·131·3) follows at once.

The initial modulus of elasticity λ is equal to the derivative of $X/(L_y^0 L_z^0)$ with respect to α_x for $\alpha_x = 1$. Hence

$$\lambda = 3KkT/V_0. \qquad (14\cdot 131\cdot 5)$$

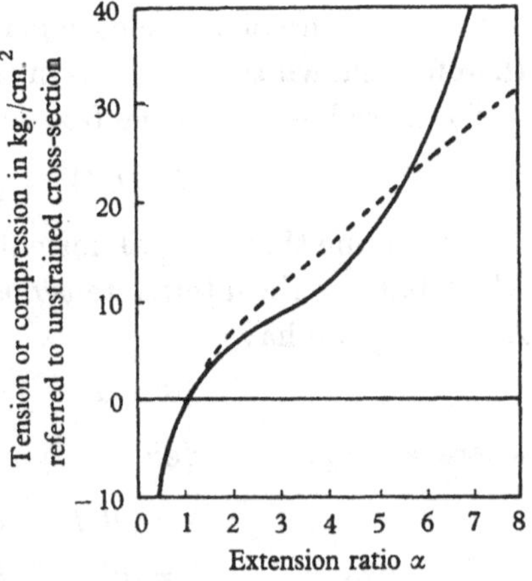

Fig. 14·2. The tension of rubber for moderate extensions.

The variation of $X/(L_y^0 L_z^0)$ with α_x is shown in fig. 14·2. It will be seen from a comparison of the observed and calculated values of $X/(L_y^0 L_z^0)$ that the theory is in accordance with the facts for compressions and for moderate elongations. For values of α_x between about 1·5 and 6, the calculated values of the tension lie slightly above the observed values, while for values of α_x greater than 6 the calculated values of the tension are much too small. This is due to the elongations of the chains becoming so large that we can no longer describe the number of configurations by a Gaussian function.

For the type of deformation known as a pure shear the results are somewhat different. In a pure shear we have $\alpha_z = 1$ and $\alpha_y = 1/\alpha_x$, the shear being measured by $\theta = \alpha_x - 1/\alpha_x$. In this case

$$F = U(V) + \tfrac{1}{2}K(\theta^2 + 3)\,kT \qquad (14\cdot131\cdot6)$$

and

$$\partial F/\partial\theta = KkT\theta. \qquad (14\cdot131\cdot7)$$

14·132. *The molecular weight.* The measured values of the initial modulus give us the quantity K/V_0, but without further assumptions we cannot relate the numerical values of K/V_0 to the properties of the network. If we assume that K is given by an expression such as (14·13·5), then K is of the order of N^*, the number of chain segments connecting the junctions. Also, if M^* is the average molecular weight of the segments and if ρ is the density of the rubber, we have $M^*N^* = \mathscr{L}\rho V$, $\mathscr{L}$ being Avogadro's number, so that

$$\lambda = 3\rho RT/M^*. \qquad (14\cdot132\cdot1)$$

Typical values of λ are 1–4 kg./cm.^{2} at $T = 300°$ K. These give $M^* = 20{,}000$ and 80,000 respectively, which correspond to segment lengths of 300 and 1200 isoprene units.

14·14. *Large elongations.* When the chains become so elongated that their configurations can no longer be described by a Gaussian distribution, the theory is much more complex, and little headway can be made unless further simplifying assumptions are made (James & Guth, 1943; Treloar, 1946). If we assume that adjacent carbon-carbon bonds can take up any direction relative to one another and are not restricted to a fixed angle of inclination, we can obtain the distribution function $P(r)$ for the distance apart of the ends of a chain by the following method.

14·141. Let $P(r)\,dx\,dy\,dz$ be the probability that the end of a chain will lie in a volume element $dx\,dy\,dz$ centred round a point distant r from the origin, which is the position of the initial atom of the chain. Then, if $p(x)\,dx$ is the probability that the x coordinate of the end of a chain will lie between x and $x + dx$, we have

$$p(x)\,dx = dx \iint P(r)\,dy\,dz.$$

We use cylindrical polar coordinates and put $r^2 = x^2 + \rho^2$, $\rho^2 = y^2 + z^2$. Then

$$p(x)\,dx = dx\int_0^\infty P(r)\,2\pi\rho\,d\rho = dx\int_x^\infty P(r)\,2\pi r\,dr,$$

and so
$$P(r) = -\frac{1}{2\pi r}\left(\frac{dp(x)}{dx}\right)_{x=r}. \tag{14·141·1}$$

We can therefore deduce $P(r)$ from $p(x)$, and the latter is the simpler to determine directly.

The function $p(x)$ can be found in various ways (see, for example, Rayleigh, 1919; Irwin, 1927), the shortest being to use Fourier transforms. (For the formulae which follow, reference can be made to Titchmarsh (1937), Chapter 2.)

We define the Fourier transform $\mathscr{F}(t)$ of a function $f(x)$ $(-\infty < x < \infty)$ by the relation

$$\mathscr{F}(t) = \frac{1}{(2\pi)^{\frac{1}{2}}}\int_{-\infty}^{\infty} f(x)\,e^{ixt}\,dx. \tag{14·141·2}$$

Then the inverse Fourier transform is

$$\frac{1}{(2\pi)^{\frac{1}{2}}}\int_{-\infty}^{\infty}\mathscr{F}(t)\,e^{-ixt}\,dt = f(x). \tag{14·141·3}$$

Also the inverse Fourier transform of $\mathscr{F}_1(t)\,\mathscr{F}_2(t)\ldots\mathscr{F}_n(t)$ is

$$\frac{1}{(2\pi)^{\frac{1}{2}}}\int_{-\infty}^{\infty}\mathscr{F}_1(t)\mathscr{F}_2(t)\ldots\mathscr{F}_n(t)\,e^{-ixt}\,dt = \frac{1}{(2\pi)^{\frac{1}{2}(n-1)}}\int_{-\infty}^{\infty} f_1(x_1)\,dx_1\int_{-\infty}^{\infty} f_2(x_2)\,dx_2\ldots$$
$$\ldots\int_{-\infty}^{\infty} f_{n-1}(x_{n-1})f_n(x - x_1 - \ldots - x_{n-1})\,dx_{n-1}. \tag{14·141·4}$$

14·142. Consider a chain consisting of n links each of length l, and let x_i be the x coordinate of the end-point of the ith link relative to the initial point of the link. Also let $f(x_i)\,dx_i$ be the probability that x_i lies in the interval x_i, $x_i + dx_i$, where $-l \leqslant x_i \leqslant l$. Then the probability $p(x)\,dx$ that the x coordinate Σx_i of the end-point of the chain lies in the interval $x, x+dx$ is given by

$$p(x)\,dx = dx\int_{-\infty}^{\infty} f(x_1)\,dx_1\int_{-\infty}^{\infty} f(x_2)\,dx_2\ldots$$
$$\ldots\int_{-\infty}^{\infty} f(x_{n-1})f(x - x_1 - \ldots - x_{n-1})\,dx_{n-1}. \tag{14·142·1}$$

Hence $p(x)$ is the inverse Fourier transform of $(2\pi)^{\frac{1}{2}(n-1)}[\mathscr{F}(t)]^n$.

In the present problem, all values of x_i lying between $-l$ and l are equally likely, and so

$$f(x_i) = 1/(2l) \quad (|\,x_i\,| \leqslant l), \quad f(x_i) = 0 \quad (|\,x_i\,| > l). \tag{14·142·2}$$

Therefore
$$\mathscr{F}(t) = \frac{1}{(2\pi)^{\frac{1}{2}}}\frac{\sin lt}{lt} \tag{14·142·3}$$

and $$p(x) = (2\pi)^{\frac{1}{2}n-1}\int_{-\infty}^{\infty} e^{-itx}[\mathscr{F}(t)]^n\,dt = \frac{1}{2\pi}\int_{-\infty}^{\infty} e^{-itx}\left(\frac{\sin lt}{lt}\right)^n dt. \quad (14\cdot142\cdot4)$$

For large values of n this must tend to the Gaussian function. To verify this we write $\xi = x/(\sqrt{n}\,l)$ and $t = s/(\sqrt{n}\,l)$. Then

$$p(x) = \frac{1}{2\pi\sqrt{n}\,l}\int_{-\infty}^{\infty} e^{-is\xi}\left(\frac{\sin(s/\sqrt{n})}{s/\sqrt{n}}\right)^n ds,$$

and for large n

$$p(x) = \frac{1}{2\pi\sqrt{n}\,l}\int_{-\infty}^{\infty} e^{-is\xi}\left(1-\frac{1}{6}\frac{s^2}{n}+\dots\right)^n ds \to \frac{1}{2\pi\sqrt{n}\,l}\int_{-\infty}^{\infty} e^{-is\xi}\,e^{-\frac{1}{6}s^2}\,ds$$

$$= \frac{\sqrt{3}}{(2\pi n)^{\frac{1}{2}}\,l}e^{-\frac{3}{2}\xi^2} = \frac{\sqrt{3}}{(2\pi n)^{\frac{1}{2}}\,l}\exp\left[-3x^2/(2nl^2)\right]. \quad (14\cdot142\cdot5)$$

The relation (14·141·1) then gives

$$4\pi P(r)\,r^2\,dr = \frac{6\sqrt{3}}{(2\pi)^{\frac{1}{2}}(nl^2)^{\frac{3}{2}}}\exp\left[-3r^2/(2nl^2)\right]r^2\,dr, \quad (14\cdot142\cdot6)$$

from which it follows that $$\overline{r^2} = nl^2. \quad (14\cdot142\cdot7)$$

In the general case of arbitrary n, we can evaluate (14·142·4) by contour integration. If $x > 0$, we deform the line of integration into any curve Γ going from $-\infty$ to ∞ and lying in the lower half of the t plane, and we write

$$p(x) = \frac{1}{2\pi}\int_{\Gamma} e^{-itx+inlt}\sum_{j=0}^{n}(-1)^j\,{}^nC_j\frac{e^{-2ijlt}}{(2ilt)^n}\,dt \quad (14\cdot142\cdot8)$$

and use the fact that

$$\frac{1}{2\pi}\int_{\Gamma}\frac{e^{iaz}}{z^n}\,dz = \begin{cases} 0 & (a<0), \\ \dfrac{i^n a^{n-1}}{(n-1)!} & (a>0). \end{cases} \quad (14\cdot142\cdot9)$$

Hence $$p(x) = \frac{1}{l}\frac{n^{n-1}}{2^n(n-1)!}\sum_{j\leqslant\frac{1}{2}n-\frac{1}{2}x/l}(-1)^j\,{}^nC_j\left(1-\frac{2j}{n}-\frac{x}{nl}\right)^{n-1} \quad (14\cdot142\cdot10)$$

and $$4\pi r^2 P(r) = \frac{r}{l^2}\frac{n^{n-2}}{2^{n-1}(n-2)!}\sum_{j\leqslant\frac{1}{2}n-\frac{1}{2}r/l}(-1)^j\,{}^nC_j\left(1-\frac{2j}{n}-\frac{r}{nl}\right)^{n-2}. \quad (14\cdot142\cdot11)$$

The general behaviour of $\log\{P(r)/P(0)\}$ as a function of $(r/nl)^2$ is shown in fig. 14·3 for $n = 100$.

14·143. It has not so far been possible to incorporate into the theory a general distribution such as (14·142·11) with the same degree either of rigour or of plausibility as a Gaussian distribution. As a tentative generalization we may take over (14·13·8) with $\exp[-\frac{1}{2}K(\alpha_x^2+\alpha_y^2+\alpha_z^2)]$ replaced by $P(u)$, where $u^2 = \frac{1}{3}Knl^2(\alpha_x^2 + \alpha_y^2 + \alpha_z^2)$, since this gives the correct expression for small values of α_x, α_y, α_z. Such a generalization cannot be quantitatively correct, and it is not easy to judge the error introduced by making this assumption, but it will give the order of magnitude of the effects due to the chains being stretched so much that they are reaching their maximum elongations.

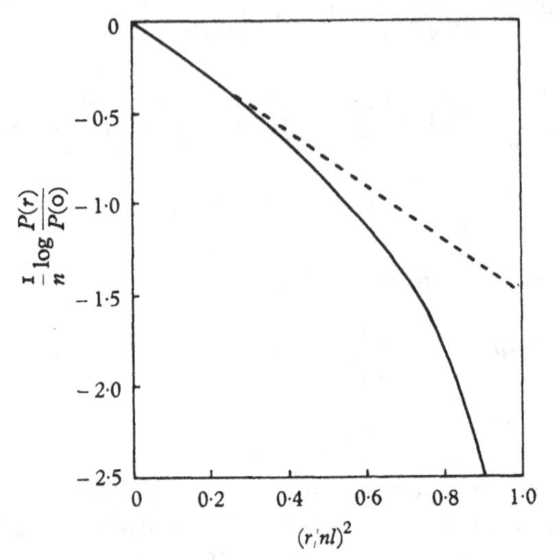

Exact distribution —— Gaussian distribution - - -

Fig. 14·3

For a linear stretch we have

$$F = \text{constant} - kT \log P(u), \tag{14·143·1}$$

where

$$u^2 = \tfrac{1}{3}Knl^2(\alpha_x^2 + 2/\alpha_x), \tag{14·143·2}$$

and so

$$\frac{X}{L_y^0 L_z^0} = \frac{1}{V_0}\frac{\partial F}{\partial \alpha_x} = -\frac{kT}{V_0}\frac{1}{uP}\frac{dP}{du}\tfrac{1}{3}Knl^2\left(\alpha_x - \frac{1}{\alpha_x^2}\right). \tag{14·143·3}$$

The general behaviour of $X/(L_y^0 L_z^0)$ as a function of α_x is shown in fig. 14·4.

14·15. The theory of the elasticity of rubber outlined in the preceding sections can only be considered as a tentative first attack on the problem. Many important features of the real assembly of macro-molecules have been entirely omitted while others have only been incorporated into the theory by making very crude approximations. Nevertheless, the results are sufficiently encouraging for us to feel that the theory is correct in its essentials, though many improvements are still possible.

Superlattices in Alloys

14·2. In a simple substitutional alloy, such as an alloy of gold in copper, each gold atom replaces an atom of copper and occupies the lattice point vacated by the copper atom. In general, the two sets of atoms are distributed at random over the lattice points, but at low temperatures and at fairly well-defined simple compositions the alloys are often in an ordered state, the different atoms being arranged in a definite way just as in an ionic compound like sodium chloride. The hypothesis that such ordered structures occur was first put forward by Tammann in (1919), but it was not until 1925 that Johansson & Linde were able to prove directly, by means of an X-ray analysis, that the CuAu alloy can exist in an ordered state.

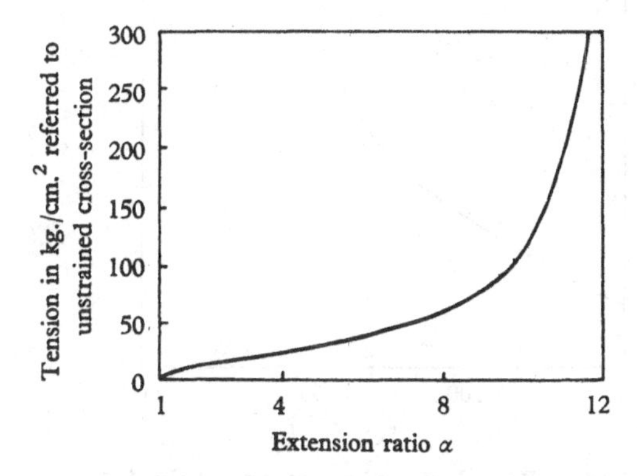

Fig. 14·4. The tension of rubber for large extensions.

It is well known that, on account of the interference of the waves scattered by the atoms at the cube corners and centres, there are no X-ray reflexions of odd orders for a body-centred cubic metal. The same is true for a disordered body-centred alloy, since there must be on the average no difference between the cube corners and the cube centres. If, however, the alloy is in an ordered state with one type of atom, called A, occupying the cube corners and the other type of atom, called B, occupying the cube centres, the odd-order X-ray reflexions must occur, since they cannot now be destroyed by the interference of the waves scattered by the atoms at the corners with those scattered by the atoms at the centres, the scattering powers of the two types of atom being different. Thus the appearance of the odd-order X-ray reflexions in a body-centred alloy is evidence that an ordered state or superlattice exists. Similarly, in a face-centred cubic alloy the appearance of the $(1, 1, 0)$ reflexions is the criterion for the existence of a superlattice.

When an orderable alloy in the disordered state is cooled, it passes at a certain temperature, known as the transition temperature, into the ordered state, sometimes gradually and sometimes abruptly. This transition is accompanied by certain characteristic changes in the properties of the alloy, which can be used to determine whether a superlattice is formed in any particular alloy.

The electrical resistance of a disordered alloy is large, owing to the scattering of the electrons by the irregularly distributed atoms, while the

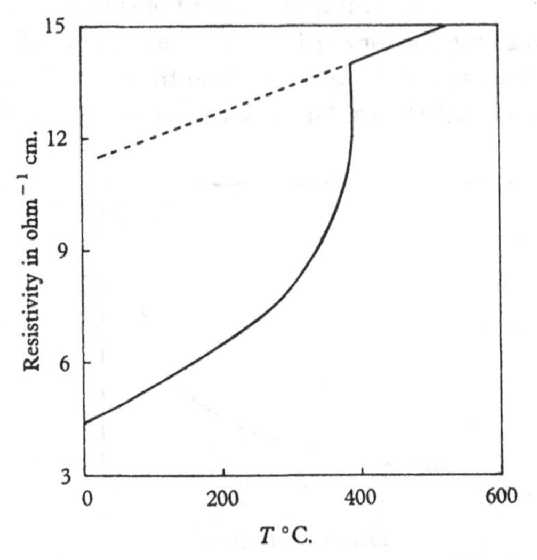

Fig. 14·5. The electrical resistance of Cu_3Au.

resistance of a perfectly ordered alloy must be of the same order as that of a pure metal, the only cause of scattering being the temperature motion of the lattice. The onset of order in an alloy as the temperature is lowered is therefore shown by there being a considerable decrease in the electrical resistance. When the ordering ceases, the resistance behaves normally, i.e. it is proportional to the absolute temperature if $T > \Theta$, where Θ is the Debye temperature. The general behaviour of the resistance is shown by fig. 14·5, which relates to Cu_3Au.

Since the ordered state must have lower energy than the disordered one, the transition from the disordered to the ordered alloy must be accompanied by the evolution of heat. Hence any latent heat or anomalous specific heat which cannot be attributed to a change in crystal structure is evidence of an ordering process.

14·21. *Elementary theory of long-distance order.* We consider first a body-centred cubic alloy consisting of equal numbers of atoms of two metals

A and B (fig. 14·6). For this structure and composition the A atoms occupy the corners of the cube cells in the ordered alloy, while the B atoms occupy the centres of the cubes. (In an infinite lattice there are as many cube centres as cube corners.) We call the cube corners the α sites and the cube centres the β sites, and we denote by r_α the fraction of the α sites occupied by A, i.e. 'right', atoms in any state of the alloy. Similarly, we denote by w_α the fraction of the α sites occupied by B, i.e. 'wrong' atoms, and by r_β and w_β the fractions of the β sites occupied by B and A respectively. We now define a degree of order s (which is such that $s=0$ for complete disorder, i.e. when $r_\alpha = w_\alpha = \frac{1}{2}$, and $s=1$ for perfect order, i.e. when $r_\alpha = 1$, $w_\alpha = 0$) by

$$s = 2r_\alpha - 1. \tag{14·21·1}$$

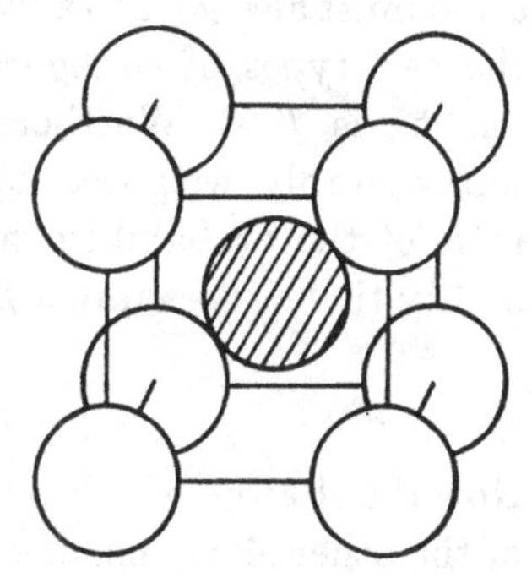

Fig. 14·6. The unit cell of an ordered body-centred cubic lattice.

A more general definition is required if the numbers of A and B atoms are not the same, as is often the case. Let N be the total number of atoms, and N_A and N_B the numbers of A and B atoms. We denote by $\nu_A = N_A/N$ the fraction of the atoms which are A atoms, and if the fraction of the sites which are α sites is also equal to ν_A we define s by

$$s = (r_\alpha - \nu_\alpha)/(1-\nu_A). \tag{14·21·2}$$

We also define ν_B as $N_B/N = 1-\nu_A$. When the number of A atoms is not equal to the number of α sites, a wider definition of s is required (see § 14·23). The order defined by (14·21·1) and (14·21·2) is called long-distance order since it depends on the possibility of distinguishing the α and β sites over macroscopic distances. There are many theories of long-distance order (Johansson & Linde, 1925; Borelius, Johansson & Linde, 1928; Gorsky, 1928; Dehlinger & Graf, 1930; Borelius, 1934; Bragg & Williams, 1934); we only deal here with the simplest aspects. A general review of the subject with extensive references will be found in the article by Nix & Shockley (1938).

Suppose that the production of two new wrong atoms by interchanging an A and a B atom, which were originally on an α and a β site respectively, requires an amount of energy $\mathscr{V}(s)$ when the alloy has order s. We can determine how the order s depends on the temperature T by the following argument. We assume that all the A atoms except one are fixed, and we consider how this one atom can be accommodated in the crystal. The possible positions in which it can be placed are those occupied by B atoms, namely, the $w_\alpha N_A$ sites which are 'right' sites for the A atom and the $r_\beta N_B$ sites which are 'wrong' sites for the A atom. Now, if the A atom goes into one of the 'right' sites, one of the B atoms must go into a β site which is 'right' for it, while if the A atom goes into one of the 'wrong' sites, a B atom must also go into a 'wrong' site. Hence the energy difference between the two types of configuration, with the particular A atom 'wrong' or 'right', is $\mathscr{V}(s)$. We therefore see that, since the 'wrong' positions are unfavourably weighted by the Boltzmann factor $\exp[-\mathscr{V}(s)/kT]$, the ratio of the probability that the A atom goes into an α site to the probability that it goes into a β site is

$$\frac{w_\alpha N_A}{r_\beta N_B \exp[-\mathscr{V}(s)/kT]}. \tag{14·21·3}$$

Now the A atom which we have considered is in no way different from any of the other A atoms, and so (14·21·3) must be the ratio of the number of A atoms on α and β sites. Hence

$$\frac{r_\alpha N_A}{w_\beta N_A} = \frac{w_\alpha N_A}{r_\beta N_B \exp[-\mathscr{V}(s)/kT]},$$

i.e.

$$\frac{r_\alpha r_\beta}{w_\alpha w_\beta} = \exp[\mathscr{V}(s)/kT]. \tag{14·21·4}$$

Now

$$r_\alpha + w_\alpha = r_\beta + w_\beta = \nu_A + \nu_B = 1, \tag{14·21·5}$$

and, since the number of A atoms on β sites must be equal to the number of B atoms on α sites, we have

$$w_\beta N_B = w_\alpha N_A. \tag{14·21·6}$$

When $\nu_A = \nu_B = \frac{1}{2}$, as for a body-centred cubic lattice, we have $w_\alpha = w_\beta$, and hence $r_\alpha = r_\beta$. In this case (14·2·4) gives

$$r_\alpha/(1-r_\alpha) = \exp[\tfrac{1}{2}\mathscr{V}(s)/kT], \tag{14·21·7}$$

and, if we eliminate r_α between this and (14·21·1), we obtain

$$s = \tanh[\tfrac{1}{4}\mathscr{V}(s)/kT]. \tag{14·21·8}$$

In the general case we have, from (14·21·3) and (14·21·6),

$$r_\alpha = \nu_A + \nu_B s, \quad w_\alpha = \nu_B(1-s), \quad r_\beta = \nu_B + \nu_A s, \quad w_\beta = \nu_A(1-s). \tag{14·21·9}$$

Hence, substituting in (14·21·4), we obtain

$$\left(\frac{1}{\nu_A(1-s)} - 1\right)\left(\frac{1}{\nu_B(1-s)} - 1\right) = \exp[\mathscr{V}(s)/kT], \tag{14·21·10}$$

which may be written as

$$\frac{\mathscr{V}(s)}{kT} = \log\left(1 + \frac{s}{\nu_A \nu_B (1-s)^2}\right). \tag{14·21·11}$$

14·211. The relation (14·21·8) or (14·21·10) gives s as a function of T when $\mathscr{V}(s)$ is known. If we treat $\mathscr{V}(s)$ as a constant then, as can be seen most easily from (14·21·8), s increases gradually from 0 to 1 as T decreases from infinity to zero; nothing like a sudden ordering of the alloy can occur. The assumption that $\mathscr{V}(s)$ is a constant is, however, untrue, since it is clear that, when the alloy is completely disordered, the interchange of two atoms does not alter the configuration at all, and hence $\mathscr{V}(0)$ must be zero. The energy $\mathscr{V}(s)$ must therefore depend upon the degree of order which exists in the alloy; it is zero when $s=0$ and it must increase as s increases. We have here to deal with a typical cooperative phenomenon in which the energy required to produce a pair of wrong atoms depends not only on the two atoms involved but also on the number of other atoms which are right or wrong.

Before we can proceed further we must know how $\mathscr{V}(s)$ depends upon s. Since it is difficult to calculate this dependence exactly, Bragg & Williams (1934) made the assumption that $\mathscr{V}(s)$ increases linearly with s, so that

$$\mathscr{V}(s)=s\mathscr{V}_0, \tag{14·211·1}$$

where $\mathscr{V}_0$ is the energy required to produce a wrong pair when the alloy is perfectly ordered. If we substitute this value for $\mathscr{V}(s)$ into (14·21·8) or (14·21·10), we obtain an equation giving s as a function of T.

14·212. We first discuss body-centred cubic lattices, for which (14·21·8) holds; an example is the CuZn alloy. The simplest procedure is to plot the curves given by (14·21·8) and (14·211·1), taking $X=\mathscr{V}_0/kT$ as the abscissa, and find the intersections of the curves. The first curve, whose equation is

$$s=\tanh \tfrac{1}{4}X, \tag{14·212·1}$$

is the same for all temperatures, while the second, whose equation is

$$s=kTX/\mathscr{V}_0, \tag{14·212·2}$$

is a straight line through the origin, whose gradient is proportional to T. The two curves always have one intersection at $s=0$, while, if the gradient of (14·212·2) is less than the gradient of (14·212·1) at the origin, there is a second intersection. This condition is that

$$kT<\tfrac{1}{4}\mathscr{V}_0. \tag{14·212·3}$$

We therefore see that the completely disordered state is a possible one at all temperatures, while if $T<T_c$, where $T_c=\frac{1}{4}\mathscr{V}_0/k$, an ordered state is possible. Now, although the disordered state is a possible one at all temperatures, yet the ordered state has lower free energy at low temperatures (see § 14·222) and hence is the stable state. We can therefore give the

following description of the behaviour of an orderable body-centred alloy which is assumed always to be in thermal equilibrium.

When $T=0$, the alloy is completely ordered and $s=1$. As the temperature is raised, the intersection of the two curves (14·212·1) and (14·212·2) moves to the left, but for not too large T the equilibrium value of s is very nearly 1. Hence, as T increases, the alloy becomes very slightly disordered until a temperature is reached for which the intersection of the curves begins to come on the part of the curve $s=\tanh\frac{1}{4}X$ which has a fairly large gradient. As T is increased past this value, the equilibrium value of s rapidly diminishes until T reaches T_c, when there is only one intersection of the curves, namely, at $s=0$. Above T_c the alloy is completely disordered. We therefore see that, although the disappearance of the order is spread over only a small temperature range below T_c, there is no discontinuity in the order. The actual form of s as a function of T is shown in fig. 14·7.

14·213. We now consider the more general case in which the number of α sites is not the same as the number of β sites, but in which the number of A atoms is still equal to the number of α sites. This situation occurs in the Cu_3Au alloy, which has a face-centred cubic structure. The A atoms are then gold atoms, while the B atoms are copper, and the α and β sites are the corners and the centres of the cube faces respectively, there being three times as many face centres as corners. We have now to find the intersections of the curves (14·21·11) and (14·211·1) with $\mathscr{V}_0$ replaced by kTX, and where, in the particular case of Cu_3Au, $\nu_A=\frac{1}{4}$, $\nu_B=\frac{3}{4}$. The curves are shown in fig. 14·8, the curve (14·21·11) being shown in a somewhat exaggerated form. (Note that (14·21·8) expresses s as a function of $\mathscr{V}(s)$, whereas (14·21·11) gives the inverse relation for $\mathscr{V}(s)$ as a function of s.)

The only difference between this case and the one discussed in the preceding section is that for a certain temperature range there are three intersections of the curves. Of these intersections the middle one corresponds to a state which is intrinsically unstable since a decrease in temperature is accompanied by a decrease in the order. The highest intersection gives a value of s greater than zero, but this disappears abruptly at the temperature T'_c, which is such that the straight line (14·211·1) touches (14·21·10) but not at the origin. We therefore see that the disordered state can exist at all temperatures but that the ordered state can only exist below T'_c; at T'_c the ordered state has $s\neq 0$. The temperature T'_c is not, however, the transition temperature. The transition temperature T_c is the temperature at which the free energy of the disordered state becomes equal to the free energy of the ordered state, and it can be shown (see § 14·224) that the ordered state becomes unstable before it ceases to exist.

It can further be shown that T_c is determined by the condition that, when $X = \mathscr{V}_0/kT$ is taken as the abscissa, the areas of the two domains bounded by the curves (14·21·10) and (14·211·1) should be equal. The proof of this is contained in equation (14·224·2) below. This equation shows that the difference between the two sides of (14·21·11) is proportional to $\partial F/\partial s$, where F is the free energy. The rule of equal areas then means that s is determined by the condition that $F(s) = F(0)$, which is the usual condition

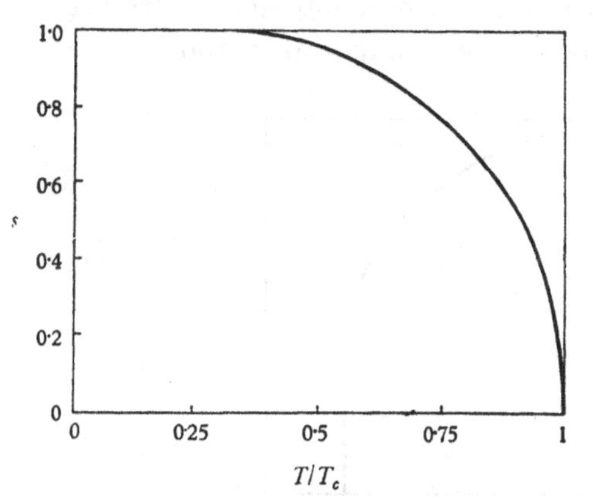

Fig. 14·7. The degree of order as a function of the temperature for body-centred cubic lattices.

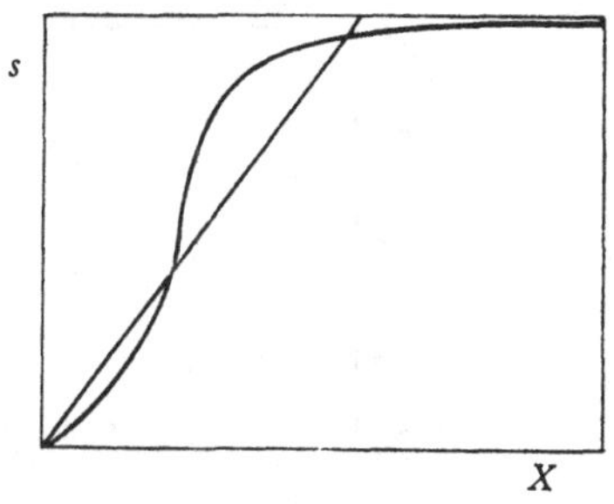

Fig. 14·8

for the heterogeneous equilibrium of two phases in contact with one another. The following is therefore a description of what happens when an ordered alloy of the type AB_3 is heated. At $T = 0$ the alloy is completely ordered; as T increases the alloy becomes more and more disordered until T_c is reached. At this point s is still not zero, but now the alloy becomes a two-phase one, since the free energies of the ordered and disordered states are equal. The transformation of the alloy into the disordered state involves a latent heat, and the temperature remains constant at T_c until the whole of the alloy is transformed into the disordered state. The transition in this case is very similar to an ordinary change of state; the only essentially new point which arises is that above the temperature T_c', which is greater than T_c, the ordered state cannot exist at all and is not merely unstable. The degree of order as a function of T is shown in fig. 14·9.

14·22. *The partition function.* We now give a more detailed analysis of the order-disorder problem, which gives a deeper insight into the assumptions that have been made implicitly, and which enables us to determine the free energy and hence all the thermodynamic properties of the alloy.

We assume that the partition function of the crystal is the product of a partition function relating to the lattice vibrations and a partition function relating to the equilibrium configuration of the two sets of atoms, and that we may confine our attention to the latter. We assume that a macroscopic state of the alloy may be defined by (among other parameters) the long-range order s as measured, for example, by the electrical resistance.

We shall also simplify the calculations by assuming that the forces between the atoms are of such short range that only interactions between nearest neighbours in the crystal lattice are of importance and we shall

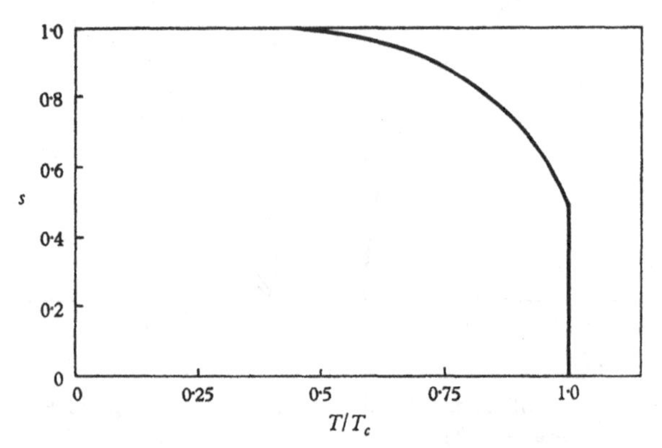

Fig. 14·9. The degree of order as a function of the temperature for face-centred cubic lattices.

assume that every lattice point has z nearest neighbours. We then denote by $\mathscr{V}_{AA}$, $\mathscr{V}_{BB}$ and $\mathscr{V}_{AB}$ the mutual potential energy of two atoms on adjacent lattice sites when both atoms are of type A, or both of type B, or one of type A and the other of type B respectively. If the numbers of nearest neighbours of the various types are characterized by the independent parameters $X, Y, \ldots$, we have to calculate the partition function Z_{system} (denoted by Z for brevity), defined by

$$Z(s, X, Y, \ldots; T) = \sum_i W(U_i; s, X, Y, \ldots) \exp[-U_i(s, X, Y, \ldots)/kT], \tag{14·22·1}$$

where $W(U_i; s, X, Y, \ldots)$ is the number of complexions of the system when its energy is $U_i(s, X, Y, \ldots)$. We ought also to sum the partition function over all values of s, but this would only introduce an unnecessary complication in the notation, since there is always one value of s (and sometimes two) for which Z is a maximum. We may therefore consider s to have a given value, which is to be determined later so as to make Z a maximum.

We use the same notation as in § 14·21, so that there are N_A atoms of type A and sites of type α, and N_B atoms of type B and sites of type β. Also r_α and w_α are the fractions of the α sites occupied by A and B atoms respectively, while r_β and w_β are the fractions of the β sites occupied by B and A atoms respectively. The average numbers of atoms on the various sites are therefore as follows:

A atoms on α sites: $r_\alpha N_A$, B atoms on α sites: $w_\alpha N_A$,

A atoms on β sites: $w_\beta N_B$, B atoms on β sites: $r_\beta N_B$.

When $N_A = N_B$, as in a body-centred cubic lattice where $z=8$, the α and β lattices are equivalent, but when $N_A \neq N_B$ the evaluation of the numbers of the different types of nearest neighbours is somewhat complex. If $N_A \neq N_B$, as in a face-centred cubic lattice where $z=12$, we take $N_A < N_B$. Then each α site has z nearest neighbours, all of which are β sites, while each β site has z nearest neighbours, of which zN_A/N_B are α sites while $z(1-N_A/N_B)$ are β sites.

Let zX denote the number of nearest AA neighbours with one A atom on an α site and the other on a β site. The numbers of the various pairs of neighbours when one atom is on an α site and the other on a β site are then as follows:

A atom on α site, A atom on β site: zX,

A atom on α site, B atom on β site: $z(r_\alpha N_A - X)$,

B atom on α site, A atom on β site: $z(w_\beta N_A - X)$,

B atom on α site, B atom on α site: $z(r_\beta N_A - r_\alpha N_A + X)$.

These expressions are obtained by noting, for example, that the total number of pairs in which one atom is an A atom on a β site and the other atom is either an A or a B atom on an α site, is equal to zN_A multiplied by the fraction w_β of the β sites that are occupied by A atoms.

Similarly, if $\frac{1}{2}zY$ is the number of AA pairs when both atoms are on β sites, the numbers of pairs of various types are as follows:

A atom on β site, A atom on β site: $\frac{1}{2}zY$,

A atom on β site, B atom on β site: $z(1-N_A/N_B)\,w_\beta N_B - zY$,

B atom on β site, B atom on β site: $\frac{1}{2}z(1-N_A/N_B)\,(r_\beta N_B - w_\beta N_B) + \frac{1}{2}zY$.

(Note that, since both atoms are on sites of the same type, we have to introduce a factor $\frac{1}{2}$ into the numbers of AA and BB pairs to avoid double counting. Note also that the total number of pairs in which one atom is an A atom on a β site and the other atom is either A or B on a β site is equal to the product of the number $w_\beta N_B$ of β sites occupied by A atoms and the number $z(1-N_A/N_B)$ of nearest neighbours of a β site that are themselves β sites.)

The numbers n_{AA}, n_{AB} and n_{BB} of pairs of the various types are therefore

$$n_{AA}=z(X+\tfrac{1}{2}Y), \tag{14·22·2}$$

$$n_{AB}=z(r_\alpha N_A+w_\beta N_B-2X-Y)=z(N_A-2X-Y), \tag{14·22·3}$$

$$n_{BB}=\tfrac{1}{2}z\{(w_\alpha-r_\alpha)N_A+(r_\beta-w_\beta)N_B+2X+Y\}=\tfrac{1}{2}z(N_B-N_A+2X+Y), \tag{14·22·4}$$

where we have used (14·21·9) to eliminate r_α, w_α, r_β, w_β. The configurational energy is then

$$U(s,X,Y)=(X+\tfrac{1}{2}Y)\mathscr{V}_0+N_A\mathscr{V}_{AB}+\tfrac{1}{2}(N_B-N_A)\mathscr{V}_{BB}, \tag{14·22·5}$$

where
$$\mathscr{V}_0=\mathscr{V}_{AA}-2\mathscr{V}_{AB}+\mathscr{V}_{BB}. \tag{14·22·6}$$

This is as far as we can proceed in general, and to make any further progress it is necessary to make some approximations, the effect of which cannot easily be estimated.

14·221. We obtain the simplest but least exact theory if we replace the exact partition function (14·22·1) by the expression

$$Z=W(s)\exp\{-\overline{U(s)}/kT\}, \tag{14·221·1}$$

where $\overline{U(s)}$ is an average of the configurational energy obtained by substituting average values of X and Y into (14·22·5), and where $W(s)$ is the total number of complexions for a given degree of order s. W is then the number of ways of dividing the α sites between the A and B atoms multiplied by the number of ways of dividing the β sites between the A and B atoms, i.e.

$$W(s)=\frac{N_A!}{(r_\alpha N_A)!\,(w_\alpha N_A)!}\,\frac{N_B!}{(r_\beta N_B)!\,(w_\beta N_B)!}, \tag{14·221·2}$$

which, by using Stirling's theorem, can be written as

$$\log W=-N_A(r_\alpha\log r_\alpha+w_\alpha\log w_\alpha)-N_B(r_\beta\log r_\beta+w_\beta\log w_\beta). \tag{14·221·3}$$

Also the average value $\overline{X}$ of X is the average number of A atoms on α sites multiplied by the average occupation number of β sites by A atoms, i.e.

$$\overline{X}=r_\alpha N_A w_\beta=\nu_A N_A(1-s)(\nu_A+\nu_B s). \tag{14·221·4}$$

Similarly
$$\tfrac{1}{2}\overline{Y}=\tfrac{1}{2}w_\beta N_B w_\beta(1-N_A/N_B)=\tfrac{1}{2}\nu_A^2(N_B-N_A)(1-s)^2. \tag{14·221·5}$$

Hence
$$\overline{U(s)}=\tfrac{1}{2}N\nu_A^2(1-s^2)\mathscr{V}_0+N_A\mathscr{V}_{AB}+\tfrac{1}{2}(N_B-N_A)\mathscr{V}_{BB} \tag{14·221·6}$$

$$=\tfrac{1}{2}N(\nu_A^2\mathscr{V}_{AA}+2\nu_A\nu_B\mathscr{V}_{AB}+\nu_B^2\mathscr{V}_{BB})-\tfrac{1}{2}N\nu_A^2 s^2\mathscr{V}_0, \tag{14·221·7}$$

and after some simplification we find that the free energy

$$F(s) = -kT \log W + \overline{U(s)}$$

is given by

$$F(s) - F(0) = NkT[\nu_A(\nu_A + \nu_B s) \log(\nu_A + \nu_B s) + \nu_A \nu_B(1-s) \log\{\nu_A \nu_B(1-s)^2\}$$
$$+ \nu_B(\nu_B + \nu_A s) \log(\nu_B + \nu_A s) - \nu_A \log \nu_A - \nu_B \log \nu_B] - \tfrac{1}{2}N\nu_A^2 s^2 \mathscr{V}_0, \quad (14\cdot221\cdot8)$$

where

$$F(0) = NkT(\nu_A \log \nu_A + \nu_B \log \nu_B) + \tfrac{1}{2}N(\nu_A^2 \mathscr{V}_{AA} + 2\nu_A \nu_B \mathscr{V}_{AB} + \nu_B^2 \mathscr{V}_{BB}). \quad (14\cdot221\cdot9)$$

We can now obtain all the thermodynamic functions by the appropriate differentiations.

14·222. *Body-centred cubic lattices.* If we put $\nu_A = \nu_B = \frac{1}{2}$ we have

$$F(s) - F(0) = \tfrac{1}{2}NkT\{(1+s)\log(1+s) + (1-s)\log(1-s) - 2\log 2\} - \tfrac{1}{8}Ns^2\mathscr{V}_0. \quad (14\cdot222\cdot1)$$

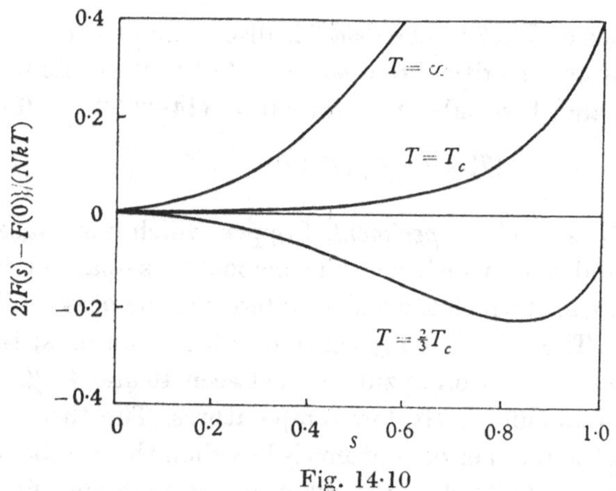

Fig. 14·10

The general behaviour of this function is shown in fig. 14·10 for various values of $\mathscr{V}_0/kT$. The positions of the maxima and minima of $F(s)$ are given by

$$\partial F/\partial s = \tfrac{1}{2}NkT\{\log(1+s) - \log(1-s)\} - \tfrac{1}{4}Ns\mathscr{V}_0 = 0, \quad (14\cdot222\cdot2)$$

i.e. by

$$s = \tanh(\tfrac{1}{4}s\mathscr{V}_0/kT), \quad (14\cdot222\cdot3)$$

which is the same relation (14·21·8) as was found by more elementary methods. Further,

$$\partial^2 F/\partial s^2 = -\tfrac{1}{4}N\mathscr{V}_0 + NkT/(1-s^2), \quad (14\cdot222\cdot4)$$

and so F has a minimum at $s=0$ so long as $kT > \frac{1}{4}\mathscr{V}_0$. When $kT < \frac{1}{4}\mathscr{V}_0$, (14·222·2) has a non-zero root $s^*(T)$ for s, and F is a minimum at $s = s^*(T)$.

The stable state, therefore, is the disordered one for temperatures greater than T_c, where $kT_c = \frac{1}{4}\mathscr{V}_0$, while for $T < T_c$ the ordered state is the stable one with a degree of order $s^*(T)$. Also, both $\partial F/\partial s$ and $\partial^2 F/\partial s^2$ are zero at $s = 0$ for $T = T_c$, and $s^* \to 0$ as $T \to T_c - 0$, so that s^* varies continuously from 0 to 1 as T decreases from T_c to 0. For values of T near T_c, s is small. We may therefore obtain the solution of the equation $s = \tanh(sT_c/T)$ by expanding $\tanh(sT_c/T)$ in powers of sT_c/T, the result being that

$$s^2 = 3(1 - T/T_c) \quad (s \ll 1,\ T < T_c). \tag{14·222·5}$$

The internal energy U is given by

$$U(s,T) - U(s,0) = -T^2 \frac{\partial}{\partial T}\frac{F(s) - F(0)}{T} = -\tfrac{1}{8}Ns^2\mathscr{V}_0 = -\tfrac{1}{2}NkT_c s^2, \tag{14·222·6}$$

and the specific heat C_V is given by

$$C_V = \frac{1}{N}\frac{dU}{dT} = \frac{1}{N}\frac{dU(0,T)}{dT} - kT_c s\frac{ds}{dT}. \tag{14·222·7}$$

Now s is continuous at $T = T_c$ while ds/dT is discontinuous. There is therefore no latent heat at the critical temperature, but the specific heat is discontinuous there, and if we substitute for s from (14·222·3) we find that

$$C_V(T_c - 0) - C_V(T_c + 0) = \tfrac{3}{2}k. \tag{14·222·8}$$

14·223. *Comparison with experiment.* Copper, which has a face-centred cubic structure, and zinc, which has a hexagonal close-packed structure, form a series of alloys, of which the ones of interest in the present discussion are the β-brasses. These have body-centred cubic structures; they exist when the atomic concentration of zinc lies between 46 and 49 %, and they form superlattices at sufficiently low temperatures. The theory of order in alloys in which the number of A atoms is less than the number of α sites is discussed later in § 14·23, but the alloy at the high-zinc limit of the β-brasses is sufficiently near the equi-atomic ratio for the relevant experimental results to be discussed at this stage.

According to the experiments of Moser (1936) and of Sykes & Wilkinson (1937), the transition from the ordered to the disordered state takes place in the β-brasses at 469° C. without the appearance of a latent heat. The experimental value for the discontinuity in the specific heat is about $5k$ per atom. The observed variation in the specific heat over a wide range of temperatures is shown in fig. 14·11, in which is also shown the specific heat calculated from (14·22·3) and (14·222·6) by assuming that $dU(0,T)/dT$ is the mean of the atomic heats of copper and zinc. It will be seen that, while theory and experiment agree in predicting a discontinuity in the specific heat but no latent heat, there is a considerable quantitative discrepancy not only in the magnitude of the discontinuity but also in the temperature variation of the configurational specific heat. Many attempts have been made, along the lines sketched in § 14·24, to improve the quantitative aspects of the theory, but on the whole they have been disappointing. It must, however, be borne in mind that the temperature variation of the specific heat provides a most stringent test of the theory and that it is very satisfactory that the general qualitative features of the transition phenomena can be so well described by a relatively simple theory.

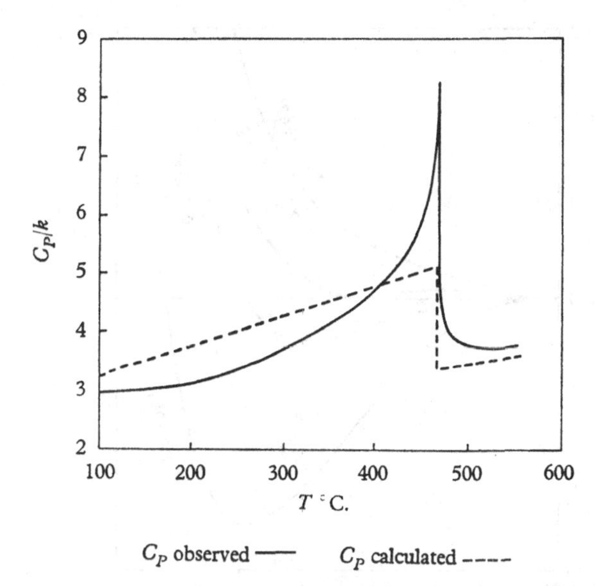

C_p observed —— C_p calculated ----

Fig. 14·11. The observed and calculated specific heats of CuZn.

14·224. *Face-centred cubic lattices.* For a face-centred cubic lattice we have $\nu_A = \frac{1}{4}$, $\nu_B = \frac{3}{4}$, and (14·221·8) becomes

$$F(s) - F(0) = \tfrac{1}{16}NkT\{(1+3s)\log(1+3s) + 6(1-s)\log(1-s) + 9(1+\tfrac{1}{3}s)\log(1+\tfrac{1}{3}s)\} - \tfrac{1}{32}Ns^2\mathscr{V}_0, \quad (14\cdot224\cdot1)$$

while

$$\partial F/\partial s = \tfrac{1}{16}NkT\{3\log(1+3s) - 6\log(1-s) + 3\log(1+\tfrac{1}{3}s)\} - \tfrac{1}{16}Ns\mathscr{V}_0, \quad (14\cdot224\cdot2)$$

$$\frac{\partial^2 F}{\partial s^2} = \tfrac{1}{16}NkT\left(\frac{9}{1+3s} + \frac{6}{1-s} + \frac{1}{1+\frac{1}{3}s}\right) - \tfrac{1}{16}N\mathscr{V}_0. \quad (14\cdot224\cdot3)$$

The general behaviour of this function is shown in fig. 14·12 for various values of $\mathscr{V}_0/kT$. It will be seen that for sufficiently large values of T there is only one turning point (at $s=0$), where $F(s)$ is a minimum. As T decreases below a value T_1, a second minimum and a maximum develop at $s = s^\dagger$ and $s = s^*$ respectively, and at $T = T_c$ the second minimum is such that $F(s^*) = F(0)$. For $T < T_c$ the minimum at $s = s^*$ is the smallest value of $F(s)$, while at a value T_2 of T the maximum at $s = s^\dagger$ reaches $s = 0$, and, for smaller values of T, $F(s)$ is a maximum at $s = 0$. We therefore see that, as T decreases from a high value where the stable state is the disordered one, we first of all reach the temperature T_1 where a metastable ordered state with $s \neq 0$ can exist. At a lower temperature T_c, both the ordered and disordered states are stable, while for temperatures between T_c and T_2 the

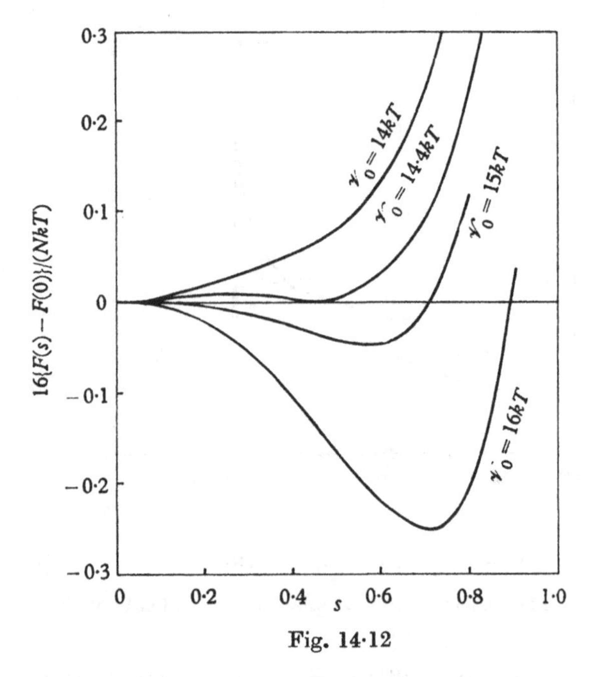

Fig. 14·12

disordered state is metastable and the ordered state is stable. For $T < T_2$ the disordered state is unstable. The degree of order is therefore a discontinuous function of T, and it jumps from $s = 0$ to $s = s^*(T_c)$ at $T = T_c$ if we only consider stable states.

The various temperatures can be found numerically from (14·223·1), (14·223·2) and (14·223·3). T_1 is the temperature for which

$$\partial F/\partial s = \partial^2 F/\partial s^2 = 0$$

for a non-zero value of s. T_c is such that $F(s) = F(0)$, $\partial F/\partial s = 0$ $(s \neq 0)$, while T_2 is such that $\partial F/\partial s = \partial^2 F/\partial s^2 = 0$ at $s = 0$. Their values are as follows:

$$kT_1 = 0{\cdot}069\mathscr{V}_0, \quad s^*(T_1) = 0{\cdot}345;$$

$$kT_c = 0{\cdot}0685\mathscr{V}_0, \quad s^*(T_c) = 0{\cdot}463; \quad kT_2 = 0{\cdot}0625\mathscr{V}_0.$$

The internal energy U is given by

$$U(s,T)-U(0,T)=-T^2\frac{\partial}{\partial T}\frac{F(s)-F(0)}{T}=-\tfrac{1}{32}Ns^2\mathscr{V}_0, \quad (14\cdot224\cdot4)$$

and the specific heat C_{Γ} is given by

$$C_{\Gamma}=\frac{1}{N}\frac{dU}{dT}=\frac{1}{N}\frac{dU(0,T)}{dT}-\tfrac{1}{16}s\frac{ds}{dT}\mathscr{V}_0. \quad (14\cdot224\cdot5)$$

Since s is discontinuous at $T=T_c$ there is a latent heat equal to $\frac{1}{32}s^*(T_c)^2\mathscr{V}_0$ per atom, i.e. $0\cdot0067\mathscr{V}_0=0\cdot098kT_c$ per atom.

14·225. *Comparison with experiment.* Gold and copper both have face-centred cubic structures and they are completely miscible in all proportions. Superlattices are formed at various compositions, but for the moment we

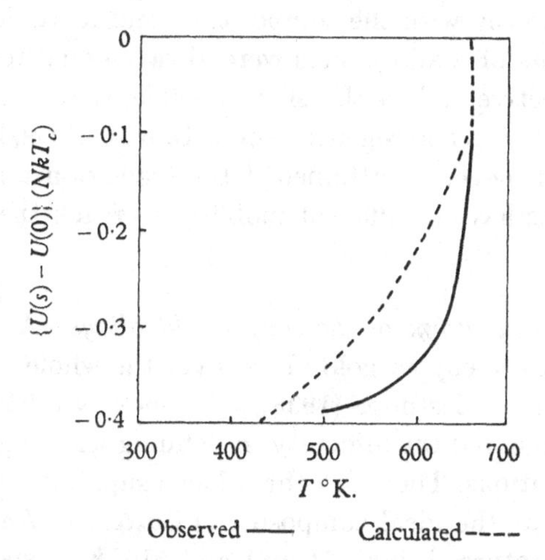

Fig. 14·13. The observed and calculated configurational energies of Cu_3Au.

shall confine our attention to the alloy with the composition Cu_3Au. This alloy undergoes a phase change at 664° K. which is accompanied by a latent heat of about 5·4 joules/gram. This is qualitatively in agreement with the preceding theory, but there are considerable quantitative discrepancies as shown by the curves in fig. 14·13, which give the observed and calculated values of the configurational energy as a function of temperature. In particular, the observed specific heat just below the transition temperature is very much greater than the calculated value.

14·23. *The effect of composition on the degree of order.* We have so far assumed that the number of A atoms present is equal to the number of α sites, and correspondingly that the number of B atoms is equal to the number of β sites. This is not necessarily so, and a more general theory is required when the alloy has an arbitrary composition. It is, however, difficult to know how to define the degree of order. If there are more B atoms than β sites, then in a state of perfect order we could fill all the β sites with B atoms, but, if the numbers of A and B atoms are xN and $(1-x)N$ respectively, we can distribute all the A atoms and the remaining B atoms over the α sites in $N_A!/\{(xN)!(N_A-xN)!\}$ ways. Thus in the supposed state of perfect order there would be a large degree of arbitrariness in the arrangement of the atoms and, by the third law of thermodynamics, there must be a more stable arrangement at sufficiently low temperatures. This might take the form of a heterogeneous mixture of two alloys with different superstructures, or alternatively a 'hyper-structure' might be formed having a larger unit cell. But it is difficult to imagine a hyper-structure which varied continuously with the composition, and it would have to be such that the numbers of α and β sites were always equal to numbers of A and B atoms respectively. It is therefore possible, that in some circumstances the stable state is a heterogeneous one. In real alloys, however, the most stable state will never be attained if the transition temperature is so low that the atoms have insufficient mobility to reach their true equilibrium positions.

14·231. *The phase diagram of the copper-gold alloys.* A discussion of the phase diagram of the copper-gold alloys over the whole concentration range has been given by Easthope (1938), by Shockley (1938) and by Li (1949), all of whom ignored the difficulty in defining the long-range order for arbitrary compositions. There are three basic superlattices to be considered which occur at the ideal compositions Cu_3Au, CuAu and $CuAu_3$, the transition temperatures being 664, 681 and 516° K. respectively. The Cu_3Au and $CuAu_3$ alloys have face-centred cubic structures and can be treated by the equations given in § 14·224. The CuAu alloy, however, cannot be treated as being basically a face-centred cubic structure, and we have to redefine the α and β sites so as to make them equal in number. The ordered CuAu structure is in fact face-centred tetragonal with a unit cell with sides a, a, c, where $a/c = 1{\cdot}08$. The gold and copper atoms lie on separate planes, the positions of the gold atoms being congruent to $(0, 0, 0)$, $(\frac{1}{2}a, \frac{1}{2}a, 0)$ and those of the copper atoms to $(\frac{1}{2}a, 0, \frac{1}{2}c)$, $(0, \frac{1}{2}a, \frac{1}{2}c)$. We may therefore, if we please, designate the alloy as Cu_2Au_2.

The change in the ratio of the lattice constants can be calculated approximately as follows. The lattice constant a is determined partly by the distance apart of the gold atoms and partly by the distance apart of the copper atoms in their respective planes, and it should therefore be equal to or slightly less than the greater of a_{Cu} and a_{Au}, the lattice constants of pure copper and gold. Actually $a = 3{\cdot}96 \times 10^{-8}$ cm. when the lattice is perfectly ordered, while $a_{Cu} = 3{\cdot}61 \times 10^{-8}$ cm. and $a_{Au} = 4{\cdot}07 \times 10^{-8}$ cm. On the other hand, c is determined by the distance apart of the gold and copper planes, and if we assume that the atoms are in contact, we have

$$\tfrac{1}{2}(a^2 + c^2)^{\frac{1}{2}} = \tfrac{1}{2}(a_{Au} + a_{Cu})/\sqrt{2}.$$

This gives $c = 3{\cdot}71 \times 10^{-8}$ cm., which is in fair agreement with the observed value $3{\cdot}67 \times 10^{-8}$ cm. On account of the change in the ratio of the lattice constants, there is necessarily a latent heat of transition for the Cu_2Au_2 structure.

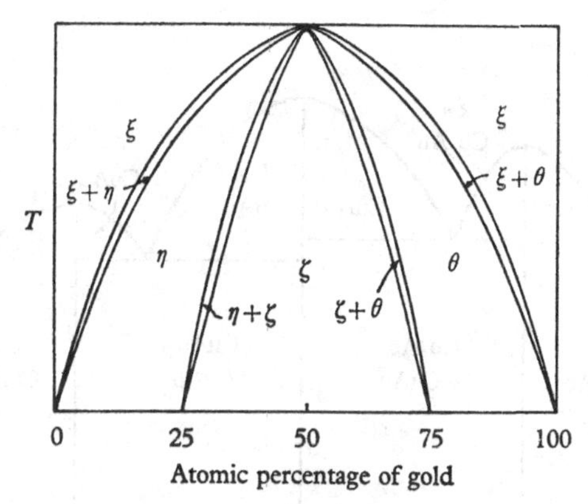

Fig. 14·14. Shockley's phase diagram for the copper-gold alloys.

Shockley treated the various superlattices as being derived from one or other of the basic Cu_3Au, Cu_2Au_2 and $CuAu_3$ superlattices by considering the number of A atoms to be different from the number of α sites, and he found it necessary to subdivide the face-centred cubic lattice into four simple cubic lattices and to define a degree of order for each sublattice. His theory is a tentative one and gives rise to the phase diagram shown in fig. 14·14. The phase ξ is completely disordered, the phases η and θ are ordered face-centred cubic structures while the phase ζ is an ordered face-centred tetragonal structure. There are also heterogeneous structures which are denoted by $\xi + \eta$, $\eta + \zeta$, $\zeta + \theta$ and $\xi + \theta$ in the diagram.

Shockley's phase diagram cannot be correct at low temperatures since it does not satisfy the third law of thermodynamics. It may, however, be qualitatively correct at high temperatures, but, if so, its predictions are in conflict with experiment on at least one point. According to the experimental results of Haughton & Paine (1931), the transition temperature for alloys derived from the Cu_3Au structure is a maximum at the 3 : 1 atomic ratio, whereas, according to the theory, a maximum can only occur at a 1 : 1 ratio. (Li's theory, on the other hand, which assumes that super-lattices can only exist for restricted concentration ranges, avoids this particular difficulty.)

In view of this discrepancy and more particularly since the ordered states are not in accordance with the third law of thermodynamics, it is possible that, except for the compositions Cu_3Au, CuAu and $CuAu_3$, the ordered states are heterogeneous mixtures. If this hypothesis is correct, the phase

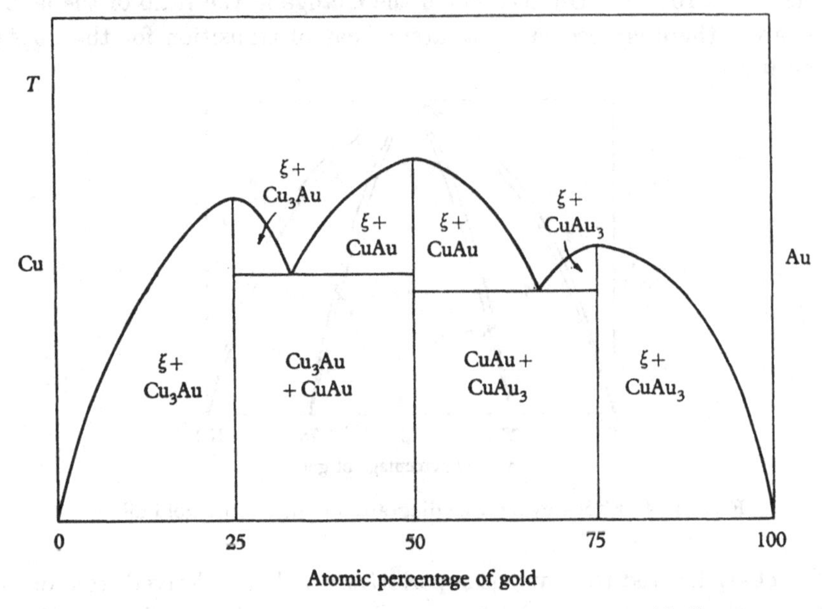

Fig. 14·15. A possible phase diagram for the copper-gold alloys.

diagram would have the general character shown in fig. 14·15. The transition temperature would have maxima at the 3 : 1, 1 : 1 and 1 : 3 atomic ratios, and at $T = 0$ the equilibrium alloy would consist of mixtures of Cu and Cu_3Au, or of $CuAu_3$ and Au, for atomic percentages of gold less than 25 and greater than 75 respectively. For intermediate compositions the alloys would be mixtures of Cu_3Au and CuAu or of CuAu and $CuAu_3$. The alternative hypothesis that a hyper-structure is formed would also lead to a stable state which is in accordance with the third law of thermodynamics, but no detailed calculations have been carried out of the other consequences of the hypothesis.

It should be emphasized once more that the theory of ordered structures with an arbitrary ratio of atoms to lattice sites is by no means in a satisfactory state, and a more profound analysis is desirable.

14·24. *Other theories of order.* The preceding calculations involve a very crude method of averaging, and much more elaborate theories have been proposed. In all of these the order is defined by more parameters than the long-distance order s. In describing the structure of the partition function in § 14·22, we introduced various parameters $X, Y, \ldots$ which denoted the numbers of nearest neighbours of various types, and to obtain a more general theory than that given in § 14·221 it is necessary to avoid identifying these parameters with their mean values $\bar{X}, \bar{Y}, \ldots$. This effectively means introducing a short-distance order which differs from the long-distance order, and which may not be zero even when the alloy is macroscopically disordered.

One method of attempting to set up a more general theory is to use the exact expression (14·22·5) for the configurational energy $U(s, X, Y)$ and an approximate value for the number of complexions $W(U, s, X, Y)$. One plausible choice of $W(U, s, X, Y)$ leads to the following equation for X (Fowler & Guggenheim, 1939, 1940):

$$\frac{(r_\alpha N_A - X)(w_\beta N_A - X)}{X(r_\beta N_A - r_\alpha N_A + X)} = \exp(\mathscr{V}_0/kTz). \qquad (14\cdot24\cdot1)$$

This method of constructing an approximate partition function is known as the quasi-chemical method, since the formula (14·24·1) can be derived by formally considering the various pairs of neighbouring atoms (enumerated in § 14·22) to 'react' with one another according to the 'quasi-chemical equation':

$$\begin{aligned}&-(A \text{ atom on } \alpha \text{ site}, A \text{ atom on } \beta \text{ site})\\ &\qquad\qquad -(B \text{ atom on } \alpha \text{ site}, B \text{ atom on } \beta \text{ site})\\ &+(A \text{ atom on } \alpha \text{ site}, B \text{ atom on } \beta \text{ site})\\ &\qquad\qquad +(B \text{ atom on } \alpha \text{ site}, A \text{ atom on } \beta \text{ site}) = 0. \qquad (14\cdot24\cdot2)\end{aligned}$$

Since $\mathscr{V}_0/z$ is the energy required to change an AA and a BB pair into an AB and a BA pair, the energy change in the 'reaction' is $\Delta H = -\mathscr{V}_0/z$, and, if we assume that the 'equilibrium constant' K is given by

$$\log K = -\Delta H/kT$$

and apply the law of mass action, we arrive at (14·24·1). We may therefore assume that (14·24·1) represents a plausible guess, which may lead to a more exact theory. It should be noted that if we replace the right-hand side of (14·24·1) by unity we obtain the value (14·221·4) for X which leads to the simplified theory of § 14·221. An entirely different method of obtaining equation (14·24·1) has been given by Bethe (1935).

An alternative approach by Kirkwood (1938) is to obtain the exact partition function as a power series in $\mathscr{V}_0/kT$. In principle the calculations can be carried out to any desired accuracy, but in practice the calculations are long and tedious, and the series may not converge or may only converge very slowly.

None of these more elaborate theories greatly improves the agreement between theory and experiment, but they do give rise to a configurational energy which, due to the short-range order, persists above the transition temperature. In view of the comparatively slight improvements introduced by these theories and in view of the fact that complete accounts of one or other of them are available in the books by Fowler & Guggenheim (1939) and Guggenheim (1952), we shall not pursue them here.

Some Exact Solutions of the One-dimensional Order-disorder Problem

14·3. None of the order-disorder problems which we have so far considered has yet been solved exactly, and the methods of solution which have to be employed consist essentially in obtaining a number of successive approximations to the partition function. It is, however, by no means certain that the approximations form a convergent series, and empirically the first approximation often seems to give a more physically plausible result than the second approximation. There is therefore considerable doubt as to the theoretical validity of the methods given above of obtaining approximate partition functions in order-disorder problems. This doubt is increased by a consideration of certain simple one- and two-dimensional problems for which exact solutions exist, which are different from those which would be obtained by approximate methods. We therefore conclude this chapter by a brief discussion of those order-disorder problems which have been solved exactly, in order to demonstrate some of the points which still require elucidation. The problems themselves have no particular physical importance.

14·31. *A finite linear chain.* We shall consider the simplest possible magnetic body, consisting of a chain of N equidistant atoms each possessing a spin and a magnetic moment which can point in either of two opposite directions. The configurations of the chain are then determined by the distribution of the positively and negatively directed spins over the lattice points. We shall assume that each atom interacts with its two neighbours only, the interaction energy being $-J$ or J according as the two spins are parallel or antiparallel. We shall also assume that there is a magnetic field present, and that the magnetic energy of a positively or negatively directed spin is $-\mu H$ or μH $(\mu > 0)$. We shall show that a chain with these properties cannot be ferromagnetic.

We group together the various configurations that have given numbers ν_1 and ν_2 of positively and negatively directed spins, the positive spins being arranged in $s+1$ distinct blocks. To calculate the number of such configurations we proceed as follows (Ising, 1925). We assume that the chain begins with a positive spin, and define δ as equal to 0 or 1 according as the chain ends with a positive or with a negative spin. Now the number of ways of dividing ν_1 indistinguishable objects up into $s+1$ sets each containing at least one object‡ is ${}^{\nu_1-1}C_s$, and this is the number of ways of arranging the positive spins on the lattice with exactly s gaps between them. Correspondingly, the ν_2 negative spins can be distributed over the $s+\delta$ gaps in the lattice in ${}^{\nu_2-1}C_{s+\delta-1}$ distinct ways, and the $\nu_1+\nu_2$ spins can therefore be distributed in ${}^{\nu_1-1}C_s\,{}^{\nu_2-1}C_{s+\delta-1}$ ways. There is a similar expression for the number of configurations in which the first spin is negative, and the total number of complexions $W(\nu_1,\nu_2,s)$ for given ν_1, ν_2 and s is

$$W(\nu_1,\nu_2,s)={}^{\nu_1-1}C_s\,{}^{\nu_2-1}C_{s+\delta-1}+{}^{\nu_2-1}C_s\,{}^{\nu_1-1}C_{s+\delta-1}, \tag{14·31·1}$$

where $\delta=0$ or 1, and where ν_1, ν_2 and s are such that $\nu_1+\nu_2=N$ with

$$\nu_1\geqslant s+1, \quad \nu_2\geqslant s+\delta \tag{14·31·2}$$

in the first term, and with

$$\nu_1\geqslant s+\delta, \quad \nu_2\geqslant s+1 \tag{14·31·3}$$

in the second term. The total number of pairs of nearest neighbours is $N-1$, of which $2s+\delta$ have antiparallel and $N-2s-\delta-1$ have parallel spins. The energy of this particular configuration is therefore

$$E(\nu_1,\nu_2,s)=-(N-4s-2\delta-1)J-(\nu_1-\nu_2)\mu H, \tag{14·31·4}$$

and so the partition function is

$$Z=\sum_{\nu_1+\nu_2=N}\sum_s\sum_{\delta=0,1}W(\nu_1,\nu_2,s)\exp[-E(\nu_1,\nu_2,s)/kT], \tag{14·31·5}$$

where ν_1, ν_2 and s must satisfy the conditions (14·31·2).

To evaluate this we denote the expression (14·31·5) by $Z(N)$ and form the generating function

$$\Phi(x)=\sum_{N=1}^{\infty}x^N Z(N). \tag{14·31·6}$$

If we write $\Phi(x)=\Phi_1(x)+\Phi_2(x)$ corresponding to the two terms in the expression (14·31·1) for $W(\nu_1,\nu_2,s)$, we can carry out the summations by inverting the order of summation. Since $\nu_2=N-\nu_1$, and since N now

‡ The number required is the coefficient of x^{ν_1} in the expression

$$(x+x^2+\ldots)\,(x+x^2+\ldots)\,\ldots,$$

there being $s+1$ factors. It is therefore the coefficient of x^{ν_1} in the expansion of $x^{s+1}(1-x)^{-s-1}$, which is ${}^{\nu_1-1}C_s$.

ranges from one to infinity, there are no restrictions on ν_1 and ν_2 except those imposed by (14·31·2), and we have

$$\Phi_1(x) = \sum_{s=0}^{\infty} \sum_{\delta=0,1} \sum_{\nu_1=s+1}^{\infty} \sum_{\nu_2=s+\delta}^{\infty} \frac{(\nu_1-1)!}{(\nu_1-s-1)!\,s!} \frac{(\nu_2-1)!}{(\nu_2-s-\delta)!\,(s+\delta-1)!}$$

$$\times x^{\nu_1+\nu_2} \exp[(\nu_1+\nu_2-4s-2\delta-1)J/kT] \exp[(\nu_1-\nu_2)\mu H/kT] \quad (14\cdot31\cdot7)$$

$$= \sum_{s=0}^{\infty} \sum_{\delta=0,1} \exp[(1-\delta)\mu H/kT]\, x^{2s+\delta+1}$$

$$\times \sum_{r_1=0}^{\infty} \frac{(r_1+s)!}{r_1!\,s!} x^{r_1} \exp\{[r_1\mu H + (r_1-s)J]/kT\}$$

$$\times \sum_{r_2=0}^{\infty} \frac{(r_2+s+\delta-1)!}{r_2!(s+\delta-1)!} x^{r_2} \exp\{[-r_2\mu H + (r_2-s-\delta)J]/kT\}. \quad (14\cdot31\cdot8)$$

Now
$$(1-y)^{-s-1} = \sum_{r=0}^{\infty} \frac{(r+s)!}{r!\,s!} y^r, \quad (14\cdot31\cdot9)$$

and hence, on summing (14·31·8) with respect to r_1 and r_2, we have

$$\Phi_1(x) = \sum_{s=0}^{\infty} \sum_{\delta=0,1} \exp[(1-\delta)\mu H/kT]\, x^{2s+\delta+1} \exp[-(2s+\delta)J/kT]$$

$$\times \{1 - x\exp[(J+\mu H)/kT]\}^{-s-1} \{1 - x\exp[(J-\mu H)/kT]\}^{-s-\delta}$$

$$= \frac{x\, e^{\mu H/kT} - 2x^2 \sinh(J/kT)}{1 - 2x\, e^{J/kT} \cosh(\mu H/kT) + 2x^2 \sinh(2J/kT)}. \quad (14\cdot31\cdot10)$$

The corresponding expression for $\Phi_2(x)$ is obtained by changing the sign of μ, and we finally obtain

$$\Phi(x) = \frac{2x\cosh(\mu H/kT) - 4x^2\sinh(J/kT)}{1 - 2x\, e^{J/kT}\cosh(\mu H/kT) + 2x^2\sinh(2J/kT)}. \quad (14\cdot31\cdot11)$$

If we split $\Phi(x)$ up into partial fractions and expand the denominators we obtain the partition function in the form

$$Z(N) = \left[\cosh\beta + \frac{e^\alpha \sinh^2\beta + e^{-\alpha}}{\sqrt{(e^{2\alpha}\sinh^2\beta + e^{-2\alpha})}}\right] [e^\alpha\cosh\beta + \sqrt{(e^{2\alpha}\sinh^2\beta + e^{-2\alpha})}]^{N-1}$$

$$+ \left[\cosh\beta - \frac{e^\alpha \sinh^2\beta + e^{-\alpha}}{\sqrt{(e^{2\alpha}\sinh^2\beta + e^{-2\alpha})}}\right] [e^\alpha\cosh\beta - \sqrt{(e^{2\alpha}\sinh^2\beta + e^{-2\alpha})}]^{N-1}, \quad (14\cdot31\cdot12)$$

where
$$\alpha = J/kT, \quad \beta = \mu H/kT. \quad (14\cdot31\cdot13)$$

If N is large, the second term is negligible compared with the first.

We may now obtain any of the thermodynamic functions. In particular, the magnetic moment $M\mu$ is given by

$$M\mu = \mu\,\partial \log Z/\partial\beta,$$

and in the limit of large N we have

$$M\mu = N\mu \frac{e^{\alpha} \sinh \beta}{\sqrt{(e^{2\alpha} \sinh^2 \beta + 1)}}. \qquad (14\cdot31\cdot14)$$

The magnetic moment vanishes when $\beta = 0$, and this model therefore cannot have a permanent magnetic moment. The differential susceptibility is

$$\chi = \mu \frac{\partial M}{\partial H} = \frac{N\mu^2}{kT} \frac{\cosh(\mu H/kT)\, e^{J/kT}}{\sqrt{\{\sinh^2(\mu H/kT)\, e^{2J/kT} + 1\}}}. \qquad (14\cdot31\cdot15)$$

The particular case $T = 0$ is singular. We then have $M\mu = N\mu$ for all H.

14·32. *A cyclic chain.* If we consider the last atom in the chain to be the nearest neighbour of the first atom, we can avoid the 'end-effects', and the results are somewhat simplified. The number of different spin configurations is the same whether we consider an open or a closed chain, but the energy levels are slightly different. The total number of pairs of nearest neighbours is now N, and if the initial atom has a positive spin there are $2(s+\delta)$ pairs with antiparallel spins and $N - 2s - 2\delta$ pairs with parallel spins. The expression (14·3·4) for the energy is therefore replaced by

$$E = -(N - 4s - 4\delta) J - (\nu_1 - \nu_2) \mu H. \qquad (14\cdot32\cdot1)$$

On repeating the calculation of the preceding section we find that the generating function is

$$\Phi(x) = \frac{2x\, e^{J/kT} \cosh(\mu H/kT) - 4x^2 \sinh(2J/kT)}{1 - 2x\, e^{J/kT} \cosh(\mu H/kT) + 2x^2 \sinh(2J/kT)} \qquad (14\cdot32\cdot2)$$

and that the partition function is

$$Z(N) = [e^{\alpha} \cosh \beta + \sqrt{(e^{2\alpha} \sinh^2 \beta + e^{-2\alpha})}]^N + [e^{\alpha} \cosh \beta - \sqrt{(e^{2\alpha} \sinh^2 \beta + e^{-2\alpha})}]^N, \qquad (14\cdot32\cdot3)$$

which is somewhat simpler than (14·3·12) but indistinguishable from it for large N.

14·33. *Approximate evaluation of the partition function.* The partition function (14·31·5) contains the two independent parameters ν_1 and s which characterize the various configurations. Partition functions of this type are of common occurrence, and on various occasions (explicitly in § 14·221 and implicitly in § 10·551) we have evaluated them approximately by inserting average values of $W(\nu_1, \nu_2, s)$ and $E(\nu_1, \nu_2, s)$, which are obtained by using the value of s which maximizes $W(\nu_1, \nu_2, s)$. We shall now show that this procedure can lead to fallacious results.

The logarithm of the general term of the partition function (14·31·5) can be written as

$$\nu_1 \log \nu_1 + \nu_2 \log \nu_2 - (\nu_1 - s) \log (\nu_1 - s) - (\nu_2 - s) \log (\nu_2 - s)$$
$$- 2s \log s + \{(n - 4s - 2\delta - 1) J + (\nu_1 - \nu_2) \mu H\}/kT, \quad (14·33·1)$$

and the value of s which maximizes this is given by

$$(\nu_1 - s)(\nu_2 - s)/s^2 = e^{4J/kT}. \quad (14·33·2)$$

If we replace the exponential by unity, which might be expected to be a valid approximation at high temperatures, the solution of this equation is

$$s = \nu_1 \nu_2 / (\nu_1 + \nu_2), \quad (14·33·3)$$

and, with this value of s, (14·33·1) becomes

$$N \log N - \nu_1 \log \nu_1 - \nu_2 \log \nu_2 + \{(N - 4\nu_1 \nu_2 / N) J + (\nu_1 - \nu_2) \mu H\}/kT. \quad (14·33·4)$$

Now $\nu_2 = N - \nu_1$, and for variable ν_1 the maximum value of (14·33·4) is given by

$$\log \{\nu_1/(N - \nu_1)\} = \{(-4 + 8\nu_1/N) J + 2\mu H\}/kT. \quad (14·33·5)$$

Then, since $M = \nu_1 - \nu_2 = 2\nu_1 - N$, we have

$$\frac{M}{N} = \tanh\left(\frac{\mu H}{kT} + \frac{MJ}{NkT}\right). \quad (14·33·6)$$

This is the same result (with $z = 2$) as was obtained in §10·551 by elementary methods, and, since it is in contradiction with the exact formula (14·31·14), the validity of the theory given in §10·551 is extremely doubtful.

14·34. *The matrix method.* An alternative method of calculating the partition function is the following. If we consider a cyclic chain and treat the spin μ_i on any atom as an operator with two values μ and $-\mu$, we can write the partition function as

$$Z = \sum_{\mu_i = \pm \mu} \exp [(\mu_1 \mu_2 + \mu_2 \mu_3 + \ldots + \mu_N \mu_1) K + (\mu_1 + \mu_2 + \ldots + \mu_N) L], \quad (14·34·1)$$

where

$$K = J/(\mu^2 kT), \quad L = H/kT, \quad (14·34·2)$$

since by taking all possible values for the μ's we obtain each configuration of the chain once and once only. Now the terms which involve μ_2 can be written as

$$\sum_{\mu_2 = \pm \mu} \exp [\mu_1 \mu_2 K + \tfrac{1}{2}(\mu_1 + \mu_2) L] \exp [\mu_2 \mu_3 K + \tfrac{1}{2}(\mu_2 + \mu_3) L], \quad (14·34·3)$$

and this is the product of the row matrix

$$(\exp [\mu_1 \mu K + \tfrac{1}{2}(\mu_1 + \mu) L], \exp [-\mu_1 \mu K + \tfrac{1}{2}(\mu_1 - \mu) L]) \quad (14·34·4)$$

and of the column matrix

$$\begin{pmatrix} \exp[\mu\mu_3 K + \frac{1}{2}(\mu+\mu_3)L] \\ \exp[-\mu\mu_3 K + \frac{1}{2}(-\mu+\mu_3)L] \end{pmatrix}. \tag{14·34·5}$$

Since μ_1 is itself an operator with the two values μ and $-\mu$ we may consider (14·34·4) as a 2×2 matrix $A(\mu_1, \mu_2)$, whose rows are characterized by the values of μ_1 and whose columns are characterized by the values of μ_2. Then, if the elements of $A(\mu_1, \mu_2)$ are $A(\mu_1, \mu_2)_{k,l}$ $(k, l = 1, 2)$, we have

$$A(\mu_1, \mu_2)_{k,l} = \exp[\mu_{1,k}\mu_{2,l}K + \frac{1}{2}(\mu_{1,k}+\mu_{2,l})L], \tag{14·34·6}$$

where
$$\mu_{i,1} = \mu, \quad \mu_{i,2} = -\mu \quad (\text{all } i). \tag{14·34·7}$$

Similarly, we may write (14·34·5) as the 2×2 matrix $A(\mu_2, \mu_3)$, and the partition function (14·34·1) can be written as the matrix product

$$Z = \sum_{\mu_1=\pm\mu} A(\mu_1, \mu_2)\,A(\mu_2, \mu_3)\ldots A(\mu_{N-1}, \mu_N)\,A(\mu_N, \mu_1). \tag{14·34·8}$$

(We do not need to write the summations with respect to $\mu_2, \ldots, \mu_N$, since these are implied by the matrix multiplication.) Now, if we consider all value of μ_1 and μ_1', the expression $A(\mu_1, \mu_2)\,A(\mu_2, \mu_3)\ldots A(\mu_N, \mu_1')$ is the 2×2 matrix representing $\mathbf{A}^N$, and Z is therefore its diagonal sum, usually denoted by spur or trace. Hence

$$Z = \operatorname{spur} \mathbf{A}^N. \tag{14·34·9}$$

This expression is easily evaluated by transforming $\mathbf{A}$ into its diagonal form. $\mathbf{A}$ is a symmetric matrix, and it can be transformed into the diagonal form by means of a symmetric matrix $\mathbf{S}$, i.e. $\mathbf{S}$ exists such that

$$\mathbf{S}^{-1}\mathbf{A}\mathbf{S} = \begin{pmatrix} \lambda_1 & 0 \\ 0 & \lambda_2 \end{pmatrix}. \tag{14·34·10}$$

The spur is invariant under such a transformation since

$$(\mathbf{S}^{-1}\mathbf{A}\mathbf{S})^N = (\mathbf{S}^{-1}\mathbf{A}\mathbf{S})\,(\mathbf{S}^{-1}\mathbf{A}\mathbf{S})\ldots(\mathbf{S}^{-1}\mathbf{A}\mathbf{S}) = \mathbf{S}^{-1}\mathbf{A}^N\mathbf{S},$$

and $\operatorname{spur}(\mathbf{S}^{-1}\mathbf{A}\mathbf{S})^N = \operatorname{spur}\mathbf{S}^{-1}\mathbf{A}^N\mathbf{S} = \operatorname{spur}\mathbf{A}^N\mathbf{S}\mathbf{S}^{-1} = \operatorname{spur}\mathbf{A}^N$, the spur of a product being independent of the order of the factors in the product $(\sum_{m,n} A_{mn}B_{nm} = \sum_{m,n} B_{mn}A_{nm})$. Hence (14·34·9) can be written

$$Z = \lambda_1^N + \lambda_2^N, \tag{14·34·11}$$

where λ_1 and λ_2 are the roots of the characteristic equation $|\mathbf{A} - \lambda\mathbf{1}| = 0$, which is

$$\begin{vmatrix} e^{K+L} - \lambda & e^{-K} \\ e^{-K} & e^{K-L} - \lambda \end{vmatrix} = 0. \tag{14·34·12}$$

Therefore $$\lambda^2 - 2\lambda\, e^K \cosh L + e^{2K} - e^{-2K} = 0, \tag{14·34·13}$$

and $$Z = [e^K \cosh L + \sqrt{(e^{2K} \sinh^2 L + e^{-2K})}]^N + [e^K \cosh L - \sqrt{(e^{2K} \sinh^2 L + e^{-2K})}]^N, \tag{14·34·14}$$

in agreement with (14·32·3)

14·35. *Two-dimensional lattices.* It was shown by Lassettre & Howe (1941), by Montroll (1941), and more particularly by Kramers & Wannier (1941) that the matrix method could be extended to deal with two-dimensional lattices in the absence of a magnetic field. Kramers & Wannier showed that an order-disorder transition can take place in a two-dimensional lattice; they located the exact transition point and obtained an approximate value of the partition function by a variation method.

In 1944 Onsager obtained an exact solution of the problem, and since that time there have been many contributions to the theory in attempts to simplify and extend it. A general review of the methods employed has been given by Newell & Montroll (1953), and the reader is referred to their article for reference to the most important papers.

It has not been possible to extend Onsager's method of solution to three-dimensional lattices, and his exact solution of the two-dimensional problem is therefore of little practical importance. It is, however, of very considerable theoretical interest, as being the first exact solution of a problem in which an order-disorder transition can take place. The full theory is too long and complex to be given here, and a general sketch of the method without a discussion of the details is not illuminating. We shall therefore merely quote the results obtained for a square lattice in which each atom interacts with its four neighbours only.

The free energy of a square lattice containing N atoms is expressible in terms of elliptic functions, and is given by

$$F = -NkT \log\{2\cosh(2J/kT)\} - \frac{NkT}{2\pi}\int_0^{\pi} \log\{\tfrac{1}{2} + \tfrac{1}{2}(1 - k_1^2 \sin^2\theta)^{\frac{1}{2}}\}\, d\theta, \tag{14·35·1}$$

where $$k_1 = \frac{2\sinh(2J/kT)}{\cosh^2(2J/kT)}. \tag{14·35·2}$$

Considered as a function of $x = 2J/kT$, k_1 is zero for $x = 0$ and $x = \infty$ and reaches its maximum value of unity when $\sinh(2J/kT) = 1$. Now the integral in (14·35·1) is singular for $k_1 = 1$, and there is therefore a real transition temperature T_0, given by the solution of the equation

$$\sinh(2J/kT_0) = 1, \tag{14·35·3}$$

at which some of the thermodynamic functions are discontinuous. Explicitly we have

$$J/kT_0 = \tfrac{1}{2}\log(1+\sqrt{2}) = \tfrac{1}{2}\log\cot\tfrac{1}{8}\pi = 0{\cdot}4407. \qquad (14{\cdot}35{\cdot}4)$$

The singularity in the partition function is such that the energy is continuous and the specific heat is discontinuous at $T = T_0$.

To determine the nature of the discontinuity we differentiate F/T, which gives

$$U = -NJ\coth(2J/kT)\,[1 + 2\pi^{-1}k_2K(k_1)] \quad (k_2^2 = 1 - k_1^2), \qquad (14{\cdot}35{\cdot}5)$$

where $K(k_1)$ is the complete elliptic integral of the first kind, namely,

$$K(k_1) = \int_0^{\frac{1}{2}\pi} (1 - k_1^2\sin^2\theta)^{-\frac{1}{2}}\,d\theta. \qquad (14{\cdot}35{\cdot}6)$$

Now $K(k_1)$ has only a logarithmic singularity at $k_1 = 1$, and so U is continuous. The specific heat per atom, however, is given by

$$C_V = \frac{1}{N}\frac{\partial U}{\partial T} = \frac{2J^2}{\pi^2 kT^2}\coth^2\frac{2J}{kT}\,[2K(k_1) - 2E(k_1) - (1-k_2)\{\tfrac{1}{2}\pi + k_2K(k_1)\}], \qquad (14{\cdot}35{\cdot}7)$$

where $E(k_1)$ is the complete elliptic integral of the second kind, namely,

$$E(k_1) = \int_0^{\frac{1}{2}\pi} (1 - k_1^2\sin^2\theta)^{\frac{1}{2}}\,d\theta, \qquad (14{\cdot}35{\cdot}8)$$

and C_V becomes logarithmically infinite at $T = T_0$. Since

$$K(k_1) \sim \log(4/k_1) \qquad (14{\cdot}35{\cdot}9)$$

for small values of $1 - k_1$, the specific heat in the neighbourhood of $T = T_0$ is given by

$$C_V = \frac{2}{\pi}(\log\cot\tfrac{1}{8}\pi)^2 k\left[\log\frac{\sqrt{2}\,kT_0^2}{J} - \log|T - T_0| - 1 - \tfrac{1}{4}\pi\right]. \qquad (14{\cdot}35{\cdot}10)$$

While the discovery of the exact solution of the elementary two-dimensional order-disorder problem has not so far led to a rigorous solution of the three-dimensional problems in which we are primarily interested, it constitutes a great step forward, in that it enables us to apply some tests to any approximate method which is proposed. (It does not, however, necessarily follow that because an approximate method gives a poor answer to the two-dimensional problem, it will also give an equally poor answer if applied to the physically important three-dimensional problems.) Most of the approximate methods which have been proposed result in partition functions for the one- and two-dimensional lattices which have singularities of a higher order than those which occur in the exact partition functions. It is therefore necessary to use the results so far obtained by approximate methods with reasonable but not with undue caution.

REFERENCES

Bethe, H. (1935). Statistical theory of superlattices. *Proc. Roy. Soc.* A, **150**, 552.

Borelius, G., Johansson, C. H. & Linde, J. O. (1928). Crystallographic transformations in metallic mixed crystals. *Ann. Phys., Lpz.*, (4), **86**, 291.

Borelius, G. (1934). The theory of transformations of metallic mixed phases. *Ann. Phys., Lpz.*, (5), **20**, 57.

Bragg, W. L. & Williams, E. J. (1934/5). The effect of thermal agitation on atomic arrangement in alloys. I and II. *Proc. Roy. Soc.* A, **145**, 699; **151**, 540.

Busse, W. F. (1932). The physical structure of elastic colloids. *J. Phys. Chem.* **36**, 2862.

Dehlinger, U. & Graf, L. (1930). Transformations of solid metal phases. I. The tetragonal gold-copper alloy CuAu. *Z. Phys.* **64**, 359.

Easthope, C. E. (1938). The critical ordering temperature in alloys. II. The existence of a two-phase region. *Proc. Camb. Phil. Soc.* **34**, 68.

Eyring, H. (1932). The resultant electric moment of complex molecules. *Phys. Rev.* **39**, 746.

Flory, P. J. & Rehner, J. (1943). Statistical mechanics of cross-linked polymer networks. *J. Chem. Phys.* **11**, 512, 521.

Fowler, R. H. & Guggenheim, E. A. (1939). *Statistical thermodynamics.* Cambridge.

Fowler, R. H. & Guggenheim, E. A. (1940). Statistical thermodynamics of superlattices. *Proc. Roy. Soc.* A, **174**, 189.

Gorsky, W. (1928). X-ray investigations of transformations in the CuAu alloy. *Z. Phys.* **50**, 64.

Guggenheim, E. A. (1952). *Mixtures.* Oxford.

Guth, E. & Mark, H. (1934). Internal molecular statistics, especially in chain molecules. *Mh. Chem.* **65**, 93.

Haughton, J. L. & Paine, R. J. M. (1931). Transformations in the gold-copper alloys. *J. Inst. Metals*, **46**, 457.

Irwin, J. O. (1927). The frequency distribution of the means of samples from a population having any law of frequency with finite moments. *Biometrika*, **19**, 225.

Ising, F. (1925). A contribution to the theory of ferromagnetism. *Z. Phys.* **31**, 253.

James, H. M. (1942). Statistical properties of networks of flexible chains. *J. Chem. Phys.* **15**, 651.

James, H. M. & Guth, E. (1943). Theory of the elastic properties of rubber. *J. Chem. Phys.* **11**, 455.

James, H. M. & Guth, E. (1949). Simple presentation of the network theory of rubber, with a discussion of other theories. *J. Polym. Sci.* **4**, 153.

James, H. M. & Guth, E. (1953). Statistical thermodynamics of rubber elasticity. *J. Chem. Phys.* **21**, 1039.

Johansson, C. H. & Linde, J. O. (1925). The X-ray determination of the atomic arrangement in the mixed-crystal series gold-copper and palladium-copper. *Ann. Phys., Lpz.*, (4), **78**, 439.

Kirkwood, J. G. (1938). Order and disorder in binary solid solutions. *J. Chem. Phys.* **6**, 70.

Kramers, H. A. & Wannier, G. H. (1941). Statistics of the two-dimensional ferromagnet. I and II. *Phys. Rev.* **60**, 252.

Kuhn, W. (1934). The shape of fibrous molecules in solutions. *Kolloidzschr.* **68**, 2.

Lassettre, E. N. & Howe, J. P. (1941). Thermodynamic properties of binary solid solutions on the basis of the nearest neighbour approximation. *J. Chem. Phys.* **9**, 747.

Li, Y.-Y. (1949). Quasi-chemical theory of order for the copper-gold system. *J. Chem. Phys.* **17**, 447.

Meyer, K. H., von Susich, G. & Valko, E. (1932). The elastic properties of organic high polymers and their kinetic significance. *Kolloidzschr.* **59**, 208.

Montroll, E. W. (1941). Statistical mechanics of nearest neighbour systems. *J. Chem. Phys.* **9**, 706.

Moser, H. (1936). Measurements of the true specific heat of silver, nickel, β-brass and quartz between 50 and 700° C. *Phys. Z.* **37**, 737.

Newell, G. F. & Montroll, E. W. (1953). The theory of the Ising model of ferromagnetism. *Rev. Mod. Phys.* **25**, 353.

Nix, F. C. & Shockley, W. (1938). Order-disorder transformations in alloys. *Rev. Mod. Phys.* **10**, 1.

Onsager, L. (1944). Crystal statistics. I. A two-dimensional model with an order-disorder transition. *Phys. Rev.* **65**, 117.

Rayleigh, Lord (1919). The problem of random vibrations, and of random flights in one, two or three dimensions. *Phil. Mag.* (6), **37**, 321.

Shockley, W. (1938). Theory of order for the copper-gold alloy system. *J. Chem. Phys.* **6**, 130.

Sykes, C. & Wilkinson, H. (1937). The transformation in the β-brasses. *J. Inst. Metals*, **61**, 223.

Tammann, G. (1919). The chemical and galvanic properties of mixed crystals and their atomic structure. *Z. anorg. Chem.* **107**, 1.

Titchmarsh, E. C. (1937). *The theory of Fourier integrals.* Oxford.

Treloar, L. R. G. (1943). The elasticity of a network of long-chain molecules. I and II. *Trans. Faraday Soc.* **39**, 26, 241.

Treloar, L. R. G. (1946). The statistical length of long-chain molecules. *Trans. Faraday Soc.* **42**, 77.

Treloar, L. R. G. (1949). *The physics of rubber elasticity.* Oxford.

Wall, F. T. (1942). Statistical thermodynamics of rubber. I and II. *J. Chem. Phys.* **10**, 132, 485.

Wall, F. T. (1943*a*). Statistical lengths of rubber-like hydrocarbon molecules. *J. Chem. Phys.* **11**, 67.

Wall, F. T. (1943*b*). Statistical thermodynamics of rubber. III. *J. Chem. Phys.* **11**, 527.

Williams, E. J. (1935). The effect of thermal agitation on atomic arrangement in alloys. III. *Proc. Roy. Soc.* A, **152**, 231.

INDEX OF SUBJECTS

INDEX OF NAMES